Non-destructive Elemental Analysis

Also of interest

Zeev B. Alfassi
Statistical Treatment of Analytical Data
0-632-05367-4

Non-destructive Elemental Analysis

Edited by

Zeev B. Alfassi

Blackwell
Science

Editorial Offices:
Osney Mead, Oxford OX2 0EL
25 John Street, London WC1N 2BS
23 Ainslie Place, Edinburgh EH3 6AJ
350 Main Street, Malden
MA 02148 5018, USA
54 University Street, Carlton
Victoria 3053, Australia
10, rue Casimir Delavigne
75006 Paris, France

Other Editorial Offices:

Blackwell Wissenschafts-Verlag GmbH
Kurfürstendamm 57
10707 Berlin, Germany

Blackwell Science KK
MG Kodenmacho Building
7–10 Kodenmacho Nihombashi
Chuo-ku, Tokyo 104, Japan

Iowa State University Press
A Blackwell Science Company
2121 S. State Avenue
Ames, Iowa 50014-8300, USA

First published 2001

Set in 10/12 Times
by DP Photosetting, Aylesbury, Bucks
Printed and bound in Great Britain by
MPG Books Ltd, Bodmin, Cornwall

DISTRIBUTORS

Marston Book Services Ltd
PO Box 269
Abingdon
Oxon OX14 4YN
(*Orders:* Tel: 01235 465500
Fax: 01235 465555)

USA
Blackwell Science, Inc.
Commerce Place
350 Main Street
Malden, MA 02148 5018
(*Orders:* Tel: 800 759 6102
781 388 8250
Fax: 781 388 8255)

Canada
Login Brothers Book Company
324 Saulteaux Crescent
Winnipeg, Manitoba R3J 3T2
(*Orders:* Tel: 204 837-2987
Fax: 204 837-3116)

Australia
Blackwell Science Pty Ltd
54 University Street
Carlton, Victoria 3053
(*Orders:* Tel: 03 9347 0300
Fax: 03 9347 5001)

A catalogue record for this title
is available from the British Library

ISBN 0-632-05366-6

Library of Congress
Cataloging-in-Publication Data
is available

For further information on
Blackwell Science, visit our website:
www.blackwell-science.com

Contributors

M. Aboudy	Department of Nuclear Engineering, Ben Gurion University of the Negev, Beer-Sheeva 84120, Israel
Z.B. Alfassi	Department of Nuclear Engineering, Ben Gurion University of the Negev, Beer-Sheva 84120, Israel
D.J. Cherniak	Department of Earth and Environmental Sciences, Rensselaer Polytechnic Institute, Troy, New York, NY 12180, USA
G. Demortier	LARN, Facultés Universitaires Notre-Dame de la Paix, 61 rue de Bruxelles, B-5000 Namur, Belgium
J. Injuk	Department of Chemistry, University of Antwerp (UIA), B-2610, Wilrijk-Antwerp, Belgium
T.E. Jeffries	Department of Mineralogy, Natural History Museum, Cromwell Road, London SW7 5BD, UK
W.A. Lanford	Physics Department, University at Albany, Albany, New York, NY 12222, USA
H.J. Mathieu	Swiss Federal Institute of Technology Lausanne (EPFL), Materials Science Department, Lausanne, Switzerland
J. Räisänen	Department of Physics, University of Jyväskylä, Finland
M. Talianker	Department of Materials Engineering, Ben Gurion University of the Negev, Beer-Sheva 84120, Israel
G. Terwagne	LARN, Facultés Universitaires Notre-Dame de la Paix, 61 rue de Bruxelles, B-5000 Namur, Belgium
R. Van Grieken	Department of Chemistry, University of Antwerp, B-2610, Wilrijk-Antwerp, Belgium
R.D. Vis	Department of Physics and Astronomy, Vrije Universiteit, Amsterdam, The Netherlands
C. Yonezawa	Department of Environmental Sciences, Tokai Research Establishment, Japan Atomic Energy Research Institute (JAERI), Tokai-Mura, Naka-gun, Ibaraki 319-1195, Japan

This book is dedicated to Pascal A. Mathieu who died at the age of 27, just before this book was published. He was the son of one of the contributors, Professor Hans Jörg Mathieu.

Contents

Introduction

Z.B. ALFASSI

Most chemical analyses are destructive, completed usually by dissolving the material being analysed and measuring the concentration of the various elements in the resulting solution. This preliminary step may involve contamination due to either the dissolving reagents or the atmosphere in the preparation room. Where analysis is non-destructive the contamination is lower than that found in destructive methods and consequently such analysis has lower limits of detection and also higher precision as contamination due to the atmosphere is random. Another important advantage of the non-destructive method is its ability to analyse samples that are either too precious or unique and must be kept intact rather than destroyed by dissolution. The number of non-destructive assays is now increasing in the fields of materials, archaeology, arts and biology.

Most chemical methods consume the analysed sample and hence are destructive. In some cases, however, the amount of consumed material is very small and the particular method of analysis can be regarded as almost non-destructive. An example of this type of analysis is laser ablation ICP-MS. Although some chemists regard ICP-MS together with XRF, SEM, XPS and AES as chemical methods, others treat them as nuclear methods of analysis. As such, they are borderline non-destructive methods making use of high-energy electrons, though these electrons are derived from a simpler source than a particle accelerator. Since true nuclear methods deal with the atomic nucleus rather than electrons, nuclear methods are considered to be non-destructive.

Chemical analysis by nuclear methods is usually elemental, i.e. it determines the quantity of the various elements contained in the analysed sample, but is unable to reveal their chemical form (e.g. compounds, valence states). The analysis is based on the reaction of the analysed element with a nuclear projectile (neutrons, accelerated charged particles, e.g. electrons, protons, heavy ions, gamma photons), or on the scattering of accelerated charged particles by the elemental nucleus. The reaction can be written in the same form as a chemical reaction, namely

Target + projectile → light product + heavy product

or in the more concise form of nuclear physics:

Target (projectile, light product) heavy product

For example, the first production of an artificial radionuclide by Joliot and Curie was through the reaction of alpha particles with aluminium metal,

$^{27}Al + \alpha \rightarrow {}^{30}P + n$ (α is the ^{4}He nucleus)

or in the physics notation, $^{27}Al\ (\alpha, n)\ ^{30}P$.

The light product is similar to the projectile, be it a neutron, a small charged particle or a photon. The basis of chemical analysis via a nuclear reaction (nuclear activation) therefore, is the measurement of the amount formed of either the light or the heavy product in a known flux of projectiles for a known length of time (the irradiation time). The amount produced is proportional to the number of target atoms in the element being analysed, and hence the measurement of the product quantity yields the number of target atoms. The concentration of product formed is too small to be measured chemically (except in very rare cases), and the only way to measure the amount of product is by pulse counting.

The light product can be measured by virtue of its energy if it is a photon or a high-kinetic-energy charged particle, or by its reaction if it is a neutron. Whatever the method of measurement, it must be completed very shortly after the reaction otherwise the light product will lose either energy or identity by interaction with the surrounding media. Consequently, the measurement of light particles must take place during the bombardment of the target element with projectiles. This type of measurement is called *Prompt Analysis*. This analysis can be undertaken in both situations of nuclear reaction (activation) and projectile scattering. In the cse of nuclear reaction, the qualitative analysis is based on the identity of the products, while in scattering, qualitative analysis concerns kinematics, i.e. the energy and direction of the scattered particle. The quantitative measurement involves counting the number of products or scattered particles.

In situations where the heavy product is radioactive, i.e. it undergoes a spontaneous nuclear transformation, its identity and amount are revealed by its radioactivity. Since radioactivity can be measured after completion of the interaction between the target element and the beam of projectiles, this method of analysis is called *Delayed Activation Analysis*. The heavy product must be radioactive in this type of analysis and therefore not all elements can be measured with any projectile. However, by choosing the correct bombarding projectile almost every element can be determined.

The scattering of charged particles is measured in both the forward (ERDA or elastic recoil detection analysis) and backward directions (RBS or Rutherford backward scattering). In the early years of research, light-ion scattering was studied. However, more recently the scattering studies of heavier ions have become more frequent. The widespread availability of 2–3 MeV accelerators has made the use of heavier ions more popular. However, in forward scattering (ERDA), such accelerators are restricted in their use to the analysis of hydrogen and deuterium. Heavier elements can be analysed with the beams from backward scattering or the forward scattering of heavier ions. [In a recent Ion Beam Analysis conference in Lisbon (1997) there were two papers on He-induced ERDA compared to 26 papers on heavy-ion ERDA. However, for routine analysis, the actual situation is different owing to the availability of the low-energy He accelerators.]

Activation by neutrons is carried out mainly in the delayed mode, where very low detection limits can be obtained. Prompt activation is used primarily for in situ and on-line analysis and for elements that cannot be determined by the delayed mode because the product is non-radioactive or has half-life which is too short or too long. An example is the determination of hydrogen and nitrogen atoms by PGNAA (prompt gamma neutron activation analysis), mainly in food samples.

The slowing of high-kinetic-energy charged particles when they pass through matter allows not only the determination of the total concentration of some elements but also their distribution as a function of distance from the sample surface, i.e. the depth profile of the various elemental concentrations. The slowing down can alter the energy of the emitted charged particle as, for example, in the reaction $^{10}B\ (n, \alpha)^7Li$. The energy loss of the alpha particle traversing the surface of the sample to reach the detector when compared with the original energy of the alpha particle, indicates the distance that the alpha particle has travelled through the sample.

The slowing down of the impinged high-energy projectile can reduce its energy to the range where a resonance reaction can occur. Thus, by changing the original energy of the projectile, the depth profile of the elemental concentration can be measured.

1 Nuclear activation analysis

Z.B. ALFASSI

1.1 Introduction

Chemical analysis by nuclear activation is an elemental analysis, i.e. it determines the contents of the various elements in the analysed sample, but cannot reveal what chemical form (compounds, valence states) they are. Nuclear activation is based on the reaction of the the element being analysed with nuclear projectiles (neutrons, accelerated charged particles e.g. protons, or gamma photons). The reaction can be written in the same form as a chemical reaction, namely

Target + projectile $\rightarrow$ light product + heavy product

or in the more concise form of nuclear physics notation,

Target (projectile, light product) heavy product

Thus, for example, the first production of an artificial radionuclide by Joliot and Curie followed the reaction of alpha particles with aluminium metal,

$^{27}Al + \alpha \rightarrow {}^{30}P + n$ (α is the ^4He nucleus)

or in the physics notation, $^{27}Al\ (\alpha, n)^{30}P$.

The light product is similar to the projectile, e.g. a neutron, a small charged particle (protons, deuterons, tritions, ^3He, ^4He) or a photon. Fundamentally, the nuclear activation method for chemical analysis involves the measurement of the amount formed of either the light or the heavy product in a known flux of projectiles for a known length of time (the irradiation time). The amount produced is proportional to the number of target atoms, and hence the measurement of this quantity yields the target atom concentration. The quantity of product is usually too small to be measured chemically (except in very rare cases); the only way to measure it, therefore, is by the nuclear physics method of pulse counting. Here, the light product can be measured by virtue of its energy, if it is a photon or a high-kinetic-energy charged particle, or by its interaction with the projectile, if it is a neutron. The common denominator in all of these processes is that measurements must be performed very soon after the formation of the product, otherwise the light product will lose either its energy or its identity, by reacting with the surrounding media. Consequently, the measurement of the light particle must be done during the bombardment of the target with the projectiles. This kind of measurement is called *prompt activation analysis*.

When the heavy product is radioactive, i.e. it undergoes a spontaneous nuclear transformation, its quantity can be measured by its radioactivity. This is the more common method in analytical chemistry, but not in the case of in situ and on-line analysis. Since the radioactivity can be measured after the end of the interaction between the target and the beam of projectiles, this method of analysis is called *delayed activation analysis*. Since the heavy product must be radioactive, not all elements can be measured with every available projectile. However, through the choice of the bombarding projectile, every element can be determined. Since many elements are activated simultaneously the radioactivity measurement must yield both the identity of the radioactive nuclides and their amounts.

The term *radioisotope*, which is occasionally used, is incorrect, since when referring to isotopes mention of the element should be made. *Nuclide* is the nuclear physics term that parallels the atom in chemistry. It therefore consists of a nucleus (composed of neutrons and protons) and orbiting electrons. As the identity of *radionuclides* (a compound term for radioactive nuclides) in a mixture cannot be determined from β^- emission (unless it is a single radionuclide or in special cases it is one of very few in the mixture), only radionuclides that emit γ-rays either with or without β^- rays can be used in activation analysis. Radionuclides used for identification purposes are called *indicator radionuclides* (IRN). The chemical analysis by nuclear activation in which chemical separation precedes the radioactivity measurement is named *radiochemical activation analysis* (RAA), whereas analysis by nuclear activation with direct measurement of the nuclear activity is named *instrumental activation analysis* (IAA). Since this book deals only with non-destructive methods this chapter will concentrate on IAA with delayed measurement only.

Most nuclear activation analyses are done with thermal neutrons as projectiles for the following reasons:

(1) Thermal neutrons react with most elements by one reaction only, namely a (n, γ) reaction and hence any activated product (IRN) can be formed only from one source, indicating unequivocally the identity of the element present in the sample.
(2) Since there is no Coulombic repulsion between the target nucleus and the projectile the cross-section for activation by neutrons is larger than that found with energetic positive ions.
(3) Thermal neutrons do not lose their ability to interact with the target while passing through it and hence can be used for the determination of bulky samples. Energetic positive ions lose their kinetic energy while penetrating the target. Since low-energy positive ions cannot react with the target, owing to Coulombic repulsion, the activation with energetic charged ions permits analysis of the surface only rather than the bulk sample. As the activation cross-section depends on the particle energy, the calculation of concentration involves the integral of the flux and the cross-section over the whole energy range, leading to a higher error than that in neutron activation analysis.

For these reasons this chapter will deal mainly with instrumental neutron activation analysis. However, since some elements cannot be determined by neutron activation owing to either an inappropriate half-life (too short or too long) of the activated nuclide, or through being a pure β^- emitter, some elements are determined by charged-particle activation analysis as discussed elsewhere.

1.2 Basic nuclear physics

This section provides only a brief basic knowledge. Readers interested in further resources can refer to some common textbooks (Harvey 1969; Choppin & Rydberg

1980; Friedlander *et al.* 1981; Vertes & Kiss 1987; Krane 1988; Ehmann & Vance 1991).

1.2.1 *Nuclides*

The nucleus of an atom is composed of protons and neutrons. A general term to describe a particle in the nucleus (either proton or neutron) is *nucleon*. The total number of nucleons in the nucleus is called the mass number, A ($A = N + Z$, where N is the number of neutrons in the nucleus, and Z is the number of protons in the nucleus). Z is called the atomic number. A nucleus is defined unequivocally by its A and Z values. The usual way to indicate a nucleus is $^{Z}_{A}X$. Nuclides of constant Z but different values of A are called *isotopes*. The writing of Z is analogous to the use of the chemical symbol for the element. For example, the oxygen nucleus has eight protons and consequently the use of the symbol for oxygen is the same as writing $Z = 8$. If for the nucleus the symbol O is written, it is not necessary to give the subscript 8, although in some cases this will be done for purposes of clarity.

Oxygen has three stable isotopes containing 8, 9 and 10 neutrons. These are written as ^{16}O, ^{17}O and ^{18}O. A common error is the designation of any nuclear species, such as ^{14}C, ^{13}N, or ^{17}O, by the word isotope. The word isotope should be reserved in connection with a particular element only, as in saying that ^{16}O, ^{17}O and ^{18}O are the three stable isotopes of oxygen or that ^{34}Cl, ^{35}Cl, ^{36}Cl are isotopes of the same element (in this case chlorine). However, ^{19}F and ^{19}O are not isotopes. The general term used to designate any nuclear species is *nuclide*. The general term for a radioactive nuclide should be *radionuclide* and not radioisotope, although it is correct to say, for example, that ^{128}I is a radioisotope of iodine.

1.2.2 *Radioactive decay*

Neutrons and protons are held together in the nucleus through nuclear forces. These forces along with the Coulombic force interact in certain ways resulting in the fact that only special combinations of neutrons and protons are stable. Too many protons increase Coulombic repulsion, while neutrons are unstable species and require screening by protons to effect stability in the nucleus. Thus, the hydrogen atom with a single-proton nucleus is infinitely stable; however, a bare neutron will decay with a half-life of 10.25 minutes. The requirement for special numerical combinations of protons and neutrons is evident from the fact that each element has only limited numbers of stable isotopes. Stating that some nuclides are unstable does not mean that these cannot be formed, it just claims that such nuclides can spontaneously transform (some of them very rapidly, while others more slowly) to other nuclides that have lower energy (i.e. lower mass according to Einstein's equation of mass and energy). This property of certain nuclides occurs via the transformation of a proton to a neutron or vice versa, thus changing their ratio. The transformation of a nuclide can also occur by emission of an alpha particle (^{4}He nuclide), although this emission takes place only for the very heavy nuclides and it is an almost unimportant phenomenon in activation analysis. Very few radionuclides are transformed by spontaneous emission of a neutron or a proton or by a spontaneous fission. Such examples of nuclear decay are also mostly unimportant in activation analysis.

All transformations neutron → proton and proton → neutron are called β decays. They are characterized by the mass number (A) remaining unchanged, whereas the atomic number is changed by ± 1. There are three transformations of this type. A neutron is transformed to a proton by a β^- process:

$$n \rightarrow p^+ + \beta^- + \bar{\nu}$$

The proton remains in the nucleus, while the electron (β^-) and the antineutrino $\bar{\nu}$ are emitted from the nucleus, both with high kinetic energies. The electron is written as β^- (and not e^-) to demonstrate that it is not an atomic electron. A proton can be transformed into a neutron by two processes. One process is β^+ decay in which a proton is transformed into a neutron and a positron (i.e. $\beta^+ = e^+$), a particle similar in mass to an electron, but possessing opposite charge:

$$p^+ \rightarrow n + \beta^+ + \nu$$

In this process there is also an emission of a neutrino (ν). The second process of proton → neutron is the reaction of a proton with one of the orbiting atomic electrons to yield a neutron. This process is called electron capture (EC). The neutrino and antineutrino are very small particles having no charge, and consequently are very difficult to detect. These also are not important to activation analysis, except for their effect on β energies, which will be explained later.

Most radionuclides are not pure β emitters as they emit simultaneously a γ photon (though more accurately, the γ emission is a very short time of less than 10^{-10} s after the β^- emission). Gamma rays (photons) are electromagnetic waves, as are light and radiowaves, but they have much higher energies (much shorter wavelengths). They have usually higher energies than X-rays, although the main difference between X-rays and γ-rays are their sources and not necessarily their energies. X-rays arise from atomic transitions (transitions between different energy levels of electrons), whereas γ-rays are due to nuclear transitions (transitions between different energy levels of the nucleons). The β^- decay does not yield, in most cases, the ground state of the product nuclide, but leaves it in an excited state. The excited state nuclide decays very rapidly (in most cases) to the ground state by the emission of either γ-rays or atomic electrons called conversion electrons. The emitted conversion electrons are not relevant to activation analysis and can be neglected. Reference only need be made to the emission of the γ-rays.

A very important difference between β processes and γ decay (besides the entities of the emitted particles, electrons vs photons, and the difference in ΔA, ± 1 in the β process and 0 in the γ decay) is the fact that in β processes there is simultaneous emission of two particles, either β^- and $\bar{\nu}$, or β^+ and ν, while in γ decay there is emission of only one photon at a time. (There are examples of a few photons being emitted from the same nuclide, but these are successive emissions and not a simultaneous emission as in the β processes.) The importance of this difference lies in the fact that each nuclear decay is a transformation between two discrete energy states, resulting in a definite energy being released in the process. If only one particle is emitted from the nucleus, this particle has a defined energy. When two particles are emitted, the released energy is distributed between them, and each particle can have a spectrum of energies ranging from zero up to a maximum energy, equal to the total energy released. *Since emitted γ photons have definite energies, they can be used to identify their emitters by measuring the energy of the photons.* As β particles do not have definite energies, they cannot usually be used to identify their emitters.

In nuclear activation analysis, different radionuclides need to be identified (in order to determine from which nuclide they were formed), in addition to measuring their activities. It is for this reason that nuclear activation analysis employs almost exclusively the measurement of spectra from emitted γ photons. In a few instances, where the IRN (indicator radionuclide) is a pure β^- emitter, the IRN can be measured either by chemical separation of the emitting element (radiochemical activation analysis, RAA) or, for long-lived IRNs, by the long waiting periods between the end of irradiation and starting the activity count. This waiting time is called usually the 'cooling time'. The end of the irradiation is called 'pile out' in the case of nuclear reactor irradiations, or EOB (end of bombardment) in the case of particle accelerators. Many β^+ emitters do not have de-exciting γ photons. They are measured by the 511 keV γ photons produced in the annihilation process. When a positron collides with an electron, the masses of both particles annihilate and are transformed into two photons with energy of 511 keV each. Since each β^+ emitter emits the 511 keV γ line, this line can be used for quantitative measurement, but not for identification. Thus, in many cases NAA with IRNs that are β^+ emitters is conducted only after chemical separation of the element. If there are only a few (2–3) non-specific γ emitter IRNs in the sample with quite different half-lives, their separate activities can be measured even without chemical separation by measuring the activities at different times (measuring the decay curve), and extracting the various activities from the time dependence of the measured activity.

It was stated that the emitted γ photons are derived from the de-excitation of the nuclide produced in the β process. However, this de-excitation does not always take place by one photon only. Moreover, not all β decays lead to the same level of excitation. These two facts together with the de-excitation by emission of conversion electrons (mainly from low-lying levels) offer possibilities that one radionuclide can have more than one kind of photon (energy), and that the number of photons must not be equal to the number of nuclides that has decayed (disintegrated). The number of photons of specific energy emitted per 100 disintegrated nuclides is called the *intensity* of that γ line (in %). Different occurrences of γ line intensity in four examples of radionuclides are set out below.

(1) ^{28}Al decays completely to the 1.778 MeV excited state of ^{28}Si. This level decays to the ground state by only one photon. Thus, it means that ^{28}Al has only one γ line of 1.778 MeV with intensity of 100%. This fact is illustrated in Fig. 1.1 by a so-called *decay scheme* of decreasing energies.

(2) ^{24}Na decays almost completely ($>$ 99.8%) to one level of ^{24}Mg. However, this level decays subsequently to the ground state by emission of two successive photons. The first photon of 2.75 MeV is followed by a second photon of 1.39 MeV. The decay scheme of Fig. 1.2 explains the ^{24}Na γ lines of 2.75 and 1.39 MeV, both possessing almost 100% intensity.

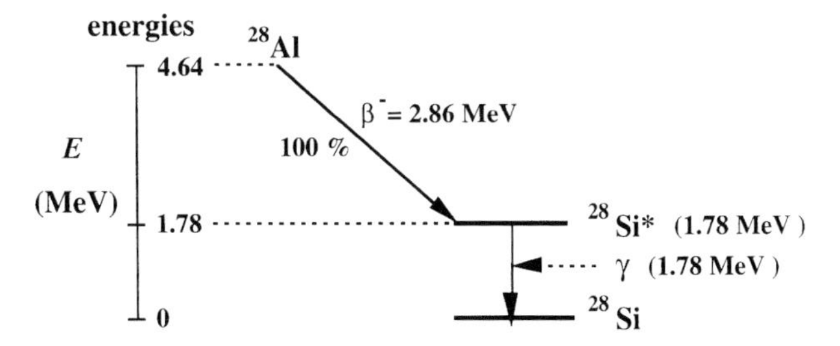

Fig. 1.1 Decay scheme illustrating ^{28}Al decaying completely to the excited state of ^{28}Si by β^- emission.

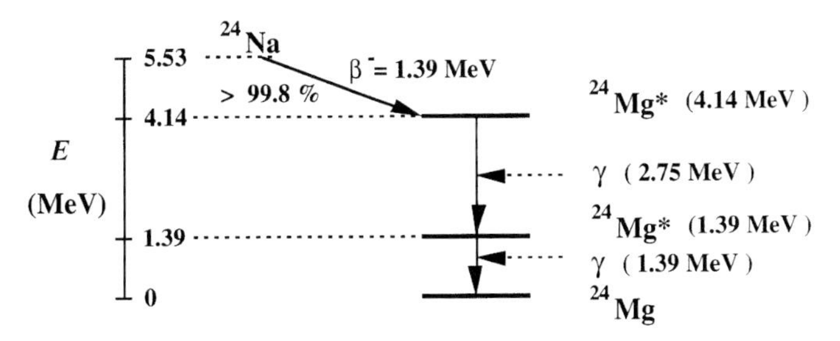

Fig. 1.2 Decay scheme illustrating ^{24}Na decaying almost completely to one level of ^{24}Mg by β^- emission. Successive γ emissions follow.

(3) ^{27}Mg decays to two different excited levels, resulting in two β^- with different energies: 1.59 MeV (29%) and 1.75 MeV (71%). The higher excited level (1.014 MeV = 1014 keV) decays by 97% directly to the ground state while 3% ($0.29 \times 0.03 = 0.008 = 0.8\%$ from the total nuclides disintegrated) decays to the lower excited state. The lower excited level decays completely to the ground state by one photon. Consequently, there are intensity values of 71.8% for the photon of 843 keV, 28% for the 1014 keV photon and 0.8% for the 170 keV photon. The decay scheme is shown in Fig. 1.3. Q represents the total energy of decay ($\beta + \gamma$).

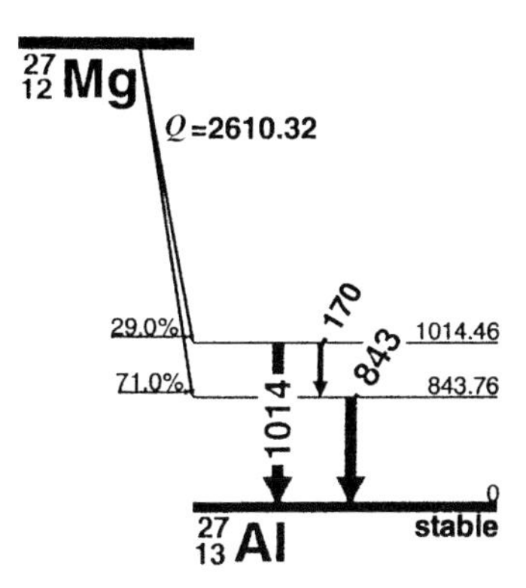

Fig. 1.3 Scheme illustrating ^{27}Mg decay to two different excited levels by two β^- emissions of differing energy.

(4) ^{38}Cl decays directly to three different states of ^{38}Ar, namely two excited states of ^{38}Ar and the ground state of ^{38}Ar (see Fig. 1.4). This is the reason for ^{38}Cl having three different β^- energies. (Remember, each is only a maximal energy, since β^- comprises a spectrum of β^- energies up to E_{maximum}.) The decay energies are 4.81 MeV (53%), 2.77 MeV (9%) and 1.77 MeV (38%). The higher excited state (3.77 MeV) decays almost completely to the lower excited state, while only very little (0.06% of the 38%) decays directly to the ground state. Consequently, ^{38}Cl has three γ lines: 3.77 MeV (with intensity 0.03%); 2.10 MeV (38%); and 1.60 MeV (38% + 53% = 91%).

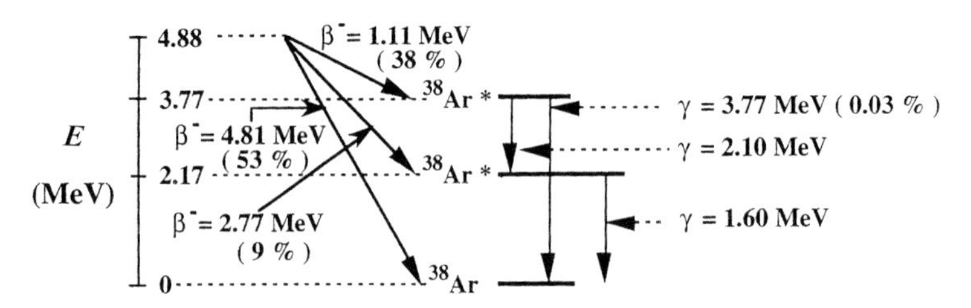

Fig. 1.4 Decay scheme illustrating direct ^{38}Cl decay to three different states of ^{38}Ar with various γ emissions.

These values together with additional physical data can be found in literature collections of nuclide decay schemes (Firestone *et al.* 1996). However, for the application of radionuclides in general, using collections of data for all γ lines without their detailed decay schemes should be sufficient (Erdtmann & Soyka 1979; Reus & Westmeier 1983; Browne *et al.* 1986). These data collections are for all known radionuclides, but for most specific cases of activation analysis it is better to use smaller collections of radionuclide data, i.e. only for those formed potentially by specific types of activation analysis (Yule 1981; Alfassi 1990; Glascock 1996).

1.2.2.1 *Kinetics of decay of radioactive nuclides* The decay of radioactive nuclides is a statistical process. Thus, the number of atoms decaying per unit time (the rate of decay) is proportional to the number of that specific radionuclide present in the sample. If the number of nuclei (atoms) of a specific radionuclide at time t is N^*, the rate of their disappearance (decay or disintegration) is given by:

$$-\frac{dN^*}{dt} = \lambda N^* \tag{1.1}$$

The constant, λ, which is different for every radionuclide, is called the *decay constant* of the radionuclide. Integration of Equation (1.1) leads to the *decay equation of radionuclides*. This is the equation describing the time dependence of the number of atoms of the specific radionuclide,

$$N^*(t) = N_0^* \, e^{-\lambda t} \tag{1.2}$$

where N_0^* is the number of atoms at time $t = 0$. Equation (1.2), which is the same as the integrated equation of any first-order chemical process, indicates the existence of a constant life for half of the atoms present. This is called the *half-life*. The half-life means that irrespective of the value of N_0^*, it takes the same time (for a specific radionuclide) for disintegration of half of the atoms to occur, leaving $0.5 \times N_0^*$ atoms of that radionuclide present. Substituting $N^*(t)$ with $N_0^*/2$ in Equation (1.2) leads to correlation between the half-life ($t_{1/2}$) and the decay constant (λ):

$$N^*(t) = N_0^*/2 \quad \Rightarrow \quad t_{1/2} = \ln 2/\lambda \quad \text{or} \quad t_{1/2} = 0.0693/\lambda \tag{1.3}$$

Equation (1.2), the decay equation, can be written also with $t_{1/2}$ instead of λ:

$$N^*(t) = N_0^*(0.5)^{t/t_{1/2}} \tag{1.4}$$

The advantage of Equation (1.4) over Equation (1.2) is that it gives an easier 'feeling' for the extent of the decay. Thus, for example, if a half-life value, $t_{1/2}$, is 2.5 days, then after 2.5 days only half of the original atoms of a radionuclide sample remains, and after 5 days the number of original atoms present reduces to one quarter. Equation (1.2) is easier to use when referring to tables of exponents. In tables of data, $t_{1/2}$ values only are given. In order to use Equation (1.2), λ should be calculated from Equation (1.3).

1.2.2.2 *Kinetics of chain decays* In some situations a radionuclide decays not to a stable nuclide but to another unstable radionuclide in a chain of decays. Each chain, though, is terminated with formation of a stable nuclide:

$$R_1 \xrightarrow{\lambda_1} R_2 \xrightarrow{\lambda_2} R_3 \xrightarrow{\lambda_3} \cdots \xrightarrow{\lambda_n} R_n \longrightarrow S$$

Here R denotes a radionuclide and S stands for a stable nuclide. The number of radionuclide atoms, given by R_k, at time t assuming that at time zero $R_1 \neq 0$ (represented by R_1^0 is given by the Bateman equation:

$$R_n = C_1 \times e^{-\lambda_1 t} + C_2 \times e^{-\lambda_2 t} \cdots + C_n \times e^{-\lambda_n t} \qquad \text{or} \qquad R_n = \sum_{i=1}^{n} C_i\, e^{-\lambda_i t} \tag{1.5}$$

$$C_1 = \frac{\lambda_1\lambda_2\lambda_3\ldots\lambda_{n-1} \times N_1^0}{(\lambda_2 - \lambda_1)(\lambda_3 - \lambda_1)\ldots(\lambda_n - \lambda_1)};\; C_2 = \frac{\lambda_1\lambda_2\ldots\lambda_{n-1} \times N_1^0}{(\lambda_1 - \lambda_2)(\lambda_3 - \lambda_2)\ldots(\lambda_n - \lambda_2)}$$

and generally

$$C_k = \frac{\prod_{i=1}^{n-1} \lambda_i}{\prod_{i=1,\, i\neq k}^{n} (\lambda_i - \lambda_k)} \times N_1^0 \tag{1.6}$$

Equation (1.6) is the solution of the n simultaneous differential equations:

$$\frac{dR_1}{dt} = -\lambda_1 R_1\,;\; \frac{dR_k}{dt} = \lambda_{k-1} \times R_{k-1} - \lambda_k R_k \qquad (k = 2, 3 \ldots n) \tag{1.7}$$

1.2.2.3 *Kinetics of formation of radioactive nuclides by irradiation* When a thin target is bombarded with a beam of projectiles the rate of nuclear transformation (the number of nuclides produced per unit time) is proportional to the beam intensity (I), which is the number of incident particles per unit time and also proportional to the target nuclei density (n) (number of target nuclei per unit volume) and the thickness of the target ($\mathrm{d}x$). The proportonality to the thickness of the target is limited to sufficiently small $\mathrm{d}x$ such that both the intensity of the beam and the energy of the projectiles remain practically unchanged. The proportionality constant is given by σ and is called the *reaction cross-section*. This name is derived from a simple model that assumes that each geometrical collision leads to a nuclear reaction and that the nuclei of the atoms occupy only a small fraction of space. In this case, σ is the geometrical cross-section of the collision pair, namely $\pi\,(r_1 + r_2)^2$, where r_1 and r_2 are the radii of the projectile and the target nucleus, respectively. Thus:

$$\frac{\mathrm{d}N^*}{\mathrm{d}t} = \sigma I n\,\mathrm{d}x \tag{1.8}$$

Since the nuclear radius has dimensions in terms of femtometres (10^{-13} cm), the cross-sections are in the order of 10^{-24} cm^2. For this reason it has been customary to express cross-sections in *barns*, where 1 barn $= 10^{-24}$ cm^2.

Equation (1.8) assumes that the beam cross-section is smaller than the target size and consequently each projectile particle is transversing through the target. In the situation where the target size is smaller than the beam cross-section, the beam intensity, I, should be replaced by the product ϕA, where ϕ is the beam flux (the number of projectile particles per unit area and per unit time), and A is the area of the target facing the beam:

$$\frac{\mathrm{d}N^*}{\mathrm{d}t} = \sigma\phi A n\,\mathrm{d}x$$

$A \times \mathrm{d}x$ is the volume of the target and consequently $nA \times \mathrm{d}x$ is the total number of target atoms, N:

$$\frac{\mathrm{d}N^*}{\mathrm{d}t} = \sigma\phi N \tag{1.9}$$

As Equation (1.9) refers to the reaction of a specific nuclide, N is the total number of atoms of that nuclide. Equation (1.9) is used mainly for irradiation in a nuclear reactor, where the target is located inside a uniform flux of neutrons. Equation (1.8) is used for narrow beams from charged-particle accelerators.

In the case of prompt activation analysis, the only important number is the number of nuclei transformed by the nuclear reaction. Hence, the number of nuclei transformed is given by the integration of Equations (1.8) or (1.9) over the irradiation time. However, for delayed activation analysis, the measured value is the number of nuclei decaying during the measurement period, which is delayed and completed a little time after the irradiation. Radionuclides are not only formed during the irradiation, but are also decaying. Combining Equations (1.9) and (1.1) leads to the complete equation for the formation rate of the number of radioactive nuclei during irradiation in a constant flux (ϕ) of projectiles:

$$\frac{\mathrm{d}N^*}{\mathrm{d}t} = \sigma\phi N - \lambda \times N^* \tag{1.10}$$

Integration of Equation (1.10) with the initial condition $N_0^* = 0$, leads to the equation for the number of radioactive nuclei at the end of irradiation (EOI) where t_i is the irradiation time:

$$N_{EOI}^* = \frac{\sigma\phi N}{\lambda}(1 - e^{-\lambda t_i}) \tag{1.11}$$

In the product λt_i, λ should have the same unit as t_i, whereas in the term before the parentheses in Equation (1.11), the unit should be the same as the time unit of ϕ. For cases where $\lambda t_i \ll 1$ ($t_i \ll t_{1/2}$), the parenthetical term is approximately equal to λt_i ($e^{-\lambda t_i} \sim 1 - \lambda t_i$ for $\lambda t_i \ll 1$) and N_{EOI}^* is equal to $\sigma\phi N t_i$, i.e. the rate of formation multiplied by the irradiation time, since the decay is negligible.

For long irradiation times relative to the half-life of decay ($t_i \gg t_{1/2}$; $\lambda t_i \gg 1$), the term in parentheses is equal to 1, which is its maximum value. This is *the maximum number of nuclei which can be formed.* Longer irradiation does not increase the number of radioactive nuclei, since the rate of formation by the nuclear reaction is equal to the rate of disappearance by radioactive decay. This statement assumes that N does not change considerably during irradiation, which is common for the usual values of ϕ ($\leq 10^{15}$ n s^{-1} cm^{-2}), σ (1 barn) and t_i (< 10 days). For extreme circumstances, $\Delta N/N$ is still usually smaller or equal to 10^{-3}. This maximal number of formed radionuclei is known as the *saturation value* or the number at saturation, N_{sat}^*:

$$N_{sat}^* = \frac{\sigma\phi N}{\lambda} \tag{1.12}$$

The saturation value can be increased by increasing the flux of the bombarding particles or the number of irradiated nuclei (the mass of the target), but not by longer irradiation. Usually the mass of the specific element in the target and not the number of atoms of this element is considered. It is desirable, therefore, to transform N (the number of atoms) into m (the mass of the specific element in the activated samples). The number of atoms in mass m of element with atomic weight M is $N = m/M \times N_A$, where N_A is the Avogadro number.

In contrast to the usual chemical measurements, where all natural isotopes of the same element react in almost the same way, nuclear activation does not show this. For example, consider chlorine. Chlorine has two stable naturally occurring isotopes: ^{35}Cl with a 75.77% abundance and ^{37}Cl with a 24.23% abundance. Irradiation of its thermal neutrons leads to a (n, γ) reaction forming the two radionuclides ^{36}Cl and ^{38}Cl. ^{36}Cl cannot be used as an IRN for activation analysis owing to the absence of γ radiation (^{36}Cl is a pure β emitter) and its long half-life. (Note that an IRN, indicator radionuclide, is a radionuclide used for identification and quantification of an element in activation analysis.) Consequently, the total number of chlorine atoms that should be used in Equations (1.9)–(1.12) equates with the number of ^{37}Cl atoms only. This number is obtained by multiplying N by the *abundance* of the parent IRN (0.2423 for ^{37}Cl). Using f for the abundance, Equation (1.12) assumes the form:

$$N_{sat}^* = \frac{m}{M} N_A \times f\frac{\sigma\phi}{\lambda} \tag{1.13}$$

The number of IRN nuclei at the end of irradiation is obtained from Equation (1.8) as:

$$N^*_{\mathrm{EOI}} = N^*_{\mathrm{sat}}\left(\mathrm{i} - \mathrm{e}^{-\lambda t_{\mathrm{i}}}\right) \tag{1.14}$$

Measurement of the γ activity (the rate at which γ-rays are emitted by the activated sample) is not performed immediately after the end of irradiation, but rather is started after some time. This is called the *decay time*, t_{d}. The decay time (also called 'cooling' time) can be quite short, e.g. the time required to transfer the sample from the irradiation port to the measurement station, or it can be quite a long time. Long decay times are used when there is interest in IRNs with a medium to long half-life (known as *medium to long-lived* IRNs). The high activity of the short-lived radionuclides at short times after irradiation obscures the activity of the longer-lived radionuclides. Consequently, to measure the activity of medium and long-lived IRNs, measurement is delayed. In order to determine the concentrations of many elements in the sample, the γ activity in the irradiated sample is measured at several times (at least three) after permitting different decay times. Since the number of atoms in the IRN at the beginning of the decay time is N_{EOI}, the number at the end of the decay period, time t_{d}, is given, according to Equation (1.2), by:

$$N^*_{\mathrm{EOD}} = N^*_{\mathrm{SOC}} = N^*_{\mathrm{EOI}} \times \mathrm{e}^{-\lambda t_{\mathrm{d}}} = N^*_{\mathrm{sat}}(1 - \mathrm{e}^{-\lambda t_{\mathrm{i}}}) \times \mathrm{e}^{-\lambda t_{\mathrm{d}}} \tag{1.15}$$

Subscript EOD indicates end of decay, while subscript SOC denotes the start of counting. The measurement of γ activity is referred to as counting since the procedure involves counting the pulses that form in the γ-ray detection system as a result of the interaction of the γ photons with the detector. The counting takes place during time t_{c}, the count time.

The number of radionuclide atoms that remains undecayed at the end of the counting time, namely N^*_{EOC}, is given, according to Equation (1.2), by:

$$N^*_{\mathrm{EOC}} = N^*_{\mathrm{SOC}} \times \mathrm{e}^{-\lambda t_{\mathrm{c}}} = N^*_{\mathrm{sat}}\left(1 - \mathrm{e}^{-\lambda t_{\mathrm{i}}}\right) \times \mathrm{e}^{-\lambda t_{\mathrm{d}}} \times \mathrm{e}^{-\lambda t_{\mathrm{c}}} \tag{1.16}$$

The number of atoms that decay during the counting period (ΔN_{c}) is given by the difference between the number of atoms at the beginning of the counting period, N^*_{SOC}, and the number of atoms at the end of this period N^*_{EOC}:

$$\Delta N_{\mathrm{c}} = N^*_{\mathrm{SOC}} - N^*_{\mathrm{EOC}} = N^*_{\mathrm{sat}}\left(1 - \mathrm{e}^{-\lambda t_{\mathrm{i}}}\right) \times \mathrm{e}^{-\lambda_{\mathrm{d}}}\left(1 - \mathrm{e}^{-\lambda t_{\mathrm{c}}}\right) \tag{1.17}$$

The actual number of decay events recorded by the detection system is actually smaller than ΔN_{c} owing to three different factors. (1) The intensity of the specific γ line (the number of photons emitted per 100 atoms decayed, denoted by I_{γ}) is in many cases smaller than 100%. (2) Not every γ photon emitted will reach the detector. Photons are emitted isotropically and only those travelling in the direction of the detector will be recorded. (3) Some of the photons reaching the detector will not necessarily interact with it. Others will transfer only part of their energy to the detector, and consequently will not be recognized as having originated from a specific IRN.

Factors (2) and (3) are combined in a term called the geometric efficiency of the detector (ε). This quantity depends on the detector, the energy of the γ line and the distance of the sample from the detector. For a large sample, ε will also depend on the sample size. The geometric efficiency is determined experimentally by counting the activity of calibrated standards (which can be bought commercially), for which the rate of disintegrations is known accurately. The rate of decay of a standard is given in its certificate at a specific date and time. Its current activity at the time of calibrating a system is calculated using Equation (1.2). Note that Equation (1.2) applies to both the number of atoms in the sample and the activity of the sample since the rate is $-\mathrm{d}N^*/\mathrm{d}t$, which is given by λN^*. Thus, the number of counts, C, recorded by the detection system is given by the equation:

$$C = \varepsilon I_\gamma \times \Delta N_\mathrm{c} = \varepsilon I_\gamma \times N^*_\mathrm{sat}\left(1 - \mathrm{e}^{-\lambda t_\mathrm{i}}\right) \times \mathrm{e}^{-\lambda t_\mathrm{d}}\left(1 - \mathrm{e}^{-\lambda t_\mathrm{c}}\right)$$

For activation in a uniform flux the number of measured counts is given by the equation:

$$C = \frac{m}{M} \times N_\mathrm{A} f \sigma \varepsilon I_\gamma \phi \times \frac{1 - \mathrm{e}^{-\lambda t_\mathrm{i}}}{\lambda} \times \mathrm{e}^{-\lambda t_\mathrm{d}}(1 - \mathrm{e}^{-\lambda t_\mathrm{c}}) \qquad (1.18)$$

For a beam that is smaller than the target, ϕ is replaced by I/A where A is the area of the target. A is expressed usually by:

$$A = \frac{m_\mathrm{sample}}{\rho\, \mathrm{d}x}$$

It should be noted that here the mass is the total mass of the sample and not the mass of the element being determined. The term $\rho\, \mathrm{d}x$ is the usual way in which nuclear physicists express the sample thickness (its units are $\mathrm{gr \cdot cm^{-2}}$).

Equation (1.18) shows that the number of recorded counts is proportional to the mass of the element responsible for a specific γ line, and that this is the basis of the activation method since the measurement of the specific γ line yields the mass of the specific original nuclide. As an example, consider 10 mg of NaCl, irradiated for 20 minutes in a nuclear reactor with a beam of thermal flux of $5 \times 10^{12}\ \mathrm{n\,s^{-1}\,cm^{-2}}$. After 60 minutes of decay the sample was counted on a Ge detector for 15 minutes. In calculating the number of γ photons emitted during the measurement sequence the following data are required: (1) atomic mass of Na (22.99) and Cl (35.45); thus the number of molecules of NaCl and hence the number of atoms of either Na or Cl is then given by $10 \times 10^{-3} \times 6.02 \times 10^{23}/(22.99 + 35.45) = 1.03 \times 10^{20}$; (2) the natural isotopes of these elements, their abundances and cross-sections for the (n, γ) reaction. For the radionuclides produced by the (n, γ) reaction knowledge of their half-life, and the intensity of their γ lines is needed. For Na and Cl these data are:

$$\underset{100\%}{^{23}\mathrm{Na}} \xrightarrow{\sigma=0.53\ \mathrm{barn}} {}^{24}\mathrm{Na} \xrightarrow{15.02\ \mathrm{hours}} 1368\ \mathrm{keV}\ (100\%),\ 2754\mathrm{keV}\ (99.9\%)$$

$$\underset{75.77\%}{^{35}\mathrm{Cl}} \xrightarrow{\sigma=43\ \mathrm{barn}} {}^{36}\mathrm{Cl} \xrightarrow{3\times10^5\ \mathrm{years}} \mathrm{no}\ \gamma$$

$$\underset{24.23\%}{^{37}\mathrm{Cl}} \xrightarrow{\sigma=0.428\ \mathrm{barn}} {}^{38}\mathrm{Cl} \xrightarrow{37.24\ \mathrm{minutes}} 1642\ \mathrm{keV}\ (31.6\%),\ 2168\ \mathrm{keV}\ (42.4\%)$$

^{35}Cl is not important here since the ^{36}Cl that is produced from it is a pure β^- emitter and no γ photons are observed. For ^{24}Na in Equation (1.18), $f(m/M)N_\mathrm{A}$ is replaced by the value of 1.03×10^{20}, while for ^{38}Cl the term is substituted with $0.2423 \times 1.03 \times$

$10^{20} = 2.50 \times 10^{19}$. The unit of λ should be s^{-1} in order to agree with the flux data. Thus, $\phi = 5 \times 10^{12}$ and λ is $0.693/(15.02 \times 3600) = 1.282 \times 10^{-5}\ s^{-1}$ for ^{24}Na and $0.693/(37.24 \times 60) = 3.102 \times 10^{-4}\ s^{-1}$ for ^{38}Cl. Since the magnitude of various values of t is given in minutes it is preferable to express λ in min^{-1} rather than change all the times to units of second. Thus, $\lambda(^{24}Na)$ is $7.691 \times 10^{-4}\ min^{-1}$ and $\lambda(^{38}Cl)$ is $1.861 \times 10^{-2}\ min^{-1}$. The cross-sections are given in barns, while ϕ is given in cm^{-2} and hence the values of σ have to be multiplied by 10^{-24}. Thus, the number of photons emitted at each energy during the counting time is:

For 1368 keV:

$$\frac{10^{12} \times 1.03 \times 10^{20} \times 0.53 \times 10^{-24}(1 - e^{-20 \times 7.691 \times 10^{-4}}) \times e^{-60 \times 7.691 \times 10^{-4}}(1 - e^{15 \times 7.691 \times 10^{-1}})}{1.282 \times 10^{-5}}$$

For 1642 keV:

$$\frac{10^{12} \times 2.50 \times 10^{9} \times 0.428 \times 10^{-24}(1 - e^{-20 \times 1.861 \times 10^{-2}}) \times e^{-60 \times 1.861 \times 10^{-2}}(1 - e^{15 \times 1.861 \times 10^{-2}})}{3.102 \times 10^{-4}} \times 0.316$$

For 2168 keV:

$$\frac{10^{12} \times 2.50 \times 10^{9} \times 0.428 \times 10^{-24}(1 - e^{-20 \times 1.861 \times 10^{-2}}) \times e^{-60 \times 1.861 \times 10^{-2}}(1 - e^{15 \times 1.861 \times 10^{-2}})}{3.102 \times 10^{-4}} \times 0.424$$

For 2754 keV:

$$\frac{10^{12} \times 1.03 \times 10^{20} \times 0.53 \times 10^{-24}(1 - e^{-20 \times 7.691 \times 10^{-4}}) \times e^{-60 \times 7.691 \times 10^{-4}}(1 - e^{15 \times 7.691 \times 10^{-4}})}{1.282 \times 10^{-5}} \times 0.999$$

It should be emphasized that this is the number of photons emitted and not those actually detected. To obtain the number of actual counts the above number is multiplied by ε, which itself depends on the size of the detector, the detector–sample distance and the energy of the photons. The real spectra obtained will also include single and double escape peaks, and should the sample be too close to the detector sum peaks as well (the meaning of these peaks will be explained later).

For some elements, principally gold and tantalum, the radionuclides produced by the (n,γ) reaction with the reactor neutrons have exceptionally high cross-sections (e.g. for ^{198}Au the cross-section is 2.5×10^4 b). As a result of the high cross-section and despite the limited concentration of ^{198}Au, the stable nuclide forms radionuclides from the absorption of two neutrons (in two consecutive steps and not via one interaction). In the example of gold, where ^{197}Au is the only one stable isotope, the reaction scheme is:

$$\begin{array}{ccccc} ^{197}Au & \xrightarrow{(n,\gamma)} & ^{198}Au & \xrightarrow{(n,\gamma)} & ^{199}Au \\ & & \downarrow \beta^- & & \downarrow \beta^- \\ & & ^{198}Hg(stable) & & ^{199}Hg(stable) \end{array}$$

The rate of ^{198}Au formation is given by the equation:

$$\frac{d^{198}Au}{dt} = {}^{197}Au \times \sigma_1\phi - (\lambda_1 + \sigma_2\phi)\,{}^{198}Au \qquad (1.19)$$

Equation (1.19) is the same as Equation (1.10) when substituting λ for $\alpha_1 = \lambda_1 + \sigma_2\phi$. Thus the number of ^{198}Au radionuclides at the end of irradiation is given by:

$$N(^{198}\mathrm{Au})_{\mathrm{EOI}} = \frac{\sigma_1 \phi(^{197}\mathrm{Au})}{\lambda_1 + \sigma_2 \phi}[1 - \mathrm{e}^{-(\lambda_1 + \sigma_2 \phi)t}] = \frac{\sigma_1 \phi(^{197}\mathrm{Au})}{\alpha_1} \times (1 - \mathrm{e}^{-\alpha_1 t}) \tag{1.20}$$

The rate of ^{199}Au formation is given by:

$$\frac{\mathrm{d}^{199}\mathrm{Au}}{\mathrm{d}t} = \sigma_2 \phi^{198}\mathrm{Au} - \lambda_2\,(^{199}\mathrm{Au}) \tag{1.21}$$

This equation is similar to the set of differential equations given in Equation (1.7), but differing in that the constants in the positive term in one equation varies from the negative of the next, as not all of the R_i disappearances lead to R_{i+1}. However, the same method of Bateman can be used here. The solution of Equation (1.21) when substituting it in Equation (1.20) yields:

$$^{197}\mathrm{Au} = \frac{\sigma_1\,\sigma_2\,\phi_2}{\alpha_1 \lambda_2} \times {}^{197}\mathrm{Au}^0 \left(1 - \frac{\lambda_2\,\mathrm{e}^{-\alpha_1 t} - \alpha_1\,\mathrm{e}^{-\lambda_2 t}}{\lambda_2 - \alpha_1}\right) \tag{1.22}$$

Equation (1.22) neglects the disappearance of ^{199}Au due to the absorption of neutrons. An equation that includes this process is given by Heydorn (1964), but its contribution is seen to be negligible for all fluxes used in activation analysis. The general treatment can be found in Friedlander *et al.* (1981, p. 201).

1.2.3 *The chart of the nuclides*

As in the periodic table of the elements, which classifies all known elements by their main properties in a systematic way, the known nuclides have been described in the chart of the nuclides (as suggested by Segre). Figure 1.5 illustrates a small part of the chart of the nuclides. The chart is constructed such that the nuclides in the same row have the same value of Z (i.e. they are isotopes of the same element), while nuclides in the same column have the same number of neutrons. The number at the left of each row is therefore the Z value of all nuclides in that line; the number at the bottom of each column indicates the neutron count in the nuclides present in that column. The darkened squares represent the naturally occurring nuclides. It can be seen from the chart that stable isotopes form an approximately ascending diagonal. This line of the naturally stable nuclides is called the stability line. Nuclides above this line (to the left) have an excess of neutrons and decay by the β^- process. Nuclides below the stability line (to the right) have an excess of protons and decay by either β^- or EC (written as ε) processes or by both. Some of the nuclides within the stability line are unstable (those having an odd number of either neutrons or protons, and mainly those with odd numbers of both). These nuclides decay sometimes by both β^- and β^+ processes. All β processes occur along the descending diagonal. (See Fig. 1.6 for a schematic representation of these processes.)

Some of the squares are divided into two (or in some cases into three). These divisions indicate where the excited states of the nuclides are relatively long-lived. In most circumstances, excited nuclides decay extremely rapidly (in less than 10^{-11}s). However, some excited states live longer. Usually, if the half-life of an excited state is longer than 1 ms (this is an arbitrary measure as some use a limit of 1 s while others use 1 μs) the excited state is called a metastable state or an isomer of the nuclide. This

	16	S 32,06 σ 0,520		S 29 187 ms β⁺ βp 5,44; 2,13...	S 30 1,18 s β⁺ 4,4; 5,1... γ 678...	S 31 2,58 s β⁺ 4,4... γ 1266...	S 32 95,02 σ 0,53 σn, α 0,004	S 33 0,75 σn, p 0,002 σn, α 0,140	S 34 4,21 σ 0,240
	15	P 30,97376 σ 0,180	P 27	P 28 268 ms β⁺ 11,5... γ 1779; 4497... βp0,680;0,956... βα2,105;1,434...	P 29 4,1 s β⁺ 3,9... γ 1273...	P 30 2,50 m β⁺ 3,2... γ (2235...)	P 31 100 σ 0,180	P 32 14,3 d β⁻ 1,7 no γ	P 33 25,3 d β⁻ 0,2 no γ
Si 28,0855 σ 0,16	Si 24 103 ms β⁺ βp 3,913	Si 25 218 ms β⁺ βp 4,08; 1,87; 3,33...	Si 26 2,21 s β⁺ 3,8... γ 829; 1622... m	Si 27 4,16 s β⁺ 3,8... γ (2210...)	Si 28 92,23 σ 0,17	Si 29 4,67 σ 0,28	Si 30 3,10 σ 0,107	Si 31 2,62 h β⁻ 1,5... γ (1266) σ 0,48	Si 32 101 a β⁻ 0,2 no γ
Al 26,98154 σ 0,230	Al 23 470 ms β⁺ βp 0,83	Al 24 129 ms | 2,07 s Iγ 426 β⁺ 13,3... γ 1369... βα 1,42; 1,79... | β⁺ 4,4; 8,7... γ 1369; 2754; 7069... βα 1,98...	Al 25 7,18 s β⁺ 3,3... γ (1612...)	Al 26 6,35 s | 7,16 · 10⁵ a β⁺ 3,2 | β⁺ 1,2 γ 1809; 1130...	Al 27 100 σ 0,230	Al 28 2,246 m β⁻ 2,9 γ 1779	Al 29 6,6 m β⁻ 2,5... γ 1273; 2426; 2028...	Al 30 3,60 s β⁻ 5,1; 6,3... γ 2235; 1263; 3498...	Al 31 644 ms β⁻ 5,6; 7,9... γ 2317; 1695...
Mg 21 122,5 ms β⁺ βp 1,94; 1,77...	Mg 22 3,86 s β⁺ 3,2... γ 583; 74...	Mg 23 11,3 s β⁺ 3,1... γ 440...	Mg 24 78,99 σ 0,052	Mg 25 10,00 σ 0,180	Mg 26 11,01 σ 0,0382	Mg 27 9,46 m β⁻ 1,8... γ 844; 1014...	Mg 28 20,9 h β⁻ 0,5; 0,9... γ 31; 1342; 401; 942...	Mg 29 1,09 s β⁻ γ 2224; 1398; 960...	Mg 30 325 ms β⁻ γ 443; 243...
Na 20 446 ms β⁺ 11,2... βα 2,15; 4,44... γ 1634...	Na 21 22,48 s β⁺ 2,5... γ 351...	Na 22 2,602 a β⁺ 0,5; 1,8 γ 1275 σ 29000	Na 23 100 σ 0,40+0,13	Na 24 20 ms | 14,96 h Iγ 472 β⁻ ~ 6 | β⁻ 1,4... γ 2754; 1369...	Na 25 59,6 s β⁻ 3,8... γ 975; 390; 585; 1612...	Na 26 1,07 s β⁻ 7,4... γ 1809...	Na 27 304 ms β⁻ 8,0... γ 985; 1698... βn 0,46...	Na 28 30,5 ms β⁻ 13,9... γ 1474; 2389... βn	Na 29 42,9 ms β⁻ 13,4... γ 2560; 1474; 1586; 1638... βn 1,70; 0,19...
Ne 19 17,22 s β⁺ 2,2... γ (1357...)	Ne 20 90,51 σ 0,037	Ne 21 0,27 σ 0,692	Ne 22 9,22 σ 0,048	Ne 23 37,2 s β⁻ 4,4... γ 440; 1639...	Ne 24 3,38 m β⁻ 2,0... γ 874 m	Ne 25 602 ms β⁻ 7,3... γ 90, 980...	Ne 26	Ne 27	Ne 28
F 18 109,7 m β⁺ 0,6 no γ	F 19 100 σ 0,0095	F 20 11,0 s β⁻ 5,4 ... γ 1634...	F 21 4,16 s β⁻ 5,3; 5,7... γ 351; 1395...	F 22 4,23 s β⁻ 5,5 ... γ 1275; 2083; 2166...	F 23 2,23 s β⁻ γ 1701; 2129; 1822; 3431...	F 24	F 25	F 26	18
O 17 0,038 σn, α 0,235	O 18 0,200 σ 0,00016	O 19 27,1 s β⁻ 3,3; 4,7... γ 197; 1357...	O 20 13,5 s β⁻ 2,8 γ 1057	O 21 3,4 s β⁻ 6,4... γ 1730; 3517; 280; 1787...	O 22	O 23	O 24		
N 16 5,3 µs | 7,13 s Iγ 120 β⁻... | β⁻ 4,3; 10,4... γ 6129; 7115... βα 1,3...	N 17 4,17 s β⁻ 3,2; 8,7... βn 1,17; 0,38... γ 871; 2184	N 18 0,63 s β⁻ 9,4 γ 1982; 1650; 817; 2467	N 19	N 20	N 21	N 22	16		
C 15 2,45 s β⁻ 4,5; 9,8... γ 5298...	C 16 0,747 s β⁻ βn 0,79; 1,72	C 17	C 18	C 19	C 20				
B 14 16,1 ms β⁻ 14,0... γ 6090; 6730	B 15		B 17		14				
	Be 14		12						
	10								

Fig. 1.5 A part of the chart of nuclides.

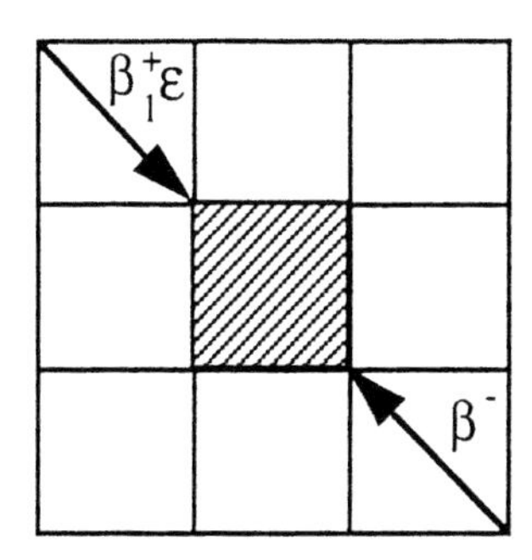

Fig. 1.6 Schematic representation of β processes occurring along the diagonal of the chart of the nuclides.

situation exists both for stable and radioactive nuclides. For example, the stable nuclide ^{77}Se (natural abundance of 7.7%) has an isomer which is written as ^{77m}Se with a half-life of 17.5 s. The radionuclide ^{82}Br ($t_{1/2}$ = 35.34 h) has an isomer ^{82m}Br with half-life of 6.1 min. In some radionuclides the isomer occasionally has a longer half-life than the ground state species, as for example ^{80}Br ($t_{1/2}$ = 17.6 min) and ^{80m}Br ($t_{1/2}$ = 4.42 h). Such radionuclides possess what is called an isometric state with higher energy which is lost on decaying to the ground state. Most isomeric states decay by emission of γ photons when decaying to their ground state. However, there are metastable states of radioactive nuclides that decay by β processes also, or by both β^- emission and isomeric transition (photon emission). The isomeric transition is denoted usually by IT.

For stable nuclides the chart of the nuclides usually gives their natural abundance as a percentage of that isotope in the natural mixture of isotopes for the specific element and the cross-section of interaction with thermal neutrons. The chart also presents radionuclide half-lives (in seconds, minutes, hours, days or years), main decay modes, and main γ lines (the energies of the γ lines). However, not all γ lines are given, and where they are their intensities are not quoted. These data as well as complete decay schemes can be found in tables of isotopes or various compilations of γ lines.

1.3 Gamma detection systems

In activation analysis a detection system for the γ photons which will measure the total number of photons together with their energy distributions is needed. Fortunately, one exists in the shape of a γ-ray spectrometer. The type of measurement undertaken is influenced by which of two solid-phase detectors is used.

1.3.1 *NaI(Tl) scintillation detector*

A scintillation detector comprises a *scintillator* or *phoshor* which is optically coupled to a photomultiplier tube. The most common scintillator for γ photon measurements is a large crystal of NaI activated with 0.1–0.2% Tl. The γ photon reacts with the detector, ejecting electrons. These electrons excite or ionize the scintillator crystal. De-excitation of the scintillator then occurs via fluorescence in about 0.2 μs as a result of the Tl^+ activator (visible light). It is the small percentage of Tl, the 'activator', which is added that shifts the emitted radiation to a longer wavelength from the ultraviolet that is normally due to emission from NaI. This visible (fluorescent) emission from the excited Tl is required for two reasons. First, to reduce the self-absorption of the emitted de-excitation light from the NaI crystal, and second, the shift from ultraviolet to visible light increases the sensitivity of the photomultiplier to the emitted photons. The most popular size of NaI(Tl) detector for routine γ spectrum measurements is a 7.5 cm diameter by 7.5 cm high cylinder. Approximately 30 eV of energy deposited in the NaI crystal produces one light photon. The light photon ejects an electron from the photocathode, and this electron is accelerated towards the dynodes in the photomultiplier by an electric voltage. In each dynode the electron ejects more electrons (typically 3–4), thus multiplying the electron current. There are usually 10 dynodes and each electron ejected from the photocathode produces about 10^6 electrons. It takes, on average, about 10 light photons to release one photoelectron at the photocathode of the multiplier. Therefore, approximately 300 eV of γ energy must be deposited in the NaI to release one photoelectron. Since the number of photoelectrons is proportional to the γ energy, the final number of electrons in the pulse (corresponding to the voltage of the electric pulse) is proportional to the energy of the γ photon. The pulse voltage obtained from the photomultiplier is quite high (50–1000 mV) and only modest amplification by a pulse amplifier is required. The electric pulses are sorted according to their energies by a multichannel pulse height analyser, which usually sorts the pulses into ranges of 0–10 V. Each energy range is fitted to one channel. The spectrum of counts per channel is actually an energy spectrum of the γ photons.

1.3.2 *Solid-state ionization detector*

The simplest idea that can be applied to measuring radioactive decay is by using the common property of ionizing radiation emission in its various forms (particles, photons). When ionizing radiation strikes a non-conducting or semiconducting material, it ejects free (conducting) electrons from the material and leaves holes or cations. The number of electrons ejected or holes formed is proportional to the energy of the striking photon or charged particle. The amount of energy required to raise an electron from the valence band to the conduction band in a semiconductor is considerably smaller than the energy needed in an insulator for the same process, or the energy required to form an electron–ion pair in the liquid or gaseous phase. Thus, for the same amount of energy absorbed in a detector by the photon, more electrons are formed in a semiconductor than in an insulator.

The larger number of electrons reduces statistical fluctuation and hence the width of the energy peak in the detector (measured as FWHM, full width at half maximum of the peak) is decreased. Figure 1.7 shows the peaks obtained from the same radionuclide source by using NaI(Tl) and Ge semiconductor detectors. The smaller FWHM in the case of the Ge ionization detector leads to much higher resolution. However, the currents associated with the ions produced in the Ge detector by the

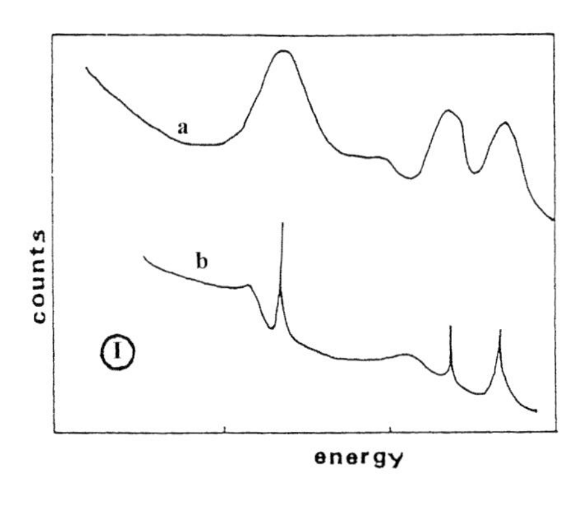

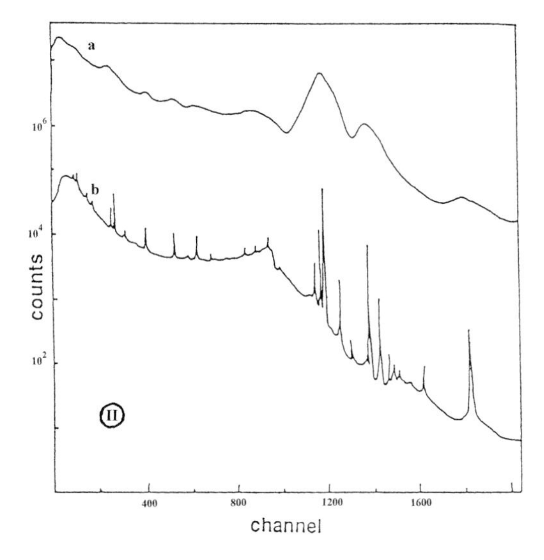

Fig. 1.7 (I) The γ-ray spectra of a mixed ^{137}Cs + ^{60}Co source: (a) taken with an NaI(Tl) detector; (b) taken with a HPGe detector.
(II) The γ-ray spectra of a neutron activated biological sample: (a) taken with a 3″ × 3″ NaI(Tl) detector; (b) taken with a Ge(Li) detector.

ionizing radiation are smaller than those formed in the scintillation detector, owing to the high multiplication factor in the photomultiplier of the NaI(Tl) set-up. Hence, before any measurement of the current from ionization detectors the signal must be amplified. In order to reduce losses in the pulse current and to reduce the rise time of the pulse (limiting the rate of pulses that can be counted accurately), a preamplifier located as close as possible to the detector is used for initial amplification of the detector output signal. The voltage pulse produced at the output of the preamplifier is proportional to the collected charge and independent of detector capacitance. However, the output pulse from the preamplifier is too low for sorting by the multichannel analyser (MCA). Further amplification is completed by the main amplifier, which also serves to shape the pulse. In order to reduce the noise arising from the leakage current of the electrons in the conduction band, which are agitated by thermal excitation, *the semiconductor is cooled to liquid nitrogen temperatures during the actual measurement of the γ spectrum.*

Of all semiconductor materials, germanium and silicon are almost exclusively used for modern γ-ray spectrometry, since only for these elements can adequate pure material be prepared. This is important as any impurities which are not four-valent will produce either free electrons in the conduction band or 'holes', leading to increased leakage current, which is not due to the ionizing radiation being measured. Even silicon and germanium could not be prepared to sufficient purity for use as γ-ray detectors in the early years of semiconductor technology. However, the impurities were not so great that they could not be compensated by the lithium ion drifting method developed in 1960. The method is used to effectively compensate p-type (i.e. acceptor type containing an excess of trivalent impurities) grown crystals of silicon and germanium. The small lithium ions are pulled into the crystals by an electric field and high temperature. Owing to its very low ionization potential, lithium acts as an electron donor impurity compensating for the excess of holes. The lithium ions, having a high mobility, drift under the influence of local fields in such a way that the number of lithium donor atoms compensate everywhere in the crystal exactly for the acceptors of the original semiconductor material. These lithium-drifted detectors are called extrinsic detectors and denoted Si(Li) (pronounced 'silly') and Ge(Li) (pronounced 'jelly').

Since lithium ions have high mobility in germanium at room temperature, Ge(Li) detectors must be maintained at the liquid nitrogen temperature of (77 K), immediately after the desired level of impurity compensation is obtained. A schematic diagram of a measurement system incorporating a semiconductor detector is given in Fig. 1.8. Note that in silicon the mobility of lithium ions at room temperature is low enough that Si(Li) detectors can be stored temporarily at room temperature without cooling. It should be remembered, however, that when measuring γ-ray spectra, both semiconductor detectors should be cooled to 77 K.

In the mid-1970s, advances in germanium purification technology made available high-purity germanium, with impurity levels as low as 10^{10} atoms cm^{-3}, permitting use in γ-ray spectrometry detection without lithium drifting. These so-called *intrinsic germanium detectors* are usually denoted HPGe (high-purity germanium). The outstanding feature of these detectors is that they do not have to be kept constantly at liquid nitrogen temperature. Nevertheless, HPGe detectors must still be cooled to 77 K during measurements to reduce the leakage current arising from the thermal excitation of electrons to the conducting band. HPGe and Ge(Li) detectors are virtually identical from the point of view of measurement. However, owing to the more convenient use of HPGe, it has completely replaced the Ge(Li) detector in contemporary γ-ray spectroscopy. Silicon cannot be prepared to such high purities and Si(Li) detectors still remain in use.

Owing to the higher atomic number of germanium, the cross-section (probability of interaction) for interaction with a γ photon is higher than it is for Si. However, Si(Li) detectors are used for the measurement of X-rays (up to 40–60 keV). Another type of detector for X-rays, or low-energy γ-rays, is the planar (thin) Ge detector. Although Si(Li) has a better discrimination for γ-rays, the planar Ge detector has higher measurement efficiency in the range 40–150 keV.

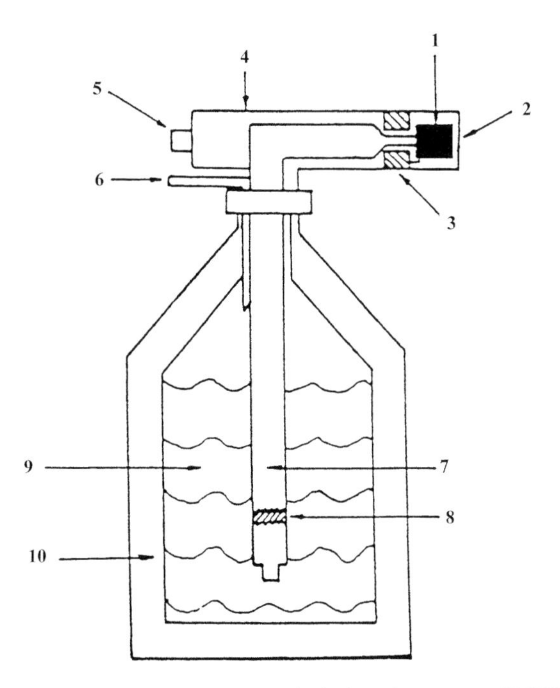

Fig. 1.8 A schematic diagram of a HPGe, Ge(Li) or Si(Li) detection system: (1) HPGe, Ge(Li) or Si(Li) crystal; (2) end cap; (3) field effect transistor preamplifier; (4) metal housing; (5) vacuum/pressure release; (6) liquid nitrogen filling tube; (7) cold finger to cool the crystal; (8) gas absorbant, in order to maintain vacuum; (9) liquid nitrogen; (10) vacuum heat insulation.

The main performance characteristic of a nuclear detector for activation analysis is its *resolution*, usually expressed as the full width of the peak at half of its maximum (abbreviated FWHM). The narrower the peak (lower FWHM), the better is the ability of the detector to separate two close peaks, i.e. it has better resolution. The Ge detectors have very high resolution compared to NaI(Tl) scintillators. At 1332 keV (a value corresponding to one of the ^{60}Co γ-rays used to characterize the FWHM of a detector) the resolution of a good HPGe detector is about 1.8 keV while the resolution of a 3″ × 3″ NaI(Tl) detector is about 60 keV. Figure 1.6 gave spectrum examples of an activated sample and radioisotope standards (^{60}Co + ^{137}Cs) measured with both NaI(Tl) (line a), and HPGe (line b) detectors to observe different resolutions. However, the detector *efficiency* (the fraction of γ photons hitting the detector that appear in the photopeak of the spectrum) of NaI(Tl) is greater, and the efficiency of a Ge detector is given usually relative to NaI(Tl).

The first Ge(Li) detectors were small and their efficiency relative to NaI(Tl) was less than 10% (at 1332 keV). However, present HPGe detectors of diameter 450 mm have higher efficiency than the NaI(Tl) 3″ × 3″ (130% at 1332 keV). The relative efficiency of Ge detectors decreases with increasing energy owing to the higher Z value for iodine. For most applications of activation analysis where there is a multitude of γ lines, resolution is more important than efficiency and therefore Ge detectors are mostly used. However, the NaI(Tl) detector is sometimes used for very high-energy γ-rays. Further advantages of the NaI(Tl) detector is its lower price, simpler electronics and operation at room temperature.

A method used to improve the counting efficiencies of both NaI(Tl) and Ge detectors is the incorporation of well detectors. These are specially built where a cylindrical well has been drilled at the centre of a cylindrical semiconductor or crystal. Alternatively, the crystal or semiconductor has been modelled around the well. The sample is placed in the well. The detector surrounds the sample in almost all directions and the geometric factor for the radiation–detector interaction is almost 4π. However, in all directions the thickness of the detector is smaller than that found in usual detectors, and this contributes to reduced efficiency at higher energy. However, for energies that are not too high, the use of well counters does increase the efficiency of counting.

1.4 The shape of the γ spectrum

In optical (absorption and emission) spectrometry of a species that has only one excited state the resulting spectrum will appear as a single line without any background. This profile is due to the 'yes/no' quantum characteristic of the energy transfer. The energy difference between the ground and excited states is such that energy is either transferred or not transferred; the energy cannot be transferred in parts. Thus, a photon of light in the infrared, visible or ultraviolet can be absorbed totally or not at all – it cannot lose only part of its energy. *The situation is different for high-energy γ-rays.* Low-energy γ-rays interact with matter mainly by the *photoelectric process.* In this process the γ photon interacts with one electron of the material, losing all its energy to this electron. The transferred energy is higher than the binding energy of the electron, and thus the electron is ejected from the atom with kinetic energy equal to this energy difference. Since the range of electrons with high kinetic energy is considerably smaller than that of γ photons, the electrons will lose their kinetic energy within the detector material and from the point of view of the detector this means that the photon energy is absorbed completely, resulting in the same energy being absorbed for all photons (assuming that all photons are of the same energy). Thus, for low-energy X-rays, the measured spectrum will be composed of discrete peaks and negligible background, like an optical spectrum. The situation is different for γ photons of higher energy. Here, the cross-section (probability of reaction or the rate constant) for the photoelectric process decreases with increasing photon energy by more than that for a competing process – the so-called *Compton scattering.*

For higher energy photons the main interaction with matter is via Compton scattering. In germanium the photoelectric effect is the dominant process for γ interaction up to energies of about 200 keV, whereas from 200 keV and more, Compton scattering assumes greater importance. (It should be stressed that below 200 keV, Compton scattering also occurs, but its contribution to photon interaction with germanium is less than that of the photoelectric process). The contribution from Compton scattering increases with photon energy despite the fact that its cross-section decreases with the energy. This occurs because this cross-section decrease is less than that observed in the photoelectric process.

In Compton scattering the γ photon interacts with what might be termed a free electron (mainly electrons from the outer shells of atoms) which is in contrast to the photoelectric process where inner-shell electrons, mainly from the innermost K shell, are subject to the interaction. Only part of the photon energy is transferred in Compton scattering. The energy the electron receives, however, is more than sufficient to eject it from the atom and it moves with the excess energy as kinetic energy which is lost in collisions over a very short distance. In retaining part of its energy the photon changes its frequency since $E = h\nu$. In the Compton process therefore, the energy of the original photon is shared between two particles (the ejected electron and the scattered photon), and consequently there is a continuous distribution of scattered photon energies. In order to conserve both energy and momentum, the γ photon cannot lose all of its energy. (Note that in the photoelectric effect, momentum conservation is compensated for by atom recoil from where the electron was ejected, while in Compton scattering there is an interaction with a 'free' electron). The maximum energy that the photon can lose is given by

$$E_\gamma/(1 + m_0c^2/2E_\gamma) \tag{1.23}$$

where E_γ is the energy of the initial photon and m_0 is the electron rest mass. The term m_0c^2 is the rest mass of an electron in energy units and is equal to 0.511 MeV.

The difference between the photoelectric and Compton processes with respect to their responses in the detector material is that the ejected electron (in both processes) will lose all its energy in the detector, owing to its short range, while the Compton scattered photon might escape from the detector without further interaction. In γ-ray spectrometry measurement is made of the *full-energy peak*, called the *photopeak*, as the associated energy will be the only quantity to appear if the interaction process is photoelectric absorption and no other. The photopeak will result from γ photons losing all their energy by a photoelectric absorption, or alternatively from a Compton scattering process which is followed by photoelectric absorption of the scattered lower energy γ photons in the detector. (The photoelectric absorption can occur after one or several scatterings, all within the detector material.) However, the scattered γ photon from the Compton process (or from the several consecutive Compton processes) might escape from the detector crystal, thus leaving the detector with less energy than the full-energy peak.

The escape of any scattered photons will reduce the number of counts in the photopeak. This escape also has a major disadvantage in that formation of a background of lower energy peaks occurs. Since a fraction of the energy (given to the electron) is absorbed in the detector, the detector registers this as a count of lower γ energy than the actual energy of the γ photon that was emitted by the IRN. The scattered photons do not have discrete energies and can therefore lie between a minimum, given by Equation (1.23), and a maximum, given by the original energy of the photon (minus a small value for the binding energy of the electron, which is not a completely free electron). For a monoenergetic γ source, the Compton scattering produces a background continuum ranging from zero up to a maximum called the *Compton edge* given by Equation (1.23). This results from the fact that the minimum energy that the scattered photon can have is not zero but rather a value given by Equation (1.24):

$$E_{\text{minimum of scattered}\,\gamma} = E_\gamma - \frac{E_\gamma}{1 + m_0c^2/2E_\gamma} = \frac{E_\gamma\, m_0c^2}{2E_\gamma + m_0c^2} \qquad (1.24)$$

For $E_\gamma \gg m_0c^2$, E_{minimum} approaches a value of $m_0c^2/2$, i.e. 256 keV. Thus for high-energy γ photons the Compton edge will be separated from the photopeak by about 256 keV. For monoenergetic sources the range between the photopeak and the Compton edge is almost free of any counts. However, most actual samples contain many IRNs having different γ energies and the Compton continuum stretches from zero up to the Compton edge of the highest γ energy.

If the energy of the measured photon is above 1.022 MeV, a third process is possible in the interaction of photon with detector – a process called *pair production*. In this process the photon energy is transformed, under the influence of a nucleus, into matter occurring as an electron–positron pair (the transformation of energy into a matter–antimatter pair). Since the rest mass of each electron or positron is 0.511 MeV, the threshold of this reaction to conserve energy is 1.022 MeV. Although the threshold energy is 1.022 MeV the cross-section for pair production is very low for energies below 1.5–1.6 MeV and can be ignored. In pair production excess energy is shared between the kinetic energies of the electron and the positron. Since the respective ranges of positrons and electrons are very short, this excess energy (E_γ − 1.022 MeV) will be deposited in the detector. When the positron loses all of its kinetic energy, it reacts with an electron in an *annihilation reaction* to form two γ photons each with energy 0.511 MeV (i.e. transformation of matter–antimatter into electromagnetic energy). Each of these 511 keV photons can either escape from the detector without any interaction, or lose part of its energy by Compton scattering, or alternatively lose all energy in the detector (either by photoelectric absorption or by successive Compton scatterings and photoelectric absorption). If one of the 511 keV photons deposits all of its energy in the detector, while the other escapes totally, the energy absorbed by the detector will be 511 keV less than the full-energy peak. With γ photons of energy above 1.6 MeV the peak of $E_\gamma - 511$ keV is usually observed, where E_γ is the photopeak energy. This peak is called the *single-escape peak*, as one 511 keV photon escapes from the detector. Another peak in the spectrum is the *double-escape peak*, arising when two 511 keV photons escape from the detector without any interaction with the detector material. The energy of this peak is $E_\gamma - 1.022$ MeV. When analysing γ-rays spectrum to find the radionuclides in the sample, if a peak with energy $E_\gamma > 1.6$ MeV is known to belong to some radionuclide, it must be remembered that in the list of the established peaks there are some that do not belong to other radionuclides, but rather are either single- or double-escape peaks. We saw that ^{28}Al has a γ energy of 1.778 MeV. Thus we would expect to see in the ^{28}Al γ-ray spectrum peaks of 1.778, 1.269 (single-escape = 1.778 − 0.511) and 0.758 MeV (double-escape peak). ^{24}Na has two γ lines of 2.75 and 1.39 MeV. In the γ-ray spectrum four lines are expected at 2.75, 2.24 (single-escape), 1.73 (double-escape) and 1.39 MeV. Although the 1.39 MeV line is above the threshold of 1.022 MeV for pair production, its cross-section for this process is quite low and pair production contributes very little to the interaction of the 1.39 MeV photon with the detector. Hence, in most cases, we will not see the single- and double-escape peaks of this photon. Figure 1.9 gives the measured γ-ray spectra of ^{24}Na. The ratio of escape peaks to the photopeak depends on the detector size (influencing the probability of the γ escape) and on the energy of the γ photon (affecting the chance that the photon will interact by pair production). In order to determine if a peak is a photopeak or SE/DE (single-escape, double-escape), it is preferable to start the analysis of the spectrum from the highest energy peak. For each peak we can remove from the list of peaks its SE and DE peaks.

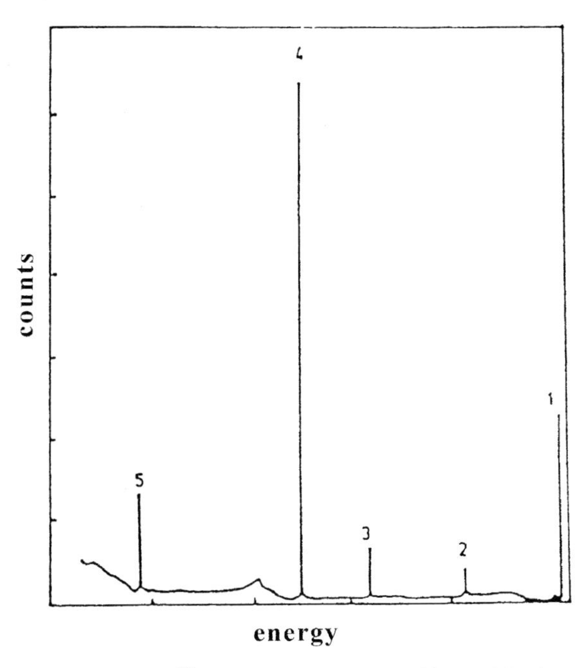

Fig. 1.9 The γ-ray spectrum of the ^{24}Na radioisotope taken with a HPGe detector: (1) photopeak at 2754 keV; (2) single-escape peak at 2243 keV (2754–511 keV); (3) double-escape peak at 1732 keV (2754–1022 keV); (4) photopeak at 1369 keV; (5) 511 keV positron annihilation.

An important artefact in many γ-ray spectra is the sum-peak, a peak which is due to the summation of two γ photons. If two photons hit the detector within a time interval smaller than the time characteristic of the detector, the detector treats both photons as a single photon and the energy deposited is the sum of the energies of the two photons. The two photons can be derived from different radionuclides, or the same radionuclide having more than one γ line. In the case of different radionuclides the resultant sum-peak is referred to as 'pile-up'; for the same nuclide the phenomenon is called coincidence summing. The sum-peaks can be reduced (relative to the photopeaks) by removing the measured sample further from the detector or by analysing a smaller sample. A lower rate of counting increases the average time between successive photons reaching the detector, and thus decreases the chances of pile-up.

In the spectrum of only one γ line there is also a peak with energy given by Equation (1.24). This peak is due to 180° backscattering (mainly from the shielding of the detector) of the original photon. The backscattered Compton photon reacts with the detector giving a peak of full energy. For a source with various γ-ray energies, this backscatter peak can range from 70 to 256 keV, thus adding to the Compton continuum rather than forming a distinct peak.

The Compton background is an intrinsic background owing to the physical characteristics of the detector and the interaction of radiation with matter. Another source of background signal is external due either to radioactive sources other than our sample or to cosmic radiation. This external background can be minimized by the use of shielding, mainly through the use of lead due to its high Z and high density. Usually, the detector is located inside a lead 'castle'. Lead shielding configurations suitable for a germanium detector are available commercially, usually in the form of a hollow cylinder with a sliding lid or revolving door for access.

In many cases where the Compton continuum interferes with the measurements of small peaks it can be reduced by using the Compton suppression method. In this method the germanium detector is located inside an annular scintillation detector or inside a ring of scintillation detectors. Many of the scattered Compton γ-rays which escape from the germanium detector (in the case of the Compton continuum), are transferring some energy to the surrounding detectors, whereas those photons that lose all their energy in the germanium detector do not cause any signal in the scintillation detector(s). The system is operated in the anti-coincidence mode, i.e. the multichannel analyser (MCA) is recording only the events occurring in the germanium detectors only. If within a fixed time the pulse in the germanium detector is followed by a pulse in the scintillation detector (indicating that the germanium pulse is in the Compton continuum and not part of a photopeak) this event is discarded. The scintillation detector(s) are usually of the NaI(Tl) type, although in some systems it is a crystal of bismuth germanate ($Bi_4Ge_3O_{12}$ – abbreviated BGO – because of its higher efficiency for the absorption of γ-rays (since it has a high-Z element). However, BGO crystals are more expensive and are therefore not used in most systems. A schematic diagram of a Compton suppression system is given in Fig. 1.10.

1.5 Neutron sources

Three main sources of neutrons are available: (1) research nuclear reactors (nuclear reactors used as neutron sources); (2) ion and electron accelerators, including neutron generators; and (3) radioactive sources.

Research nuclear reactors have the highest neutron fluxes, but are limited by their price and availability. Consequently, nuclear reactors are used for neutron activation analysis of minute amounts of sample where the main interest is laboratory chemical analysis, but not in situ or on-line activation analysis. Research nuclear reactors are usually large devices in which fissionable material, almost exclusively ^{235}U, is fissioned into two nuclides with the simultaneous emission of neutrons that induce further fission in a chain reaction. The fission-produced neutrons are very energetic. The cross-section for neutron-induced fission of fissionable nuclides increases with decreasing energy of the neutrons, and in order to increase the neutron activity, moderators that slow down the neutrons are inserted into the reactor. To reflect back

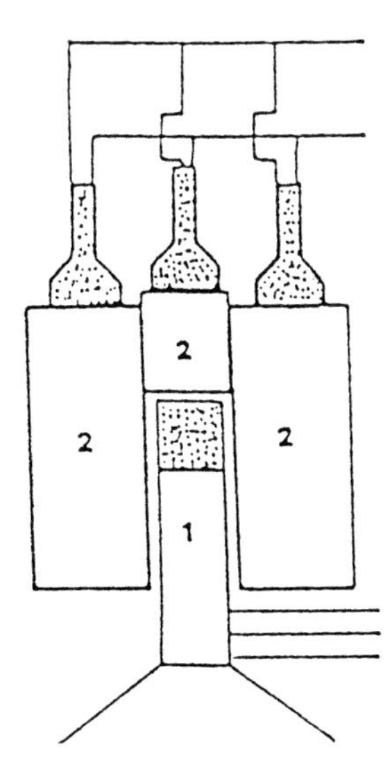

Fig. 1.10 Schematic diagram of an anti-coincidence Compton suppression γ spectrometer: (1) a HPGe detector with liquid nitrogen cooling; (2) NaI(Tl) detectors with photomultipliers.

some of the neutrons that leak from the reactor core, reflectors are used. The fission process releases large amounts of energy, mainly due to halting the recoil of the two fissioned particles, and the system must be cooled using a coolant (either liquid or gas). The nuclear reactors are categorized according to their fuel, moderator, coolant, reflector and configuration. Almost all research nuclear reactors (neutron sources) are heterogeneous reactors in which fuel rods are used. The fuel is enriched ^{235}U (natural uranium has only 0.7% of ^{235}U, the fissile material). Most research reactors contain 93–99% ^{235}U. Many reactors have rods comprising U–Al alloys. However, some of the newer designs (mainly those converted to 20% ^{235}U) are of the uranium–silicide type. TRIGA reactors operate with uranium–zirconium hydride fuel, which owing to its large negative temperature coefficient of reactivity permits operation of the reactor in pulses.

In the light water reactor (LWR), ordinary water (H_2O) is used both as a moderator and as a coolant. The reflector is mainly graphite but there are also Be- or H_2O-reflected reactors. There are either pool-type or tank-in-pool-type LWRs. Due to the relatively high cross-section for capturing thermal neutrons by hydrogen atoms, and therefore the relatively small amount of moderator ability, the flux of neutrons in light water reactors invariably contains a large fraction of fast and epithermal neutrons. The available power is in the range of 10–5000 kW with neutron fluxes of $5 \times 10^{10}-1.5 \times 10^{14}$ n cm^{-2} s^{-1}.

Many reactors have a unique design, though there are some common commercial types, for instance the American TRIGA and the Canadian Slowpoke. The TRIGA reactor is a popular multi-purpose research reactor. About 50 are in operation with power levels of 18 kW–3 MW and fluxes of $7 \times 10^{11}-3 \times 10^{13}$ n cm^{-2} s^{-1}. The most common types are the 250 kW (Mark II) and 1 MW (Mark I). They are pool-type LWRs, graphite reflected with uranium–zirconium hydride fuel with ^{235}U enrichment of 10–70%. The Slowpoke is a low-power (20 kW) reactor designed specifically as a teaching aid with additional uses in activation analysis and production of small amounts of radioisotopes. The system is designed to operate remotely. It can be provided with up to five irradiation sites in the core with flux of 10^{12} n cm^{-2} s^{-1} and five further tubes outside the reflector with half that flux.

Heavy water research reactors are of the tank closed-type. They usually contain enriched uranium fuel. Heavy water is used as both moderator and cooler, and neutrons are reflected from the heavy water and graphite. Owing to the low cross-section for absorption of thermal neutrons by D and O, and owing to the higher mass of moderator, these reactors are characterized by a well-thermalized neutron flux (very few epithermal and fast neutron fluxes, except inside the core). Because of the lower moderation power of D compared to H, the physical size of a heavy water reactor is larger and hence it has available a large irradiation volume. The reactor power is usually between 10 and 26 MW (fluxes of up to 2×10^{14} n cm^{-2} s^{-1}).

1.5.1 *Samples introduction*

The method of introducing a sample into a neutron flux depends on the physical structure of the reactor. It is essential that the introduction of the sample will not affect the operation of the reactor. The irradiation site may be within the reactor core or outside in the moderator/reflector region. If the reactor is an open-pool type with access from above, vertical tubes or ropes can be installed to lower the samples into the core or close to its side. The samples are contained in sealed ampules in order to prevent contact with the water surrounding the core. The ampules are usually made of aluminium because of its corrosion-resistance and short-lived activation products. For short irradiations, a polyethylene capsule can also be used. The manual loading of samples is neither quick nor reproducible in time, even when performing short irradiations. In order to obtain rapid and reproducible sample introductions, mechanical systems are used. Two types of mechanical system are used: chain-driven racks and pneumatic devices. The latter is the more common – it has shorter loading and unloading times, requires less maintenance and has fewer failures during operation. With the pneumatic device the sample is pushed along a tube by pressurized gas (air or nitrogen). The sample transfer time depends on the pressure and the distance over which the sample is transferred. In many systems the transit time is a second or less. The pneumatic device can be automated very easily. Figure 1.11 depicts schematically the pneumatic system. In the case of the closed-tank reactor the irradiation is achieved by either pneumatic device or with neutron beams.

1.6 Instrumental neutron activation analysis (INAA)

1.6.1 *Techniques*

INNA is one of the simplest techniques for trace element analysis. Samples are inserted into polyethylene or polypropylene cans for short irradiation or into a quartz vial, which is inserted into an aluminium can, for long irradiation. PVC cans or glass

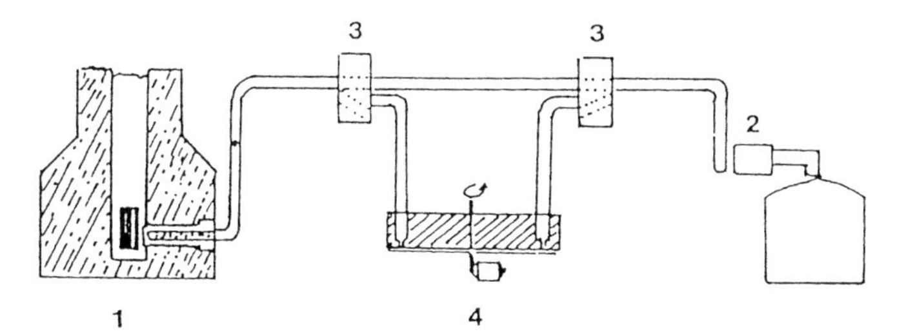

Fig. 1.11 Schematic layout of a fast transfer system for INAA: (1) reactor; (2) γ spectrometer; (3) valves; (4) sample changer.

vials cannot be used since they will be become radioactive owing to the formation of ^{38}Cl or ^{24}Na, respectively. For short irradiation times each sample is 'sent' alone to the reactor by means of a pneumatic-operated tube – the rabbit system. For long irradiation times many samples can be lowered into the reactor and left there for the required time. After the samples are removed from the reactors they are either counted (means form the γ-ray spectra) immediately or the γ-ray spectrum is taken after a 'cooling' (decay) time when the high radioactivity level due to short-lived radionuclides has reduced. Before counting the vials are washed to remove surface contaminants. In some cases the samples are counted inside the vial, while in others the samples are removed from the vial before the measurement to avoid interference from trace elements in the vial material.

1.6.2 *Calculation*

In general the first calculation step involves *peak identification*, i.e. identifying which peaks are photopeaks (discarding single-escape, double-escape and sum-peaks) and assigning the radioactive species producing them. Tables of γ-ray energies listed in order of increasing energies are available for ease of identification (Erdtmann & Soyka 1979; Yule 1981; Reus & Westmeier 1983; Browne *et al.* 1986; Alfassi 1990; Glascock 1996). Some tables give the energies of all known radionuclides while others give only those for radionuclides that are formed in nuclear activation, e.g. in a nuclear reactor. For optimum identification the tables give the half-life of the radionuclide, its probability of formation and other γ lines for the same radionuclide. Thus, if we find one γ peak of some radionuclide, but not its accompanying line or lines, it suggests an incorrect identification. The different lines must also be checked that they occur in the ratios reported, taking into account the relative intensities of the γ lines and the efficiency of the detector at those energies. Nowadays multichannel analysers (MCAs) are PC-based and have in their database a nuclide library. However, most libraries check only for the energy of the photopeak with a pre-set tolerance of 1–2 keV. Some commercial computer programs also check for the presence (in the right ratios) of other lines of the same radionuclide. Most programs do not attempt identification through half-life measurements. It should be remembered that single- and double-escape energies are not listed in most libraries and this also can lead to misidentifications. Figure 1.12 gives examples of γ-ray spectra measured from an activated sample.

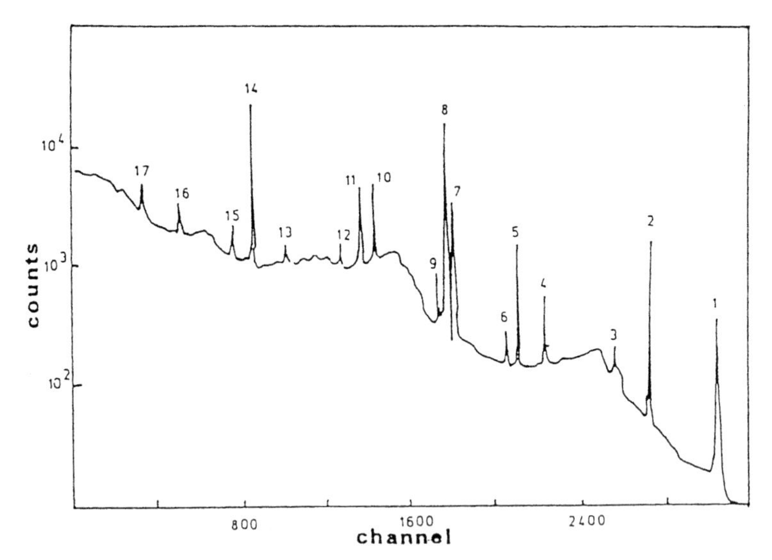

Fig. 1.12 The γ-ray spectrum of a neutron activated geological sample measured with a HPGe system 15 min after the end of irradiation: (1) 3084 keV ^{49}Ca; (2) 2754 keV ^{24}Na; (3) 2573 keV SE of ^{49}Ca; (4) 2243 keV SE of ^{24}Na; (5) 2113 keV ^{56}Mn; (6) 2062 keV DE of ^{49}Ca; (7) 1811 keV ^{56}Mn; (8) 1779 keV ^{28}Al; (9) 1732 keV DE of ^{24}Na; (10) 1434 keV ^{52}V; (11) 1369 keV ^{24}Na; (12) 1268 keV SE of ^{28}Al; (13) 1014 keV ^{27}Mg; (14) 846 keV ^{56}Mn + ^{27}Mg; (15) 757 keV DE of ^{28}Al; (16) 511 keV β^+ annihilation; (17) 320 keV ^{51}Ti.

The second step in the calculation is ascertaining the net counts under the peak (plotted as the number of counts in each channel per channel number). Various methods exist for the evaluation of the net peak 'area' (actually the net peak count). The simpler methods define the channels belonging to the peak and sum all counts in the peak. The total is $\sum_{i=L}^{u} C_i$, where C_i is the count in channel number i, and L and u are the lower and upper channels belonging to the peak. However, this count also contains a background (Compton and external) which should be subtracted. The background is estimated by taking the average counts of 3–5 channels at both sides of the peak, and multiplying by the number of channels in the peak:

$$\text{Background} = \frac{(u - L + 1)}{2 \times 3} \left[\sum_{i=L-3}^{L-1} C_i + \sum_{i=u+1}^{u+3} C_i \right]$$

The value of 3 in the previous equation can be increased or decreased, but it is considered an optimal value – lower numbers will increase the statistical error in the evaluation of the background, while larger numbers will increase the possibility that the following channels already belong to other peaks.

Radioactive decay is not an absolute process but rather is of a statistical nature. Consequently the larger the number of counts the more accurate the result. The standard deviation of each decay count (as long as the counting time is small compared to the half-life of the measured radionuclide) is equal to the square-root of the number of counts. The number of net counts is given by:

N = net counts = Total − Background = $T - B$

The standard deviation of the net counts σ_N and the relative standard deviation, $\%\sigma_N$, is given by the equation:

$$\sigma_N \sqrt{T+B} = \sqrt{N+2B}; \quad \%\sigma_N = 100 \cdot \sqrt{T+B}/N = 100 \cdot \sqrt{N+2B}/N$$

It should be remembered that the rate of counting is limited by both the rate of processes occurring in the detector (which cause 'pile-up' or sum-peak in instances of high count rates) and the rate of sorting the pulses by the multichannel analyser (MCA) and the electronic system. During the 'busy' period of the electronic system further acquisition of signals is not possible, so that each signal reaching the system during this 'busy' period will be discarded. The 'busy' time is called *dead time* and the actual measurement time (live time) is shorter than the real time (measured by the clock). Most MCA systems contain a device for correction of dead time. However, these devices are only accurate up to about 10% dead time. Therefore, it is usually recommended not to take measurements when the dead time is above 10%. Despite this recommendation, there are situations where measurement with a high dead time is inevitable, e.g. the measurement of a small amount of short-lived IRN in the presence of considerably greater activity of a longer-lived radionuclide (Alfassi & Rietz 1994). In this case the correct live time can be measured (Anders 1960, 1961) using a pulse generator of known rate. The generator pulses are introduced to the same electronic system (the height of the pulse is chosen in a range where there are no γ signals). The live time is then calculated from the ratio of the measured and expected counts from the pulse generator. Instead of using a pulse generator the longer-lived radionuclide can be measured after partial decay or preferably at a larger distance (Alfassi & Lavi 1995).

1.6.2.1 *Calibration and calculation* The amount of a nuclide in a sample can be calculated from the counts of the radionuclide produced from its neutron activation by using Equation (1.18) if the physical constants σ, f, I_γ, ϕ and λ are known. The geometry factor, ε, is measured by using calibrated radioactive standards, which can be bought commercially. These standards are calibrated in a 4π detector, i.e. a detector that surrounds the radioactive source from all directions such that any β particle or photon interacts completely with the detector. The valve of ε depends on the distance of the source from the detector and the energy of the γ line. For a fixed distance, ε is a maximum function of the energy, the maximum is at 100–150 keV. For higher energies than the maximum, log ε is approximately a linear function of the log (energy), although for more accurate interpolations a quadratic dependence of log ε on log (energy) is used.

Standard sources often used for calibration are ^{241}Am (60 keV), ^{131}Ba (81, 302, 356 and 383 keV), ^{137}Cs (661 keV), ^{60}Co (1773 and 1332 keV), ^{22}Na (1275 keV) and ^{88}Y (898 and 1836 keV). These standard radioactive sources are used also for the calibration of the number of channels versus the appropriate γ energy. The standard radioactive sources are supplied with a certificate, certifying their activity at a fixed date. The activity is given either in μCi (3.7×10^4 dps = disintegrations per second) or in kBq (1 kBq = 10^3 Bq, 1 Bq = 1 dps $\Rightarrow$ 1 μCi = 37 kBq). The activity of the source at the time of measurement is calculated using Equation (1.2), i.e. $A(t) = A_0 \exp(-\lambda t)$, where $A(t)$ is the activity at the measurement time, A_0 is the certified activity and t is the time elapsed from the date of certification to the measurement. The quantities f, I_γ and λ are tabulated and known very accurately. However, σ is a function of the energy of the neutrons and the real term in Equation (1.18) should not be $\sigma\phi$ but rather the integral, $\int_0^\infty \sigma(E) n(E)\, dE$, where $n(E)\, dE$ is the energy-flux density or the flux distribution function (number of neutrons per second per cm^2 with energy in the range from $E + dE$ to E), $\phi = \int_0^\infty n(E)\, dE$. These functions $n(E)$ and $\sigma(E)$ are not always known and occasionally vary with time and place in the reactor. In order to overcome this uncertainty problem the results of neutron activation analysis have been calculated using comparators. A comparator is a standard which contains a known amount of the element to be determined in the sample. If the sample (denoted by subscript s) and the comparator (denoted by subscript c) are irradiated simultaneously and counted under the same conditions, then Equation (1.15) gives

$$m_s = m_c \frac{C_s}{C_c} \frac{D_c}{D_s} \frac{M_c}{M_s}$$

where D is the decay factor $[D = \exp(-\lambda t_d)]$, and M is the measurement time factor $[M = 1 - \exp(-\gamma t_c)$ where t_c is the counting time].

This method of comparators is the most accurate available, and it is the best one to use if the concentration of a single (or even a few) element(s) in the sample is required. Attention should be paid to the sample and the comparators being of similar size and composition to compensate for unequal self-absorption of the measured γ photons in the sample or comparator. This problem of self-absorption is almost negligible if both sample and comparators are small. The main problem in the use of a comparator for each element is when INAA is used for multielement analysis, which is the common method. In some samples more than 25 elements are determined simultaneously (more accuately, by 2–3 irradiations for different periods of times followed by 3–6 countings at different times after the irradiations for various periods of times; see Dams *et al.* 1970; Simons & Landsberger 1985; Cunningham & Stroube 1987; Hancock *et al.* 1987; Wyttenbach *et al.* 1987; Zhuk *et al.* 1988). For these multielement determinations it is difficult to prepare comparators for all the elements; besides that it becomes impractical owing to the large storage space required for all the comparators. Thus for multielement analysis either multielemental comparators have to be used, or 1–2 element comparators are used in the calculations for other elements.

Some multielements standards (comparators) of different types (geological, biological, etc.) have been prepared by either groups of scientists or governmental or international organizations. However, these standards (so-called SRM, standard reference material) are not recommended as primary standards because some large uncertainties in some elements and problems with homogeneity of small samples (of the SRMs) used in NAA do exist.

The use of one or two elements as comparators for all elements has become more widely adopted in the last few years. The advantage of using two elements, or one element with two isotopes which are activated rather than only one IRN, is based on the possibility that two activated nuclides, if chosen well, can give more information on the energy distribution of the neutron flux in the reactor (DeCorte *et al.* 1969, 1984; Simonits *et al.* 1976). The most popular method is the so-called k_0 method, in which k_0 values are determined for most elements according to the equation:

$$k_0 = \frac{B_c}{B_s} \quad \text{where} \quad B = \frac{C}{f \times I_\gamma \times \sigma_{th}}$$

The subscripts c and s stand for comparator and sample/element, respectively, while σ_{th} is the cross-section for the reaction with thermal neutrons. The most popular comparator is Zr, which has two active isotopes, used to monitor the thermal and epithermal flux simultaneously. The calculation of the different quantities of elements in a sample involves both the thermal and epithermal neutron fluxes.

1.6.3 *Nuclear interferences*

There are two principal interferences in the calculation of trace element concentrations by INAA. The first is the formation of the same radionuclide from two different elements. This is almost impossible if all neutrons are thermal since then the only possible reaction is (n, γ). However, all reactors have some higher energy neutrons, which can also induce other reactions. Examples of this are: the determination of Mg in the presence of Al; Cr in the presence of Fe; or Al in the presence of Si or P. Magnesium can only be determined through its less abundant isotope reaction ^{26}Mg (n, γ) ^{27}Mg. The cross-section for this reaction is rather low, making measurement not too sensitive – it suffers interference because ^{27}Mg can also be formed from non-thermal neutrons reacting with Al in a (n, p) reaction, ^{27}Al (n, p) ^{27}Mg. Similarly, the determination of chromium by the ^{50}Cr (n, γ) ^{51}Cr is interfered with by the non-thermal neutron reaction ^{54}Fe (n, α) ^{51}Fe. Al is determined by ^{28}Al, namely ^{27}Al (n, γ) ^{28}Al, which can be produced also by the reactions ^{31}P (n, α) ^{28}Al, and ^{28}Si (n, p) ^{28}Al. In tissue samples it was found that the contribution of P to ^{28}Al is more than ten times the ^{28}Al arising directly from Al in most reactors (Alfassi & Lavi 1984, 1991; Glascock *et al.* 1985; Kratochvil *et al.* 1987; Landsberger *et al.* 1988).

For pure thermal neutrons only a few cases show that a specific IRN can be formed from two elements. This happens when the IRN can also be formed by consecutive reactions with two neutrons. Platinum (Pt) has several IRNs, but all of them except one are short-lived and suffer from interferences from the major elements in the background. Thus, minute amounts of Pt are determined by ^{199}Au formed in the process:

$$^{198}\text{Pt} \xrightarrow{(n,\gamma)} {}^{199}\text{Pt} \xrightarrow{\beta^-} {}^{199}\text{Au}\ (3.139\,\text{d},\ 158.3\,\text{keV}\ \gamma\ \text{line})$$

However, ^{199}Au is also formed by a two-neutron reaction with ^{197}Au, since ^{198}Au has a very high cross-section for neutron capture (2.65×10^4 barn) (see Lin *et al.* 1992; Alfassi *et al.* 1998):

$$^{197}\text{Au} \xrightarrow{(n,\gamma)} {}^{198}\text{Au} \xrightarrow{(n,\gamma)} {}^{199}\text{Au}$$

It will be seen later how the first kind of interferences can sometimes be removed.

The second type of interference derives from two different radionuclides having very close γ lines. An example of this kind of interference involves the 846.8 keV line of ^{56}Mn and the 843.8 keV line of ^{27}Mg. Nowadays very good spectrometers are able to separate these two peaks if they have comparable activities, but this is impossible for low-resolution germanium detectors or when one activity is much higher. If the lines cannot be separated by the spectrometer there are two methods available to overcome the interference:

(1) Since the half-life of ^{56}Mn (2.56 h) is longer than that of ^{27}Mg (9.45 min), decay of two hours will leave practically only ^{56}Mn (the activity of ^{27}Mg will decrease by a factor of 6647 whereas ^{56}Mn activity will be decreased by less than a factor of two). Measurements of the sample soon after the irradiation and after a decay time of 2–3 h will give the activities of both ^{56}Mn (from the delayed measurement) and ^{27}Mg (by subtracting the corrected activity of ^{56}Mn from the first measurement).
(2) The activities of both radionuclides can be calculated using the fact that ^{27}Mg also has another γ line at 1014 keV. The ratio of the activities of the two lines can be measured in a pure sample of Mg, and thus the 843.8 keV activity of ^{27}Mg can be calculated from the 1014 keV activity. Subtracting this activity from the combined (843.8 + 846.8 keV) peak yields the activity of the ^{56}Mn 846.8 keV peak.

1.6.4 *A special case – INAA of trace elements in silicon*

Silicon has three naturally stable isotopes: ^{28}Si (92.21%), ^{29}Si (4.70%) and ^{30}Si (3.09%). Of these only ^{30}Si produces a radionuclide by the usual (n, γ) reaction of thermal neutrons. The isotope ^{31}Si produced by the ^{30}Si (n, γ) ^{31}Si reaction has a 2.62 h half-life and very few γ-rays, showing an intensity of only 0.07% (1.26 MeV). Considering its low abundance, the relatively low cross-section (0.28 barn) and mainly the low intensity of the γ-rays, ^{31}Si does not present a major problem in instrumental analysis by γ-ray spectrometry. However, owing to the presence of fast and epithermal neutrons in the reactor the main activity induced in silicon is due to ^{28}Al, ^{29}Al and ^{27}Mg formed by the reactions ^{28}Si (n, p) ^{28}Al, ^{29}Si (n, p) ^{29}Al and ^{30}Si (n, α) ^{27}Mg, respectively. As the half-lives of these radionuclides are short, (2.3, 6.6 and 9.5 min, respectively), instrumental activation analysis is possible if it is applied to radionuclides with half-lives longer than 1–2 h. In any case, the activity induced in nuclides with shorter half-lives is usually insufficient to measure the purity required for electronic grade semiconductors owing to the small number of radioactive atoms that can be formed until saturation.

As there is no background from the Si matrix after about a day or two and since the concentrations of the trace elements are low, the way to increase the sensitivity (decrease the lower limit of detection) is to irradiate with a larger total number of neutrons (higher flux, longer irradiation times, or both). In the HFR reactor at Petten, The Netherlands, a special irradiation facility was constructed by Philips for irradiation of silicon samples, in which silicon wafers of up to 15 cm diameter can be irradiated with 4×10^{13} thermal n cm^{-2} s^{-1} for periods from 72 to 96 h

(Theunissen *et al.* 1987; Verheijke *et al.* 1987). Researchers also used large Ge(Li) detectors (100–150 cm^3) and counted for 8–16 h. Thus, they achieved the lowest detection limits. Table 1.1 summarizes some of the limits of detection given in the literature for the Si matrix.

Table 1.1 Limit of detection of trace element in Si by instrumental neutron activation analysis (ppb by weight)

Element	Verheijke *et al.* (1987)[a]	Lindstrom (1986)[b]	Revel *et al.* (1984)[c], Revel (1988)[c]	Fujingawa & Kudo (1979)[d]
Na	0.5	0.1	0.3	0.1
Mg		3000		
Al		500		
Cl		5		
K	0.15	100	0.6	
Ca	10	2000		
Sc	0.00003	0.003	< 0.002	
Ti	3	1000	30	
V			10	
Cr	0.002	1	0.02	0.1
Mn	15	0.1		
Fe	0.3	50	4	1.6
Co	0.0005	0.1	0.1	0.001
Ni	0.15	2000	3	
Cu	< 0.56	0.5	0.1	0.1
Zn	0.015	1.0	0.2	0.3
Ga	0.0015		0.05	
Ge	1			
As	< 0.033	0.02	0.001	0.008
Se	0.002		0.1	
Br	< 0.0024	0.02	0.01	
Rb	0.01			
Sr	0.15		4	
Y	200			
Zr	0.15	500	2	
Nb	1.5			
Mo	0.006		0.01	
Ru	0.0015		0.1	
Pd	0.06			
Ag	0.003	0.01	0.08	
Cd	0.015		0.3	
In	0.004		0.3	
Sn	0.2			
Sb	< 0.18	0.01	0.02	0.006
Te	0.004		0.2	
I	8			
Cs	0.0005	0.05		
Ba	0.06	10	1.5	
La	0.00015	0.01	0.0007	
Ce	0.0009		0.1	
Pr	0.006			
Nd	0.01			
Sm	0.00003	0.01	0.0004	
Eu	0.00007	0.05	0.002	
Gd	0.006			

Contd

Table 1.1 *Contd*

Element	Verheijke *et al.* (1987)[a]	Lindstrom (1986)[b]	Revel *et al.* (1984)[c], Revel (1988)[c]	Fujingawa & Kudo (1979)[d]
Tb	0.0001		0.006	
Dy	300			
Ho	0.003			
Er	0.04			
Tm	0.003			
Yb	0.0002		0.001	
Lu	0.0004			
Hf	0.003	20	0.005	
Ta	0.005	0.05	0.01	
W	< 0.025	0.02	0.0001	
Re	0.0003			
Os	0.0007			
Ir	0.000004		0.0002	
Pt	0.004		0.005	
Au	< 0.0016	0.0005	0.0001	0.0008
Hg	0.0006		0.7	
Tl	3			
Pb	300			
Th	0.0002	0.1	0.0015	
U	0.001	0.3	0.02	

[a] Irradiated wafers of 15 cm diameter and 0.5 mm thickness (20.6 g) with flux of 4×10^{13} n cm^{-2} s^{-1} for 72–96 h and counted for 6–8 h.
[b] Irradiated samples of 100 mg for up to 6 h at 5×10^{13} n cm^{-2} s^{-1}.
[c] Si (1 g) irradiated for 72 h at 2.3×10^{14} n cm^{-2} s^{-1}.
[d] Irradiated 5–10 g samples for 11 d at 2×10^{13} n cm^{-2} s^{-1}.

For some elements, e.g. Cl and Mn, shorter irradiation times were used, between 0.5 and 2.0 h. Special correction was performed in the determination of Na from the reaction ^{28}Si (n, pα) ^{24}Na, as was shown by Niese (1977), by irradiating very pure Si in a reactor core with and without a cadmium cover to absorb thermal neutrons. The contribution of the ^{28}Si (n, pα) ^{24}Na reaction to ^{24}Na analysis in a silicon matrix depends on the spectrum of the neutrons in the reactor core, since a very high energy is required for this reaction and the Q value exceeds that of a 14 MeV neutron generator (Niese 1977). Niese (1977) also found that the contribution of the ^{28}Si (n, pα) ^{24}Na is equal to 0.7 ppb. For the HFR reactor at Petten, The Netherlands, the contribution is 0.55 (Verheijke *et al.* 1987) or 0.6 ppb (Theunissen *et al.* 1987). Revel *et al.* (1984) found the contribution of ^{28}Si (n, pα) ^{24}Na to be less than 0.4 ppb. Haas *et al.* (1982) warned against the common use of wrapping the sample for irradiation in Al foil, as it was shown that there is recoil of ^{24}Na from the Al wrapping to the sample, owing to the high recoil energy from the ^{27}Al (n, α) ^{24}Na reaction. In many other uses it is not important, but because of the very low concentration of Na in pure Si, this contribution might be considerable.

Several important elements cannot be determined using this method, including the light elements H, Li, Be, B, C, N, O and F, as they do not form radioisotopes or their isotopes are too short-lived. Phosphorus also cannot be determined because of its lack of γ emission from the only isotope produced, ^{32}P. However, ^{32}P β-rays can be determined destructively by the extraction of the phosphomolybdic complex in the presence of hold-back carriers such as Ta and Au (Jaskolska & Rowlinska 1975). The chemical yield of the extraction was found to be 83.8 $\pm$ 4% and the decontamination factors were 79–6500 for the different elements studied. The limit of detection was found to be 3×10^{-11} g. Jaskolska and Rowlinska (1975) performed their study on phosphorus-doped silicon and Alfassi and Yang (1989) showed that for this situation there is no need for chemical separation and 10–20 days of cooling are sufficient to ensure that the only β emitter is ^{32}P (with an accuracy of 1–2%, which is much better that the reproducibility of the extraction yield). The limit of detection should be about the same, as both methods use liquid scintillation counting for the measurement of the β activity, although Alfassi and Yang (1989) used a discriminator which reduced the detection limit by a factor of two.

1.6.5 *A suggested procedure for a general case*

Greenberg *et al.* (1984) described the procedures adopted by the NAA group of the National Institute of Standards and Technology of the USA. The sample is first packaged in a cleaned container. The container, according to the type of sample and the length of irradiation, can be a polyethylene bag or vial, sealed by heating, or flame-sealed high-quality quartz (to eliminate as much as possible the active sodium) or an aluminium can. The samples are irradiated in a well-moderated (by D_2O) neutron flux of 10^{14} n cm^{-2} s^{-1} using a pneumatic tube for transferring the sample to the reactor. The length of the irradiation, decay and counting depends on the analysed elements. For one or few elements the timings are optimized for these elements. For determination of mercury in urine, for instance, the samples (in quartz ampules) are irradiated for 1 h and counted both after decay of 5 days and 20 days in order to measure the Hg content via the formation of both ^{197}Hg ($t_{1/2}$ = 64.1 h, E_γ = 77 and 191 keV) and ^{203}Hg ($t_{1/2}$ = 46.59 d, E_γ = 279 keV). The long decay is used to reduce the activities of Na, Cl and Br.

To measure Cl and Br in various oils, Greenberg *et al.* (1984) used short irradiations of a few minutes and measured the activity of ^{38}Cl ($t_{1/2}$ = 37.2 min, E_γ = 1642.2 and 2167.6 keV) after decay of several minutes. The concentration of Br was determined via ^{82}Br (35.3 h, 554.3 and 776.5 keV) after a decay time of 24 h. For lower detection limits of bromine, irradiation times of several hours up to 24 h are allowed.

For a complete multielement analysis (20–40 elements, which can be determined by NAA), the elements are divided into three groups according to the half-lives of their IRNs: 2 min–15 h; 0.5–5 days; and $>$ 5 days. Some elements appear in more than one group. The same sample can thus be used for the two irradiations. The usual scheme of irradiations, decays and counting is illustrated in Fig. 1.13.

The accuracy and precision of Greenberg *et al.*'s (1984) determinations can be found in Table 1.2, which give their results for two NBS standard reference materials (SRMs) together with the certified values. The uncertainties listed represent the estimated overall analytical uncertainties at the 95% confidence level. SRM 1648 is of Urban Dust and SRM 1572 is of Citrus Leaves.

Cunningham and Stroube (1987) applied INAA to 240 food composites. They irradiated samples with neutron flux of $4\text{–}9 \times 10^{13}$ n cm^{-2} s^{-1}. Two irradiations were performed on each sample: one of 15 s and one for 4–5 h. For the short irradiation, a 2

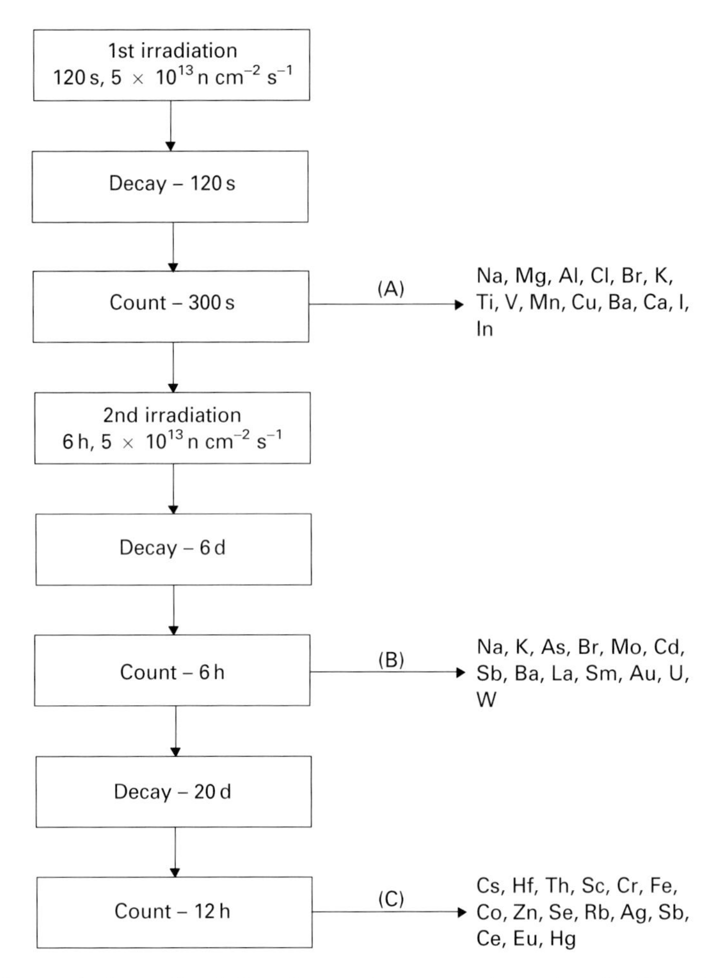

Fig. 1.13 A general scheme of irradiation, decay and count procedures adopted in the nuclear activation analysis of trace elements.

min decay time and a 10 min counting time were used. For the longer irradiation, decay times varied from 5 days to 4 weeks and counting times were 6–8 h. In the short irradiation they determined Ca, Cl, Cu, Ga, I, In, K, Mg, Mn, Na, S, Sr, Tl and V. In the long irradiation they determined Ag, As, Au, Ba, Br, Cd, Ce, Co, Cr, Cs, Eu, Fe, Hf, La, Lu, Mo, Rb, Sb, Sc, Se, Sm, Ta, Th, W, Yb and Zn.

Sun and Jervis (1985) used a Slowpoke reactor, which is a nuclear reactor specially designed for teaching, with activation analysis and production of small amounts of radioisotopes in mind. The reactor operates at a low power of 20 kW and has a maximum flux of 10^{12} n cm^{-2} s^{-1} (Sun and Jervis used flux of 2.5×10^{11} n cm^{-2} s^{-1}). An advantage of this type of reactor is its high stability (1–2% over long periods), which avoids having to irradiate standards with each batch of samples. Sun and Jervis (1985) determined 35 elements in rocks and soils using two irradiation and three counting times: Al, Ba, Ca, Cl, Dy, I, Mg, Mn, Sr, Ti, U and V were determined by 1

Table 1.2 Elemental concentrations (ppm, unless % is indicated) determined by INAA versus NBS (NIST) certified values.

Element	SRM 1648 INAA	SRM 1648 Certified value	SRM 1572 INAA	SRM 1572 Certified value	Method (see Fig. 1.13)
Ag	6.5 ± 0.5				C
Al(%)	3.5 ± 0.1	3.42 ± 0.11			A
As	119 ± 5	115 ± 10	31 ± 0.3	3.1 ± 0.3	B
Ba	750 ± 60		21.4 ± 1.2		A
Br	515 ± 30		8.2 ± 0.3		A, B
Ca(%)	5.9 ± 0.5		3.19 ± 0.10	3.15 ± 0.10	A
Cd	72 ± 6	75 ± 7			B
Ce	54 ± 3		0.28 ± 0.08		C
Cl(%)	0.46 ± 0.02				A
Co	17.9 ± 0.5		0.02 ± 0.006		C
Cr	410 ± 10	403 ± 12	0.74 ± 0.18	0.8 ± 0.2	C
Cs	3.5 ± 0.2		0.098 ± 0.010		C
Eu	0.81 ± 0.08		0.010 ± 0.003		C
Fe	39200 ± 800	39100 ± 1000	92 ± 13	90 ± 10	C
Hf	4.3 ± 0.3				C
Hg			0.079 ± 0.016	0.08 ± 0.02	A
I	20 ± 5				A
In	1.00 ± 0.07				A
K(%)	1.00 ± 0.06	1.05 ± 0.01	1.76 ± 0.03	1.82 ± 0.06	A, B
La	43 ± 2		0.19 ± 0.03		B
Mg(%)	0.85 ± 0.08				A
Mn	800 ± 20				A
Na(%)	0.41 ± 0.02	0.425 ± 0.002			B
Rb	53 ± 9		5.0 ± 0.3	4.84 ± 0.06	C
Sb	46 ± 2		0.041 ± 0.010		B, C
Sc	6.3 ± 0.2		0.011 ± 0.001		C
Se	27 ± 2	27 ± 1			C
Sm	4.2 ± 0.4		0.052 ± 0.006		B
Sr			94 ± 3	100 ± 2	C
Th	7.5 ± 0.3		0.015 ± 0.004		C
Ti(%)	0.41 ± 0.02				A
V	133 ± 7	140 ± 3			A
W	4.9 ± 0.6				B
Zn	4800 ± 200	4760 ± 140	28.3 ± 0.8	29 ± 2	C

min irradiation followed by 5 min counting after a decay of 10–15 min; a longer irradiation of 16 h was followed by 30 min counting after 4–9 d decay for As, Br, K, La, Na, Sb, Sm, W and Yb, and 100 min counting after a decay of 15–20 d for Ce, Co, Cr, Cs, En, Fe, Hf, Ln, Nd, Rb, Sb, Sc, Se, Sn, Ta, Tb, Th, Yb and Zn. Relatively large standard deviations between several samples were found for Cl (32%), Sb (24%), Ti (18%), Ba (16%) and Br (14%), but for most elements the average relative standard deviation was less than 10%. A similar scheme was used by the same authors for samples of coal and ashes (Sun & Jervis 1987), except that the short irradiation time for coal samples was increased to 5 min.

Iskander (1985) measured the concentrations of 28 elements in tobacco and cigarette paper. Samples were irradiated in two irradiation periods – the first for three minutes and the second for eight hours with a neutron flux of 2×10^{12} n cm^{-2} s^{-1} from a TRIGA Mark I reactor. The samples irradiated for 3 min were counted for 3 min after a 1 min decay (to measure Al, V, Ti, Ca and Mg) and after an additional decay time of 26 min (total decay time 30 min) they were counted again for 1000 s (to determine Cl and Mn). Additionally, the longer irradiated samples were counted twice: 4000 s after a decay time of 12 h (to determine K, Na, As, Br and La) and 40 000 s after a decay time of 21 d (Ba, Rb, Th, Cr, Ce, Hf, Fe, Sb, Sr, Ni, Sc, Se, Zn, Cs, Co, Eu).

McOrist and coworkers (1994, 1997) studied the trace element concentration in various Australian opals. They irradiated their samples twice, for 1 min at a thermal flux of 2.5×10^{13} n cm^{-2} s^{-1}, and for 18 h at a thermal flux of 5×10^{12} n cm^{-2} s^{-1}. Each irradiated sample was counted twice. The short-time irradiated samples were counted for 10 min after 20 min of decay, and then for 45 min after a decay time of 24 h. After the longer irradiation time the samples were counted for 2 and 4 h after 7 and 28 days of decay, respectively.

Wallner and Katzlberger (1997) studied the concentration of 14 elements in spruce tree rings and needles, taken from an area where an aluminium refinery was located. Al, Mn, Ca and Cl were determined by 3–7 min of irradiation at a thermal neutron flux of 5×10^{10} n cm^{-2} s^{-1}. They calculated that for Al (assuming that the main interference for its 1779 keV line is from the Compton of the 2113 keV line of ^{56}Mn) the optimal irradiation time is 4 min. Na, K, Rb, Ba, Zn, Fe, Sc, Cr, Co and Br were determined by 10 h of irradiation with a thermal neutron flux of 8×10^{13} n cm^{-2} s^{-1}. The samples were counted on three occasions, after decay of 4–5 d, (Na and K), 7 d (Br and Ca) and decay of 14 d (Ba, Co, Cr, Fe, Rb, Sc and Zn).

Oddone *et al.* (1997) studied the elemental content of Anatolian obsidians (volcanic glasses) by INAA using both thermal and epithermal neutrons. Mg, K, Ca, Ti, Mn, As and Dy were determined by a 2 min irradiation time at a thermal neutron flux of 1×10^{12} n cm^{-2} s^{-1} (TRIGA Mark II reactor at 250 kW). The sample was counted six times after decay of 0.5, 10, 30, 120, 480 and 960 min. The longer the decay time the longer the count. Al and Si were determined by repeating the short-time irradiation for a sample wrapped in a 10 mm thick Cd foil. Both Al and Si were determined from the 1779 keV line of ^{28}Al, using the epithermal neutron activation ^{28}Si(n,p)^{28}Al (see Section 1.6.8 for more details). Na, Ca, Sc, Cr, Fe, Co, Zn, Se, Rb, Zr, Nb, Sb, Ba, La, Ce, Nd, Sm, Eu, Gd, Tb, Ho, Tm, Yo, Lu, Hf, Ta and Th were determined following an irradiation time of 20 h. The samples were counted on five occasions following decay times of 3, 6, 24, 48 and 96 d. Epithermal neutrons INAA (see Section 1.6.7) were incident for long irradiation times also (using samples placed inside a 1.5 mm thick Cd box) in order to reduce interference from elements having low resonance integrals and high thermal cross-section (small I_0/σ ratio). The samples were irradiated for 40 h and counted five times after decay of 3, 6, 12, 24 and 48 d. The epithermal-neutron INAA improves the sensitivity and accuracy of the measurement of Zn, Se, Rb, Zr, Nb, Sb, Cs, Ba, Gd, Tb, Ho, Tm, Yb, Hf, Ta and Th. Ni, Sr and U, which could not be measured by thermal INAA, could be determined by the epithermal INAA.

Sato (1990) suggested four different timing schemes for the determination of various elements in biological samples (three irradiation times with the last one counted twice, after different decay times). His suggestion for various elements is detailed in Table 1.3. Table 1.4 provides three examples from the literature for the

Table 1.3 INAA timings for various elements in biological samples

Element	IRN	Half-life of IRN	γ-ray (keV) of IRN	Timings system	Interfering reactions
Na	^{24}Na	15.02 h	1369	2,3	$^{24}Mg(n,p)$, $^{27}Al(n,\alpha)$
Mg	^{27}Mg	9.46 min	1014	1	$^{27}Al(n,p)$
Al	^{28}Al	2.24 min	1779	1	$^{28}Si(n,p)$, $^{31}P(n,\alpha)$
Si	^{31}Si	2.62 h	1266	2	$^{31}P(n,p)$, $^{34}S(n,\alpha)$
S	^{37}S	5.0 min	3103	1	
Cl	^{38}Cl	37.3 min	1642	1,2	
K	^{42}K	12.36 h	1525	2	$^{42}Ca(n,p)$
Ca	^{49}Ca	8.72 min	3084	1	
Sc	^{46}Sc	83.8 d	889	4	$^{46}Ti(n,p)$
V	^{52}V	3.76 min	1434	1	
Cr	^{51}Cr	27.70 d	320	4	$^{54}Fe(n,\alpha)$
Mn	^{56}Mn	2.58 h	847	2	$^{56}Fe(n,p)$
Fe	^{59}Fe	44.6 d	1099	4	
Co	^{60}Co	5.27 yr	1173	4	$^{63}Cu(n,\alpha)$
Ni	^{58}Co	70.8 d	811	4	
Cu	^{66}Cu	5.10 min	1039	1	$^{66}Zn(n,p)$
Zn	^{65}Zn	244.1 d	1116	4	
As	^{76}As	26.3 h	559	3	$^{79}Br(n,\alpha)$
Se	^{75}Se	118.5 d	265	4	
Br	^{82}Br	35.34 h	776	3	
Rb	^{86}Rb	18.8 d	1077	4	
Sr	^{87m}Sr	2.80 h	388	2	
Mo	^{99}Mo–^{99m}Tc	66.02 h	141	3	
Cd	^{115}Cd–^{115}In	53.4 h	336	3	
Sb	^{122}Sb	2.68 d	564	3	
Sb	^{124}Sb	60.20 d	1691	4	
I	^{128}I	24.99 min	443	2	
Cs	^{134}Cs	2.06 yr	796	4	
Ba	^{139}Ba	82.9 min	166	2	
Hg	^{203}Hg	46.8 d	279	4	

The timing systems are:

System:	1	2	3	4
Time of irradiation:	1–10 min	2–20 min	10 h	10 h
Time of counting:	1–2 min	1 h	5–8 d	30 d
Time of decay:	2–5 min	10 min	1–2 h	3–5 h

irradiation of biological samples together with timings and the limit of detection of some elements via INAA.

1.6.6 *An example of interferences removal in INAA – determination of Pt in aerosols* (see Alfassi *et al.* 1998)

Platinum is emitted into the air by corrosion of catalytic converters used in cars to reduce the emission of CO, NO_x and unburnt hydrocarbons. In order to determine the concentration of Pt in air, a high-volume air-pump is used to pump air through a filter and the amount of Pt deposited on the filter (together with aerosols) is determined. Pt could be assessed using several radionuclides. However, ^{191}Pt suffers from very low abundance (0.01%) compared with the parent stable nuclide, i.e. ^{190}Pt. Additionally, ^{197}Pt and ^{199}Pt suffer from relatively short half-lives (18.3 h and 30.8

Table 1.4 Detection limits found for INAA of various biological samples

Biological sample	Irradiation time[a]	Decay time	Counting time	Detection limits (ppb)
(1) Human blood[b]	5 d	7 mon	2 h	Co(4), Cs(4), Fe(10000), Se(20).
	24 h, E	21 d	2 h	Br(1000), Fe(1500), Rb(300), Se(100), Zn(500).
(2) Hair[c]	30 s	10 s	20 s	Ag(140), Cl(13000), F(23000), Se(160).
	10 m	5 m	5 m	Al(1600), Ba(7600), Ca(52000), Cu(3800), I(260), K(16000), Mn(750), Na(4100), S(0.28%), Zn(3500).
	16 h	2 d	50 m	As(46), Au(2), Sb(45)
	16 h	21 d	50 m	Co(97), Hg(200), Ag(140), Ba(7600), Se(160), Zn(3500).
(3) Meat, fish and poultry[d]	15 s	2 m	10 m	Ca(82000), Cl(2400), Cu(8300), K(0.1%), Mg(0.013%), Mn(1900), Na(260), V(81).
	4 h	5–28 d	7 h	Ag(120), As(65), Br(130), Cd(170), Co(23), Cr(350), Cs(39), Eu(3.7), Fe(6800), Rb(430), Sb(32), Sc(1.4), Se(300), Zn(460).

[a] d = days; h = hours; s = seconds; m = minutes; mon = months; E denotes irradiation with epithermal neutrons.
[b] Chisela and Bratter (1986).
[c] Chatt *et al.* (1985).
[d] Cunningham and Stroube (1987).

min). If measurement is completed shortly after irradiation, these IRNs will not be observed because of the large quantity of Comptons from the major elements of Na, Cl and Br. Thus, Pt can be determined in minute amounts only via ^{199}Au [3.14 d, 158.4 keV (36.9%)] which is formed in the sequence of reactions:

$$^{198}\mathrm{Pt}\,(7.2\%) \xrightarrow{(\mathrm{n},\gamma)} {}^{199}\mathrm{Pt} \xrightarrow{\beta^-} {}^{199}\mathrm{Au}$$

Now ^{199}Au suffers from both of the interference types mentioned before:

- it can be formed also from other elements besides Pt
- another radionuclide formed from other elements has a proximate γ line

In the following it will be shown how these interferences can be overcome.

(1) ^{199}Au can also be formed from the successive absorption of two neutrons by ^{197}Au due to the very high cross-section for the (n,γ) reaction of ^{198}Au (26 500 barn) and the high cross-section for the (n,γ) reaction of ^{197}Au (98.7 barn):

$$^{197}\mathrm{Au} \xrightarrow{(n,\gamma)} {}^{198}\mathrm{Au} \xrightarrow{(n,\gamma)} {}^{199}\mathrm{Au}$$

Since ^{198}Au is a radionuclide with half-life close to ^{199}Au (2.69 d) it can be measured simultaneously with ^{199}Au, and can be used to calculate the concentration of Au in the sample and hence to calculate the contribution of Au to the measured peak of ^{199}Au. Subtracting this contribution reveals the counts of ^{199}Au formed from ^{198}Pt and hence the concentration of Pt.

(2) Another problem is the very close signals of the γ lines for ^{199}Au (158.4 keV) and ^{47}Ca (159.3 keV). If similar counts are obtained for these two radionuclides the γ spectra show two clearly distinct peaks although the lower one (158.4 keV) does not reduce to zero, and hence there is less accuracy in the calculation of the area under each peak (i.e. the counts due to each radionuclide). The situation is worse when the number of counts of one of the radionuclides is considerably higher (ratio >4), in which case the spectrum does not show two distinct peaks but one peak with a shoulder more or less visible. In order to separate this composite peak into its two signals two methods can be used. An important note concerning this measurement is that a computer convolution program to calculate the peak area should not be used as the fitting function code is characteristic of pure photopeaks only. In this situation it is better to calculate the area by a simple integration of the counts under the peak, and then subtract a trapezoid for the background correction (Alfassi & Rietz 1994). The two methods for separation of the radioisotope contributions are based on (a) different temporal behaviour of the two contributions and (b) another γ line of the interferant.

(a) *Temporal separation* – Although ^{199}Au and ^{47}Sc have close half-lives which usually will not allow the temporal separation of their contributions, their parent isotopes (as neither is produced directly from the (n, γ) reaction and both are products of β decay of the (n, γ) products) have considerably different half-lives and hence the two contributions can be separated by measuring twice and solving the two equations (obtained by writing the equations for the expected counts using x and y for the masses of Pt and Ca) with the two unknowns m_{Pt} and m_{Ca} (or x and y, $x = m_{\mathrm{Pt}}$ and $y = m_{\mathrm{Ca}}$):

$$^{198}\mathrm{Pt} \xrightarrow{(n,\gamma)} {}^{199}\mathrm{Pt} \xrightarrow{\beta^-} {}^{199}\mathrm{Au}$$

$$^{46}\mathrm{Ca} \xrightarrow{(n,\gamma)} {}^{47}\mathrm{Ca} \xrightarrow{\beta^-} {}^{47}\mathrm{Sc}$$

The disadvantage of this method is the need to make two measurements, and thus the counting time required is more than doubled, since because of the decay between the two measurements the counting time of the second decay must be longer than the first in order to obtain the same statistical accuracy.

(b) ^{47}Sc does not have an uninterfered γ peak to calculate its contribution to the composite peak; however, its parent, ^{47}Ca, has a good uninterfered γ line at 1297 keV and this γ line can be used to measure the concentration of Ca and consequently its contribution via ^{47}Sc to the 158–159 keV peak.

The above-mentioned interferences are the only ones present in a well-moderated nuclear reactor (heavy-water moderated), while in most research reactors (light-water moderated) the high-energy neutron flux will lead to additional interferences by the ^{47}Ti (n, p) ^{47}Sc and the ^{199}Hg (n, p) ^{199}Au reactions. These interferences can be removed by additional irradiation inside a box constructed from B or Cd, which are thermal neutron absorbers, as explained in Sections 1.6.7 and 1.6.8. However, this correction is not mandatory in most cases. The reaction ^{199}Hg (n, p) ^{199}Au can be neglected as the cross-section is lower than that for (n, γ) and the fast flux is smaller than the thermal flux and the Hg concentration is small. The ^{47}Ti (n, p) ^{47}Sc reaction can be a greater problem owing to the higher concentration of the more abundant Ti. However, since Ti has several isotopes the measurement of ^{48}Sc, which can be formed only by the ^{48}Ti (n, p) ^{48}Sc reaction, can be used to calculate the contribution from the ^{47}Ti (n, p) ^{47}Sc reaction.

1.6.7 *Epithermal INNA*

In usual instrumental neutron activation analysis (INAA), the neutron energy spectrum of the whole reactor is used. However, in some situations the use of a portion of the neutron spectrum only is preferable; such systems are characterized by large differences in activation cross-sections for the desired and the interfering nuclides in the various parts of the energy spectrum. It might be that the required trace elements are activated within one part of the neutron spectrum while the interfering major elements are activated more strongly within other parts of the spectrum. Thus, it is better to avoid these other parts.

The neutron energy spectrum in a nuclear reactor is usually divided, for convenience, into three portions, and their relative abundances are dependent on the reactor structure. The most abundant fraction is that containing thermal neutrons, i.e. those neutrons which are in thermal equilibrium with the moderator atoms. Their most expected energy is equal to kT (where k is Boltzmann's constant and T is the moderator temperature), which at room temperature is equal to about 0.025 eV. Neutrons with energy above that of the thermals are divided into fast neutrons (those directly formed from fission having not been moderated at all, with energy mainly above 1 MeV), and epithermal neutrons (i.e. partly moderated and having energy between tenths of eV and 1 MeV). When the whole reactor neutron energy spectrum is used for activation, the main contribution is that from the thermal neutrons, both because of their higher abundance and their usually higher cross-section for (n, γ) reactions.

A special situation where advantage can be taken of the epithermal and fast neutrons is the case where required trace elements are activated more strongly relative to the major elements by epithermal or fast neutrons. Most major elements in geology and biology follow the $1/v$ cross-section rule (i.e. their activation cross-section is inversely proportional to the square-root of the neutron energy) throughout the whole energy spectrum and hence have considerably lower (n, γ) cross-sections for epithermal neutrons. In contrast, many of the less-abundant elements have, in addition to their thermal activation, large activation cross-section resonances in the epithermal energy region and consequently can be activated preferentially in this region. Similarly, several of the less-common elements can be activated by other neutron reactions besides the common (n, γ) reaction. These (n, p) and (n, α) reactions require higher energy than the thermal activation (and in most cases, they do not occur in the epithermal region and are induced only by fast neutrons).

A simple example demonstrating the advantage of using neutron filters (thermal neutron absorbers) in activation analysis is seen in Fig. 1.14, which shows the γ-ray spectra of a sample of blood serum, activated with reactor neutrons once within a cadmium wrapping and once without any absorber (bare irradiation) (Lavi & Alfassi 1984; Etzion *et al.* 1987). In the instance of activation without the Cd absorber, the Comptons of the major elements Na and Cl cover the peaks of bromine and iodine. (Bromine can be determined only after a delay of several days while the shorter-lived iodine cannot be determined instrumentally and can be determined only after chemical separation.) The activation with epithermal neutrons (using a Cd cover) shows clearly the peaks of Br and I. The spectra of a more complicated sample (geological) irradiated with and without a Cd cover is shown in Fig. 1.15.

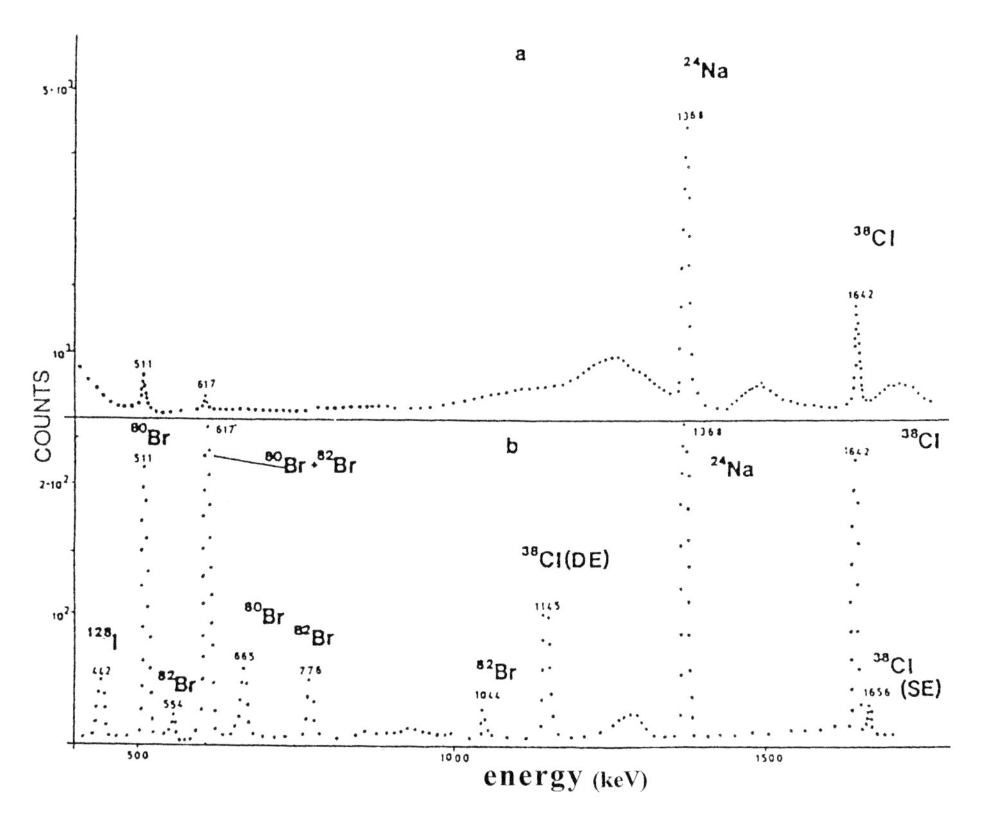

Fig. 1.14 Two γ-ray spectra of a neutron activated blood sample taken with a Ge(Li) detector: (a) irradiated bare with reactor neutrons; (b) irradiated in a Cd shield.

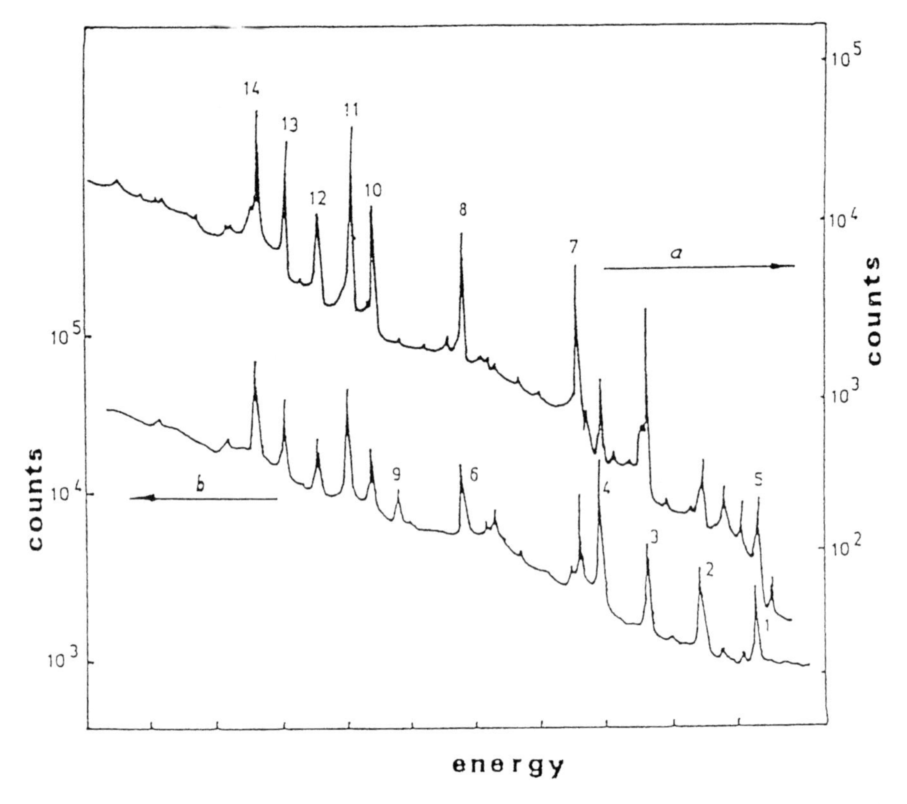

Fig. 1.15 Two γ-ray spectra of a neutron activated air-dust after 10 days cooling: (a) inside a Cd sheet; (b) irradiated bare. The peaks are: (1) 1732 keV DE of ^{24}Na; (2) 1595 keV ^{140}La; (3) 1474 keV ^{82}Br; (4) 1369 keV ^{24}Na; (5) 1691 keV ^{124}Sb; (6) 1099 keV ^{59}Fe; (7) 1317 keV ^{82}Br; (8) 1044 keV ^{82}Br; (9) 889 keV ^{46}Sc; (10) 828 keV ^{82}Br; (11) 777 keV ^{82}Br; (12) 690 keV ^{82}Br; (13) 619 keV ^{82}Br; (14) 554 keV ^{82}Br.

Owing to the presence of both thermal and epithermal neutrons the product $\sigma\phi$ in Equation (1.15) should be replaced by the term $\phi_{th}\,\sigma_{th} + \phi_{epi}\,I_0$, where ϕ_{th} and σ_{th} are the flux and cross-section of the reaction with thermal neutrons, while, ϕ_{epi} and I_0 are defined by the following equations [where I_0 is called the resonance integral of the (n, γ) reaction]:

$$I_0 = \int_{0.55\,\text{eV}}^{1\,\text{MeV}} \frac{\sigma(E)}{E}\,\mathrm{d}E; \qquad \phi_{epi} = \text{total epithermal flux} \Big/ \int_{0.55\,\text{eV}}^{1\,\text{MeV}} \mathrm{d}E/E$$

The 1 MeV upper boundary was chosen here because neutrons with energies above 1 MeV are categorized as fast neutrons rather than epithermal neutrons. The 0.55 eV lower boundary was chosen as this is the cut-off in the neutron absorption by Cd, the main filter used to capture thermal neutrons.

The epithermal activation properties of a nuclide can be expressed by the ratio I_0/σ_{th} or by means of the absorber ratio. An absorber ratio is the ratio of the activities of the nuclide irradiated once without absorber cover and once covered by an absorber of thermal neutrons. Thus the cadmium ratio is given by

$$R_{Cd} = \frac{\phi_{th} \times \sigma_{th} + \phi_{epi} \times I_0}{\phi_{th} I_0} = 1 + \frac{\phi_{th} \times \sigma_{th}}{\phi_{epi} I_0}$$

The most important fact is not only how the insertion of an absorber of thermal neutrons influences the activity of the specifically measured nuclide, but also how it influences the interfering nuclides. Thus, it would be advantageous to analyse an element by epithermal neutron activation rather than using the whole spectrum of reactor neutrons (ENAA versus RNAA as it is most usually written, i.e. epithermal and reactor neutron activation analysis), only if the ratio of the resonance integral to the thermal neutron cross-section (I_0/σ_{th}) is larger for this element than the same ratio for the interfering elements. Several criteria have been suggested and tabulated to measure the advantage of ENAA over RNAA (Brune & Jirlow 1964; Bem & Ryan 1971; Gladney *et al.* 1980; Parry 1980, 1984; Tian & Ehmann 1984; Alfassi 1985; Currie 1986).

1.6.7.1 *Thermal neutron absorbers* The main absorbers for thermal neutrons are cadmium and boron due to their high cross-sections for the reaction with thermal neutrons. The variation of cross-section with energy is different for Cd and for B, and hence an intelligent choice of absorber (sometimes also called a filter) will lead to the optimized detection of nuclides (Rossitto *et al.* 1972). In some circumstances, one absorber is used and in others a combination of two absorbers, e.g. Cd + B (Rossitto *et al.* 1972; Parry 1980; Williamson *et al.* 1987). In some experiments, filters that absorb the epithermal neutrons in other regions of the neutron spectrum are used, allowing more selectivity for some elements (Tokay *et al.* 1985).

Since the main absorbers are boron and cadmium, it is very important to compare them from a technical point of view. The absorber is used as: a covering sheet to wrap the sample, mainly cadmium from which metallic sheets are commercially available; as a capsule built from these materials; as a mixture with the sample, usually with B_2O_3 (Rosenberg 1981); or as a permanently lined installation within one of the irradiation ports of the reactor. The use of absorber-lined ports has the disadvantages that: (1) scattered thermal neutrons can come from angles not covered by the lining of the absorber leading to lower absorber ratios, while when the absorber is used as a capsule or wrapping, it is covered from all angles; (2) the capsule in which the sample is held during the pneumatic transfer and which is usually made from polyethylene leads to partial thermalization of the epithermal neutrons; and (3) the lining reduces the total flux of the neutrons in the reactor and causes excessive use of the nuclear fuel. In contrast, the use of an absorber-built capsule or wrapping suffers from the disadvantages of the absorption reactions.

Cadmium is activated to short- and long-lived nuclides and therefore unloading and unpacking of the sample for medium- and long-lived radionuclide measurements involves radiation safety problems due to the high radiation dose. Short-lived ($t_{1/2}$ < 20–30 s) radionuclides cannot be measured at all in a cadmium capsule since the short half-life prohibits the safe unpacking of the vessel and the activity of the absorber is too high to allow measurement together with the filter.

While the absorption of neutrons by ^{10}B does not lead to radioactive products, the reaction ^{10}B (n, α) ^{7}Li is very exoergic (Q = 2.792 Me V) and the samples are heated considerably. Stroube *et al.* (1987) found that in the 20 MW reactor at the National Bureau of Standards, thermal heating of the boron nitride vessel limited the length of irradiation for freeze-dried foods to 4 s and prevented completely safe irradiation of wet food. When biological samples are irradiated, the heating accelerates the thermal decomposition of organic compounds producing a high pressure situation in the sample container, especially when it is airtight, which may lead to explosion and contamination or even destruction of the irradiation port. In other cases, elements may be volatilized and lost. In order to avoid these effects, the time of irradiation should be limited.

A possible advantage of boron over cadmium is the reuse of the same filter in subsequent irradiations. Cadmium filters cannot usually be reused, at least not immediately, owing to the long-lived radioactivity produced in cadmium during irradiation. The activity produced in boron filters is small and is due only to the contamination in boron. However, the use of boron filters is still limited in many cases when the total dose of neutrons is high: structural failure of the capsule can occur probably resulting from the excessive formation of helium in the ^{10}B (n, α) ^{7}Li reaction. Cadmium filters are easily manufactured from metallic cadmium sheets of about 1 mm thickness. Boron, however, is a difficult material to machine and Stuart and Ryan (1981) prepared boron shields from forming a mixture of boron carbide powder and paraffin wax. Parry (1982, 1984) used the same method, but instead of using boron carbide, she used boron powder. (It should be mentioned that paraffin causes a small amount of thermalization of the epithermal neutrons.) The most easily machinable refractory boron compound is BN (Soreq & Griffin 1983), and consequently many of the studies with boron filters have involved the use of BN capsules. Ehmann *et al.* (1980) suggested not using BN because of its relatively high cross-section of 1.81 barn for the ^{14}N (n, p) ^{14}C reaction which leads to the formation of an appreciable amount of the long-lived radioactive ^{14}C. Instead, they claimed boron carbide should be used, another refractory compound of boron. Unfortunately boron carbide is extremely hard and not machinable, and a capsule can be made only by the hot-pressing process.

One of the principal disadvantages of boron filters is the amount of impurities found in boron powder, as discussed in detail by Bem and Ryan (1971). However, if a boron capsule is used together with a permanently installed Cd lining, the interferences due to the activities of ^{28}Al, ^{56}Mn and ^{38}Cl from the boron contaminants are significantly reduced (Parry 1982, 1984).

Both cadmium and boron have high absorption cross-sections for low-energy neutrons; however, the energy dependence of the cross-sections differs considerably. Cadmium approaches a perfectly sharp filter for the thermal region and has some resonance in the epithermal range, whereas boron behaves as an almost perfect $1/v$ absorber with no sharp energy cut-off. Although the cross-section for neutron capture by boron is lower than the cross-section for absorption by cadmium in the lower energy range of 0.01–1 eV, this can be compensated for by using thicker boron absorbers. A 0.25 cm thick boron shield is sufficient to stop practically all thermal neutrons. The effective cut-off energy is almost independent of the thickness of the Cd absorber while it increases considerably with the thickness of the boron absorber.

1.6.8 *Fast-neutron INAA*

The reactions that occur with fast neutrons of energy usually in the MeV range should be considered in two ways: (1) the use of these reactions for the determination of some elements; and (2) the possible interference of these reactions in the determination of some elements by the (n, γ) reaction, owing to the formation of the same nuclide, as was discussed earlier in Section 1.6.3 dealing with nuclear interferences. These interferences can only be solved by the use of double irradiation, i.e. irradiation with a bare core and then irradiation inside a Cd or B filter followed by calculation of the contribution from each element. The same treatment is usually adopted when using (n, p) and (n, α) reactions in the determination of certain elements.

The main advantage of these reactions is that they produce nuclides that are different from those produced by (n, γ) reactions. Consequently, this may lead to faster determinations when producing a short-lived nuclide rather than the long-lived one normally produced in the (n, γ) reaction (Alfassi & Lavi 1984a). In other cases, these reactions may enable the determination of elements that cannot be measured via (n, γ) reactions because the produced radionuclide is only a β emitter.

1.6.8.1 *Determination of phosphorus and silicon* Thermal neutron activation cannot be used for the determination of phosphorus and silicon. For instance, the radiative capture (n, γ) reaction with the only stable isotope of P leads to formation of ^{32}P, which is a pure β emitter. In the case of silicon [stable isotopes ^{28}Si (92.2%), ^{29}Si (4.7%) and ^{30}Si (3.1%)], the only radionuclide produced by the (n, γ) reaction is ^{31}Si, which is almost wholly a β emitter. Its very low intensity of γ-rays (1266 keV; 0.07%) together with the low abundance of ^{30}Si and low cross-section for radiative capture (0.11 barn) enables only the determination of relatively large amounts of silicon. In contrast, activation with fast neutrons leads to the formation of ^{28}Al via both ^{31}P (n, α) ^{28}Al and ^{28}Si (n, p) ^{28}Al, and of ^{29}Al by the ^{29}Si (n, p) ^{29}Al reaction. ^{28}Al is also produced by the ^{27}Al (n, γ) ^{28}Al reaction. This leads to a procedure for the determination of Si from the activity of ^{29}Al, which is produced from silicon only, using the activities of ^{28}Al from its activation with a Cd filter and also without a filter to determine the concentration of both aluminium and phosphorus. However, the activity of ^{29}Al produced from Si is almost two orders of magnitude less than the activity of ^{28}Al produced from the same element. Thus, the use of ^{29}Al will both limit the minimal amount of silicon that can be determined and reduce the accuracy of the measurement.

Another problem associated with the measurement of ^{29}Al is that its main γ line at 1273 keV suffers from interference due to the single-escape peak at 1268 keV of the more abundant ^{28}Al. However, when the concentration of Si is high, this element can be determined by the ^{29}Si (n, p) ^{29}Al reaction as was done by Hancock (1982) for the measurement of silicon in pottery using a Cd shield to decrease the formation of ^{28}Al. The samples were allowed to decay for 17–20 min before counting, ensuring further decrease of the ^{28}Al activity. Ördögh *et al.* (1974) measured in a similar method the concentration of silicon in very small inhomogeneous lymph node samples.

The concentration of P was determined spectrophotometrically by the molybdenum blue method and hence the concentration of Si and Al can be found from the

^{28}Al activity induced by both epithermal activation (Cd cover) and reactor neutron irradiation. Another method, as suggested by Alfassi and Lavi (1984b), involved using the simultaneous determination of ^{27}Mg and ^{28}Al, each of them for both reactor activation and irradiation with non-thermal neutrons, to measure concurrently Mg, Al, Si and P. Each of these radionuclides can be formed by three reactions as illustrated in Fig. 1.16. The concentration of the four elements Mg, Al, Si and P can be obtained from a solution of the four equations for the four measured activities (the 844 keV line due to ^{27}Mg and the 1778 keV line due to ^{28}Al, each without a filter and with a cadmium absorber).

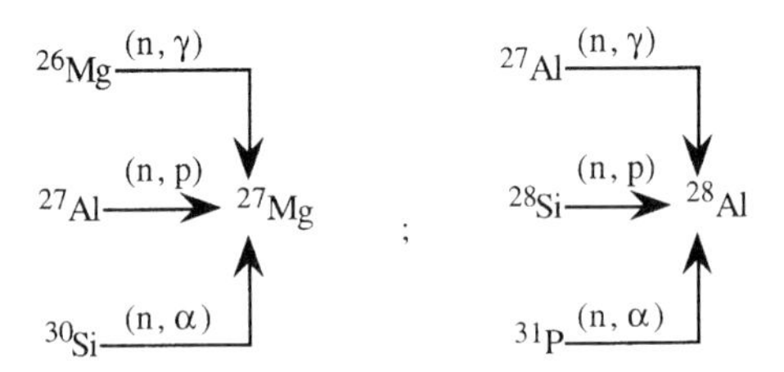

Fig. 1.16 Scheme illustrating formation of ^{27}Mg and ^{28}Al by three simultaneous decay reactions (Alfassi & Lavi 1984b).

1.6.9 *Neutron activation analysis with emissions other than γ emission*

As mentioned before, in most delayed activation analyses only the γ emission is measured, since β^- emitters are difficult to distinguish owing to their extended spectra rather than their monoenergetic emission, X-ray emission is less penetrating, and neutron emission is very rare. However, all of these emissions can be used in special situations.

1.6.9.1 *β^-* It has been shown for both biological samples (Steinnes 1967, 1971) and most geological samples that owing to the higher concentration of P the principal and relatively long-lived β^- emitter is ^{32}P, and after 10 days more than 95% of the β^- emission is due to ^{32}P, and hence the total β^- measurement can be used for the determination of P.

1.6.9.2 *X-rays* (Mantel 1990) X-rays suffer from two disadvantages, which can be turned to advantage in special cases. First, the range of X-rays is quite short and only those emitted from the first micrometres of a sample can be detected. This is a disadvantage when the whole bulk concentration is needed. However, it is advantageous when only the first few microns of a sample are important and when within them the concentrations of the elements to be measured are higher. Second, fewer nuclides emit X-rays (mainly after electron capture, or internal conversion) than those emitting γ-rays. This is disadvantageous since fewer elements can be determined by this method. However, if the major elements of the sample, which interfere with the measurement of the trace elements, do not emit X-rays the few elements that do emit X-rays can be determined even at small concentrations because of the lower background.

Owing to the low penetration of X-rays, the samples used for X-ray measurement should be very thin. X-rays can arise from internal conversion (X-rays of the Z of an IRN), electron capture (X-rays of the $Z-1$ element) or internal conversion after a β decay ($Z+1$ element). In γ-ray spectrometry the electrons emitted by the active sample are stopped by a thin perspex or aluminium foil as the electrons emitted by the active sample are less penetrating than the γ-rays. In the case of low-energy X-rays, which are less penetrating than some of the electrons, the electrons should be removed by a magnetic field (Alfassi *et al.* 1978; Mantel *et al.* 1978; Graman 1986).

X-ray measurement are used mainly for determination of rare earths in geological samples (Hertogen & Gijbels 1971; Mantel & Amiel 1973; Baedecker *et al.* 1977), copper (Allen & Brookhart 1974) and bromine (Rupaport *et al.* 1979) in biological and geological samples, but also other elements can be analysed such as Nb in steel (Mantel *et al.* 1982) and also Co and Hg in sea-water (Stiller *et al.* 1984).

1.6.9.3 *Neutrons* (Alfassi 1990) Delayed neutrons are emitted from only very few radionuclides. In many cases the radionuclide decays by both β^- and neutron emission. For example, ^{87}Br decays 2.8% by neutron emission while ^{88}Br decays 6.6% by neutron emission. However, ^{88}Se decays by only 0.8% and ^{90}Br decays 25% by neutron emission. At least 57 radionuclides are known to emit neutrons, but none is formed by the (n, γ) reaction since this requires the nuclides to be more than one neutron further removed from the line of stability. However, such radionuclides can be formed in a neutron-induced fission of a fissile nuclide such as ^{235}U, since the heavy elements have a higher percentage of neutrons in the nucleus. This is the reason that activation analysis followed by measurement of neutrons is used only for the determination of U and Th.

Since other elements do not produce neutron emitters the background from other elements is very small. The neutrons are measured (after their moderation in a paraffin layer) by BF_3 counters. These are gaseous ionization chambers that measure the α particles formed by the ^{10}B(n, α)^{7}Li reaction. Any α particles beyond the chamber, however, are not able to penetrate the counter walls. The delayed neutron emitters, formed in the neutron-induced fission, have half-lives ranging from 0.08 s to 55.6 s. In order to remove delayed neutrons of the interfering ^{17}N (4.16 s) and also to reduce the error through incorrect time determinations (which is more prominent for short-lived radionuclides), the determination of the emitted neutrons is completed only for those nuclides with half-lives longer than about 10 s. This is achieved by delaying the counting procedure for 30–60 s (Amiel 1963; Shenberg *et al.* 1988). During this delay time some short-lived radionuclides can be measured (Shenberg *et al.* 1988).

1.6.10 *Depth profiling by INAA*

Depth profiling involves the measurement of concentration of a trace element as a function of its distance from the surface of the sample. This can be done by a destructive method in which the whole sample is activated and then divided into a number of sections of varying distances from the surface, either by gradual slow dissolution or by gradual milling and grinding. (For each layer, its thickness and γ or β activity of the various elements present are measured.) However, many samples are too important to be destroyed in the measurement. By using (n, α) or (n, p) reactions the depth profiling can be completed non-destructively. The emitted charged particle (α or p) has a definite kinetic energy. This particle, while transversing the material from its place of formation inside the sample to the detector positioned in front of the sample, loses part of this energy, the amount lost depending on the distance travelled by the particle in the sample. From the measured energy, the distance from the surface can be calculated. Thus, the energy spectra measured for this particle is related to the concentration depth profile of the element, which is responsible for this nuclear reaction.

In order to enable the measurement of low concentrations, the nuclear reaction should have a high cross-section, usually of the order of at least 10 barn. These high cross-sections rule out most elements and only very few elements can be determined.

The method was developed in 1972 by Ziegler *et al.* for the determination of boron because of its high cross-section in the ^{10}B (n, α) ^{7}Li (σ = 3837 barn). This cross-section is five orders of magnitude larger than that of almost any other stable nuclide for the (n, α) reaction (except ^{6}Li). The measured sample is placed in a vacuum chamber (pressure less than 5×10^{-6} torr). In the original study only about 10^{11} neutrons were required to detect a concentration of 100 ppm boron in a 2 cm Si wafer.

1.7 Charged-particle activation analysis (CPAA)

An extensive review of CPAA can be found in Blondiaux *et al.* (1995) and Vandecasteele (1988). The most important application of CPAA is in the analysis of trace concentrations of light elements such as C, N and O, which are difficult to measure using other methods. However, these elements produce β^+ emitters, which do not form γ-rays in addition to the 511 keV photons from the annihilation of the positron. These 511 keV photons are not only produced from any β^+ emitters but derive also from annihilation of the positrons produced in the surrounding material from high-energy photons. Consequently the emitter element can be determined only after chemical separation.

Following chemical separation, carbon can be determined to the ppb level. However CPAA can be used non-destructively for larger concentrations. In contrast to neutron activation, which does not require the kinetic energy of the neutrons, charged-particle activation requires kinetic energy, even for exorgic reactions, in order to overcome the Coulombic repulsion between the particles and the nucleus of the target. This Coulombic barrier (repulsion) increases with Z of the target nucleus. This barrier can be used to determine an energy range, which will be above the barrier for C, N and O, but below the barrier for the major elements in the matrix, not allowing for activation of the matrix itself.

Vandecasteele *et al.* (1979) analysed carbon in Zr, Nb, Ta and W without chemical separation, using the ^{12}C (d, n) ^{13}N activation reaction. They used 5 MeV deuterons for Zr and 7 MeV deuterons for Nb, Ta and W. The matrix was not activated owing to the Coulombic repulsion. The samples were irradiated for 2–5 min ($t_{1/2}$ of ^{13}N is 9.96 min) and then etched with acid. During the etching ^{15}O formed from the ^{14}N (d, n) ^{15}O decay ($t_{1/2}$ of ^{15}O is 2.09 min). The two 511 keV annihilation photons were measured in coincidence between two 3″ × 3″ NaI(Tl) detectors. Concentrations of a few tens of ppm in Zr and Nb, 1 ppm in Ta and less than 15 ppb in W were measured.

Nitrogen is determined by either the ^{14}N (p, n) ^{14}O or the ^{14}N (p, α) ^{11}C reactions. ^{14}O has a half-life of 70.95 s and can be determined by either its γ energy of 2313 keV or via the 511 keV annihilation photons. ^{11}C (20.38 min) has no γ photons and must be determined via the 511 keV annihilation photons. The disadvantage of the ^{14}N (p, n) ^{14}O reaction is the relatively low cross-section (Nozaki & Iwamoto 1981). However, this is compensated for by the characteristic γ-rays. The sensitivity via this reaction is about 1 ppm.

Oxygen is determined mostly by ^{3}He activation via the reaction ^{16}O (^{3}He, p) ^{18}F. The activation by protons or α particles suffers from higher interference than ^{3}He activation. The best reaction for the analysis of oxygen at the trace level is the ^{16}O (t, n) ^{18}F (Valladon & Debrun 1977; Revel *et al.* 1977). However, only a few accelerators have the ability to work with tritium beams.

CPAA is important mostly for the analysis of C, N and O at very low levels, although other elements can be determined by CPAA. In some cases CPAA leads to the formation of γ emitters such as ^{7}Be (53.4 d, 477.6 keV photons) used for the determination of B via ^{10}B (p, α) ^{7}Be and Li via and ^{7}Li (n, p) ^{7}Be. Distinguishing between B and Li can be achieved by using two irradiations, either with different protons energies or once with protons and then subsequent irradiation with deuterons (Alfassi & Sellschop 1989).

A matrix that is completely unsuitable for neutron activation analysis, but which can be analysed for trace elements by CPAA, is uranium metal and uranium compounds. Neutrons will induce fission of the uranium nuclei leading to formation of tens of radionuclides. Owing to the high Z of uranium, however, the Coulombic repulsion of charged particles will prevent fission or activation up to high energies (see Erasmus *et al.* 1986).

Concentration measurements of trace elements by non-destructive proton activation analysis have been carried out using several matrices, e.g. Al (Barrandon *et al.* 1976; Debrun *et al.* 1976; Shibata *et al.* 1979), Si (Barrandon *et al.* 1976; Debrun *et al.* 1976), Fe (Goethals *et al.* 1986), Co (Krivan 1975a; 1976; Barrandon *et al.* 1976; Debrun *et al.* 1976), Mo (Krivan 1975b; Barrandon *et al.* 1976; Debrun *et al.* 1976), Rh (Barrandon *et al.* 1976; Debrun *et al.* 1976), Ag (Barrandon *et al.* 1976; Debrun *et al.* 1976), coal (Landsberger 1984), and environmental samples (Wauters *et al.* 1987).

References

Alfassi, Z.B., *J. Radioanal. Nucl. Chem.* **90**, 151 (1985).
Alfassi, Z.B., in *Activation Analysis*, ed. Z.B. Alfassi. CRC Press, Boca Raton, FL, Vol. I, p.97 (1990).
Alfassi, Z.B., in *Activation Analysis*, ed. Z.B. Alfassi. CRC Press, Boca Raton, FL, Vol. II, p.35 (1990).
Alfassi, Z.B., Biran-Izak, T. and Mantel, M., *Nucl. Instrum. Meth.* **151**, 227 (1978).
Alfassi, Z.B. and Lavi, N., *Analyst* **109**, 959 (1984).
Alfassi, Z.B. and Lavi, N., *J. Radioanal. Nucl. Chem.* **84**, 363 (1984a).
Alfassi, Z.B. and Lavi, N., *Analyst* **109**, 959 (1984b).
Alfassi, Z.B. and Lavi, N., *J. Radioanal. Nucl. Chem.* **150**, 183 (1991).
Alfassi, Z.B. and Lavi, N., *Nucl. Instrum. Meth. Phys. Res.* **A365**, 190 (1995).
Alfassi, Z.B., Probst, T.U. and Rietz, B., *Anal. Chim. Acta* **360**, 243 (1998).
Alfassi, Z.B. and Rietz, B., *Analyst* **119**, 2407 (1994).

Alfassi, Z.B. and Sellschop, J.P.F., *J. Radioanal. Nucl. Chem.* **135**, 341 (1989).
Alfassi, Z.B. and Yang, M.H., *J. Radioanal. Nucl. Chem.* **132**, 99 (1989).
Allen, D.R. and Brookhart, W., *Anal. Chem.* **46**, 1297 (1974).
Amiel, S., *Anal. Chem.* **34**, 1683 (1963).
Anders, O.U., *Anal. Chem.* **33**, 1368 (1960); **33**, 1706 (1961).
Baedecker, P.A., Rowe, J.J. and Steinnes, E., *J. Radioanal Chem.* **40**, 115 (1977).
Barrandon, J.N., Benaben, P. and Debrun, J.L., *Anal. Chim. Acta* **83**, 157 (1976).
Bem, H. and Ryan, D.E., *Anal. Chim. Acta* **124**, 373 (1971).
Blondiaux, G., Debrun, J.L. and Maggiore, C.J., in *Handbook of Modern Ion Beam Materials Analysis*, eds. J.R. Tesmer and M. Nastasi. Materials Research Society, Pittsburgh, 1995, p. 205.
Browne, E., Firestone, R.B. and Shirley, V.S., *Table of Radioactive Isotopes*. Wiley, New York (1986).
Brune, D. and Jirlow, K., *Nukleonik* **6**, 242 (1964).
Choppin, G.R. and Rydberg, J., *Nuclear Chemistry*. Pergamon Press, Oxford (1980).
Currie, L.A., *Anal. Chem.* **40**, 587 (1986).
Cunningham, W.C. and Stroube Jr, W.B., *Sci. Total Environment* **63**, 29 (1987).
Dams, R., Robbins, J.A., Rahn, K.A. and Winchester, J.W., *Anal. Chem.* **42**, 861 (1970).
Debrun, J.L., Barrandon, J.N. and Benaben, P., *Anal. Chem.* **48**, 167 (1976).
De Corte, F., Jovanovic, S., Simonits, A., Moens, L. and Hoste, J., *Atomkernenergie* (Suppl.) **44**, 641 (1984).
De Corte, F., Speecke, A. and Hoste, J., *J. Radioanal. Chem.* **3**, 205 (1969).
Ehmann, W.D., Brückner, J. and McKown, D.M., *J. Radioanal. Chem.* **57**, 491 (1980).
Ehmann, W.D. and Vance, D.E., *Radiochemistry and Nuclear Methods of Analysis*. Wiley, New York (1991).
Erasmus, C.S., Sellschop, J.P.F. and Alfassi, Z.B., *Nucl. Intrum. Methods* **B15**, 5469 (1986).
Erdtmann, G. and Soyka, W., *The Gamma Rays of the Radionuclides*. Verlag Chemie, Weinheim (1979).
Etzion, Z., Alfassi, Z., Lavi, N. and Yagil, R., *Biol. Trace Elem. Res.* **12**, 411 (1987).
Firestone, R.B., Shirley, V.S., Baglin, C.M., Chu, Y.S. and Zipkin, J., *Table of Isotopes*, 8th edn. Wiley, New York (1996).
Friedlander, G., Kennedy, J.W., Macias, E.S. and Miller, J.M., *Nuclear and Radiochemistry*, 3rd edn. Wiley, New York (1981).
Fujingawa, K. and Kudo, K., *J. Radioanal. Chem.* **52**, 411 (1979).
Gladney, E.S., Perrin, D.R., Balagna, J.P. and Warner, C.L., *Anal. Chem.* **52**, 2128 (1980).
Glascock, M.D., *Tables for Neutron Activation Analysis*, 4th edn. University of Missouri Press, Columbia (1996).
Glascock, M.D., Tian, W.Z. and Ehmann, W.D., *J. Radioanal. Nucl. Chem.* **92**, 379 (1985).
Goethals, P., Bosman, B., Krivan, V. and Vandecasteele, C., *Bull. Soc. Chim. Belg.* **95**, 331 (1986).
Graman, L.B., *J. Radioanal. Nucl. Chem.* **99**, 75 (1986).
Greenberg, R.R., Fleming, R.F. and Zeisler, R., *Environ. International* **10**, 129 (1984).
Haas, E.W., Hofmann, R. and Richter, F., *J. Radioanal. Chem.* **69**, 219 (1982).
Hancock, R.G.V., *J. Radioanal. Chem.* **69**, 313 (1982).
Hancock, R.G.V., Grynpas, M.D. and Alpert, B., *J. Radioanal. Nucl. Chem.* **110**, 283 (1987).
Harvey, B.G., *Introduction to Nuclear Physics and Chemistry*, 2nd edn. Prentice Hall, Englewood Cliffs, NJ (1969).
Hertogen, J. and Gijbels, R. *Anal. Chim. Acta* **56**, 61 (1971).
Heydorn, K., *Radiochim. Acta* **3**, 161 (1964).
Iskander, F.Y., *J. Radioanal. Nucl. Chem.* **89**, 511 (1985).
Jaskolska, H. and Rowlinska, L., *J. Radioanal. Chem.* **26**, 31 (1975).
Krane, K.S., *Introductory Nuclear Physics*. Wiley, New York, 1988.
Kratochvil, B. Motkosky, N., Duke, M.S.M. and Ng, D., *Can. J. Chem.* **65**, 1047 (1987).
Krivan, V., *Anal. Chim. Acta* **79**, 161 (1975a).
Krivan, V., *Anal. Chem.* **47**, 469 (1975b).
Krivan, V., *Talanta* **23**, 621 (1976).
Landsberger, S., *Int. J. Environ. Anal. Chem.* **16**, 337 (1984).
Landsberger, S., Arendt, A., Keck, B. and Glascock, M., *Trans. Am. Nucl. Soc.* **56**, 230 (1988).
Lavi, N. and Alfassi, Z.B., *Analyst* **109**, 361 (1984).
Lindstrom, R.M., in *Microelectronic Processing: Inorganic Materials Characterization*, ed. L.A. Casper, AES Symposium Series No. 295. American Chemical Society, Washington, DC, 1986, p. 294.
Lin, X., Heydorn, K. and Rietz, B., *J. Radioanal. Nucl. Chem.* **160**, 85 (1992).
Mantel, M., in *Activation Analysis*, ed. Z.B. Alfassi. CRC Press, Boca Raton, FL, 1990, Vol. I, p. 111.
Mantel, M. and Amiel, S., *J. Radioanal. Chem.* **16**, 127 (1973).
Mantel, M., Amiel, S. and Alfassi, Z.B., *Anal. Chem.* **50**, 441 (1978).

Mantel, M., Shenberg, C. and Rapaport, M.S., *J. Radioanal. Chem.* **75**, 145 (1982).
McOrist, G.D. and Smallwood, A.S., *J. Radioanal. Nucl. Chem.* **223**, 9 (1997).
McOrist, G.D., Smallwood, A. and Fardy, J.J., *J. Radioanal. Nucl. Chem.* **185**, 293 (1994).
Niese, S., *J. Radioanal. Chem.* **38**, 37 (1977).
Nozaki, T.M. and Iwamoto, M., *Radiochim. Acta* **29**, 57 (1981).
Oddone, M., Yegingil, Z., Bigazzi, G., Ercan, T. and Ozdogan, M., *J. Radioanal. Nucl. Chem.* **224**, 27 (1997).
Ördögh, M., Orban, E., Miskovitz, G., Appel, J. and Szabo, E., *Int. J. Appl. Radiat. Isot.* **25**, 61 (1974).
Parry, S.J., *J. Radioanal. Chem.* **59**, 423 (1980).
Parry, S.J., *J. Radioanal. Chem.* **72**, 195 (1982).
Parry, S.J., *Radioanal. Nucl. Chem.* **81**, 143 (1984).
Rapaport, M.S., Mantel, M. and Nothman, R., *Anal. Chem.* **51**, 1356 (1979).
Reus, U. and Westmeier, W., *Atomic Data and Nuclear Data Tables* **29**, 1–192 and 193–406 (1983).
Revel, G., *Euroanalysis VII Conference Proceeding 1987*, ed. E. Roth. Editions Physiques, Paris, 1988, p. 167.
Revel, G., Da Cunha Belo, M., Linck, I. and Kraus, L., *Rev. Appl. Phys.* **12**, 81 (1977).
Revel, G., Deschamps, N., Dardenne, C., Pastol, J.L., Hania, B. and Dinh Ngugen, H., *J. Radioanal. Nucl. Chem.* **85**, 137 (1984).
Rosenberg, R.J., *J. Radioanal. Chem.* **62**, 145 (1981).
Rossitto, F., Terrani, M. and Terrani, S., *Nucl. Instrum. Meth.* **103**, 77 (1972).
Sato, T., in *Activation Analysis*, ed. Z.B. Alfassi. CRC Press, Boca Raton, FL, 1990, Vol. II, pp. 323–358.
Shenberg, C., Nirel, Y., Alfassi, Z. and Shiloni, Y., *J. Radioanal. Nucl. Chem.* **114**, 207 (1988).
Shibata, S., Tanaka, S., Suzuki, T., Umezawa, H., Lo, J.G. and Yeh, S.J., *Int. J. Appl. Rad. Isot.* **30**, 563 (1979).
Simonits, A., De Corte, F and Hoste, J., *J. Radioanal. Chem.* **31**, 467 (1976).
Simons, A. and Landsberger, S., *J. Radioanal. Nucl. Chem.* **110**, 555 (1985).
Soreq, H. and Griffin, H.C., *J. Radioanal. Chem.* **79**, 135 (1983).
Steinnes, E., *Talanta* **14**, 753 (1967).
Steinnes, E., *Anal Chim. Acta* **57**, 457 (1971).
Stiller, M., Mantel, M. and Rapaport, M.S., *J. Radioanal Nucl. Chem.* **83**, 345 (1984).
Stroube Jr., W.B., Cunningham, W.C. and Lutz, G.J., *J. Radioanal. Nucl. Chem.* **112**, 341 (1987).
Stuart, D.C. and Ryan, D.E., *Can. J. Chem.* **59**, 1470 (1981).
Sun, J.X. and Jervis, R.E., *J. Radioanal. Nucl. Chem.* **89**, 553 (1985).
Sun, J.X. and Jervis, R.E., *J. Radioanal. Nucl. Chem.* **114**, 89 (1987).
Theunissen, H.J.J., Jaspers, M.J.J., Hansen, J.M.G. and Verheijke, M.L., *J. Radioanal. Nucl. Chem.* **113**, 391 (1987).
Tian, W.Z. and Ehmann, W.D., *J. Radioanal. Nucl. Chem.* **84**, 89 (1984).
Tokay, R.K., Skalberg, M. and Skarnemak, G., *J. Radioanal. Nucl. Chem.* **96**, 265 (1985).
Valladon, M. and Debrun, J.L., *J. Radioanal. Chem.* **31**, 385 (1977).
Vandecasteele, C., *Activation Analysis with Charged Particles*. Ellis Horwood, New York, 1988.
Vandecasteele, C., Strijckmans, K. and Hoste, J., *Anal. Chim. Acta* **108**, 127 (1979).
Verheijke, M.L., Jaspers, H.J.J., Hansen, J.M.G. and Theunissen, M.J.J., *J. Radioanal. Nucl. Chem.* **113**, 397 (1987).
Vertes, A. and Kiss, I., *Nuclear Chemistry*. Elsevier, Amsterdam (1987).
Wallner, G. and Katzlberger, Ch., *J. Radioanal. Nucl. Chem.* **223**, 139 (1997).
Wauters, G., Vandecasteele, C. and Hoste, J., *J. Trace Microprobe Tech.* **5**, 169 (1987).
Williamson, T.G., Penneche, P.E., Hostica, B., Brenizer, J.S. and Ngugen, T.L., *J. Radioanal. Nucl. Chem.* **114**, 387 (1987).
Wyttenbach, A., Bajo, S. and Tobler, L., *J. Radioanal. Nucl. Chem.* **114**, 137 (1987).
Yule, H.P., in *Nondestructive Activation Analysis*, ed. S. Amiel. Elsevier, Amsterdam, p. 113 (1981).
Zhuk, L.I., Kist, A.A., Mikholskaya, I.N., Osinskaya, N.S., Tilayev, T., Tursunbayev, S.I. and Agzamova, S.V., *J. Radioanal. Nucl. Chem.* **120**, 369 (1988).
Ziegler, J.F., Cole, G.W. and Baglin, J.E.E., *J. Appl. Phys.* **43**, 3809 (1972).

2 Prompt gamma neutron activation analysis with reactor neutrons

C. YONEZAWA

2.1 Introduction

Prompt gamma neutron activation analysis (PGNAA) (Alfassi 1994, 1995) is an isotopic and elemental analysis method for measuring prompt γ-rays emitted by capture or inelastic scattering reactions when a sample is irradiated with neutrons. Since PGNAA uses large penetrative radiation (neutrons) for exciting and γ-rays for detecting, it suffers fewer matrix effects than other non-destructive analytical methods that use either X-rays or charged particles as their applications are limited to analysis of surface regions only. Consequently, PGNAA permits bulk sample analysis. It is an 'in beam' spectroscopic method based on nuclear properties alone and is therefore independent of the chemical form of the elements in the sample. Furthermore, since PGNAA can analyse light elements such as H, B, N, S, Si and so on with negligibly low induced radioactivity, it is particularly suitable for analysis of archaeological samples.

Four types of neutron source are used for PGNAA: nuclear reactors; radioisotope neutron sources such as ^{252}Cf, ^{241}Am–Be, ^{238}Pu–Be, ^{239}Pu–Be and so on; 14 MeV neutron generators; and accelerators. The neutron sources, except for nuclear reactors and ^{252}Cf, emit neutrons of energy higher than 4 MeV and therefore, in these cases, along with capture reactions, threshold inelastic scattering reactions (n, n$'\gamma$) are also used in analysis. Nuclear reactor and ^{252}Cf neutron sources use only capture reactions because these sources emit neutrons of energy less than 2.5 MeV, which is the threshold energy for an inelastic scattering reaction. The higher energy sources can be used to analyse O and C using inelastic scattering, while nuclear reactor and ^{252}Cf neutrons are not energetic enough to analyse these elements. The nuclear reactor is used primarily as the neutron source for PGNAA owing to the fact that it has the highest neutron flux and hence allows the highest analytical sensitivities to be achieved. Analytical sensitivity of reactor-based PGNAA has been improved drastically by using a cold neutron-guided beam that was developed in late 1960s. Therefore, only PGNAA methods using reactor neutrons are described in this chapter. In spite of their lower intensity neutron flux, radioisotope sources, 14 MeV neutron generators and accelerators have all been used for in situ element analysis during geological exploration, on-line bulk analysis of coal and mineral ore, *in vivo* medical diagnosis and explosives detection at airports (Alfassi 1994, 1995).

2.2 History of reactor-based PGNAA

After the discovery of neutrons by Chadwick in 1932, emission of γ-rays during the capture of thermal neutrons was observed almost simultaneously by two separate research efforts in 1934 (see Gladney 1979). One of these observations was made by Chadwick's student Lea while the other was made by Fermi *et al.* The capture γ-rays at 2223 keV from the ^{1}H (n, γ) ^{2}H reaction was observed in their experiments. Owing to the development of the nuclear reactor, high flux neutrons had become available. Therefore, measurements and the compilation of prompt γ-rays from elements for the purposes of nuclear physics as well as elemental analysis were carried out using such reactors during the 1950s and 1960s in association with detectors such as the magnetic pair spectrometer and NaI(Tl) γ-ray scintillation spectrometer.

The first reactor-based PGNAA experiment for elemental analysis was carried out in 1966 by Isenhour and Morrison (1966a) using a NaI(Tl) scintillation spectrometer. In order to improve analytical selectivity at low resolution of the NaI(Tl) detector, they developed a new method called the 'modulation technique' in which γ-ray spectra were measured with and without neutron irradiation using a rotating neutron chopper and applied it to the determination of boron (Isenhour & Morrison 1966b).

Drastic improvements in the analytical selectivity of PGNAA were obtained by developing a high resolution γ-ray detector, Ge(Li), in the late 1960s (Greenwood 1967). The development of the Compton suppression and pair spectrometer with both NaI(Tl) and Ge(Li) detectors (Orphan & Rasmussen 1967) improved further the selectivity of PGNAA. At almost the same time, basic examination of B determination with a reactor-derived neutron beam, using NaI(Tl) and Ge(Li) detectors independently, had been carried out in Japan (Furukawa 1967).

A neutron guide tube and a cold neutron source, both of which generate ideal neutrons for PGNAA, were developed in the late 1960s and the first PGNAA experiment using a thermal neutron guided beam was carried out on biological samples in 1968 by Comar *et al.* (1969). Following this, a highly sensitive PGNAA experiment for multielement determination and its application to geological samples were completed by Henkelmann and Born (1973).

The PGNAA studies until this moment had been performed using temporally constructed systems or as supplements to nuclear physics. Research on the high-sensitive determination of B and Cd using an 'internal target type' PGNAA facility (Gladney *et al.* 1976) triggered the construction of permanent PGNAA facilities at the National Institute of Standards and Technology (NIST) by scientists at the University of Maryland (Anderson *et al.* 1981) and at the University of Missouri (Hanna *et al.* 1981) in the United States. Many basic studies on PGNAA and its applications have been carried out using these systems.

Although early studies using low-energy guided neutron beams were reported before the mid-1970s, not many experiments had been carried out until the low-energy guided neutron beam facilities became widely available in the 1990s. Permanent and stand-alone PGNAA systems for analytical purposes with cold and thermal neutron-guided beams have been constructed in the 1990s and much PGNAA research has been undertaken using these systems.

2.3 Principles and characteristics

2.3.1 *Principles*

A schematic representation of a neutron capture reaction utilized in a reactor-based PGNAA is shown in Fig. 2.1. When a nucleus in the sample is bombarded with

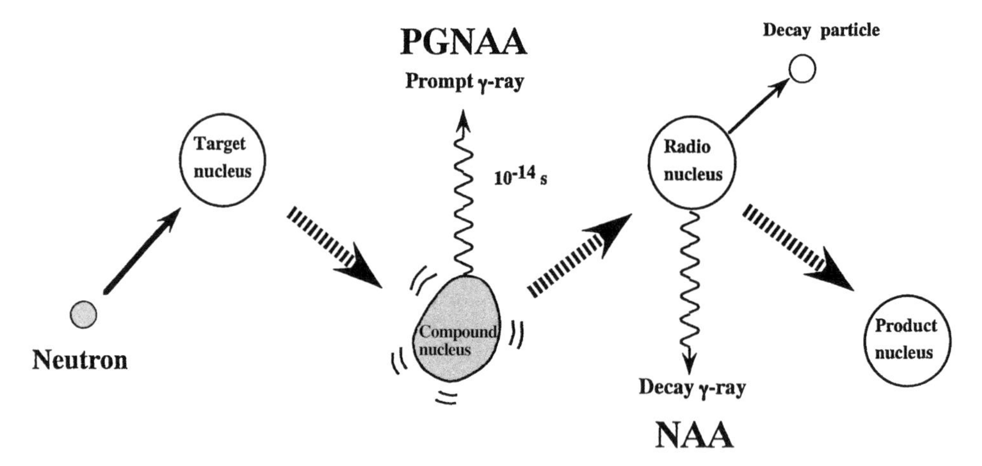

Fig. 2.1 Schematic representation of the neutron capture reaction and its decay.

neutrons it may capture the incident neutron and subsequently form a compound nucleus. The excited compound nucleus is de-excited by emitting prompt (capture) γ-rays within 10^{-14} s and is changed to a radioactive or stable nucleus. When the de-excited nucleus is radioactive the nucleus disintegrates to a product nucleus by emitting decay particles (β^-, β^+ and α, etc.) and γ-rays (decay γ-rays) with the half-life characteristics of a radionucleus. Both the neutron activation analysis (NAA) and the prompt gamma neutron activation analysis (PGNAA) use neutron capture reactions. In NAA, the γ-rays, electrons and positrons emitted from decaying radionuclides are measured, whereas in PGNAA the prompt γ-rays are detected. In both methods, the elements or isotopes in the sample are analysed from the energy and the intensity of the measured γ-ray photons or emitted particles. The energy of the prompt γ-rays is characteristic of the emitting nucleus, and the number of photons of particular energy is directly proportional to the concentration of this element. In NAA the radiation measurement is carried out after neutron irradiation while in PGNAA it is completed simultaneously during the neutron irradiation.

2.3.2 *Neutron capture reaction and prompt γ-rays*

Neutrons, which possess no electric charge, have been shown to be very effective in penetrating positively charged nuclei, thereby producing nuclear reactions. Neutrons are usually classified into three groups according to their energy or velocity (see Alfassi 1994, 1995): thermal neutrons (0.025 eV); epithermal neutrons (0.5 eV to 1 MeV); and fast neutrons (above 1 MeV). Thermal neutrons are those neutrons in equilibrium at ordinary room temperature and their most probable energy is 0.025 eV. Neutrons which are in equilibrium at temperatures much lower than room temperature are called cold neutrons. Although (n, p), (n, α), (n, 2n) and (n, n′γ) reactions are predominant with fast neutrons, capture reactions, i.e. (n, γ), are predominant with thermal and epithermal neutrons. A schematic representation of the capture reaction cross-section versus neutron energy is shown in Fig. 2.2. The cross-sections for many nuclides are low for fast neutrons, while epithermal neutrons are

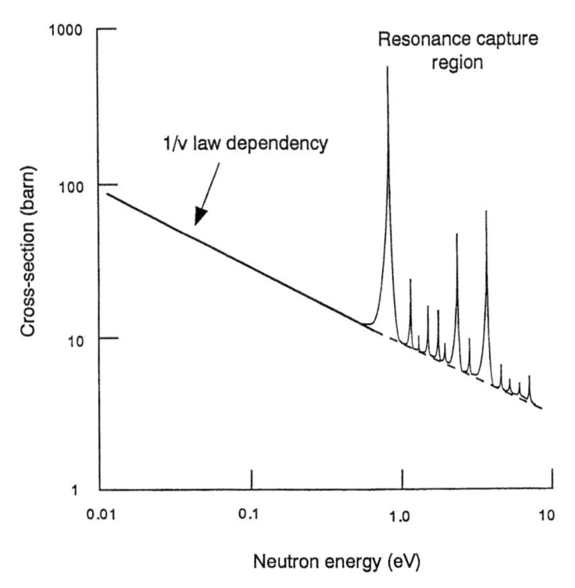

Fig. 2.2 Schematic representation of the capture reaction cross-section versus neutron energy.

characterized by a number of resonance peaks. The cross-sections for thermal neutrons are proportional to $1/v$, where v is the neutron velocity, i.e. the cross-section of the capture reaction becomes larger at lower neutron energies.

When a nucleus in a sample is bombarded with thermal neutrons the incident neutrons are captured by the nucleus and form a compound nucleus which is at an excited state. The excitation energy (E^*) is provided by the kinetic energy of the bombarding neutron (KE_n) and the binding energy of the neutron in the compound nucleus (B_n):

$$E^* = KE_n + B_n \tag{2.1}$$

For thermal neutrons $KE_n \ll B_n$.

$$^1H + ^1n_{th} \longrightarrow ^2H^*, E^* \cong B_n \tag{2.2}$$

One way to complete the reaction process is for ^2H to return to its ground state by emitting a prompt γ-ray of equivalent energy E^*:

$$^2H^* \longrightarrow ^2H + E_\gamma \tag{2.3}$$

This process, also known as an (n, γ) reaction, has been observed in most elements (except Li and B).

The excitation energy can be evaluated according to the mass difference. For example, in the thermal neutron capture reaction of Equations (2.2) and (2.3), the total energy of prompt γ-rays becomes:

$$E^* = B_n = \{[m(^1H) + m_n] - m(^2H)\}c^2 = \Delta mc^2 \text{ (MeV)} \tag{2.4}$$

where $m(^1H)$ is the mass of the ^1H nuclide, $m(^2H)$ is the mass of the ^2H, m_n is the rest mass of the neutron and c is the velocity of light. Using the experimental data of the atomic mass unit [$m(^2H)$ = 2.0141017 amu, $m(^1H)$ = 1.007825 amu, m_n = 1.0086649 amu, and 1 amu = 931.5 MeV/c^2] one can obtain E^* = 0.1163 amu·c^2 = 2.2246 MeV. The only available de-excitation mode of this compound nucleus is the emission of γ-rays of 2.22323 MeV energy with the balance of the reaction energy being dissipated through deuteron recoil. The $^2H^*$ nuclide de-excites in one shot, utilizing all E^*. However, many other compound nuclides de-excite through numerous excited states and eject many low-energy prompt γ-rays. These prompt γ-rays serve for identification and quantitative determination of the element being analysed in the sample.

2.3.3 *Characteristics of PGNAA*

Owing to the above-mentioned properties of the neutron capture reaction and prompt γ-rays, PGNAA has the following advantages:

(1) Multielements can be determined simultaneously and non-destructively.
(2) It is capable of analysing light elements which are difficult to analyse by other analytical methods.
(3) Reactions producing stable nuclides can be used.
(4) Analysis depends only on cross-section, specimen mass and incident flux, not on the rate of decay.
(5) Samples have lower residual radioactivity after irradiation.
(6) The combination of a long range of neutrons in matter and minimal absorption of high-energy γ-rays enables analysis of bulk samples.
(7) The prompt γ-rays are emitted before the product has any possibility of losses of volatile species as a result of the Sziland–Chalmers processes of elements such as Hg or Br prior to off-line counting.

PGNAA has also the following disadvantages:

(1) A large facility (nuclear reactor) is required as the neutron source.
(2) Lower neutron flux and detection efficiency result in lower analytical sensitivity.
(3) Higher initial excitation energy increases the number of nuclear levels available during de-excitation and so prompt γ-ray spectra are more complex than the decay γ-ray spectra.
(4) Analysis is limited to a single sample at a time and a relatively long time is required for the analysis, hence the number of samples has to be small.
(5) Scattered neutrons cause detector degradation and more shielding is required.
(6) Owing to low analytical sensitivities, longer irradiation and relatively large samples are required.

2.4 Facilities

2.4.1 *Types of reactor-based facilities*

The reactor-based PGNAA facilities that have been utilized for analytical purposes are listed in Table 2.1. Irradiation can be conducted in the core or external to the

Table 2.1 Representative reactor-based PGNAA facilities around the world

Organization (country)	Reactor (power)	Type	Neutron flux ($n\ cm^{-2}\ s^{-1}$)	γ-Ray detector	Others	Reference
University of California Los Alamos (USA)	OWR (5 MW)	Internal	4×10^{11}	Ge(Li)–NaI(Tl)		Jurney *et al.* (1967)
GKSS-Research Centre Geesthacht (Germany)	FRG-1 (5 MW)	Internal	8×10^{12}	Ge–NaI(Tl)		Spychala *et al.* (1987)
University of Missouri (USA)	MURR (20 MW)	External	5×10^{8}	Ge(Li)–NaI(Tl)		Hanna *et al.* (1981)
University of Maryland–NIST[a] (USA)	NBSR (10 MW)	External	3.3×10^{8}	Ge–NaI(Tl)		Anderson *et al.* (1981)
National Tsing Hua University (Taiwan)	THOR (1 MW)	External	1.3×10^{6}	Ge–NaI(Tl)		Chung and Yuan (1995)
KFA Jülich (Germany)	FRG-2 (23 MW)	Guided beam	1.5×10^{8}	Ge	Cold neutron	Rossbach (1991)
NIST[a] (USA)	NBSR (20 MW)	Guided beam	1.5×10^{8}	Ge–BGO	Cold neutron	Paul *et al.* (1997)
Institute of Isotopes (Hungary)	BRR (10 MW)	Guided beam	4×10^{6}	Ge–BGO	Thermal neutron	Molnár *et al.* (1997)
JAERI[b] (Japan)	JRR-3M (20 MW)	Guided beam	1.4×10^{8} 2.4×10^{7}	Ge–BGO Ge–BGO	Cold neutron Thermal neutron	Yonezawa *et al.* (1993)
University of Texas Austin (USA)	TRIGA-II (1 MW)	Guided beam	5×10^{7}	Ge	Cold neutron	Ríos-Martínez *et al.* (1998)
Paul Scherrer Institute (Switzerland)	Ring-cyclotron[c] (590 MeV)	Guided beam	6.9×10^{7}	Ge–BGO Ge–NaI(Tl)	Cold neutron	Crittin *et al.* (2000)

[a] National Institute of Standards and Technology. [b] Japan Atomic Energy Research Institute. [c] Spallation neutron source using the ring-cyclotron; the value in the parentheses is the injected proton energy.

reactor by extraction of a neutron beam, and the facilities are classified accordingly as (1) an internal target type and (2) an external target type. In addition to the above a third low-energy guided neutron beam type facility, which uses cold or thermal guided neutron beams as the neutron source instead of the neutron beam commonly used in the external target type facility, is also employed.

2.4.1.1 *Internal target type facilities* In the internal target type facilities a sample is irradiated in the high neutron flux region of a nuclear reactor and the emitted prompt γ-rays are measured by a detector positioned outside the reactor. Although a high neutron flux, higher than 10^{12} n cm^{-2} s^{-1} is available in this target type, the detection efficiency is low owing to a large sample-to-detector distance, in some cases as much as 6 m, and consequently the analytical sensitivities are low. Owing to the high neutron flux irradiation, induced radioactivities in the sample are higher than those of the external target and low-energy guided neutron beam type facility, and the size and shape of sample are limited. Most facilities of this type were constructed to perform nuclear structure investigations and only a few facilities have been used for analytical purposes.

A PGNAA facility at the GKSS Research Centre in Germany (Spychala *et al.* 1987) is shown in Fig. 2.3 as an example of the internal target type facility. The facility was set up in the SMW swimming pool-type reactor of FRG-1. A sample in a graphite capsule is irradiated at the position next to the core along the tangential channel with a thermal neutron flux of 8×10^{12} n cm^{-2} s^{-1}. The emitted γ-rays are measured by a

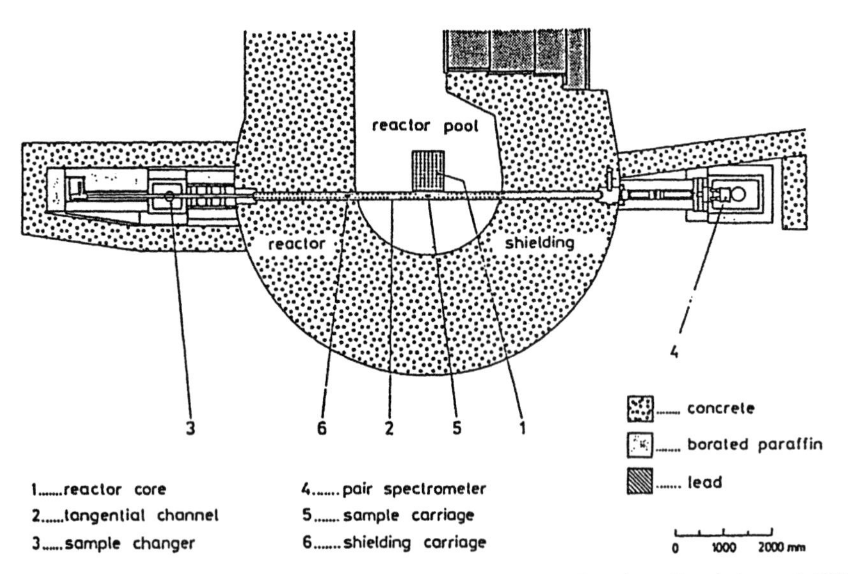

Fig. 2.3 Internal target type facility of the GKSS-Research Centre Geesthact (Spychala *et al.* 1987). A sample in a graphite capsule is irradiated at the position next to the core along the tangential channel. The emitted γ-rays are measured by the Ge–NaI(Tl) detector-pair spectrometer separated from the sample by 6 m. (Reprinted from *Nucl. Geophys. (Int. J. Radiat. Appl. Instrum.)*, **1**, M. Spychala, W. Michaelis and H.-U. Fanger, Prompt gamma-ray neutron activation analysis of river sediments, p. 309. Copyright 1987, with permission from Elsevier Science.)

Ge–NaI(Tl) detector pair spectrometer separated from the sample by a distance of 6 m. Only double-escape peaks of high-energy γ-rays by triple coincidence between the Ge and the two optically isolated NaI(Tl) detectors were measured. The PGNAA facility at the University of California at Los Alamos in the United States is also this type (Jurney *et al.* 1967).

2.4.1.2 *External target type facilities* In the external target type facilities a sample is irradiated with a neutron beam from a horizontally or vertically oriented hole in the reactor shielding and the emitted prompt γ-rays are measured by a detector placed near the sample. The neutron flux of this type is lower than that of the internal target type due to decreased neutron flux extracted from the reactor (this normally decreases the flux by a factor of 10^{-5} to 10^{-6}). However, larger γ-ray detection efficiency can be obtained than in the internal target type facility. The trade-off between high flux with low detection efficiency for the internal target type and low flux with high detection efficiency in the external target type set-up leads approximately to the same sensitivity in the two configurations.

A facility at the University of Maryland–NIST (Anderson *et al.* 1981) is shown in Fig. 2.4 as an example of the external target type. This facility was set up at the external beam port of the 20 MW NBSR reactor. The neutron beam is extracted vertically through a He gas-filled beam thimble with collimator and the rotating beam shutter from the heavy water region of the reactor. The external beam tube has lithium carbonate ($LiCO_3$)–paraffin (inside) and boron carbide (B_4C)–paraffin (outside) as neutron shielding materials. A sample is set in the sample box, made from Plexiglas, and is irradiated with a neutron beam of 4.5 cm diameter and of 2×10^8 n cm^{-2} s^{-1} neutron flux with a Cd (Au) ratio of 55 measured at an operating power of 10 MW. The sample box and the external beam tube are filled with He gas and neutron irradiation is carried out in the He atmosphere. The emitted prompt γ-rays are measured using a multi-mode γ-ray spectrometer with annulus-type NaI(Tl) and Ge(Li) detectors, set adjacent to the sample position with its axis at 90° to the neutron beam direction. The γ-ray spectrometer is equipped to measure three modes of spectrum simultaneously: single, Compton suppression, and pair modes in the energy range of 0.05–11 MeV. Other facilities of this type are at the University of Missouri (Hanna *et al.* 1981) in the United States and at the National Tsing Hua University in Taiwan (Chung & Yuan 1995). Most PGNAA studies have been carried out using this type of facility. As a unique facility, a movable PGNAA facility using a low-power mobile reactor THMER (0.1 W, Taiwan) has been reported (Chung *et al.* 1985). This facility, including the reactor, can be moved on a mobile trailer and has been applied to *in vivo* elemental analysis of the human body.

By comparing with the internal target type facilities, the external target type has several advantages as follows. Owing to the lower neutron flux and fission γ-rays from the reactor core, (1) samples are subjected to no heating, little radiation damage and have very little residual activity; (2) large or fragile samples can be irradiated and analysed; (3) neutron filters can be placed in the neutron beam to remove the γ-ray flux or neutrons of certain energy ranges; and (4) greater flexibility in the sample and detector set-up, e.g. coincidence counting, can be achieved with two or more detectors. However, the γ-ray detection efficiency is not as high as would be expected. The

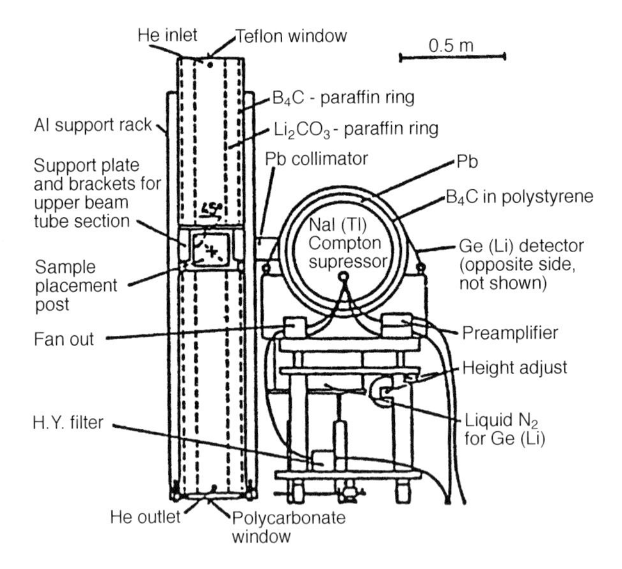

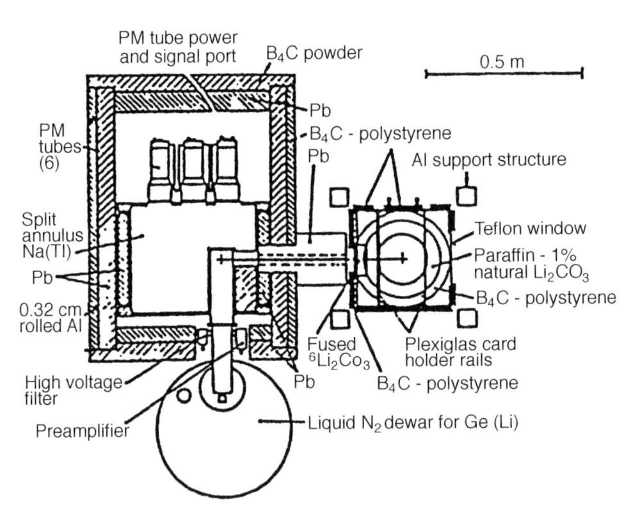

Fig. 2.4 External target type facility of the University of Malyland–NIST (Anderson *et al.* 1981). The facility was set up at the vertical beam port of the NBSR. A sample is irradiated in the sample box and the emitted prompt γ-rays are measured by the Ge–NaI(Tl) detector system. (Reprinted from *J. Radioanal. Chem.*, **63**, D.L. Anderson, M.P. Failey, W.H. Zoller, W.B. Walters, G.E. Gordon and R.M. Lindstrom, Facility for non-destructive analysis for major and trace elements using neutron-capture gamma-ray spectrometry, p. 97. Copyright 1981, with permission from Akadémiai Kiadó Rt.)

conventional thermal neutron beam contains not only γ-rays but also fast and epithermal neutrons, and large amounts of hydrogenous moderating and shielding materials are required to moderate these fast and epithermal neutrons. Because of this the γ-ray background counts become high due mainly to hydrogen. In order to reduce the γ-ray count rate the γ-ray detector cannot be positioned close to the sample – it must instead be positioned at a distance of 40–75 cm from the sample (Anderson *et al.* 1981; Hanna *et al.* 1981).

2.4.1.3 *Low-energy guided neutron beam type facilities* The external target type PGNAA facilities, which use a conventional thermal neutron beam, have the disadvantage of low analytical sensitivity owing to poor neutron beam quality. This disadvantage can be improved, however, by increasing the neutron flux and detection efficiency.

A cold and thermal neutron beam guided by a tube consisting of a Ni mirror or Ni–Ti multilayer supermirror from the reactor core to the outside is virtually free of γ-rays, fast neutrons and epithermal neutrons. The γ-rays, fast neutrons and epithermal neutrons are removed by using the differences in reflection angles (for a curvature guide) and a filter (for a straight guide) and consequently pure low-energy neutron beams are obtained. When the low-energy guided neutron beam is used for the PGNAA neutron source, some advantages can be obtained: there is no need to use bulky hydrogenous thermalizing materials and the γ-ray background count due to hydrogen can be reduced drastically; furthermore, the γ-ray background counts generated by scattered fast neutrons and γ-rays in the neutron beam assume low levels. Consequently, the γ-ray detector can be located close to the sample without creating an unacceptably high background counting rate. This leads to improved efficiency for detecting the capture reaction γ-rays. In addition, as the neutron energy distribution in the guided beam is shifted to a lower energy region, the effective capture cross-section increases according to the '$1/v$' law, which means that cold neutrons are three times as effective in producing the analytical signal as are thermal neutrons. Both the enhanced cross-section and lowered background result in better sensitivity for many elements of interest.

For examples of the low-energy guided neutron beam type facility, refer to the JAERI JRR-3M neutron guide facility and the PGNAA system (Yonezawa *et al.* 1993) as shown in Figs 2.5 and 2.6, respectively. The system can be set to either the cold or thermal neutron beam guide of the 20 MW JRR-3M reactor, the upgraded Japan Research Reactor No. 3, and is characterized by a high sensitivity and low γ-ray background system. The system is designed to achieve the lowest γ-ray background. This is obtained by using lithium fluoride (LiF) tiles for neutron shielding, placing samples in a helium atmosphere and using a Ge–BGO detector system for Compton suppression. The neutron beam is collimated to 20×20 mm^2 by using a LiF collimator. The average neutron flux at the sample position over the 20×20 mm^2 area is 1.4×10^8 (2.4×10^7) n cm^{-2} s^{-1} with a relative standard deviation of 18 (5)% for the cold (thermal) neutrons. The peak neutron energies (wavelengths) of the cold and thermal neutron distributions are 3.0 meV (0.52 nm) and 15 meV (0.23 nm), respectively. The neutron beam intensity is monitored continuously both with a ^{3}He counter and a 0.05 (0.5) mm thick Ti plate for the cold (thermal) neutrons measuring the prompt γ-rays of Ti. The flux fluctuation during one operation cycle of the reactor (26 days) is 0.8%. The multi-mode γ-ray spectrometer comprises a closed-end coaxial-type high-purity Ge detector, BGO (bismuth germanate, $Bi_4Ge_3O_{12}$) shielding detectors and a multichannel analyser (MCA) system controlled by a personal computer. Three modes of prompt γ-ray measurements, namely single mode, Compton suppression mode and pair mode, can be performed simultaneously in the energy range between 0 and 12 MeV. Owing to the low γ-ray background design, the present system has very low hydrogen background counts, which give the highest γ-ray background counts in usual systems by factors of 54–166. Due to the low energy

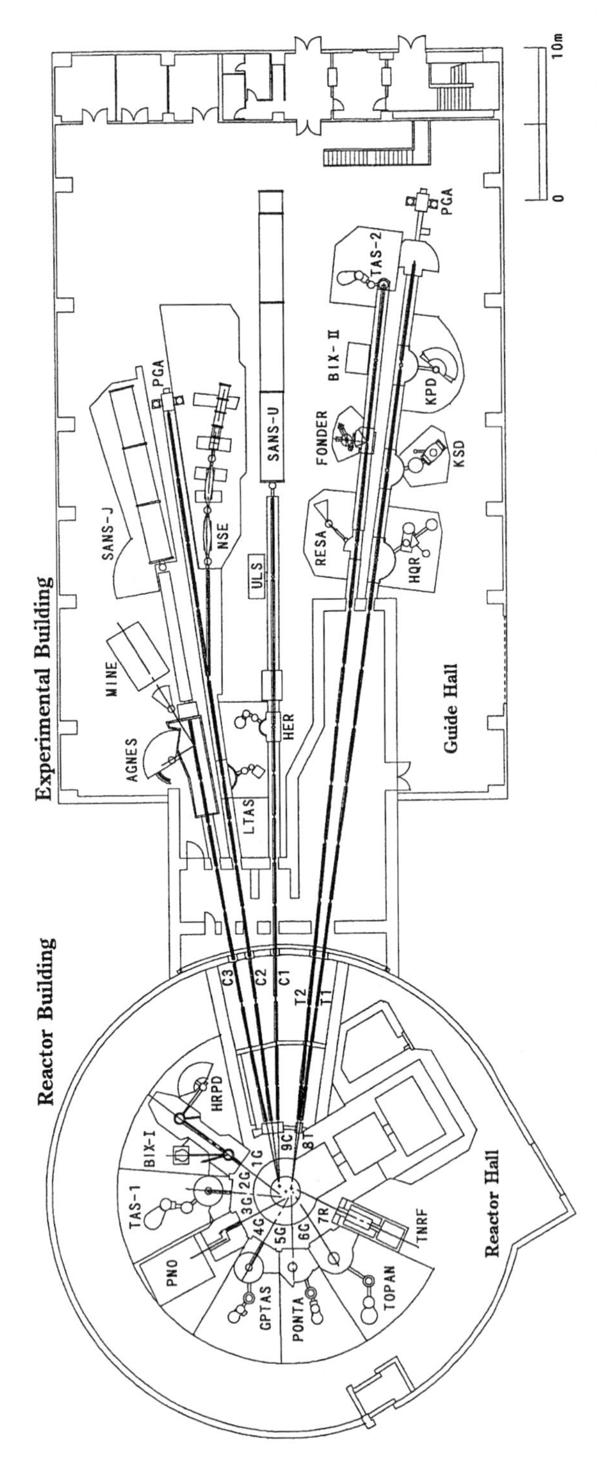

Fig. 2.5 General layout of the JRR-3M neutron guide beam facility at JAERI. The JRR-3M has a liquid hydrogen cold neutron source and cold and thermal neutron beam guides. The PGNAA facility is set at either the cold neutron beam port (C2-3-2) or the thermal neutron beam port (T1-4-1).

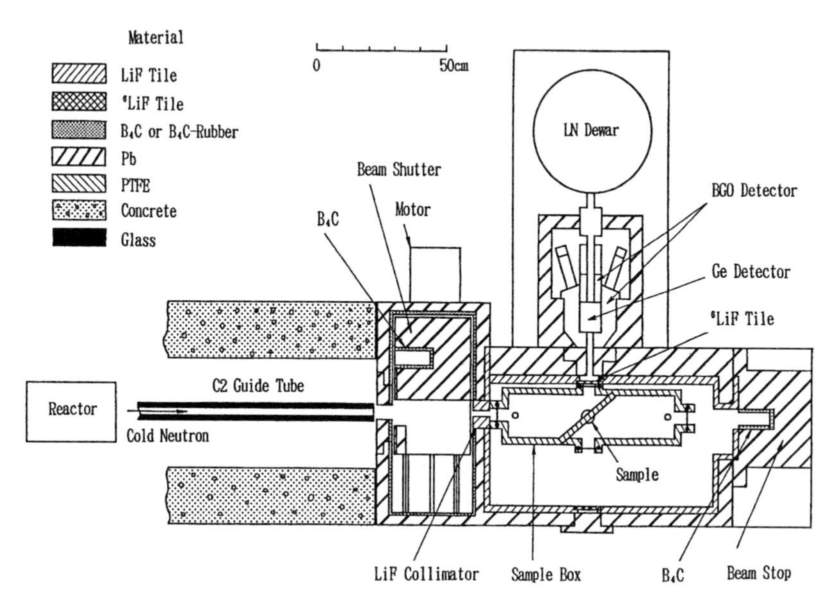

Fig. 2.6 Layout of the JRR-3M PGNAA cold neutron beam facility. A sample is placed in the sample box made from Teflon and the emitted prompt γ-rays are measured by the Ge–BGO detector system under flowing He gas (Yonezawa *et al.* 1995). (Reprinted from *J. Radioanal. Nucl. Chem.*, **193**, C. Yonezawa, M. Magara, H. Sawahata, M. Hoshi, Y. Ito and E. Tachikawa, Prompt gamma-ray analysis using cold and thermal guided neutron beams, p. 171. Copyright 1995, with permission from Akadémiai Kiadó Rt.)

of the guided neutron beams and the low background, analytical sensitivities and detection limits of elements are better than those observed in usual PGNAA systems. The JRR-3M PGNAA system using cold neutrons has higher analytical sensitivities (by factors of 13–27) and lower detection limits (by factors of 3–8) than thermal neutrons under the same geometric conditions. Other facilities of this type are at KFA Jülich in Germany (Rossbach 1991), NIST (Paul *et al.* 1997), the University of Texas at Austin (Ríos-Martínez *et al.* 1998) and the Institute of Isotopes (IKI) (Molnár *et al.* 1997) in Hungary. Although it does not use the reactor neutron, the facility at the Paul Scherrer Institute (Crittin *et al.* 2000) in Switzerland, which employs a cold neutron guided beam from a spallation neutron source using the ring-cyclotron, is also of this type.

Comparison of γ-ray spectra obtained by the external target type and the low-energy guided beam type facilities is given in Fig. 2.7 (Lindstrom & Yonezawa 1995). Figure 2.7(a) is the low-energy portion of the prompt γ-ray spectrum of coal fly ash, measured by the JRR-3M cold neutron guided beam facility at JAERI (Yonezawa *et al.* 1993). The peaks are distinct from the baseline, even down to X-ray energies. By contrast, the spectrum of the same material [Fig. 2.7(b)] measured by the external target type system using a conventional thermal neutron beam from the NBSR at NIST (Anderson *et al.* 1981) is dominated by a continuum of γ-rays originating from the reactor core and from neutron capture in the inner end of the beam tube and scattered from the sample into the detector.

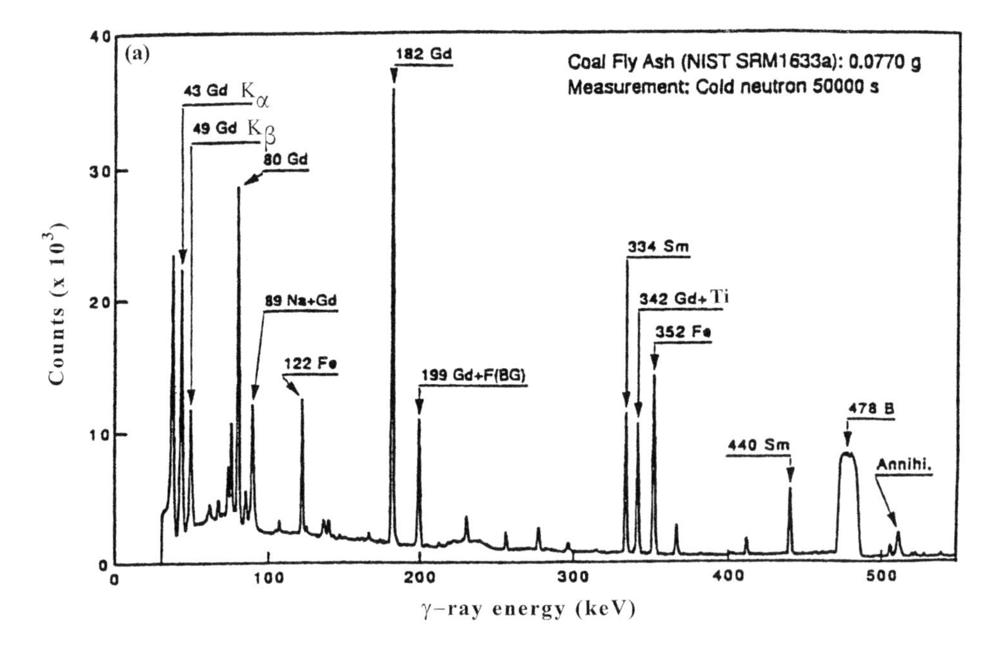

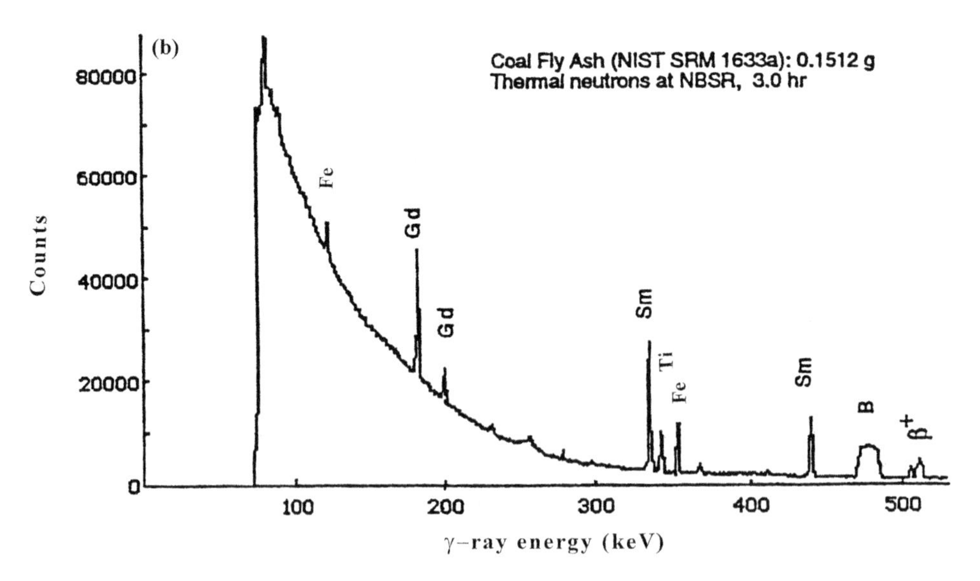

Fig. 2.7 The low-energy region of the PGNAA spectrum of NIST SRM 1633a Coal Fly Ash. In (a), the sample was irradiated with the cold neutron beam of the JRR-3M. In contrast, in (b), the conventional thermal neutron beam at NIST was used. Both spectra were measured in the Compton suppression mode (Lindstrom & Yonezawa 1995). (Reprinted from Z.B. Alfassi and C. Chung (eds.), *Prompt Gamma Neutron Activation Analysis*, R.M. Lindstrom and C. Yonezawa, Prompt gamma activation analysis with guided neutron beams, p. 93. Copyright 1995, with permission from CRC Press.)

2.4.2 *Creation of a neutron beam*

High neutron flux irradiation and high-efficiency γ-ray measurement are required for highly sensitive analysis. Fast neutrons and γ-rays in the neutron beam lead to high γ-ray background effects and this in turn leads to low γ-ray detection efficiency due to the necessarily large sample-to-detector distance required to limit the count rate. Therefore, a highly pure thermal neutron beam is desired, which has fractions of fast neutrons and γ-rays that are as low as possible. In this section the creation of a neutron beam having high purity and high flux is therefore described.

2.4.2.1 *Conventional thermal neutron beam* Extraction of neutrons from well-thermalized neutron fields such as a heavy water tank and thermal column provides high-quality neutron beams. A low γ-ray tangential type beam is more suitable than a radial type beam for PGNAA.

Various filters such as Bi, Si, silica (SiO_2) and Pb have been utilized to cut-off unwanted fast neutrons and γ-rays. The filter absorbs much of the fast neutrons and γ-ray fluxes while transmitting most of the thermal neutrons. Although single-crystal Si and silica are effective at removing fast neutrons, such crystals are not suitable for the removal of γ-rays due to low atomic number. Bismuth, however, is suitable for removing both fast neutrons and γ-rays. Four single Si crystals of 50 cm total length and an 18 cm length of single-crystal silica were used in the instruments at the Universities of Missouri (Hanna *et al.* 1981) and Washington (Lombard *et al.* 1968), respectively. A combination of single crystals of Be and Bi cooled with liquid nitrogen was found to provide a high-purity thermal neutron beam (Alam *et al.* 1993). This system was used at the NIST cold neutron beam facility (Prask 1994), though it has not been utilized in a conventional thermal neutron beam. A diffracted neutron beam from a multilayered graphite monochromator and a sapphire crystal is used as a high-quality thermal neutron beam of 0.0143 eV neutrons with a neutron flux of 6×10^6 n cm^{-2} s^{-1} at the Massachusetts Institute of Technology (Harling *et al.* 1993).

A collimator made from neutron and γ-ray absorber can be used to create a beam size fitted to a sample of particular size and to unify neutron direction. Collimators were made by making holes in Pb, graphite, B and Li contained in polyethylene/paraffin and Mansonite.

2.4.2.2 *Guided neutron beam* The cold and thermal guided neutron beams can be obtained by guiding low-energy neutrons, moderated with heavy water and a low-temperature moderator, from the reactor core to outside via the neutron guide tube. An evacuated or He-filled guide tube (generally rectangular in cross-section and made from polished glass slabs coated with a reflecting film of natural Ni or ^{58}Ni for total reflection of entering neutrons) is most often used for guiding the slow neutrons. Just as with light in an optical fibre, neutrons striking the inner surface of the Ni-coated guide at an angle of incidence less than a critical angle are totally reflected. Thus, a beam can be transported through the guide over long distances with little loss of flux. The efficiency of such a guide (i.e. the fraction of the neutrons entering the guide that are reflected and transported to the outlet) is proportional to the square of the critical angle θ_c. The critical angle is proportional to the neutron wavelength; numerically, θ_c

$= 0.0173\lambda$ for Ni (θ_c in radians, λ in nm). This is 0.18° for thermal neutrons ($\lambda = 0.18$ nm at a velocity $v = 2200$ m s^{-1}) and 0.40° at 0.4 nm, the most probable wavelength of a 30 000 Maxwellian distribution. Thus, the slower, longer wavelength neutrons are collected and transported more efficiently. The largest critical angle for a single substance suitable for mirror coating is obtained with ^{58}Ni, for which $\theta_c = 0.47°$ at 0.4 nm. Thermal neutrons can be guided as well, though with lower efficiency than cold neutrons. Since the critical angle is smaller at shorter wavelengths, a smaller fraction of the neutrons entering the guide emerges from the exit. However, with a thermal guide, the substantial engineering difficulties of operating a low-temperature source adjacent to the reactor core are avoided. Multilayer Ni–Ti supermirrors have been prepared with a critical angle more than three times greater than ^{58}Ni.

In addition to the desired slow neutrons, other radiation enters the neutron guide at the reactor's end: fast (epithermal and fission) neutrons and γ-rays from fission in the fuel and neutron capture in the reactor and beam-tube components. Two methods have been used to remove the remaining undesirable components and produce beams of pure slow neutrons. The first method (used at the JRR-3M, JAERI; see Kawabata *et al.* 1990) includes using a curved section of neutron guide between the reactor and the PGNAA system. Slow neutrons are reflected around the curve while γ-rays, fast neutrons and epithermal neutrons (which travel in straight lines) are removed from the beam. One resulting difficulty is that the flux distribution in the guide is non-uniform, with more neutrons at the outside of the curve. This gradient can be reduced (but not removed entirely) by adding a straight guide after the bend, long enough for several reflections. The second solution (used at the NBSR, NIST; see Prask 1994), is the insertion of a filter into the beam. The filter at NIST is a 127 mm Be and a 178 mm Bi, single-crystal, both at liquid nitrogen temperature. Cold Bi removes fast neutrons and γ-rays from the beam without greatly affecting the slow neutron spectrum. Nearly all neutrons with a wavelength shorter than the Be Bragg cut-off at 0.395 nm (energy $E > 5$ meV; velocity $v > 1000$ m s^{-1}) are removed. The resulting beam has a Cd ratio of 10^4 or more and only a few mR h^{-1} of direct γ radiation (Lindstrom & Yonezawa 1995).

2.4.2.3 *Cold neutron source* The thermal neutrons, most probable neutron energy being 0.025 eV, are obtained by moderating the fast neutrons produced by the fission reaction of ^{235}U with a moderator such as heavy water, light water, graphite and so on at room temperature. The cold neutrons, with energy of 0.001 eV, are obtained by moderating the thermal neutrons with a cold moderator. Most guided beams are associated with a cold neutron source, which is produced by a low-temperature cold moderator placed adjacent to the core of the reactor. The most popular cold moderators are liquid hydrogen (H_2) and deuterium (D_2); however, heavy water ice (D_2O), cooled solid organic compounds, super-critical hydrogen and solid deuterium have been used or proposed. Owing to the larger neutron moderating effect of hydrogen rather than deuterium, the cold neutron source of liquid hydrogen (20 K) can be smaller and thinner than that of liquid deuterium. A larger quantity of moderator and a larger scale cooling system are necessary for the liquid deuterium (25 K) cold source than are needed for the liquid hydrogen due to its lower moderation efficiency. However, the obtained cold flux is higher owing to reduced absorption by the deuterium. Cold source systems using heavy water ice (30–40 K) or cooled organic compounds (liquid methane and solid mesitylene [$(CH_3)_3C_6H_3$]) are only of small size in order to make the cooling simpler. However, the heavy water ice source has low moderation efficiency and the cooled organic compound source has safety problems arising from unexpected hydrogen and organic gas evolution and the production of tarry substances by radiolysis. Therefore, cooled organic compounds are unsuitable for a high flux reactor. The super-critical hydrogen and solid deuterium are expected to be the moderator materials of the future. Liquid hydrogen is used in the JRR-3M at JAERI, while liquid deuterium is used in the spallation neutron source of PSI and cooled mesitylene is used in the TRIGA-II of the University of Texas. Although the NBSR at NIST used heavy water ice initially, the cold moderator was changed to the liquid hydrogen in 1995.

2.4.3 *Radiation shielding and sample support*

The PGNAA technique demands on-line irradiation and counting. However, while neutrons are striking the sample, scientists are working close to the sample and are exposed to the neutron and γ-ray fields. Sustained exposure of a Ge detector to fast neutrons risks reduction of its high-resolution performance and may even destroy it due to a distortion hole-trapping ability. Therefore, a neutron and γ-ray shield surrounding the sample and spectrometer is mandatory in order to protect workers from radiation exposures and to suppress the neutron scatter to the detectors.

Different combinations of shielding materials are utilized for fast and thermal neutrons in neutron shielding. The thermal neutrons can be absorbed by an element which has a large neutron absorption cross-section such as Li, B or Cd. In addition to the neutron absorber, an hydrogenous moderator, such as polyethylene, paraffin, water, etc., is utilized to lower the neutron energy since thermalized neutrons are absorbed more efficiently. Lithium compounds such as LiF or Li_2CO_3 are the best neutron absorbers to utilize near a γ-ray detector. The low-energy neutrons can be absorbed efficiently by ^6Li through a ^6Li (n, α) ^3H reaction, which does not accompany γ-ray emission. LiF tiles and LiF or Li_2CO_3 contained in polyethylene sheets and blocks are commercially available. ^6Li-enriched ^6LiF tile and fused ^6LiF are used to absorb neutrons more efficiently than those of a natural isotopic Li composition with minimized γ photon absorption. In the case of neutron absorption with B and Cd, γ-ray emission is accompanied. Therefore, in addition to the B or Cd, compounds such as H_3BO_3, B_4C, Cd sheet, etc. and γ-ray shielding materials such as Pb and Fe are utilized together with the neutron absorber. Concrete blocks are useful for shielding both neutrons and γ-rays.

The sample irradiation atmosphere and the materials which are used for the sample support and the sample container are also sources of γ-ray background counts. Since O and N absorb and scatter neutrons, air can give a degradation neutron beam intensity and high γ-ray background. Comparison of background γ-ray spectra measured in air and He atmospheres by the JRR-3M PGNAA facility (Yonezawa *et al.* 1993) are shown in Fig. 2.8. A large number of nitrogen peaks are detected in the spectrum measured in air, while no γ-ray peaks from nitrogen are detected in that

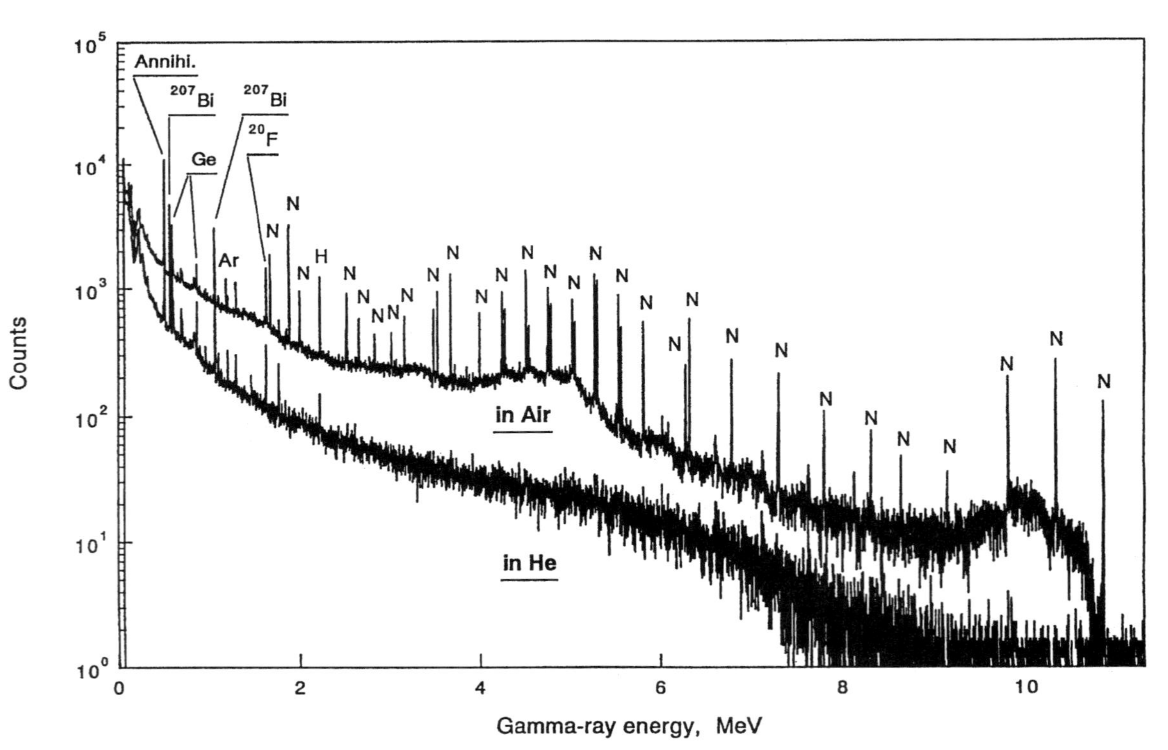

Fig. 2.8 Comparison of the γ-ray background spectra using cold neutron beam irradiation obtained in air with that obtained in a He atmosphere. Both spectra were measured in the single mode using the JRR-3M PGNAA facility (Yonezawa *et al.* 1993). (Reprinted from *Nucl. Instrum. Methods Phys. Res.*, **A329**, C. Yonezawa, A.K. Haji Wood, M. Hoshi, Y. Ito and E. Tachikawa, The characteristics of the prompt gamma-ray analyzing system at the neutron beam guides of JRR-3M, p. 207. Copyright 1993, with permission from Elsevier Science.)

spectrum measured in He. Hence, He and vacuum are utilized as irradiation atmospheres.

Low neutron absorption and low γ-ray emission materials must be used around the sample such as the sample box and sample holder. Good examples are fluorocarbon resin, Zr and Sn, Teflon, Al and Mg. These are utilized for sample box and sample holder materials owing to their good mechanical properties and low cost. A sample to be analysed must also be wrapped and suspended at the beam position, although the wrapping and suspension materials themselves provide further sources of prompt γ-ray background. Therefore, materials that emit the lowest possible level of prompt γ-rays must be selected for these functions. A fluorocarbon resin film is the most suitable material for wrapping the sample because the neutron capture cross-sections and prompt γ-ray emission probabilities of the constituent elements (C and F) are very small. Furthermore, the wrapping material must also be easy to seal. Fluorinated ethylenepropylene resin (FEP) and perfluoroalkoxy resin (PFA) films meet these requirements. However, both FEP and PFA films have difficulty in handling powdery samples because of static electric charge. With powdery samples, polyethylene film is the preferred wrapping even though it contains H, which gives γ-rays of energy 2223 keV. This material can be used only if testing for γ-rays of energy higher than 2223 keV. For liquid samples, an FEP or PFA tube is utilized. For suspending the samples, Teflon or FEP string is used.

2.4.4 *The γ-ray spectrometer*

Neutron-capture prompt γ-rays are characteristically of higher energy and possess a larger number of spectral peaks than conventional decay γ-rays. Decay γ-ray energies are mostly less than 3 MeV, whereas capture γ-rays are up to 11 MeV. During spectrometer development, a magnetic pair spectrometer and a NaI(Tl) detector scintillation γ-ray spectrometer were used for prompt γ-ray measurements. However, these spectrometers suffered from poor detection efficiency or energy resolution. After the 1960s, when the high-resolution semiconducting Ge(Li) detector was developed, the γ-ray detector at the various PGNAA facilities was replaced first by a Ge(Li) detector and then by a high-purity Ge detector.

γ-ray photons are detected in a Ge detector by three processes: photoelectric absorption, Compton scattering and pair production. The photoelectric absorption is dominant at low energy (lower than a few hundred keV) and the pair production occurs most at high energy (threshold of 1.02 MeV, but not important below 1.6 MeV), while the Compton scattering is most important in the mid-energy range. When a high-energy γ-ray ($E_\gamma > 1.6$ MeV) photon is absorbed in the detector by pair production, two photons of 511 keV each are emitted from the positron annihilation. The single-escape peak is due to escape of one annihilation photon from the detector while the double-escape peak is caused by the escape of both photons. Compton suppression and pair spectra modes are adopted not only to lower the Compton continuum but also to simplify the spectrum. While a large annulus-type NaI(Tl) scintillation detector was used for the Compton suppression and pair modes in the early stages of γ-ray spectrometry, it was replaced recently by the BGO owing to its higher detection efficiency for high-energy γ-rays. The high density of BGO and the large Z value of Bi contribute to increased detection efficiency for high-energy γ-rays. The energy resolution of the BGO is inferior to the NaI(Tl), though energy resolution is not important at all in Compton suppression and less important than efficiency in pair spectrometry.

An example of a multi-mode γ-ray spectrometer which uses a Ge–BGO detector system is shown in Fig. 2.9. The spectrometer consists of a closed-end coaxial high-purity Ge detector, BGO shielding detectors and a pulse height analyser system controlled by a personal computer. The BGO detectors are composed of a main BGO, which axially surrounds the Ge detector and two catcher BGOs. The Ge detector is mounted on a 57 cm long horizontal cryostat and 30 litre side-looking liquid nitrogen dewar. The Ge detector is set with its axis perpendicular to the neutron beam at a distance of 29.5 cm (for the cold neutron beam) and 24.5 cm (for the thermal neutron beam) from the sample position. The main BGO is separated into eight parts and the scintillation photons are collected by eight photomultiplier tubes (PMT). The two catcher BGOs are divided into two parts and the photons are collected by four PMTs. A block diagram of the electronics of the γ-ray spectrometer of the JRR-3M PGNAA facility is shown in Fig. 2.10. Single, Compton suppression and pair modes of the prompt γ-ray spectra are measured simultaneously. Each of these signals is fed into an 8192-channel analogue-to-digital converter (ADC). In the single mode, all the signals from the Ge detector are recorded, i.e. full-energy γ-ray peaks, escape peaks and Compton continuum. In the Compton suppression mode, only those signals from the Ge detector which do not have a coincidence signal in the BGO detectors are recorded. For the pair spectrum, triple coincidence between the Ge and the two halves of the main BGO crystals are required to accept a Ge pulse at the ADC. All of the signals from the ADC are fed into and sorted by a multichannel analyser.

Three modes of a prompt γ-ray spectra at the high-energy portion (3–10 MeV)

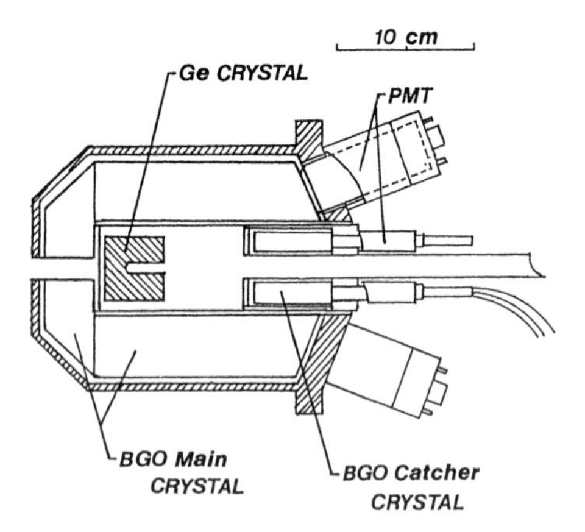

Fig. 2.9 Cross-sectional view of the Ge–BGO detector system for the JRR-3M PGNAA facility (Yonezawa *et al.* 1993). The detector system comprises a closed-end coaxial Ge detector and BGO detectors, which are composed of a main and catcher BGOs. (Reprinted from *Nucl. Instrum. Methods Phys. Res.*, **A329**, C. Yonezawa, A.K. Haji Wood, M. Hoshi, Y. Ito and E. Tachikawa, The characteristics of the prompt gamma-ray analyzing system at the neutron beam guides of JRR-3M, p. 207. Copyright 1993, with permission from Elsevier Science.)

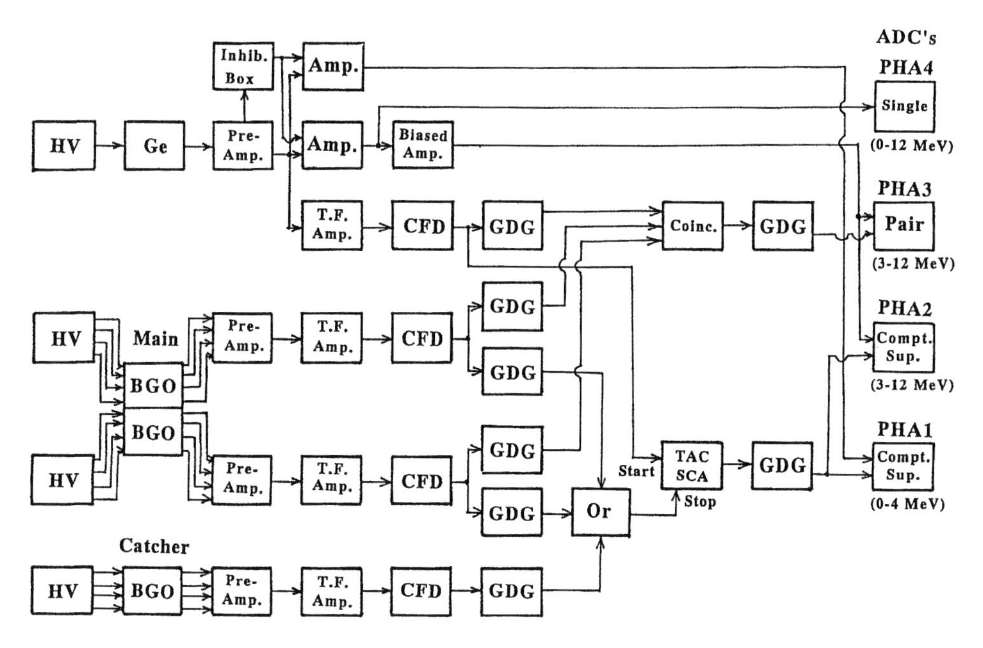

Fig. 2.10 Block diagram of the electronics for the multi-mode γ-ray spectrometer of the JRR-3M PGNAA facility (Yonezawa *et al.* 1993). T.F. Amp: timing filter amplifier; CFD: constant fraction discriminator; GDG: gate and delay generator; TACSCA: time-to-amplitude converter and single-channel analyser. (Reprinted from *Nucl. Instrum. Methods Phys. Res.*, **A329**, C. Yonezawa, A.K. Haji Wood, M. Hoshi, Y. Ito and E. Tachikawa, The characteristics of the prompt gamma-ray analyzing system at the neutron beam guides of JRR-3M, p. 207. Copyright 1993, with permission from Elsevier Science.)

using a Ge–BGO detector system are shown in Fig. 2.11. In the single mode spectrum, full-energy peaks, together with the single- and double-escape peaks, are observed. In the Compton suppression mode spectrum, the Compton continuum, single- and double-escape peaks are suppressed. Small full-energy peaks, which are not distinct in the single mode spectrum, can clearly be seen in the Compton suppression spectrum. Ideally, the pair mode is set for the detection of double-escape peaks only. However, due to incomplete optical isolation of the main BGO detector, single-escape peaks could still be detected.

2.5 Elemental analysis

2.5.1 *Basics for elemental analysis*

2.5.1.1 *Analytical sensitivity and detection limit* Analytical sensitivities (cps mg^{-1}) and detection limits depend on the neutron spectrum, neutron flux, γ-ray detection efficiencies and coexisting elements. The analytical sensitivities and detection limits without the presence of any matrix for 73 elements, except for those that are gaseous and radioactive, obtained with the cold neutron beam of the JRR-3M (Yonezawa 1997), are shown in Table 2.2. The analytical sensitivities were obtained by measuring

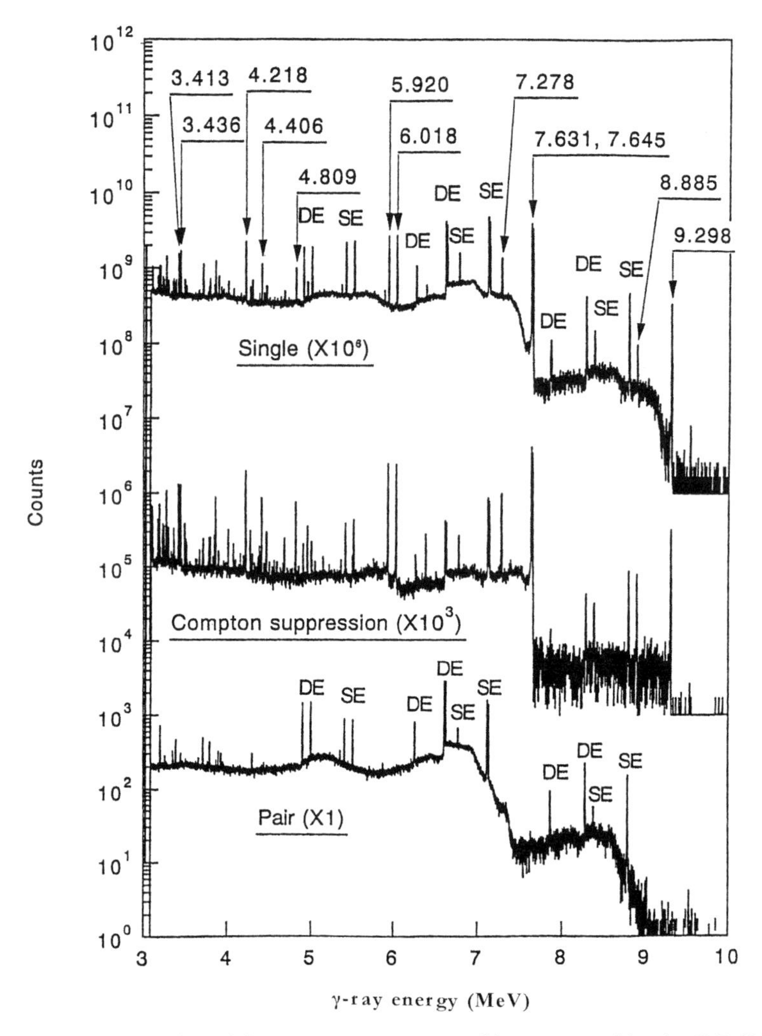

Fig. 2.11 High-energy portion of the prompt γ-ray spectra of iron measured by the JRR-3M PGNAA facility (Yonezawa *et al.* 1993). An iron sample (0.8806 g) was measured with the cold neutron beam for 2000 s. SE and DE denote single- and double-escape peaks, respectively. (Reprinted from *Nucl. Instrum. Methods Phys. Res.*, **A329**, C. Yonezawa, A.K. Haji Wood, M. Hoshi, Y. Ito and E. Tachikawa, The characteristics of the prompt gamma-ray analyzing system at the neutron beam guides of JRR-3M, p. 207. Copyright 1993, with permission from Elsevier Science.)

the Compton-suppression prompt γ-ray spectra of known amounts of high-purity elemental compounds and the detection limits were set from the Compton suppression spectrum of FEP film, which was acquired for 50 000 s. The detection limit was taken as three times the standard deviation of the background count (3σ) in the region corresponding to the prompt γ-ray lines of specified elements in the FEP spectrum. For each element the most sensitive and interference-free prompt γ-ray lines, decay γ-ray lines, and X-ray lines are given in the table. The highest analytical sensitivity at the cold neutron beam is plotted against the lowest detection limit for each element in Fig. 2.12.

Table 2.2 Analytical sensitivities and detection limits of the elements using the cold neutron guided beam of JRR-3M (data cited from Yonezawa 1997)

Element	E_γ (keV)	Analytical sensitivity (cps mg^{-1})	Detection limit (μg)
H	2223	3.14	1.3
Li	2032	0.0467	24
Be	854	0.0145	159
	6809	0.00566	49
B	478	2300	0.0025
C	1262	0.00290	807
N	1885	0.0122	119
	5269	0.00700	115
F	582	0.0198	333
	1634 D	0.0272	267
Na	92	1.04	12
	472	0.867	4.7
Mg	585	0.0350	186
	3918	0.00752	73
Al	1779 D	0.110	15
Si	3540	0.0298	23
P	513	0.0909	54
	637	0.0294	88
S	841	0.253	15
Cl	517	5.21	0.74
	1165	3.60	0.79
K	770	0.574	3.1
Ca	1942	0.0546	18
Sc	228	14.9	0.65
Ti	342	1.97	1.7
	1381	1.90	0.79
V	125	2.85	3.9
	1434 D	1.69	0.70
Cr	835	0.688	3.9
Mn	84	6.23	2.5
	212	2.67	2.7
Fe	352	0.229	28
Co	230	7.05	1.0
	556	2.92	0.94
Ni	465	0.558	5.1
Cu	160	0.811	12
	278	0.789	6.4
Zn	115	0.203	51
	1078	0.107	15
Ga	145	0.527	16
	508	0.174	28
Ge	596	0.393	13
As	165	1.13	8.7
Se	239	1.65	5.0
	614	0.796	4.4
Br	245	0.919	6.5
Sr	898	0.159	13
	1837	0.134	7.0
Y	203	0.189	30
	777	0.167	11
Zr	934	0.0261	71
Nb	99	0.201	61
	256	0.0795	53

Contd

Table 2.2 *Contd*

Element	E_γ (keV)	Analytical sensitivity (cps mg^{-1})	Detection limit (μg)
Mo	778	0.465	3.8
Ru	540	0.278	11
Pd	512	1.11	5.3
	717	0.169	19
Ag	198	5.21	1.7
Cd	558	403	0.0108
In	96	17.8	0.69
	273	13.5	0.39
Sn	1293	0.0178	110
Sb	122	0.230	49
	283	0.108	40
Te	603	0.609	11
I	134	1.04	10
Ba	628	0.0555	50
	1436	0.0311	41
La	218	0.338	22
Ce	662	0.0956	29
Pr	177	0.548	14
Nd	697	7.99	0.68
Sm	334	749	0.0071
Eu	90 D	740	0.0470
	207	17.3	0.64
	221	25.7	0.34
Gd	182	1564	0.0064
Tb	75	0.937	19
	352	0.110	24
Dy	186	67.4	0.11
Ho	137	7.20	1.8
Er	185	20.2	0.41
	816	4.55	0.35
Tm	205	3.17	1.9
Yb	515	2.53	1.8
	636	0.277	5.9
Lu	151	3.77	2.3
	458	1.10	2.7
Hf	214 D	14.0	0.57
Ta	57 X	7.06	2.5
	270	1.29	4.3
W	59 X	2.19	7.1
	146	0.583	18
Re	61 X	18.8	1.0
	208	1.20	5.7
Ir	65 X	2.57	7.1
	352	0.144	19
Pt	67 X	0.320	64
	356	1.22	3.0
Au	69 X	5.90	3.3
	215	1.36	4.5
Hg	71 X	9.52	1.5
	368	53.7	0.055
Tl	73 X	0.348	60.2
	348	0.0691	46
Pb	75 X	0.00059	36000
	7368	0.00147	240
Bi	320	0.00172	2200

D: Decay γ-ray. X: Characteristic X-ray.

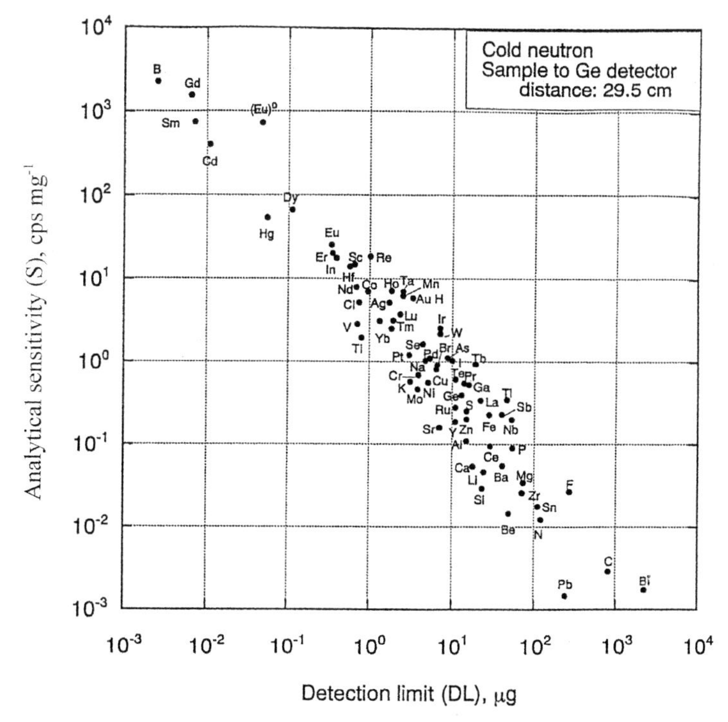

Fig. 2.12 The highest analytical sensitivity and the lowest detection limit of elements by PGNAA with the cold neutron beam of JRR-3M (Yonezawa *et al.* 1995). (Reprinted from *J. Radioanal. Nucl. Chem.*, **193**, C. Yonezawa, M. Magara, H. Sawahata, M. Hoshi, Y. Ito and E. Tachikawa, Prompt gamma-ray analysis using cold and thermal guided neutron beams, p. 171. Copyright 1995, with permission from Akadémiai Kiadó Rt.).

Boron, Sm, Gd and Cd have the highest analytical sensitivity, with detection limits down to 2.5–11 ng. Mercury, Dy, Eu, Er, Nd, In, Hf, Cl, Sc, Ti, Co and H can be detected to 0.05–1 µg. Lead, F, C and Bi have the lowest sensitivity with detection limits in the range of 0.2–2 mg. Other elements are estimated to have a medium sensitivity, such as V, Yb (and most of the other lanthanides), Mn, Ag, Ni, Na, K, S, Fe, Si, P, Sn and N etc., with detection limits ranging from 1 to 100 µg. The decay γ-rays at 1634 keV from ^{20}F ($t_{1/2}$: 11.03 s), 1779 keV from ^{28}Al ($t_{1/2}$: 2.2406 min), 1434 keV from ^{52}V ($t_{1/2}$: 3.75 min), 90 keV from ^{152m2}Eu ($t_{1/2}$: 1.600 h), and 214 keV from ^{179m1}Hf ($T_{1/2}$: 18.68 s) have much higher sensitivities than the corresponding prompt γ-rays. As shown in Table 2.2, characteristic X-rays were detected for heavy elements. The emission of the X-rays may be due to internal conversion following the emission of prompt γ-rays. The analytical sensitivity of the X-rays for the elements with atomic number larger than Ta is higher than that of the prompt γ-rays except for Pt, Hg, Pb and Bi. Measurement of the characteristic X-rays is therefore useful for the analysis of these heavy elements.

The detection limits of elements are affected by the type and amount of coexisting elements. The detection limit of the selected 21 elements in various materials including biological, geological and environmental samples (Yonezawa 1996) are shown in Fig. 2.13. Compton-suppression prompt γ-ray spectra for 100–200 mg of the samples were

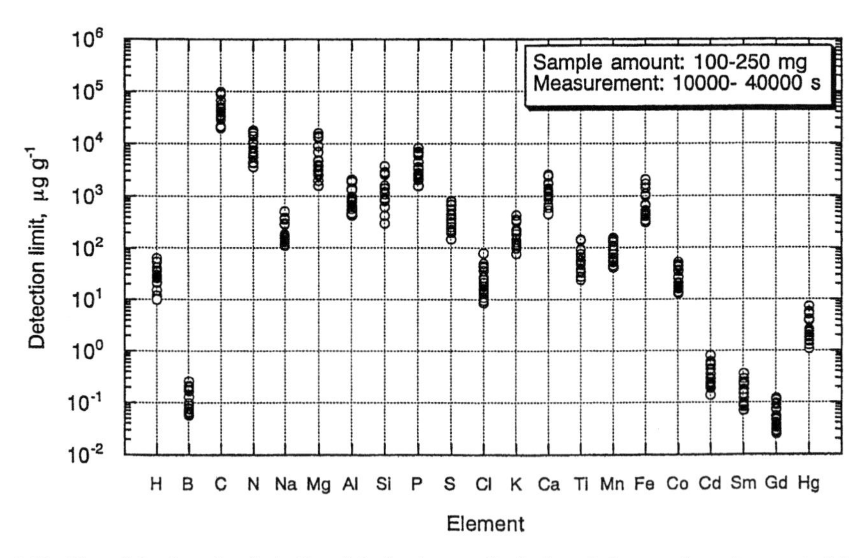

Fig. 2.13 Plot of the detection limit (3σ of the background) of selected elements in various materials with the cold neutron guided beam of JRR-3M (Yonezawa 1996). Compton-suppression prompt γ-ray spectra for 100–250 mg of the various materials including biological, geological and environmental samples were measured. (Reprinted from *Anal. Sci.*, **12**, C. Yonezawa, Multi-element determination by a cold neutron-induced prompt gamma-ray analysis, p. 605. Copyright 1996, with permission from The Japan Society for Analytical Chemistry.)

measured for 10 000–40 000 s with the cold neutron beam of the JRR-3M and the detection limits were determined as three times the standard deviation for the background counts (3σ) in the region corresponding to the prompt γ-ray lines of the specified elements. The detection limits varied by about one order of magnitude, depending on the type and amount of coexisting elements. Owing to a high concentration of Cl, the detection limits in sea samples were higher than in the other samples. The orders of the detection limits were: 25–820 ng g^{-1} for B, Cd, Sm and Gd; 1.1–820 µg g^{-1} for H, Na, S, Cl, K, Ti, Mn, Co and Hg; and 0.031–10% for C, N, Mg, Al, Si, P, Ca and Fe.

2.5.1.2 *Doppler broadening of the B 478 keV γ-ray* Boron is determined in PGNAA to have the highest analytical sensitivity. When boron is irradiated with thermal neutrons, prompt γ-rays of energy 478 keV are emitted from the excited $^{7*}Li$ nuclei produced by the ^{10}B (n, α) $^{7*}Li$ reaction with a large neutron capture cross-section. Since the produced $^{7*}Li$, with a short lifetime of 1.05×10^{-13} s, has a high kinetic energy due to the nuclear reaction, the 478 keV γ-rays are emitted from recoiling $^{7*}Li$ particles in flight. This leads to an abnormally wider peak as compared to a prompt γ peak from a stationary nuclei owing to the Doppler effect. Spectral line-shapes of the 478 keV γ-ray measured for various boron samples (Sakai *et al.* 1996) are shown in Fig. 2.14 together with that line-shape of the 472 keV prompt γ-ray from Na with no Doppler broadening. As is seen in Fig. 2.14, the 478 keV γ-ray peak is 8–10 times wider than for a normal γ-ray and the line-shape varies among the boron samples. The peak-shape of the Doppler-broadened 478 keV γ-ray lines has been studied by

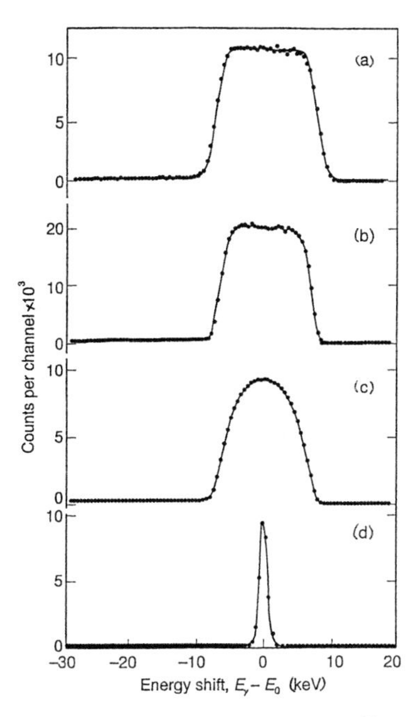

Fig. 2.14 Spectral line-shape of the prompt γ-ray at 478 keV of ^{7*}Li measured for gaseous boron trifluoride, BF_3 at 900 torr (a), aqueous solution of boric acid H_3BO_3 with 4.0 wt% (b) and elemental boron (c). For comparison, the line-shape (d) of the prompt γ-ray at 472 keV from a stationary ^{24*}Na nucleus (Na_2CO_3 target). The solid curves were obtained by spectral fitting (Sakai *et al.* 1996). (Reprinted from *Radiochim. Acta*, **72**, Y. Sakai, C. Yonezawa, H. Matsue, M. Magara, H. Sawahata and Y. Ito, Ranges of ^{7}Li produced in ^{10}B(n, α)^{7*}Li reaction, p. 45. Copyright 1996, with permission from R. Oldenbourg Verlag GmbH.)

Neuwirth and coworkers (Hauser *et al.* 1974; Neuwirth *et al.* 1975) and Sakai *et al.* (1994, 1996). The description of the spectral line-shape based on the study by Sakai *et al.* (1994, 1996) follows.

Let us assume that the 478 keV γ-ray is emitted from the moving ^{7*}Li at an angle θ to the direction of the detector. The moving ^{7*}Li (initial velocity, $v_0 = 4.8 \times 10^8$ cm s^{-1}) gradually loses its kinetic energy in the sample while simultaneously decaying with lifetime, τ, of 1.05×10^{-13} s (decay constant, $\lambda = 9.49 \times 10^{12}$ s^{-1}). However, as a first approximation we neglect the loss of kinetic energy, and the decay of ^{7*}Li.

The peak shape, $f(E, v(t))$, of the 478 keV γ-rays emitted from flying ^{7*}Li at angle θ to the detector is expressed by:

$$f(E, v(t)) = A \int_0^\pi \sin\theta \cdot g(E, \theta)\mathrm{d}\theta. \tag{2.5}$$

The line-shape $g(E, \theta)$ of the Doppler-shifted γ-ray emitted from the ^{7*}Li flying at angle θ is expressed by the Gaussian function

$$g(E, \theta) = \exp\left\{-\frac{0.693}{\Gamma^2}(E + \Delta E(\theta, v(t)))^2\right\}, \tag{2.6}$$

where Γ is the energy resolution (half of the full width at half maximum (FWHM)) of the peak with the measurement condition and $\Delta E(\theta)$ is the Doppler-shifted energy of the 478 keV γ-ray emitted from the flying ^{7*}Li at angle θ to the detector. The $\Delta E(\theta)$ value is obtained from the γ-ray energy E_0 without Doppler shift (478 keV), initial velocity v_0 of the ^{7*}Li and velocity of light, c, by means of the following equation:

$$\Delta E(\theta) = \frac{E_0 v_0 \cos\theta}{c} \tag{2.7}$$

For a more accurate calculation the decay and slowing-down processes of the ^{7*}Li have to be considered. The decay of the ^{7*}Li is given by

$$N(t) = N_0 \exp(-\lambda t), \tag{2.8}$$

where λ (s^{-1}) is the decay constant and t (s) is the decay time. The final peak-shape $F(E)$ of the 478 keV γ-rays, taking into account both the decay and the slowing-down processes of the ^{7*}Li, is expressed by

$$F(E) = A \int_0^\infty N(t)\, f(E, v(t))\mathrm{d}t, \tag{2.9}$$

where $v(t)$ is the velocity of the ^{7*}Li in the samples. The ^{7}*Li velocity $v(t)$ is assumed by applying the Lindhard–Schraff–Schiott (LSS) theory to this case as follows:

$$v(t) = v_0 \exp(-Dt) \tag{2.10}$$

where D (s^{-1}) is the so-called degradation (slowing-down) constant. Finally, the observable peak-shape of the 478 keV γ-rays can be calculated from the formula obtained by substituting Equations (2.5), (2.6), (2.7), (2.8) and (2.10) in (2.9). Thus, this implies that the peak-shape of the 478 keV γ-ray is influenced by the ^{7*}Li degradation process in the samples and the FWHM of the peak is inversely correlated to the degradation constant, D. The D values observed by analysing the measured spectra shown in Fig. 2.14 were $< 0.01 \times 10^{12}$ s^{-1} (BF_3), 1.22×10^{12} s^{-1} (H_3BO_3 solution), 1.48×10^{12} s^{-1} (H_3BO_3 crystals), and 2.57×10^{12} s^{-1} (elemental B), which all agree well with those values obtained by calculations using the LSS theory.

2.5.1.3 *Spectral interference* Because of the complexity of prompt γ-ray spectra, spectral interference often constitutes a major source of error in PGNAA. Possible spectral interferences for the γ-ray lines of the selected elements (i.e. those which are suitable for determination due to high analytical sensitivities and least spectral interference from other elements) are shown in Table 2.3 together with the equivalent amounts of the interference elements (Yonezawa & Haji Wood 1995; Yonezawa 1996). The equivalent amounts are those quantities of the studied element (mg) equivalent to the γ-ray peak counts due to 1 mg of the interfering element. Owing to the low analytical sensitivity of carbon, a large degree of interference to carbon was found, especially from elements with a large cross-section, such as Sm, Gd and Hg. Because of the Doppler-broadened wider peak of boron at 478 keV, the γ-ray lines of 11 elements such as Na, Si, P, Cl, Mn, Co, Ni, Sr, Cd, Sm and Hg, were also found in the same peak. Although many elements show spectral interference to boron, only Na

Table 2.3 Spectral interference of selected elements and their equivalent amounts

Element	E_γ (keV)	Interfering element and equivalent amount[a] (Av. $\pm \sigma_{n-1}$)
B	478	Na ((3.10 $\pm$ 0.12) $\times$ 10^{-4}), Si ((1.05 $\pm$ 0.22) $\times$ 10^{-7}), P ((3.1 $\pm$ 1.2) $\times$ 10^{-7}), Co ((2.02 $\pm$ 0.30) $\times$ 10^{-4}), Ni ((3.59 $\pm$ 0.20) $\times$ 10^{-6}), Sr ((4.73 $\pm$ 0.16) $\times$ 10^{-6}), Cd (0.00116 $\pm$ 0.00011)
C	1262	Si (0.31 $\pm$ 0.08), Fe (12 $\pm$ 0.7), Sm (2500 $\pm$ 670), Gd (20000 $\pm$ 13000), Hg (1000 $\pm$ 220)
N	1885	Na (0.16 $\pm$ 0.03), Ti (0.81 $\pm$ 0.10)
Na	92	Sr (0.0010 $\pm$ 0.0002)
Si	3540	Sr (0.019 $\pm$ 0.002)
P	637	K (0.12 $\pm$ 0.03), Ti (0.063 $\pm$ 0.014), Gd (560 $\pm$ 170)
S	841	Mg (0.011 $\pm$ 0.002), K (0.072 $\pm$ 0.002), Mn (0.041 $\pm$ 0.009), Sr (0.0031 $\pm$ 0.0012)
Cl	786 + 788	Al ((1.5 $\pm$ 0.4) $\times$ 10^{-4}), Co (0.17 $\pm$ 0.04)
K	770	Sr (0.0033 $\pm$ 0.0004), Gd (35 $\pm$ 15)
Ca	1942	P (0.015 $\pm$ 0.0009), Ti (0.024 $\pm$ 0.004)
Ti	1381	Mn ($<$0.003), Co (0.013 $\pm$ 0.003), Sr (0.0019 $\pm$ 0.0001), Sm (0.93 $\pm$ 0.34)
Mn	212	Ni (0.0011 $\pm$ 0.0003)
Fe	352	S (0.0098 $\pm$ 0.0010), Ca (0.0048 $\pm$ 0.0005), Mn (0.29 $\pm$ 0.05), Co (0.31 $\pm$ 0.06), Sr (0.0029 $\pm$ 0.00052), Gd (18 $\pm$ 7)
Co	230	Mn (0.010 $\pm$ 0.001), Fe (0.0042 $\pm$ 0.00024), Sr ((1.5 $\pm$ 0.5) $\pm$ 10^{-4})
Cd	558	Co (0.0059 $\pm$ 0.0014), Sm (0.0052 $\pm$ 0.0006)
Sm	334	Mn ((9.7 $\pm$ 0.8) $\times$ 10^{-5}), Sr ((8.2 $\pm$ 1.9) $\times$ 10^{-7}
Hg	368	Ca ((3.0 $\pm$ 0.4) $\times$ 10^{-5}), Fe ((6.4 $\pm$ 0.2) $\times$ 10^{-4}), Co ((3.2 $\pm$ 0.2) $\times$ 10^{-4}), Sr ((9.1 $\pm$ 1.2) $\times$ 10^{-6})

[a] Equivalent amounts of elements due to γ-ray interference as determined from the measurement of more than four samples (data cited from Yonezawa & Haji Wood 1995; Yonezawa 1996).

and Ni are important practically in the interference of analysis for biological, geological and environmental materials. Although not shown in Table 2.3, Ba, Co, Os and Xe interfere in the measurement of the hydrogen peak at 2223 keV (Paul *et al.* 1996). Furthermore, the single-escape peak of Ti at 2219 keV interferes with the measurement of hydrogen when it is determined in Ti matrix samples. However, the spectral interference might be corrected by computer fitting and reference peak methods using other interference-free γ-ray peaks.

2.5.1.4 *Effects of neutron absorption and scattering* When a sample is irradiated with neutrons, both neutron absorption and neutron scattering influence the analytical sensitivities by modifying the average neutron fluence rate within a sample. For in-beam PGNAA, deviation from a linear response (i.e. signal versus mass of analyte) may occur, depending on the macroscopic absorption, scattering cross-sections and the target shape. In matrices containing large concentrations of absorbing nuclides, a decrease in the overall fluence rate is due to neutron absorption within the target. This self-shielding effect is well understood theoretically for simple geometry in both isotropic and parallel monoenergetic neutron fields (Fleming 1982). The neutron absorption correction in the monoenergetic-parallel neutron beam mostly utilized in PGNAA is described here. Since in general there will be neutron self-shielding within the material, the average flux will be less than the neutron flux incident on the sample. The self-shielding factor is defined as the ratio of the average flux, $\overline{\phi}$, in the sample to the flux incident on the sample, ϕ_0:

$$f \equiv \frac{\overline{\phi}}{\phi_0} \tag{2.11}$$

The magnitude of this factor depends on a single dimensionless variable, x, which express the size of the absorber in units of the absorption mean-free-path and is therefore proportional to the macroscopic total cross-section, Σ_t. The self-shielding factors for slab-, sphere- and cylinder-shaped samples irradiated with a 2200 m s^{-1} parallel thermal neutron beam were obtained with the following equations:

Slab of thickness T; with $x = T\Sigma_t$

$$f = \frac{1 - e^{-x}}{x} \tag{2.12}$$

Sphere of radius R; with $x = 2R\Sigma_t$

$$f = \frac{3}{x^3}\left[\frac{x^2}{2} - 1 + (1 + x)\,e^{-x}\right] \tag{2.13}$$

Cylinder of radius R; with $x = 2R\Sigma_t$

$$f = \frac{2}{x}[I_1(x) - L_1(x)] \tag{2.14}$$

where I_1 and L_1 are the modified Bessel and Struve functions, respectively, and are shown in Abramowitz and Stegun (1972). The macroscopic neutron cross-sections for the calculation of the self-shielding factor are defined as the number of atoms per cubic centimetre multiplied by the microscopic cross-section

$$\Sigma = \rho N_A \sum_{i=1}^{N} \frac{w_i \sigma_i}{A_i} = \frac{\rho N_A}{A} \sum_{i=1}^{N} n_i \sigma_i \tag{2.15}$$

where Σ is the macroscopic cross-section for the compound or mixture (cm^{-1}), ρ is its density (g cm^{-3}), N_A is Avogadro's number, A is the atomic weight of the compound or mixture, σ is the microscopic cross-section for the ith component (cm^2), A_i is the atomic weight of the ith component, w_i is the weight fraction of the ith component, and n_i is the number of atoms of the ith component in a molecule of the compound. The total cross-section, Σ_t, is the sum of Σ_s and Σ_a, the cross-sections for scattering and absorption, respectively. The macroscopic thermal neutron (0.0253 eV or 2200 m s^{-1}) cross-sections for several compounds and mixtures are tabluated in the study by Fleming (1982). The macroscopic absorption cross-section (Σ_a) for practical Maxwellian-distributed neutrons is calculated using the macroscopic cross-sections of 2200 m s^{-1} monoenergetic neutrons by

$$\overline{\Sigma_a} = \frac{\sqrt{\pi}}{2}\sqrt{\frac{T_0}{T}} \cdot \Sigma_a(v_0), \tag{2.16}$$

where T is the neutron temperature (K), T_0 is 293.6 K and v_0 is 2200 m s^{-1}.

When materials containing elements which have large neutron scattering cross-sections are irradiated, elemental sensitivities and γ-ray background counts are influenced by neutron scattering in the samples. Because of the large neutron scattering cross-section of hydrogen (values for thermal neutrons are 80 and 20.5 for the bound and free hydrogen atoms, respectively), analytical sensitivities of elements are influenced by neutron scattering where hydrogen is present in biological materials. However, such effects have not been observed for matrices containing other weaker neutron-scattering elements such as C, N, or O in biological materials. In addition to the effect on analytical sensitivities, neutron scattering leads to enhanced background as a result of capturing the scattered neutrons in the sample shielding and support materials. Such background enhancements are described in Section 2.5.1.6.

The effect of neutron scattering on analytical sensitivity differs with the nature of the neutron beam characteristics. The effects in the situation of a conventional thermal neutron beam, which also contains high-energy neutrons, were examined by measuring solid or dissolved samples that contain various elements along with various concentrations of hydrogen by setting the samples at an angle of 45° to the neutron beam (Mackey *et al.* 1991, 1992). The effects of neutron scattering in the conventional thermal neutron beam of the University of Maryland–NIST facility using various quantities of disk-shaped and sphere-shaped tris-hydroxymethyl-aminomethane (THAM) is shown in Fig. 2.15. Elemental sensitivities of 1.27 cm diameter disk-shaped samples increased with increasing disk thickness over the range of 0–2 mm, but decreased beyond 2 mm while for sphere-shaped samples the sensitivities are constant. For most organic compounds and biological materials, the sensitivities observed for disks where thickness was approximately equal to diameter (i.e. targets for which the shapes were closest to spherical) agreed with those for spheres. As the target thickness approaches zero the sensitivity approaches the value that would be obtained in the absence of scattering. Results of Monte Carlo simulations (Mackey & Copley 1993) confirmed that spherical targets are less affected by elastic scattering. Agreement between simulation and experiment provides additional evidence that the observed sensitivity enhancements for conventional thermal neutron beam PGNAA are the result of elastic scattering. Elastic scattering changes the mean-free-path of the neutron within the target and therefore changes the probability for neutron absorption (Mackey 1994).

The neutron scattering effects were examined for conventional thermal neutron beams by measuring pillow-shaped (Mackey 1991, 1992) and cylindrical-shaped (Trubert *et al.* 1991) samples containing various element solutions and adjusting the hydrogen concentration with D_2O/H_2O mixtures. The analytical sensitivities of elements in the solution samples increased linearly with increasing hydrogen con-

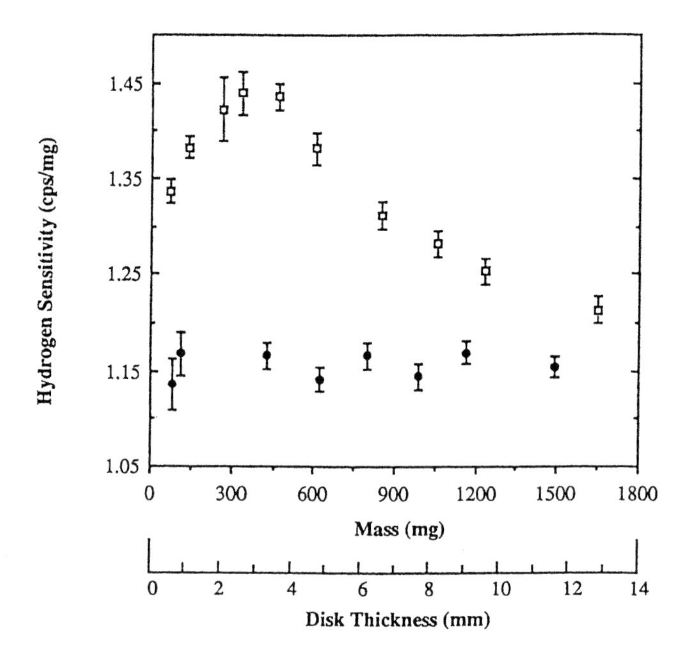

Fig. 2.15 Hydrogen sensitivities in various amounts of tris-hydroxymethylaminomethane in disks (□) and spheres (●) with the conventional thermal neutron beam of the NBSR (Mackey *et al.* 1992). (Reprinted from *Anal. Chem.*, **64**, E.A. Mackey, G.E. Gordon, R.M. Lindstrom and D.L. Anderson, Use of spherical targets to minimize effects of neutron scattering by hydrogen in neutron capture prompt gamma-ray activation analysis, p. 2366. Copyright 1992, with permission from American Chemical Society.)

centrations. In both experiments with different conventional thermal neutron beams the same sensitivity enhancements were obtained for the $1/v$ elements. However, reduced enhancement than for the non-$1/v$ elements was obtained for samarium which has a large absorption cross-section in the lower energy region of 0.094 eV. Although the same effect was expected for Cd and Gd, which have large absorption cross-sections at 0.18 eV (Cd) and 0.03 eV (Gd), this effect was not found for them. It was suggested that the lesser effect of the resonance absorption of Cd and Gd than that of Sm is due to a smaller ratio of resonance absorption cross-sections to thermal neutron absorption cross-sections than that ratio found for Sm. The neutron beams used in both experiments were well thermalized and the Cd(Au) ratios of the beams are 55 and 80. This means that the contents of the high-energy neutron components are very low. The difference between the sensitivity enhancement observed for Sm and those observed for the other elements studied provides additional evidence that the observed sensitivity enhancements are due, at least in part, to a scattering-induced decrease in the average energy of the neutrons.

The effects of neutron scattering on analytical sensitivities in the case of cold neutron guided beams are quite different from those of a conventional thermal neutron beam. As for conventional thermal neutron scattering, the effects depend on the size, shape, and scattering power of the target. In addition, scattering changes the temperature of the neutrons and hence their capture cross-section (Mackey 1994). The plot of analytical sensitivities for hydrogen versus target mass of disk-shaped and sphere-shaped THAM measured with the cold neutron beam at NIST at 300 K and 77 K is shown in Fig. 2.16. The sensitivities increased with increasing hydrogen concentration with a conventional thermal neutron beam for the disk targets, whereas the sensitivities for the same targets decrease with increasing hydrogen content with the cold neutron beam at 300 K. The analytical sensitivities remain constant with a conventional thermal neutron beam for spherical targets whereas the analytical sensitivities with the cold neutron beam at 300 K decreased with increasing target mass. Furthermore, the analytical sensitivities for both the disk and spherical targets at 77 K were higher than those measured at 300 K and the sensitivities for the spherical targets became constant with increasing target mass, the same as that observed for spherical targets with the conventional thermal neutron beam. Sensitivity changes for various elements including H, K, Br, Gd and Sm in various thicknesses of disk-shaped KBr were also examined using the cold neutron beam at NIST. Sensitivities for the '$1/v$' elements such as H, K and Br and a weak 'non-$1/v$' element (Gd) decreased by the same relative amounts when increasing target thickness, whereas the sensitivity decrease for the resonance element (Sm) was significantly smaller than those observed for the '$1/v$' elements. As can be expected the experimental results agree with those of the cold neutron scattering within a room-temperature target leading to an increase in the average energy of the neutrons, and therefore a decrease in absorption cross-sections and sensitivities (Paul & Mackey 1994).

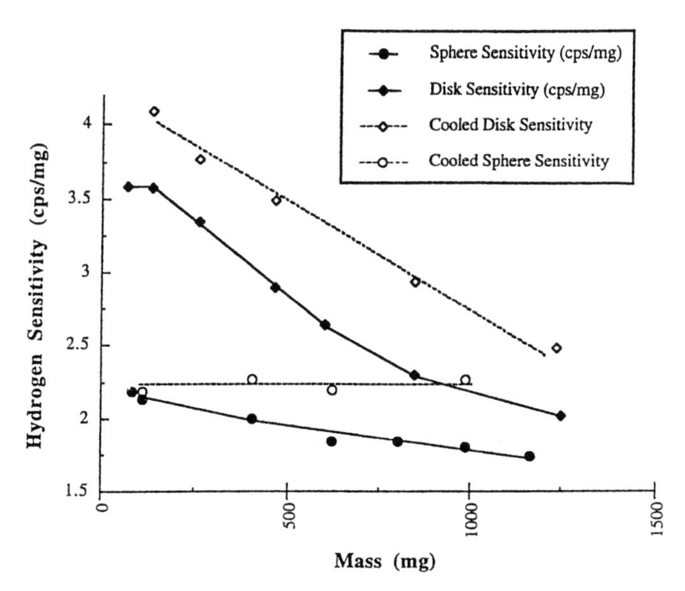

Fig. 2.16 Hydrogen sensitivities in various amounts of tris-hydroxymethylaminomethane in disks (◇) and spheres (○) with the cold neutron guided beam of the NBSR. Open symbols and filled symbols represent 77 K and 300 K targets, respectively (Mackey 1994). (Reprinted from *Biological Trace Element Research*, **43–45**, E.A. Mackey, Effects of target temperature on analytical sensitivities of cold-neutron capture prompt gamma-ray activation analysis, p. 103. Copyright 1994, with permission from Humana Press.)

2.5.1.5 *γ-ray attenuation* Attenuation of the γ-ray signal prior to detection may result in significant measurement error. The γ-ray attenuation by matter between the sample and detector, such as air and shielding materials, is the same for the various samples (and also the standards) analysed. However, attenuation within samples may also be a source of error.

The attenuation is proportional to the thickness, density and mass number (Z) of the attenuating medium and is inversely proportional to γ-ray energy. If the sample is very thick, is of high density, or contains large quantities of high-Z elements, this self-attenuation may be significant, particularly for γ-rays of low energy (< 500 keV). As an example, consider the measurement of hydrogen in a Ti cylinder of thickness 1 cm mounted at an angle of 45° to the detector with the cold neutron beam of the NBSR at NIST (Paul 1997). The γ-ray attenuation within the sample results not only in a 12% decrease in the 2223 keV H signal, but also in a 3% decrease in the ratio of the 1381 keV Ti to the 2223 keV H signal. A number of approaches may be used to minimize the γ-ray self-attenuation. First, match the elemental composition and geometry of the samples and standards in a comparative standardization. Second, correct the γ-ray attenuation by comparing the relative intensities of the sample γ-ray lines with those measured in a thin foil of the same material (where self-attenuation is minimal) when one of the elements present in the sample matrix emits a spectrum of γ-rays lines covering a range of energies. And third, if the composition and geometry of the sample are known, the γ-ray self-attenuation may be corrected using derived equations (Debertin & Helmer 1988).

The effect of γ-ray attenuation must be considered when a semi-thick γ-ray collimator is used and the sample is either larger than or not properly aligned with the collimator. Paul (1995) examined this effect by measuring prompt γ-rays from Ti foils of various diameters but with the same thickness using different apertures of a collimator made of copper. The ratio of the count rates for the 341 and 1381 keV peaks were smaller for the Ti foils with diameters larger than the aperture of the collimator than those Ti foils with diameters smaller than the aperture of the collimator. Thus the collimator should be composed of a high-Z, high-density material (e.g. Pb or W) of sufficient thickness, and in addition should be set with accurate alignment against the solid angle of the detector.

2.5.1.6 *γ-ray background* There are various background γ-ray sources in a nuclear reactor:

- capture reaction γ-rays emitted from the surrounding materials and the detectors
- γ-rays arising from inelastic scattering of the small fraction of fast neutrons
- reactor γ-rays scattered by both the sample and the surroundings

The γ-ray background counts of typical PGNAA facilities are listed in Table 2.4. Hydrogen exhibits the highest γ-ray background count for the conventional thermal neutron beam type facilities (Anderson *et al.* 1982b) owing to the utilization of hydrogenous moderators and shielding materials. However, low-energy neutron guided beam type facilities have considerably lower hydrogen background counts compared to the conventional thermal neutron beam type facilities, since in the

Table 2.4 Comparison of γ-ray background count rates under neutron irradiation

		Background count rate (cps)		
		Conventional thermal neutron beam facility		Cold neutron guided beam facility
Element or nuclide	E_γ (keV)	University of Maryland–NIST[a]	University of Missouri[a]	JAERI[b]
H	2223	1.56	0.51	0.0094
B	478	0.1	<0.1	<0.0040
C	1262	0.0014	0.0350	0.0020
N	1885	0.072	<0.008	0.0025
Na	472	0.057	0.349	<0.0035
Al	1779	0.027	0.268	0.0064
Cl	517	0.007	0.009	<0.0019
Ti	342		0.063	<0.0025
Fe	352	0.007	0.136	<0.0028
^{207}Bi	569			0.0242
^{207}Bi	1063			0.0081

[a] Cited from Anderson *et al.* (1982b). [b] Cited from Yonezawa (1997).

former no hydrogenous materials are utilized as moderating and shielding materials. For example, the hydrogen background at the University of Maryland–NIST using a conventional thermal neutron beam is 4 mg H (Mackey 1992), while the hydrogen background of the cold neutron guided beam facilities are as low as 24 ± 8 μg H for the JRR-3M JAERI (Yonezawa *et al.* 1995), 40 μg H for the NIST (Lindstrom *et al.* 1993) and 480 μg H for the Jülich Research Centre (Rossbach 1991). Additional background γ-ray peaks detected are: the capture γ-ray peaks of N and Ar from air; B, Al, Fe and Pb peaks from the construction materials of the various facilities; and Na and I peaks from γ-ray detectors. The γ-ray background peaks of the radioactive nuclide ^{207}Bi are also detected in those facilities using the BGO detector (Yonezawa *et al.* 1993). The air-originated γ-ray background components such as N, H and Ar can be reduced by irradiating the sample in a He atmosphere or evacuated sample container.

Since most of the γ-ray background counts are due to the capture of scattered neutrons by the materials surrounding the sample, the background-count levels in the conventional thermal neutron beam PGNAA facilities depend on the total scattering cross-section of the samples (Anderson & Mackey 1993). The relationship between the total scattering cross-section of the samples and background counts of Al and Pb for the University of Maryland–NIST facility is shown in Fig. 2.17. The background γ-ray counts of H, B, Na, Cl, Al and Pb were increased by increasing the total scattering cross-section of the samples analysed and the relationships were fitted by a second-order polynomial function with correlation factor of over 0.98. Therefore, the γ-ray background counts of those γ-ray lines can be estimated from the total scattering cross-section of the samples. As the peak count rates of hydrogen is proportional to the total scattering cross-sections in hydrogenous materials, the γ-ray background counts of components other than H can be estimated from the counts of the H peak. Similarly, the H background counts in the hydrogenous samples can be

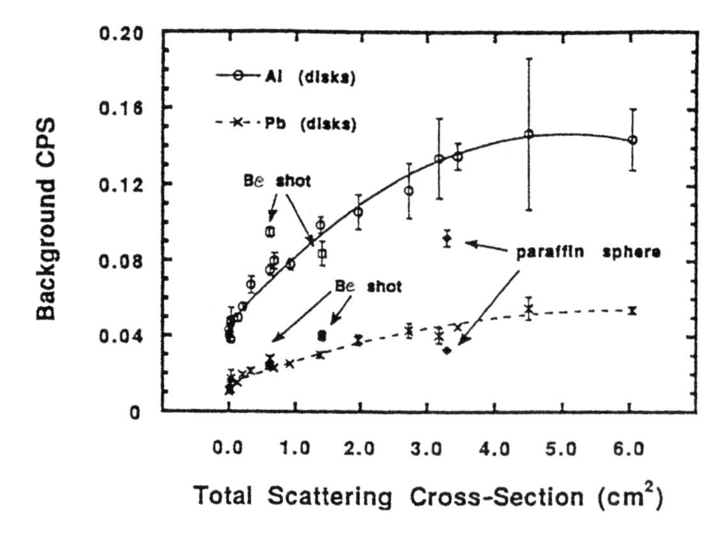

Fig. 2.17 The γ-ray background count rates (in cps) for the 1779 keV Al photopeak and the 7368 keV Pb photopeak as a function of the total scattering cross-section with the conventional thermal neutron beam of the NBSR (Anderson & Mackey 1993). (Reprinted from *J. Radioanal. Nucl. Chem.*, **167**, D.L. Anderson and E.A. Mackey, Neutron scattering-induced background enhancement in prompt gamma-ray activation analysis, p. 145. Copyright 1993, with permission from Akadémiai Kiadó Rt.)

estimated from the counts of other background components such as Al and Pb. Background enhancement was also observed in the cold neutron PGNAA at NIST (Paul 1997). Irradiation of graphite and Be targets yielded up to a factor of 10 enhancement in the H background. However, because of the lower γ-ray background counts and utilization of pure low-energy neutrons, the degree of the enhancement in low-energy neutron guided beam type facilities is smaller than that of the conventional thermal neutron beam type facilities.

Prompt γ-ray spectrum measurement of a sample is performed by sealing the sample into an FEP film and placing it in the path of the neutron beam using Teflon string. The γ-ray spectrum of an empty FEP bag is shown in Fig. 2.18. In addition to the γ-ray background counts emitted from materials surrounding the sample described previously, the γ-ray counts of F and C from the FEP film and Teflon string are detected in the spectrum. Evaluation and correction of the γ-ray background counts are therefore essential in quantitative elemental analysis.

2.5.1.7 *Prompt γ-ray data* The most reliable and complete source of information concerning thermal neutron capture γ-ray data for nuclides is the Evaluated Nuclear Structure and Decay Data File (ENSDF), produced by an international committee of expert evaluators. The updated data for mass chains are available in the journal *Nuclear Data Sheets* and for lightweight nuclei (mass number $A < 45$) in the journal *Nuclear Physics A*. An updated ENSDF is available on the Word Wide Web home page of the National Nuclear Data Center (http://www.nndc.bnl.gov/nndc/ensdf/) while the most recent (1998) compact disk version of the ENSDF is available as a supplement to the eighth edition of the *Table of Isotopes*. There is also an older, but

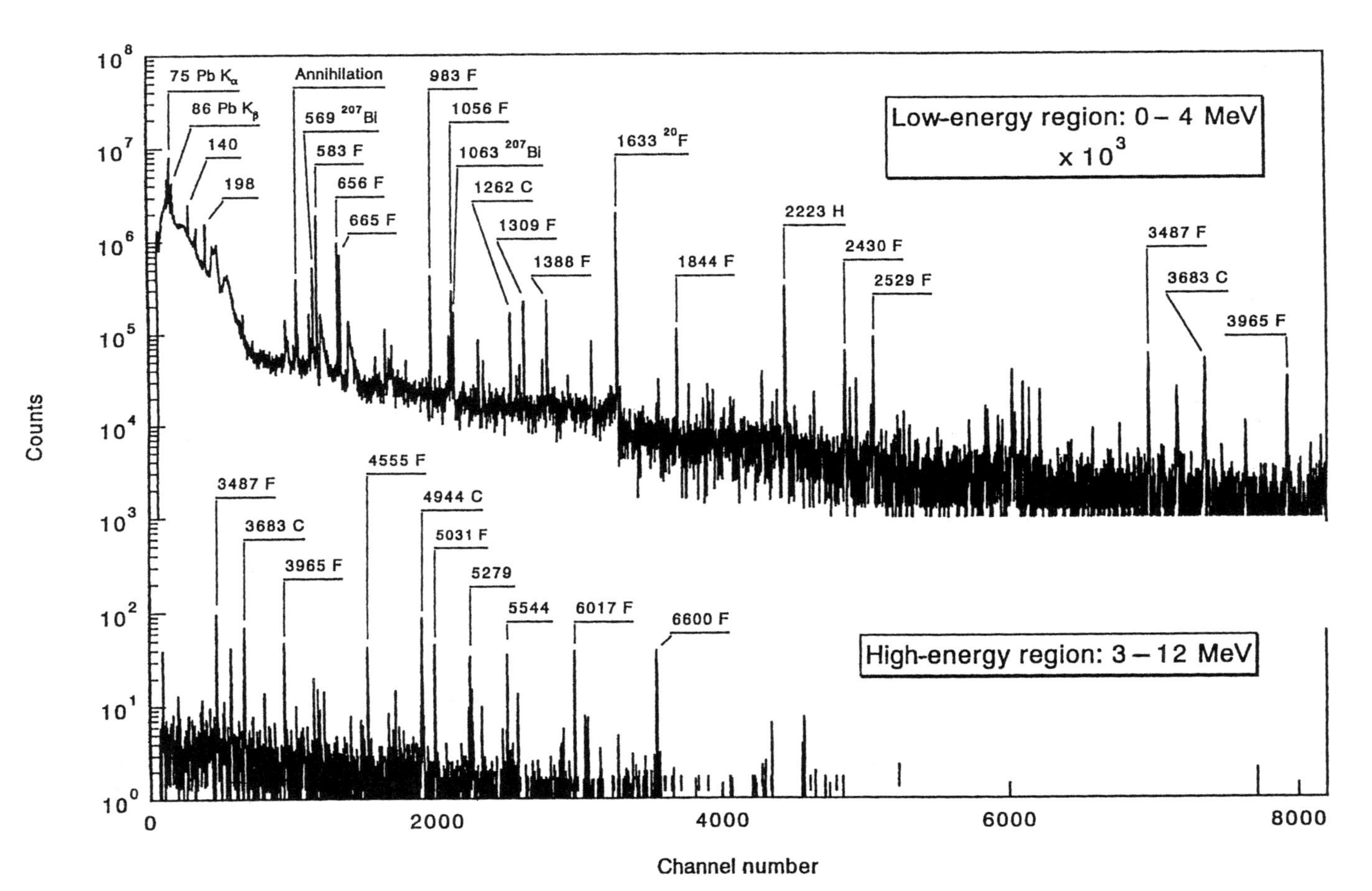

Fig. 2.18 Compton suppression PGNAA spectrum of FEP film (0.0616 g) measured with the cold neutron guided beam of the JRR-3M for 50 000 s.

quite useful compilation of thermal neutron capture γ-ray data for naturally occurring elements that was made by Lone *et al.* (1981). This compilation comprises two tables. In one the γ-ray data are arranged in order of energy and in the other the γ-rays are arranged by element. This compilation is very useful for elemental analysis, although much of the listed data are now old and require updating. Furthermore, low-energy γ-ray data below 500 keV are missing or incorrect, especially for the lanthanides. This is due to the data being obtained from historically less pure compounds. The data from Lone *et al.* (1981) for the light elements and the ENSDF data from 1991 for the heavier nuclides ($A > 44$) have recently been compiled and published by Tuli on a Word Wide Web home page (http://www.nndc.bnl.gov/wallet/tnc/capgam.shtml) and in the Appendix to his book (Tuli 1995).

Inaccuracy and incompleteness of the data available for use in PGNAA provide a significant handicap in the qualitative and quantitative analysis of complicated γ-ray spectra. An international effort to develop an accurate database for PGNAA was started in the form of the Coordinated Research Project (CRP) under the auspices of the International Atomic Energy Agency (IAEA) from 1999. Not only will accurately evaluated capture γ-ray data be made available, but also the data necessary for quantitative analysis, such as that for calculating k_0 factors and cold neutron capture cross-sections.

2.5.1.8 *γ-ray spectrum analysis* Neutron-capture prompt γ-rays have the following characteristics: higher energy; larger number of lines; and wider peak of boron γ-rays at 478 keV by Doppler broadening rather than decay γ-rays. Decay γ-ray energies are below 3 MeV, while prompt γ-ray energies rise to about 11 MeV. In addition to a larger number of prompt γ-ray lines, single- and double-escape peaks due to the high-energy γ-rays lead to more complex γ-ray spectra than those for decay γ-rays. However, the same computer programs are used for the analysis of decay spectra and prompt γ-ray spectra, as there are no other differences except for the above-described characteristics. Thus, only inherent features in prompt γ-ray spectrum analysis such as energy and detection efficiency calibrations for high-energy γ-rays and peak analysis of Doppler-broadened γ-rays will be described. Although special software for prompt γ-ray spectral analysis has not been offered commercially, Hypernet-PC has a prompt γ-ray data library and quantitative analysis function by k_0 standardization (Fezekas 1998).

Energy calibration is carried out for the low-energy region by using calibrated standard γ-ray sources and is extended to the high-energy region (up to 11 MeV) by using capture γ-rays of N and Cl whose emitted γ-ray energies are known accurately. The γ-ray detection efficiency curve is created in the low-energy region by measuring calibrated γ-ray standard sources such as $^{57,60}Co$, ^{54}Mn, ^{137}Cs, ^{152}Eu and ^{226}Ra, etc., and the curve is extended up to 2.745 MeV using ^{24}Na. The curve is then extended up to 11 MeV using capture γ-rays of C, Cl and N. As ^{24}Na and capture γ-rays of C emit cascade γ-rays (i.e. more than one γ line), the low-energy calibration can be used for the higher energy calibration, since the ratio of the cascade γ-rays is known. The capture γ-rays from N spread widely over an energy range up to 11 MeV and have accurate energy and emission probabilities. An example of the γ-ray detection efficiency curve is shown in Fig. 2.19. The curve

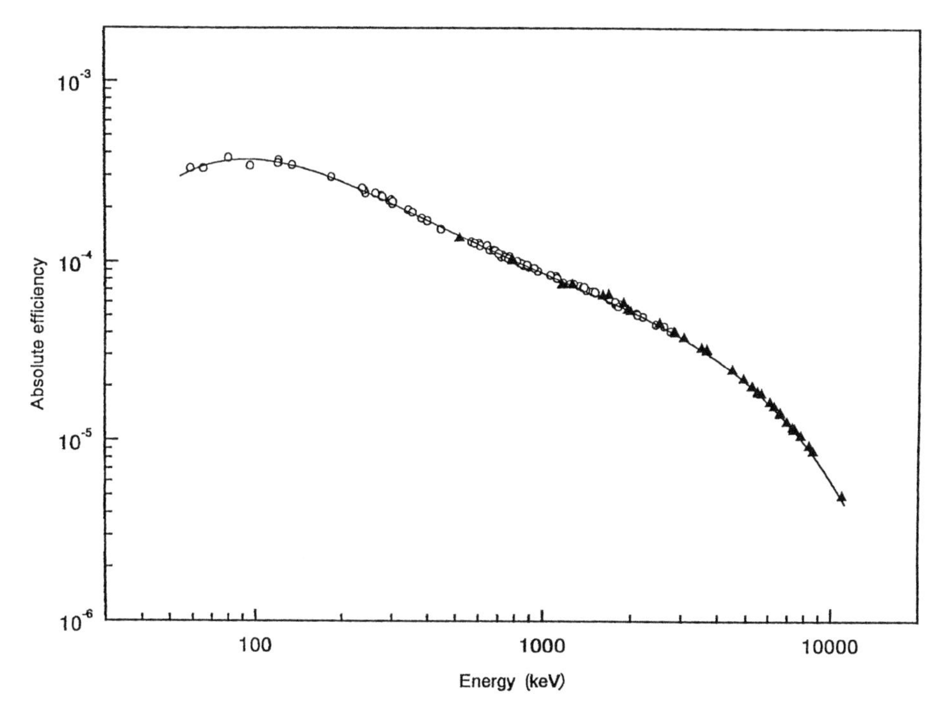

Fig. 2.19 Full-energy γ-ray detection efficiency curve of a HPGe detector. The curve was obtained by measuring radioactive γ-ray sources and prompt γ-rays from C, Cl and N.

was fitted with a fourth-order polynomial function and with an uncertainty of 2.3% (Matsue & Yonezawa 2000).

The Doppler-broadened B 478 keV peak can be analysed by fitting to the equations described in Section 2.5.1.2. A simplified method has been developed by assuming that the moving ^{7*}Li produced by the ^{10}B (n, α) reaction has a mean velocity ($\bar{v}$) instead of the velocity $v(t)$ to avoid troublesome calculation of the double integration (Magara and Yonezawa 1998). In this method, the 478 keV peak shape $F_1(E)$ is fitted to Equation (2.17), which is obtained by substituting $v(t) = \bar{v}$, $E_0\bar{v}/c = k$ and $0.693/\Gamma = 1/(2\sigma)$ in Equations (2.5)–(2.10).

$$F_1(E) = N_1 \int_0^{\pi} \exp\left\{-\frac{(E - E_0 + \kappa\cos\theta^2)}{2\sigma^2}\right\} \sin\theta \, d\theta \tag{2.17}$$

where N_1 is the peak height and σ is the peak width. The peak shape of the interfering normal peak, $F_2(E)$, can be fitted with the Gaussian function

$$F_2(E) = N_2 \exp\left\{-\frac{(E - E_0)^2}{2\sigma^2}\right\} \tag{2.18}$$

where N_2 is the peak height and σ is the peak width of the interfering peak. The baseline shape of the broadened peak $F_3(E)$ is fitted to the step function

$$F_3(E) = A \tan^{-1}\{a(E - E_0)\} + C \tag{2.19}$$

where A is the difference between the low-energy and high-energy sides, a is the slope of the baseline and C is the height of the baseline. The peak fitting is carried out by using non-linear least-squares regression with both κ and σ taken as free parameters. The peak-fitting result for a rock sample (Yonezawa *et al.* 1999a) is shown in Fig. 2.20. The Doppler-broadened 478 keV B peak was overlapped with the 472 keV Na peak and both peaks were well deconvoluted by the fitting. The program can be used also for interference correction with other elements, e.g. Cl, Ni, Co. However, large errors will be obtained for boron compounds which have higher degradation (slowing-down) constants, D (of about 2×10^{12} s^{-1}) due to the assumption that the ^{7*}Li velocity is constant. A fitting program, which includes the degradation process of ^{7*}Li in the sample, is required to analyse boron compounds, which have a high degradation constant, D.

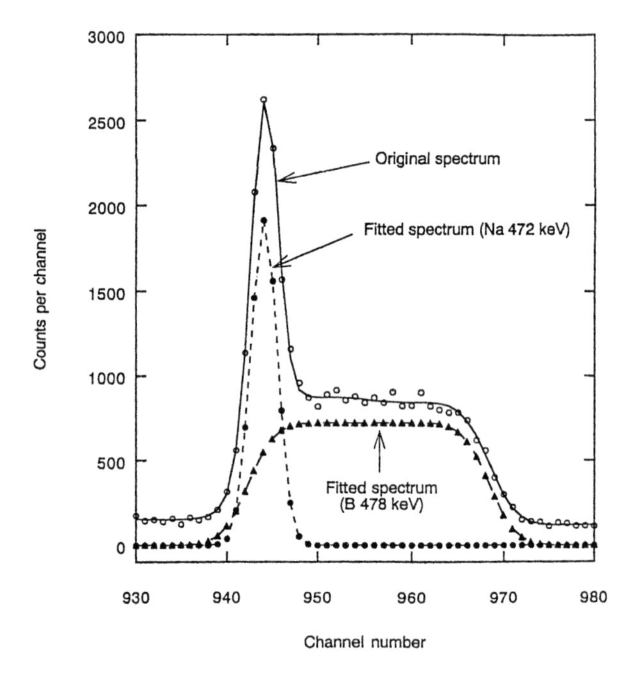

Fig. 2.20 Computer fitting of a spectrally interfered boron peak. The sample of andesite (GSJ JA-2; 0.1961 g) was measured with the thermal neutron guided beam of the JRR-3M for 35 000 s (Yonezawa *et al.* 1999a). (Reprinted from *J. Radioanal. Nucl. Chem.*, **239**, C. Yonezawa, P.P. Ruska, H. Matsue, M. Magara and T. Adachi, Determination of boron in Japanese geochemical reference samples by neutron-induced prompt gamma-ray analysis, p. 571. Copyright 1999, with permission from Akadémiai Kiadó Rt.)

2.5.2 *Standardization*

2.5.2.1 *Introduction of standardization* For a given set of experimental conditions and a homogeneous sample of volume V irradiated in a neutron beam of flux density Φ for a time t_m, the net peak area, N_γ, representing the integral number of counts due to a neutron capture γ-ray of energy E_γ from mass m_x of element x, is given in general (Molnár *et al.* 1998) by

$$N_{\gamma,x} = m_x N_\mathrm{A} \frac{\theta_x}{M_x} \gamma_x \int_V \int_{E_\mathrm{n}=0}^{\infty} \int_{t=0}^{t_\mathrm{m}} \sigma_x(E_\mathrm{n}) \Phi(E_\mathrm{n}, t, V) \varepsilon(E_\gamma, V) \mathrm{d}t \, \mathrm{d}E_\mathrm{n} \, \mathrm{d}V \tag{2.20}$$

where N_A is Avogadro's number, θ is the abundance of the capturing isotope of the element of interest, γ is the number of γ-rays emitted per capture, M is the atomic weight, $\sigma(E_\mathrm{n})$ is the capture cross-section for neutron energy E_n while $\varepsilon(E_\gamma, V)$ is the counting efficiency of the detector at energy E_γ. Elemental concentrations in a sample can be determined by an absolute method using an equation derived from Equation (2.20). However, the absolute method is not useful in practice owing to difficulties obtaining accurate neutron spectra and neutron cross-section, $\sigma_x(E_\mathrm{n})$, data at each neutron energy. In particular, spectra of filtered beam and some guided beam neutrons at downstream position are complexed as a result of absorption of certain wavelength (energy) neutrons by the filter, diffractometer, etc. Moreover, evaluation of the reaction rate for epithermal neutrons by resonance interactions is difficult too. In addition to these difficulties, insufficient accuracy of relevant nuclear data, such as the γ-ray emission rate, makes the absolute method non-compliant with the requirements for analytical work, even when studying ideal samples. Consequently, this method has not been pursued in general PGNAA work. In practice, elemental determination has been carried out using comparative and internal standard methods.

2.5.2.2 *Comparative method* Neutron-capture reaction rates, which are difficult to evaluate in the absolute method, can be obtained by measuring known amounts of element under the same conditions as the sample. In other words, the right-hand side terms of Equation (2.20), except for m_x, can be replaced by the specific count rate of the element, i.e. the so-called analytical sensitivity, A_sp (cps g^{-1}):

$$N_\mathrm{A} \frac{\theta_x}{M_x} \gamma_x \int_V \int_{E_\mathrm{n}=0}^{\infty} \sigma_x(E_\mathrm{n}) \Phi(E_\mathrm{n}, t, V) \varepsilon(E_\gamma, V) \, \mathrm{d}E_\mathrm{n} \, \mathrm{d}V = \frac{N_{\gamma,x}/t_\mathrm{m}}{m_x} = A_{\mathrm{sp},x} \tag{2.21}$$

The analytical sensitivities A_sp (cps g^{-1}) can be obtained by measuring standard samples containing known amounts of analytes. An elemental mass of the elements $(m_x)_\mathrm{smp}$ in the sample is obtained by dividing γ-ray count rates by the analytical sensitivities:

$$(m_x)_\mathrm{smp} = \frac{(N_{\gamma,x}/t_\mathrm{m})_\mathrm{smp}}{A_{\mathrm{sp},x}} \tag{2.22}$$

Relative standards and samples can be irradiated in an NAA with equal neutron fluxes by irradiating in the same capsule simultaneously. However, this procedure is difficult in PGNAA. Thus, the neutron flux difference between the standards and samples becomes a source of error in PGNAA. The neutron differences are corrected by normalizing the γ-ray peak count rates of the standards and samples with the neutron fluxes measured by a neutron monitor such as Ti and Cr (Failey *et al.* 1979; Anderson *et al.* 1990a; Yonezawa & Haji Wood 1995; Yonezawa 1996).

Elemental determination is carried out using the normalized peak count rates of the samples and normalized analytical sensitivities obtained from the standards. In general, variation of neutron intensities in a high-power reactor like the JRR-3M is as small as 0.8% during one cycle of 26 days at 20 MW full-power operation (Yonezawa *et al.* 1995).

Differences of geometrical conditions and elemental compositions between the standards and samples affect the γ-ray detection efficiencies, neutron scattering and self-absorption of γ-rays and neutrons. It is imperative, therefore, that elemental compositions and the geometry of standards and samples be as closely matched as possible.

2.5.2.3 *Standardization based on an internal standard method* The differences in neutron flux, chemical composition and geometrical condition between the sample and the standard can become sources of error in the comparative method due to the scattering and absorption effects of neutrons. However, these effects can be eliminated by using an internal standard. The determination is carried out by using a calibration curve in which the peak count ratio is plotted versus the mass ratio of the analyte to the internal standard, or by calculating relative concentrations of an element using calibration factors. Recently, a multielement determination method based on internal standardization has been developed (Lindstrom *et al.* 1992; Paul 1995; Molnár *et al.* 1998).

When a sample containing analyte x and internal standard element c is irradiated, a ratio of analytical sensitivities for both elements is obtained from Equation (2.21).

$$\frac{A_{\mathrm{sp},x}}{A_{\mathrm{sp},c}} = \frac{\theta_x \gamma_x \varepsilon(E_{\gamma,x})/M_x \int_{E_\mathrm{n}=0}^{\infty} \sigma_x(E_\mathrm{n})\mathrm{d}E_\mathrm{n}}{\theta_c \gamma_c \varepsilon(E_{\gamma,c})/M_c \int_{E_\mathrm{n}=0}^{\infty} \sigma_c(E_\mathrm{n})\mathrm{d}E_\mathrm{n}} = k_{\mathrm{r,c}}(\mathrm{x}) \tag{2.23}$$

The terms related to the volume and to the neutron flux in Equation (2.21) are cancelled as it is the same sample. The factor $k_{\mathrm{r,c}}(\mathrm{x})$ is constant under the same γ-ray detection efficiencies and neutron spectra. Relative elemental contents can be obtained by using the factors under the same experimental conditions. This method is called the 'ratio method' (Paul 1995). Absolute elemental contents are obtained by measuring the sample mixed with known amounts of the internal standard, or in the case of a multielement sample by determining at least one element in the sample by a different analytical method. Hydrogen, Cl, Ti, Cr, Ni, Sm and Hg have all been used as the internal standard element.

The $k_{\mathrm{r,c}}(\mathrm{x})$ factors are not universal and are limited to fixed conditions that were used in their measurement. Capture cross-sections for almost all elements follow the $1/v$ law for neutron temperature equal to or lower than room temperature. The capture cross-section, σ, of an element at velocity v is given by $\sigma = \sigma_0 \cdot v_0/v$ ($v_0 = 2200\,\mathrm{m\,s^{-1}}$). Substituting σ and normalizing the ratio of analytical sensitivities in Equation (2.23) with detection efficiencies (Lindstrom *et al.* 1992; Molnár *et al.* 1998), leads to a k_0 factor that is independent of the experimental conditions:

$$k_{0,\mathrm{c}}(\mathrm{x}) = \frac{A_{\mathrm{sp},x}/\varepsilon(E_{\gamma,x})}{A_{\mathrm{sp},c}/\varepsilon(E_{\gamma,c})} = \frac{\theta_x \gamma_x \sigma_{0,x}/M_x}{\theta_c \gamma_c \sigma_{0,c}/M_c} \tag{2.24}$$

This method is called 'k_0–PGNAA' because it is similar to the k_0–INAA method. Any element can, in principle, be used as an internal standard (comparator). However, Cl is used mostly owing to its high sensitivity and its simple γ-ray spectrum. The k_0 factors for various elements have been measured at several facilities.

The k_0 factors of elements obtained using Cl as a comparator have been measured with the thermal neutron guided beam at the Institute of Isotopes (IKI), Hungary; with the cold neutron guided beam at NIST (Molnár *et al.* 1998); and with the cold and thermal neutron guided beams at JAERI (Matsue & Yonezawa 2000). Some of these are listed in Table 2.5. The element k_0 factors in Table 2.5 show the highest sensitivity γ-ray line with respect to the 1951 keV Cl line. The k_0 factors of these elements were obtained by measuring chloride compounds of the element and mixtures of ammonium chloride with the single element or compounds of the element such as the oxide, carbonate, etc. For those elements where it is difficult to measure a k_0 factor using Cl as a comparator (due usually to the analytical sensitivity of an element deviating largely from that of Cl or where the element's γ-ray lines are

Table 2.5 Comparison of the k_0 factors measured by various low-energy guided beams using the Cl γ-ray peak at 1951 keV as a comparator

Element	E_γ (keV)	JAERI[a] (Cold)	JAERI[b] (Thermal)	IKI[c] (Thermal)	NIST[d] (Cold)
H	2223	1.86 ± 0.061	1.80 ± 0.061	1.981 ± 0.015	
B	478	380 ± 32	371 ± 31		
C	1262	0.000643 ± 0.000031	0.000762 ± 0.000035	0.0139 ± 0.0003	
N	1885	0.00698 ± 0.00032	0.00573 ± 0.00020		
Na	472	0.115 ± 0.0038	0.116 ± 0.0041	0.114 ± 0.001	
Si	3540	0.0211 ± 0.00037	0.0218 ± 0.00010		
P	637	0.00477 ± 0.000080	0.00470 ± 0.00077		
S	841	0.0573 ± 0.0012	0.0554 ± 0.0012		
K	770	0.128 ± 0.0042	0.127 ± 0.0044	0.130 ± 0.002	
Ca	1942	0.0464 ± 0.0016	0.0469 ± 0.0019	0.0479 ± 0.0010	
Ti	1381	0.591 ± 0.0062	0.583 ± 0.0059		
Cr	834	0.142 ± 0.0049	0.141 ± 0.0049	0.145 ± 0.002	0.149 ± 0.004
Mn	314	0.149 ± 0.0078	0.152 ± 0.0052	0.148 ± 0.002	0.138 ± 0.005
Fe	352	0.0269 ± 0.0011	0.0273 ± 0.0010	0.0269 ± 0.0006	
	7631	0.0559 ± 0.0027	0.0559 ± 0.0023	0.070 ± 0.003	0.069 ± 0.002
Co	277	0.615 ± 0.021	0.619 ± 0.021	0.606 ± 0.011	
	555	0.509 ± 0.020	0.516 ± 0.018	0.52 ± 0.02	0.460 ± 0.012
Ni	465	0.0811 ± 0.0028	0.0812 ± 0.0028	0.0804 ± 0.0010	0.081 ± 0.002
Sr	898	0.0425 ± 0.0014	0.0419 ± 0.0021		
Cu	278	0.0762 ± 0.0025	0.0766 ± 0.0027	0.0375 ± 0.0005	
	385	0.0166 ± 0.00058	0.0174 ± 0.00070	0.0084 ± 0.0003	0.0077 ± 0.0003
Cs	117	0.172 ± 0.0059	0.172 ± 0.0060		
Cd	558	62 ± 1.5	69 ± 1.3	74 ± 2	
Ba	627	0.0108 ± 0.00037	0.0111 ± 0.00040	0.00107 ± 0.00002	
Hg	368	7.0 ± 0.24	7.1 ± 0.26	6.46 ± 0.08	
Sm	334	116 ± 0.99	146 ± 4.2		
Gd	182	214 ± 1.3	255 ± 3.1		
Pb	7367	0.00329 ± 0.000033	0.00338 ± 0.000057	0.00518 ± 0.00015	

[a] The factors measured with the cold neutron beam of JRR-3M (Matsue & Yonezawa 2000). [b] The factors measured with the thermal neutron beam of JRR-3M (Matsue & Yonezawa 2000). [c] The factors measured with the thermal neutron beam of BRR (Molnár *et al.* 1998). [d] The factors measured with the cold neutron beam of NSBR (Molnár *et al.* 1998).

interfered with by those of Cl), the k_0 factor is obtained by using other elements as the comparator. The obtained k_0 factors are then converted into $k_{0,\text{Cl}}$ values. In spite of different neutron spectrum characteristics, the factors measured agree within almost 10% among all the facilities except in the cases of a few elements. However, the k_0 factors of the non-$1/v$ elements of Cd, Sm and Gd varied more among the facilities than those valves for the $1/v$ elements. The k_0 factors using Ti as a comparator have been measured with the conventional thermal neutron beam (Paul & Lindstrum 1997). The factors converted to the Cl comparator factor agreed within about 10% among the $1/v$ elements, except for Cd, Sm and Gd. Owing to the large capture cross-sections in the lower energy region, Cd, Sm and Gd are easily affected by neutron spectrum differences.

Relative elemental contents of element, x, with respect to element, y, in the same sample can be obtained using following equation:

$$\rho_{x:y} = \frac{A_x/\varepsilon_x k_{0,c}(\mathrm{x})}{A_y/\varepsilon_y k_{0,c}(\mathrm{y})} \tag{2.25}$$

By analogy with the previously described internal standard method, absolute contents cannot be obtained directly by the k_0–PGNAA. Absolute elemental contents are usually obtained by determining at least one element in the sample using other analytical methods. Analytical results of multielement samples, obtained by the present method, for the Standard Reference Materials of River Sediment are shown in Table 2.6. The absolute elemental contents given in Table 2.6 are obtained by normalizing with the certified value for Cr. The analytical results agree within less than 6% of the NIST-certified values.

Table 2.6 Analytical results for the Standard Reference Material of River Sediment (NIST SRM1645) by the k_0 standardization

Element	Unit	JAERI[a] (Av. ± error)	IKI[b] (Av. ± error)	Certified/reference value[c] (Av. ± error)
H	%	0.875 ± 0.010		
B	μg g^{-1}	33.7 ± 0.32		
Na	%	0.551 ± 0.012		0.54 ± 0.01
Si	%	26.0 ± 0.37		
S	%	1.51 ± 0.0069		(1.1)
Cl	%	0.0140 ± 0.00048		
K	%	1.31 ± 0.010	1.4 ± 0.04	(1.2)
Ca	%	2.90 ± 0.044	3.0 ± 0.1	(2.9)
Ti	%	0.0679 ± 0.00030		
Cr	%	2.96[d]	2.96[d]	2.96 ± 0.3
Mn	%	0.0772 ± 0.0029	0.077 ± 0.011	0.0785 ± 0.0097
Fe	%	11.0 ± 0.10	11.5 ± 0.3	11.3 ± 1.2
Cd	μg g^{-1}	10.1 ± 0.18	10.4 ± 0.3	10.2 ± 1.5
Sm	μg g^{-1}	1.39 ± 0.04		
Gd	μg g^{-1}	1.44 ± 0.12		

[a] Analysed using the cold neutron guided beam; errors show standard deviation of three samples (Matsue & Yonezawa 2000). [b] Analysed using the thermal neutron guided beam; errors show counting statistics of 1σ (Molnár *et al.* 1998). [c] The values in parentheses are the reference values. [d] Analytical results are normalized for the certified Cr value.

A method similar to the k_0–PGNAA, i.e. 'internal mono-standard method' (Sueki *et al.* 1996), has been developed to analyse large archaeological samples. A compound nuclear constant comprising the neutron cross-section (σ) and γ-ray intensity (γ) for analyte (x) and comparator (c), namely ($\sigma_x\gamma_x/\sigma_c\gamma_c$), is used to determine elements instead of the k_0 factor in this method. The differences between γ-ray detection efficiencies for samples where there are serious measurement problems in larger samples can be corrected using prompt γ-rays from samples such as Si, Ti, Fe, etc. in this method.

2.6 Applications

PGNAA has been applied to the determination of light elements that are difficult to analyse using other methods and in the analysis of important samples such as archaeological artefacts where its non-destructive analytical ability and low induced activity are particular advantages. Furthermore, since negligibly low radioactivity is induced after the PGNAA measurement, the same sample can be used subsequently in other elemental analyses (Zeisler *et al.* 1988). PGNAA has been applied to the analysis of environmental, geological and biological samples where it has been combined with XRF, INAA, RNAA, ICP-AES, ICP-MS, etc.

2.6.1 *Hydrogen analysis*

Although hydrogen in many materials is important and bound as moisture at the percentage or sub-percentage levels, it is known to cause embrittlement and cracking in metallic materials and bulk electrical effects in semiconductor materials at ppm levels. The detection limits for hydrogen by PGNAA using a conventional thermal neutron beam is 1 mg owing to the presence of hydrogenous compounds in shielding materials. However, detection limits of 7–20 μg H can be obtained by PGNAA using a low-energy guided neutron beam. Thus, ppm levels of H in samples can be determined by low-energy guided neutron beam PGNAA.

Moisture content in hydrodesulfurization catalysts (Heurtebise & Lubkowitz 1976b) and hydrogen contents including moisture in biological (Anderson *et al.* 1990b, 1993, 1994b), geological (Anderson *et al.* 1985) and volcanic (Anderson *et al.* 1982a) samples, and in construction materials (Anderson *et al.* 1983) were determined using conventional thermal neutron beams. Cold neutron beam PGNAA was applied to the trace determination of H in Ti and its alloys for jet-engine compressor blades, semiconductor materials, fullerenes and their derivatives and solid proton conductors (Paul 1997). In the analyses of jet-engine compressors constructed from Ti and its alloys (Paul & Lindstrom 1994a; Paul *et al.* 1996), high concentrations of H, up to 750 μg H per gram of Ti, were detected. It was confirmed by analyses that embrittlement due to hydrogen is the main cause for compressor blade failure. Although H contents in many samples are at levels less than detection limits in the analyses of semiconductor and related materials, Paul and Lindstrom (1994b) found $\leq$5–10 μg H g^{-1} in hydrothermal quartz, 50–80 μg H g^{-1} in semiconductor grade Ge, and $\leq$2% H in the 1 μm thick dielectric film on a silicon wafer. Hydrogen uptake by a solid proton conductor such as Yb-doped and undoped strontium cerate ($SrCeO_3$) was measured directly (Krug *et al.* 1995). Limits of detection (LOD) for H in various materials using the NIST cold neutron beam PGNAA were found to be 8.5 μg g^{-1} in Coal Fly Ash (NIST SRM1633), 5.0 μg g^{-1} in quartz and 15 μg g^{-1} in Ti alloy (Ti–6Al–4V) (Paul 1997). The limits of detection were calculated by Currie's equation (Currie 1968).

$$LOD = \frac{4.65(R_b/t)^{1/2}}{S} \tag{2.26}$$

where R_b is the background counting rate (cps), t is the duration of the count (s), and S is the analytical sensitivity of H (cps mg^{-1}). By comparing with the conventional H determination method using vacuum degassing and vacuum fusion techniques, PGNAA has the advantages of being non-destructive, independent of the chemical form and can be conducted in situ. Disadvantages include the time required for the analyses (several hours of counting time are generally needed to obtain acceptable counting statistics) and poorer detection limits than those found in the conventional methods.

2.6.2 *Boron analysis*

Boron has been recognized for some time as an essential element for higher plants, though its tolerance range between deficiency and toxicity is rather narrow for agriculturally essential plants. However, there is growing evidence that boron is important in human nutrition (Nielsen 1990), especially under conditions of nutritional (e.g. Mg deprivation) or metabolic (e.g. menopausal hormonal changes causing Ca loss) stress. Boron probably plays an important role in normal bone maintenance, especially under conditions favouring Ca loss from bones. Because of this, worries have been expressed about environmental effects on human health and agricultural crops where boron is leached by ash from coal-fired power stations and incinerators used for waste treatment.

Owing to its large neutron cross-section in the (n, α) reaction, boron is an important element in brain cancer treatment by boron neutron capture therapy, and also as an impurity in reactor materials. Trace determination of boron in biological and environmental samples, and in various materials is also useful. Although traces of boron can be determined by destructive analytical methods, e.g. inductively coupled plasma-atomic emission spectrometry (ICP-AES), ICP-mass spectrometry (ICP-MS) and spectrophotometry, its accurate determination is difficult by these methods owing to contamination and evaporation during chemical procedures. Anderson *et al.* (1990b, 1993) reported that significant boron losses are found in acid digestion of orange juice and in freeze-drying dietary samples. However, boron is one of the most sensitive elements measured by PGNAA and because this is a non-destructive analysis, accurate determinations can be performed without contamination and evaporation losses. As a result of boron's analytical sensitivity, most methodological studies have been carried out for boron and applied to the analysis of various reference materials to evaluate their analytical abilities (Gladney *et al.* 1976; Higgins 1984; Leeman 1988; Anderson *et al.* 1990a; Yonezawa & Haji Wood 1995; Yonezawa *et al.* 1999a). Since PGNAA can determine trace amounts of boron, which are difficult to determine by other methods accurately, it is an essential analytical method for the preparation of reference materials.

It is known that in areas with very low concentrations of boron in soil, there is increased incidence of arthritis. For example, boron contents were determined in bone and tooth samples from 18 New Zealand individuals suffering from severe rheumatoid arthritis and also in samples from 14 control individuals with no medical history of arthritis in order to study the relationship between boron content and arthritis (Ward 1987). Significant differences of boron contents in the bone samples between the rheumatoid arthritis individuals (16.13 ± 7.53 μg g^{-1}) and the control sample (19.79 ± 4.18 μg g^{-1}) were statistically confirmed by a standard *t*-test. However, although the boron content in the tooth samples of the rheumatoid arthritis individuals is lower than in the control group, significant statistical difference was not found between the results by the *t*-test.

By combining boron determination by PGNAA with other multielement determinations by INAA, the multielemental compositions in human placenta (Ward *et al.* 1987) and the brain tissue of Alzheimer's disease patients (Ward & Mason 1987) were obtained to study the relationship between the trace element and disease. Boron determination in the plasma, faeces, urine and body tissues from sheep fed with water of high boron content were carried out to elucidate boron metabolism of ruminants (Miyamoto *et al.* 2000). It was found that boron in drinking water is absorbed quickly by sheep and that levels of boron in plasma, urine and faeces increased rapidly and reached a plateau five days after the beginning of the administration. Boron was found to accumulate in tissues such as the liver, kidney, etc.

A boron neutron capture therapy (BNCT), which targets brain tumours selectively with α and $^{7*}Li$ particles emitted by the ^{10}B (n, α) $^{7*}Li$ reaction (using a nuclear reactor) is noted as an effective method for brain cancer treatment (Hatanaka 1986). PGNAA has been used for ^{10}B determination in human tissue as an essential analytical method in BNCT treatment (Kobayashi & Kanda 1983; Fairchild *et al.* 1986). When a ^{10}B-labelled organic compound, e.g. mercaptoundecahydrododecaborate ($Na_2{}^{10}B_{12}H_{11}SH$) is injected into a patient, the ^{10}B compound is concentrated in the tumour while normal tissue is left unaffected. The content of ^{10}B in the tissue changes with time and must be determined accurately to optimize neutron irradiation. Destructive methods such as ICP-MS, ICP-AES and spectrophotometry require a long time to decompose the sample, while PGNAA can measure ^{10}B rapidly without any chemical procedure. The ^{10}B content is determined using the hydrogen in the sample as an internal standard and can be determined to 0.8–1.8% precision for 14–50 μg ^{10}B g^{-1} samples within 15 min (Yonezawa *et al.* 1996). Because of this, PGNAA has been used to optimize neutron irradiation time in BNCT for the analysis of blood, tumours and other tissues.

Boron distribution in the atmosphere was studied by the PGNAA facility of the University of Maryland–NIST using a conventional thermal neutron beam (Anderson *et al.* 1994a). Atmospheric samples were collected during warm and cold weather conditions at continental, coastal, and marine sites as well as at coal-fired power stations and several volcanic sites. Atmospheric gas-phase and particulate boron as well as soluble and insoluble boron concentrations in rain and snow were determined by PGNAA. Coal-fired power stations are the major source of anthropogenic boron release into the environment owing to high contents of boron in coal. Gladney *et al.* (1978) evaluated boron release from coal-fired power stations by determining boron levels fuel coal and residual ashes (see also James *et al.* 1982).

PGNAA has been applied to the determination of boron in chondritic meteorites. Curtis *et al.* (1980) analysed large numbers of various chondrites for their boron content. Shaw *et al.* (1988) analysed boron distributions in chondritic meteorites by combining PGNAA and α-track images.

Because of its large neutron capture cross-section, the boron content in reactor materials must be controlled to be as low as possible. PGNAA is the most suitable method for ensuring this level of control. Levels of boron to parts per million in reactor-grade graphite, Be (Yonezawa & Haji Wood 1995) and high-temperature alloys (Yonezawa *et al.* 1986) have been determined. As for other applications of PGNAA in material analysis, measurements of boron in borosilicate glasses (Riley & Lindstrom 1987) and barrier-type anodic oxide films on Al (Morisaki *et al.* 1996) were also performed.

Apart from elemental determination, non-destructive chemical state analysis by PGNAA has been attempted (Sakai *et al.* 1999) based on the line-shape of the B 478 keV γ-ray, which varies with the chemical state of boron (i.e. the degradation constant, D). The method was applied to the chemical state analysis of boron and other elements in the products of Thermit reactions and was combined with Mössbauer spectroscopy and X-ray diffraction. The degradation constant of the reaction products agreed with that of metallic boron. From the analytical results of PGNAA as well as Mössbauer spectroscopy and X-ray diffraction, it was revealed that Fe and B products of the Thermit reactions are composed of intermetallic compounds with Al.

2.6.3 *Other elements*

PGNAA has been applied to the determination of S, N and Sm separately, and many elements simultaneously (other than H and B). For example, the content of sulfur in lunar samples and reference materials of environmental samples have been analysed. Analytical results for sulfur in lunar rock samples by various analytical methods differed by as much as a factor of 1.5, owing to the effect of the various chemical forms of S. PGNAA determines sulfur regardless of its chemical form. The sulfer contents in lunar rock samples from the Apollo 17 mission were determined (Islam *et al.* 1980) using a bulk Allende meteorite as a reference sample and correcting for spectral interference from Fe. Analytical results by PGNAA were intermediate. Sulfur contents in reference materials of coal, coal fly ash, cement, orchard leaves and bovine liver were determined in order to apply PGNAA for environmental samples (Jurney *et al.* 1977). Although the 841 keV γ-ray line is of the highest sensitivity for sulfur, it suffers spectral interference from K, Ca, Cr and Mn (Curtis *et al.* 1979). It was also pointed out that spectral interference from Cr and Mn must be corrected in the analysis of chondritic meteorites.

Although analytical sensitivity is low for nitrogen, PGNAA has been applied to the determination of nitrogen contents in biological samples (proteins) and explosive materials detection owing to the fact that nitrogen emits the highest energy γ-rays (10.8 MeV) among common elements. András *et al.* (1979) constructed an experimental facility for screening the production of high-protein content samples from single grain seeds and for the determination of total protein content in maize samples. Chung *et al.* (1993) determined the nitrogen content in the human body by *in vivo* prompt γ-ray activation analysis (IVPGNAA) using a mobile nuclear reactor. In both of these experiments, NaI(Tl) and BGO detectors, which have higher γ-ray detection efficiencies than a Ge detector, were used for detecting the 10.8 MeV γ-rays to improve the accuracy of nitrogen detection. Residence time of feeds in the gastrointestinal tract of ponies was measured using Sm, which is inert in the digestion process, as an intestinal marker (James *et al.* 1984).

PGNAA has been applied to multielement determinations in archaeological, environmental, biological, cosmochemical and geochemical samples (combined with other analytical methods) using its advantageous features of non-destructive analysis and low induced activity. Large numbers of analyses have been carried out on reference materials to study the multielemental analysis capabilities of PGNAA (Failey *et al.* 1979; Graham *et al.* 1982; Anderson *et al.* 1985; Vogt & Schlegel 1985; Simsons & Landsberger 1987; Rossbach 1992; Becker *et al.* 1994; Yonezawa 1996).

PGNAA has also been used in archaeological studies. Multielement determinations were performed on copper-based metallic artefacts (Glascock *et al.* 1984; Oura *et al.* 1999) and pottery bowls (Sueki *et al.* 1998). Glascock *et al.* (1984) determined Zn, Sn, Pb and Cu in Roman and Greek coins, and in Greek and Iranian bronzes. The analytical results agreed well with those obtained by destructive chemical analysis methods showing that PGNAA is a useful analytical technique for archaeological study without sample dissolution.

Multielement determination of rocks and meteorites that are difficult to decompose because of their chemical stability have been studied by PGNAA and it was shown that PGNAA is a useful analytical tool in geological, environmental and cosmochemical studies (Glascock *et al.* 1985; Tittle & Glascock 1988; Latif *et al.* 1999). PGNAA, together with INAA was used for the multielement determination of plume, stratospheric and grand-fall ash of the Mount St Helens eruption in 1980 (Vossler *et al.* 1981; Anderson *et al.* 1982a). Kitto *et al.* (1987, 1988) developed an instrument for simultaneous sample collection of atmospheric particles and gases using filters and ^{7}LiOH-impregnated filters and utilized this set-up for multi-elemental determination of atmospheric samples by PGNAA and INAA. Multi-elemental compositions of zeolite cracking catalysts, which contain light rare earth elements (REEs) and are used by petroleum refineries, were analysed and the REEs released by oil-burning including those from automobile into the environment, have also been studied (Kitto *et al.* 1992). PGNAA is highly sensitive for analysis for toxic elements such as Cd and Hg and can be determined to ppm levels in environmental samples. Because of this, PGNAA together with INAA is applied to the multielemental determination of river sediments in the study of pollution history and geochemical background (Spychala *et al.* 1987; Kuno *et al.* 1997, 1999). Kuno *et al.* (1997, 1999) determined multielemental compositions in polluted and non-polluted estuarine sediments. Large numbers of elemental compositions can be determined accurately by combining PGNAA with other elemental analysis methods, such as XRF, ICP-MS, ICP-AES, INAA, RNAA, etc. using only small amounts of sample. Such combinations of methods allowed the estimation of contamination sources in elemental compositions taken from oil-contaminated environmental samples in the Persian Gulf (Yonezawa *et al.* 1999b).

Anderson *et al.* (1990b, 1993, 1994b) applied PGNAA using a conventional thermal neutron beam to food and related materials (see also Hight *et al.* 1993; Anderson 1995; Anderson & Cunningham 1995). Owing to the high concentration of hydrogen in organic material, which leads to high neutron scattering, sensitivity and background enhancements become important in the analysis of food and related materials when using a conventional thermal neutron beam. These researchers established the multielement determination of H, B, C, N, Na, S, Cl, K, etc. in food and related samples correcting sensitivity and background enhancements (Anderson *et al.* 1990b, 1993) and applied the methods to the analysis of foods and dietary supplements (Hight *et al.* 1993; Anderson *et al.* 1994b). For instance, in a food analysis (Anderson *et al.* 1994b), B, S, K, Na, Cl and added NaCl intake were measured in 234 food items collected in the southeastern United States for a Total Diet Study (TDS) organized by the US Food and Drug Administration (FDA). In the dietary supplement analyses (Hight *et al.* 1993), a multimethod approach was used to determine 36 elements in a wide variety (42) of dietary supplements to develop a database of these products in order to identify potential health risks arising from toxic elements. A multielement determination of ceramic glaze, using PGNAA and XRF to evaluate the release of toxic elements such as Cd and Pb from ceramic ware glazes into food, was also carried out (Anderson 1995).

Medical diagnosis has been studied by *in vivo* PGNAA (IVPGNAA) using radioisotope neutron sources such as ^{238}Pu–Be and ^{252}Cf, and accelerator-type neutron sources (Chung 1995). Chung (1990) studied elemental analysis of some environmental contaminants, such as Cd and Hg, as well as vital elemental constituents, such as Ca, Cl, N and P, in the human body by IVPGNAA using a low-power mobile reactor. It is known that toxic elements, e.g. Cd and Hg, are concentrated in the liver and kidney after intake by ingestion and inhalation. The determination of Cd and Hg in the liver and kidney was performed by IVPGNAA using rats. From these studies it was found that lower Cd contents than those deemed to be medically significant can be determined by IVPGNAA of the liver and kidney, and moreover with lower radiation doses than are used at the other facilities with radioisotope and accelerator neutron sources (Chang *et al.* 1985; Chen & Chung 1989). Unfortunately, the detection limit for Hg in the above target organs, even with prolonged scan periods, is much higher than the normal safe levels for mercury in patients, therefore limiting its clinical application to screening individuals exposed to high concentrations of mercury (Chang *et al.* 1987). Further, the determination of vital constituents such as Ca, N, P and Cl in the body can also be carried out by IVPGNAA (Chung & Yuan 1988; Chung *et al.* 1993; Chung & Chen 1993).

Multielement determinations in industrial materials such as steel and various alloys (Heurtebise & Lubkowitz 1976a), hydrodesulfurization catalysts (Heurtebise *et al.* 1976; Heurtebise & Lubkowitz 1977) and construction materials (Anderson *et al.* 1983) can be performed with PGNAA alone or in combination with other methods.

2.6.4 *Isotopic analysis*

Since PGNAA detects γ-rays from nuclear transitions, it can be used for isotopic analysis. A small number of PGNAA applications have been found for the study of S, Si, Ni and Fe, which are difficult to analyse by mass spectrometry (Nakamura *et al.* 1993; Yonezawa 1995, 1999b; Yoshikawa *et al.* 1997). However, the precise analysis of the mass spectrometry method is difficult to replicate because of the statistical limit of radiation decay. The isotopic ratio of $^{32}S/^{34}S$ can be obtained to 2% precision in pure sulfur with natural isotopic composition (Yonezawa *et al.* 1999b). However, the measured precision of $^{32}S/^{34}S$ in various oil samples ranged from 6 to 28%, owing to the insufficient sensitivity of the PGNAA method. Isotopic analysis of ^{30}Si was also performed to study the diffusion behaviour of silicon in engineering barrier products used for the disposal of high-level radioactive waste glass, employing ^{30}Si as a tracer (Yoshikawa *et al.* 1997). However, this particular method was not practical because it was insufficiently sensitive. Although PGNAA is unable to perform precise isotopic analysis, it can be used for isotopic analysis when rough analytical data are required.

2.7 New techniques for PGNAA

PGNAA has been applied so far to bulk analysis where samples larger than a few tenths of a milligram are irradiated with neutron beams of about 4–20 cm^2 or placed in a reactor core. However, micro-sample analysis in which the sample size is smaller than 1 mm^2 and spatial analysis in which the spatial resolution is less than 1 mm have become possible recently through development of a neutron lens that is able to focus a neutron beam to a width of less than 1 mm. The neutron lens (Kumakhov & Komarov 1990; Kumakhov & Sharov 1992; Chen *et al.* 1992) comprises over a thousand polycapillary glass fibres made from lead silicate, each with hollow capillaries of diameter 10 μm. The lens is able to compress a neutron beam to the same focal point beyond the lens. The polycapillary glass fibres are bundled parallel at the lens entrance and bent in such a way that the exit ends of all fibres aim neutrons at a common point, i.e. the focus. The principle of the neutron lens is the multiple total external reflection of neutrons from the smooth inner walls of the capillary channels. When an incident neutron beam strikes the reflecting surface of a capillary at a grazing angle that is smaller than the critical angle of the material, it undergoes total reflection with little loss of intensity. Neutrons satisfying the critical angle condition can be transported effectively through large angles of the gently bent capillary channels. The critical angle depends on neutron energy and the higher critical angle for cold neutrons enables such beams to be more easily focused. A smaller channel diameter enables sharper bending of the fibres. The neutron lens was known as Kumakhov optics (Kumakhov & Komarov 1990) for X-ray focusing and was subsequently developed by the research group at NIST for neutron focusing (Kumakhov & Sharov 1992; Chen *et al.* 1992). Two types of neutron lenses, straight and bender, have so far been developed.

The straight type neutron lens (Xiao *et al.* 1994) focuses a neutron beam along the same axis as the incident neutron beam (see Fig. 2.21). This neutron lens was able to

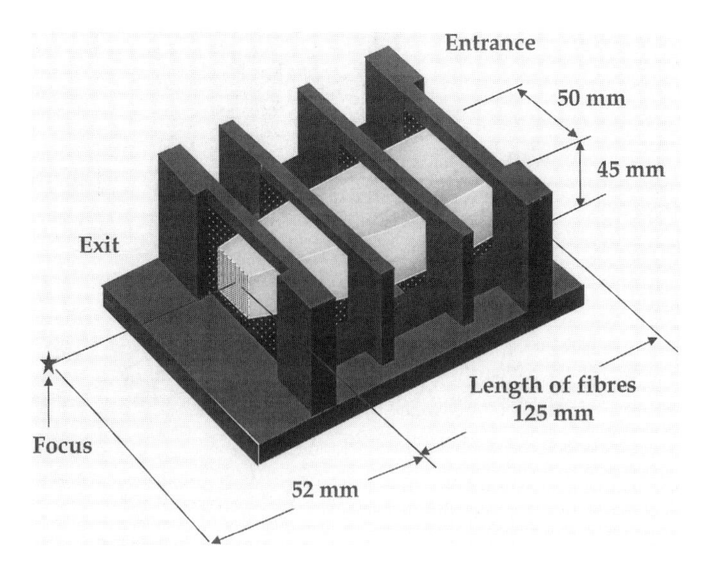

Fig. 2.21 Schematic diagram of the straight-type neutron lens. The NBSR cold neutron beam of $50 \times 45\ \mathrm{mm}^2$ was focused at the focal point 52 mm from the exit of the lens with a beam size of 0.53 mm diameter (FWHM) and intensity gain of 80 (Chen *et al.* 1995). (Reprinted from *Nucl. Instrum. Methods Phys. Res.*, **B95**, H. Chen, V.A. Sharov, D.F.R. Mildner, R.G. Downing, R.L. Paul, R.M. Lindstrom, C.J. Zeissler and Q.F. Xiao, Prompt gamma activation analysis enhanced by a neutron focusing capillary lens, p. 107. Copyright 1995, with permission from Elsevier Science.) (Courtesy Dr H.H. Chen-Mayer)

compress the cold neutron beam (cross-section $50 \times 45\ \mathrm{mm}^2$ with neutron flux $6.5 \times 10^8\ \mathrm{n\ cm^{-2}\ s^{-1}}$) of the NBSR at NIST onto a focal spot of diameter 0.53 mm (full width at half maximum, FWHM) at a distance of 52 mm from the exit of the lens with an average gain of 80 in the neutron flux density (Chen *et al.* 1995). The PGNAA experiments were carried out using the neutron lens for small samples and for spatial resolution measurements. A tapered aperture (1 mm diameter at the smaller end and 3 mm thickness) made of pressed ^{6}LiF was used to reduce the γ-ray background. In the small sample analyses, glass-bead samples containing 2 wt.% Gd ($\sim$0.23 mm $\times$ 0.385 mm cross-section) and a Cd metal particle ($\sim$0.5 mm $\times$ 0.45 mm cross-section) (Chen-Mayer *et al.* 1997a) were analysed. Although γ-ray backgrounds were enhanced through the scattering and absorption of incident neutrons by the materials used to construct the lens, the prompt γ-ray peak intensities of Cd and Gd were enhanced more than the backgrounds by a factor of 60. This resulted in an improvement of the minimum detection levels of Cd and Gd by a factor of 20. In the spatial resolution measurements (Chen *et al.* 1995), the samples consisted of rows of lead–silica and borosilicate fibres of diameter 0.4 mm (five each) and 1 mm (three each) arranged alternately (B–Pb–B–Pb . . .), side by side. The count rate of the boron peak at 478 keV was monitored as each row of sample was scanned along the x direction in the focal plane. Figure 2.22 shows an overlay of an optical microscope image of the fibres and the PGNAA count rate at the corresponding fibre positions. Although the peaks were partially resolved [(peak – valley)/(peak – background) $\sim$38%] in the 0.4 mm diameter sample, the peaks were almost completely resolved (97%) in the 1 mm sample.

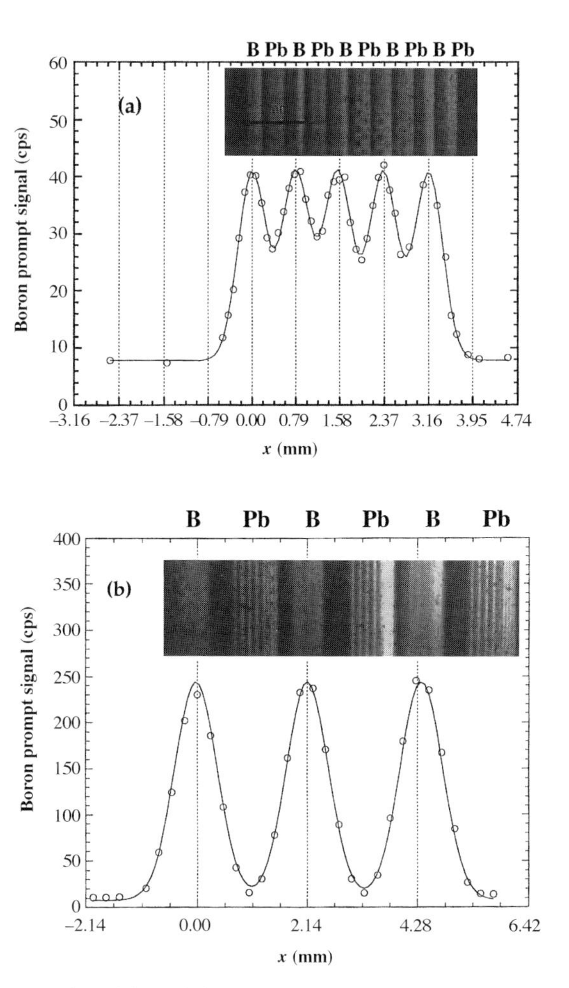

Fig. 2.22 Analytical results of spatial resolution measurements of samples consisting of alternate rows of lead–silica and borosilicate fibres using a straight-type lens. The figure shows an overlay of an optical microscope image of the fibres and the PGNAA count rate at the corresponding fibre positions (Chen *et al.* 1995). (Reprinted from *Nucl. Instrum. Methods Phys. Res.*, **B95**, H. Chen, V.A. Sharov, D.F.R. Mildner, R.G. Downing, R.L. Paul, R.M. Lindstrom, C.J. Zeissler and Q.F. Xiao, Prompt gamma activation analysis enhanced by a neutron focusing capillary lens, p. 107. Copyright 1995, with permission from Elsevier Science.) (Courtesy Dr H.H. Chen-Mayer)

Since the straight type neutron lens generates high γ-ray backgrounds due to unfocused neutrons, the bender type neutron lens, which focuses the neutron beam away from its line of sight, has been developed (Chen-Mayer *et al.* 1997b). The bender neutron lens, shown in Fig. 2.23, focuses the cold neutron beam of the NBSR (cross-section 50 × 45 mm^2) onto a focal spot at a distance of 95 mm from the lens exit and 20 mm below the bottom edge of the beam path. Although the results obtained from a focused beam of diameter 0.65 mm (FWHM) and gain of 20 in neutron flux were inferior than in the straight type, the bender type lens produced only half the background increase obtained with the straight type lens with no sample. In addition,

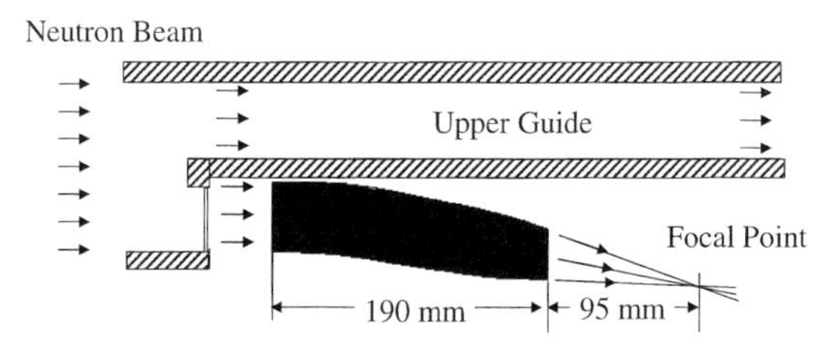

Fig. 2.23 A schematic diagram of the bender-type neutron lens set at the cold neutron beam guide of the NBSR. The lens focused the cold neutron beam of cross-section 50 × 45 mm^2 to a focal spot at a distance 95 mm from the lens exit and 20 mm below the bottom edge of the beam path. (Chen-Mayer *et al.* 1997b). (Reprinted from *Rev. Sci. Instrum.*, **68**, H. Chen-Mayer, D.F.R. Midlner, V.A. Sharov, Q.F. Xiao, Y.T. Cheng, R.M. Lindstrom and R.L. Paul, A polycapillary bending and focusing lens for neutrons, p. 3744. Copyright 1997, with permission from American Institute of Physics.) (Courtesy Dr H.H. Chen-Mayer)

when a Gd shard (1 mm^2) sample was measured the Gd γ-ray intensity-to-background ratio can be improved by a factor of as much as 6 relative to that of the straight type lens. The bender type lens has been applied to the determination of multielements in small samples and the spatial distribution measurements of coated steel samples (Chen-Mayer *et al.* 2000).

In the analyses of small samples, elemental concentrations of Fe, Cr, Mo and Ni in a small chip of stainless steel (NIST SRM160b, 2.1 mg, 0.46 mm in diameter and 1.70 mm thick) and of B in a glass (NIST SRM611, 0.7 mm thick) were determined by comparative standardization. The analytical results of both samples agreed within 10% of the NIST-certified values. In the spatial distribution measurements, the spatial distribution of Fe and Cr in a Cr-coated steel sample, with an estimated Cr thickness of 50 μm, taken from a boiler tank tube, was analysed. Although Cr distribution was detected, satisfactory spatial distribution was not obtained in the analysis owing to the beam size being larger than the Cr thickness. A much smaller beam ($\sim$0.05 mm) was needed to resolve the interface region created by the Cr coating.

Although the polycapillary-fibre neutron lens is able to create cold neutron beam sizes of diameter smaller than 1 mm, the lens still has high γ-ray background. Further development is therefore necessary to create a neutron beam which has a smaller beam size, a lower γ-ray background and a higher gain than those beams obtained with the polycapillary-fibre neutron lenses used for practical spot analysis and compositional mapping.

It is essential that a more intense source of neutrons than present nuclear reactors is used in the spot analysis and three-dimensional composition mapping of samples. A spallation neutron source (SNS), which generates neutrons by spallation reactions of high-energy charged particles such as protons in heavy nuclides such as Hg, Pb, or U using proton accelerators, is a promising alternative neutron source to the nuclear reactor. An SNS has been constructed at the Paul Scherrer Institute (PSI) in Switzerland and a high-quality PGNAA facility has recently been installed at the cold neutron beam guide (Crittin *et al.* 2000). The neutron flux at the sample position is as high as 6.9×10^7 n cm^{-2} s^{-1}, and therefore, analytical sensitivities almost equal to those obtained by a reactor can be obtained using the SNS. Construction of further high-intensity SNSs are planned in the United States, Europe and Japan.

Neutron beams of intensity higher than those obtained by a nuclear reactor will soon be available and consequently more sensitive PGNAA facilities than those at present will follow. The SNS can generate greater numbers of epithermal neutrons than a reactor and therefore, selective elemental analysis will be possible using resonance neutron reactions of epithermal neutrons. It is expected that PGNAA will advance as a high-sensitive and highly selective elemental analysis technique for micro-samples and three-dimensional composition mapping through using such new neutron sources.

References

Abramowitz, M. and Stegun, I.A., *Handbook of Mathematical Functions with Formulas, Graphs, and Mathematical Tables*, John Wiley & Sons, New York, pp. 374, 498 (1972).

Alam, S., Zaman, M.A. and Islam, S.M.A., *Nucl. Instrum. Methods Phys. Res.*, **B83**, 235 (1993).

Alfassi, Z.B., *Chemical Analysis by Nuclear Methods*, John Wiley & Sons, Chichester (1994).

Alfassi, Z.B. and Chung, C., *Prompt Gamma-Neutron Activation Analysis*, CRC Press, Boca Raton, FL (1995).

Anderson, D.L., Failey, M.P., Zoller, W.H., Walters, W.B., Gordon, G.E. and Lindstrom, R.M., *J. Radioanal. Chem.*, **63**, 97 (1981).

Anderson, D.L., Phelan, J.M., Vossler, T. and Zoller, W.H., *J. Radioanal. Nucl. Chem.*, **71**, 47 (1982a).

Anderson, D.L., Zoller, W.H., Gordon, G.E., Walters, W.B. and Lindstrom, R.M., in *Neutron-Capture Gamma-Ray Spectroscopy and Related Topics 1981*, Eds. T. von Egidy, F. Gönewein, B. Maier, The Institute of Physics, Bristol and London, p. 655 (1982b).

Anderson, D.L., Gordon, G.E. and Lepel, E.A., *Can. J. Chem.*, **61**, 724 (1983).

Anderson, D.L., Sun, Y., Failey, M.P. and Zoller, W.H., *Geostandards Newsletter*, **9**, 219 (1985).

Anderson, D.L., Cunningham, W.C. and Mackey, E.A., *Fresenius' J. Anal. Chem.*, **338**, 554 (1990a).

Anderson, D.L., Cunningham, W.C. and Mackey, E.A., in *Biological Trace Element Research*, Ed. G.N. Schrauzer, The Humana Press Inc., p. 613 (1990b).

Anderson, D.L., Cunningham, W.C. and Alvarez, G.H., *J. Radioanal. Nucl. Chem.*, **167**, 139 (1993).

Anderson, D.L. and Mackey, E.A., *J. Radioanal. Nucl. Chem.*, **167**, 145 (1993).

Anderson, D.L., Kitto, M.E., Macarthy, L. and Zoller, W.H., *Atmospheric Environment*, **28**, 1401 (1994a).

Anderson, D.L., Cunningham, W.C. and Lindstrom, T.R., *J. Food Comp. Anal.*, **7**, 59 (1994b).

Anderson, D.L., *J. Radioanal. Nucl. Chem.*, **192**, 281 (1995).

Anderson, D.L. and Cunningham, W.C., *J. Radioanal. Nucl. Chem.*, **194**, 351 (1995).

András, L., Bálint, A., Csõke, A. and Nagy, A.Z., *Radiochem. Radioanal. Lett.*, **40**, 27 (1979).

Becker, D.A., Anderson, D.L., Lindstrom, R.M., Greenberg, R.R., Garrity, K.M. and Mackey, E.A., *J. Radioanal Nucl. Chem.*, **179**, 149 (1994).

Chang, P., Chung, C., Yuan, L. and Weng, P., *J. Radioanal. Nucl. Chem.*, **92**, 343 (1985).

Chang, P., Ho, Y., Chung, C., Yuan, L. and Weng, P., *Nucl. Technol.*, **76**, 241 (1987).

Chen, H., Downing, R.G., Mildner, D.F.R., Gibson, W.M., Kumakhov, M.A., Ponomarev, I.Yu. and Gubarev, M.V., *Nature*, **357**, 391 (1992).

Chen, H., Sharov, V.A., Mildner, D.F.R., Downing, R.G., Paul, R.L., Lindstrom, R.M., Zeissler, C.J. and Xiao, Q.F., *Nucl. Instrum. Methods Phys. Res.*, **B95**, 107 (1995).

Chen-Mayer, H., Sharov, V.A., Mildner, D.F.R., Downing, R.G., Paul, R.L., Lindstrom, R.M., Zeissler, C.J. and Xiao, Q.F., *J. Radioanal. Nucl. Chem.*, **215**, 141 (1997a).

Chen-Mayer, H., Mildner, D.F.R., Sharov, V.A., Xiao, Q.F., Cheng, Y.T., Lindstrom, R.M. and Paul, R.L., *Rev. Sci. Instrum.*, **68**, 3744 (1997b).

Chen-Mayer, H.H., Mackey, E.A., Paul, R.L. and Mildner, D.F.R., *J. Radioanal. Nucl. Chem.*, **244**, 391 (2000).

Chen, W.K. and Chung, C., *J. Radioanal. Nucl. Chem.*, **133**, 349 (1989).

Chung, C., Yuan, L., Chen, K., Weng, P., Chang, P. and Ho, Y., *Int. J. Appl. Radiat. Isot.*, **36**, 357 (1985).

Chung, C. and Yuan, L., *Appl. Radiat. Isot.*, **39**, 977 (1988).

Chung, C., in *Activation Analysis Volume II*, Ed. Z.B. Alfassi, CRC Press, Boca Raton, FL, p. 299 (1990).

Chung, C., Wei, Y.Y. and Chen, Y.Y., *Appl. Radiat. Isot.*, **44**, 941 (1993).

Chung, C. and Chen, Ya-Yu, *J. Radioanal. Nucl. Chem.*, **169**, 333 (1993).

Chung, C. and Yuan, L., *Nucl. Technol.*, **109**, 226 (1995).

Chung, C., in *Prompt Gamma-Neutron Activation Analysis*, Eds. Z. Alfassi, C. Chung, CRC Press, Boca Raton, FL, p. 101 (1995).

Comar, D., Crouzel, C., Chasteland, M., Riviere, R. and Kellershohn, C., in *Proceedings of the 1968 International Conference of Modern Trends on Activation Analysis*, Ed. J.R. De Voe, P.D. LaFleur, National Bureau of Standards Special Publication 312, Vol. 1, 1969, p. 114; *Nucl. Appl.*, **6**, 344 (1969).

Crittin, M., Kern, J. and Schenker, J.-L., *J. Radioanal. Nucl. Chem.*, **244**, 399 (2000).

Currie, L.A., *Anal. Chem.*, **40**, 586 (1968).

Curtis, D.B., Gladney, E.S. and Jurney, E.T., *Anal. Chem.*, **51**, 158 (1979).

Curtis, D.B., Gladney, E.S. and Jurney, E.T., *Geochim. Cosmochim. Acta*, **44**, 1945 (1980).

Debertin, K. and Helmer, R.G., *Gamma- and X-ray Spectrometry with Semiconductor Detectors*, Elsevier, Amsterdam (1988).

Failey, M.P., Anderson, D.L., Zoller, W.H., Gordon, G.E. and Lindstrom, R.M., *Anal. Chem.*, **51**, 2209 (1979).

Fairchild, R.G., Gabel, D. Laster, B.H., Greenberg, D., Kiszenick, W. and Micca, P.L., *Med. Phys.*, **13**, 50 (1986).

Fezekas, B., Östör, J., Kiss, Z., Simonits, A and Molnár, G.L., *J. Radioanal. Nucl. Chem.*, **233**, 101 (1998).

Firestone, R.B., Shirley, V.S., Baglin, C.M., Frank Chu, S.Y. and Zipkin, J., *Table of Isotopes*, 8th edn, Vols. 1 & 2, John Wiley & Sons Inc., New York (1996).

Fleming, R.F., *Int. J. Appl. Radiat. Isot.*, **33**, 1263 (1982).

Furukawa, Y., Koyama, M. and Yuki, M., *Radioisotopes*, **16**, 499 (1967).

Gladney, E.S., Jurney, E.T. and Curtis, D.B., *Anal. Chem.*, **48**, 2139 (1976).

Gladney, E.S., Wangen, L.E., Curtis, D.B. and Jurney, E.T., *Environ. Sci. Technol.*, **12**, 1084 (1978).

Gladney, E.S., LA-8028-MS (1979).

Glascock, M.D., Spalding, T.G., Biers, J.C. and Cornman, M.F., *Archaeometry*, **26**, 96 (1984).

Glascock, M.D., Coveney, Jr., R.M., Rittle, C.W., Gartner, M.L. and Murphy, R.D., *Nucl. Instrum. Methods Phys. Res.*, **B10/11**, 1042 (1985).

Graham, C.C., Glascock, M.D., Carni, J.J., Vogt, J.R. and Spalding, T.G., *Anal. Chem.*, **54**, 1623 (1982).

Greenwood, R.C., *Trans. Am. Nucl. Soc.*, **10**, 28 (1967).

Hanna, A.G., Brugger, R.M. and Glascock, M.D., *Nucl. Instrum. Methods*, **188**, 619 (1981).

Harling, O.K., Chabeuf, J.M., Lambert, F. and Yasuda, G., *Nucl. Instrum. Methods Phys. Res.*, **B83**, 557 (1993).

Hatanaka, H., *Boron-Neutron Capture Therapy for Tumors*, Nishimura Co. Ltd., Tokyo (1986).

Hauser, U., Neuwirth, W., Pietsch, W. and Richter, K., *Z. Physik*, **269**, 181 (1974).

Henkelmann, R. and Born, H.-J., *J. Radioanal. Chem.*, **16**, 473 (1973).

Heurtebise, M. and Lubkowitz, J.A., *J. Radioanal. Chem.*, **31**, 503 (1976a).

Heurtebise, M. and Lubkowitz, J.A., *Anal. Chem.*, **48**, 2143 (1976b).

Heurtebise, M., Buenafama, H. and Lubkowitz, J.A., *Anal. Chem.*, **48**, 1969 (1976).

Heurtebise, M. and Lubkowitz, J.A., *J. Radioanal. Chem.*, **38**, 115 (1977).

Higgins, M.D., *Geostandards Newsletter*, **8**, 31 (1984).

Hight, S.C., Anderson, D.L., Cunningham, W.C., Capar, S.G., Lamont, W.H. and Sinex, S.A., *J. Food Comp. Anal.*, **6**, 121 (1993).

Isenhour, T.L. and Morrison, G.H., *Anal. Chem.*, **38**, 162 (1996a).

Isenhour, T.L. and Morrison, G.H., *Anal. Chem.*, **38**, 167 (1966b).

Islam, M.A., Kennett, T.J., Prestwich, W.V. and Rees, C.E., *J. Radioanal. Chem.*, **56**, 123 (1980).

James, W.D., Graham, C.C., Glascock, M.D. and Hanna, A.G., *Environ. Sci. Technol.*, **16**, 195 (1982).

James, W.D., Arnold, F.F., Pond, K.R., Glascock, M.D. and Spalding, T.G., *J. Radioanal. Nucl. Chem.*, **83**, 209 (1984).

Jurney, E.T., Motz, H.T. and Vegors Jr., S.H., *Nucl. Phys.*, **A94**, 351 (1967).

Jurney, E.T., Curtis, D.B. and Gladney, E.S., *Anal. Chem.*, **49**, 1741 (1977).

Kawabata, Y., Suzuki, M., Takahashi, H., Onishi, N., Shimanuki, A., Sugawa, Y., Niino, N., Kasai, T., Funasho, K., Hayakawa, S. and Okuhata, K., *J. Nucl. Sci. Technol.*, **27**, 1138 (1990).

Kitto, M.E. and Anderson, D.L., in *The Chemistry of Acid Rain: Sources and atmospheric processed*, Eds. R.W. Johnson, G.E. Gordon, W. Lalkins, A. Elzermann, American Chemical Society, Washington, DC, p. 84 (1987).

Kitto, M.E., Anderson, D.L. and Zoller, W.H., *J. Atmospheric Chem.*, **7**, 241 (1988).

Kitto, M.E., Anderson, D.L., Gordon, G.E. and Olmez, I., *Environ. Sci. Technol.*, **26**, 1368 (1992).

Kobayashi, T. and Kanda, K., *Nucl. Instrum. Methods. Phys. Res.*, **204**, 525 (1983).

Krug, F., Schober, T. Paul, R.L. and Springer, T., *Solid State Ionics*, **77**, 185 (1995).

Kumakhov, M.A. and Komarov, F.F., *Phys. Rep.*, **191**, 289 (1990).

Kumakhov, M.A. and Sharov, V.A., *Nature*, **357**, 390 (1992).

Kuno, A., Matsuo, M., Takano, B., Yonezawa, C., Matsue, H. and Sawahata, H., *J. Radioanal. Nucl. Chem.*, **218**, 169 (1997).
Kuno, A., Sampei, K., Matsuo, M., Yonezawa, C., Matsue, H. and Sawahata, H., *J. Radioanal. Nucl. Chem.*, **239**, 587 (1999).
Latif, Sk.A., Oura, Y., Ebihara, M., Kallmeyn, G.W., Nakahara, H., Yonezawa, C., Matsue, H and Sawahata, H., *J. Radioanal. Nucl. Chem.*, **239**, 577 (1999).
Leeman, W.P., *Geostandards Newsletter*, **12**, 61 (1988).
Lindstrom, R.L., Fleming, R.F., Paul, R.L. and Mackey, E.A., in *Proceedings of The International k_0 Users Workshop – Gent*, Ed. F. De Corte, Gent, Belgium, p. 121 (1992).
Lindstrom, R.M., Zeisler, R., Vincent, D.H., Greenberg, R.R., Stone, C.A., Mackey, E.A., Anderson, D.L. and Clark, D.D., *J. Radioanal. Nucl. Chem.*, **167**, 121 (1993).
Lindstrom, R.M., Yonezawa, C., in *Prompt Gamma-Neutron Activation Analysis*, Ed. Z.B. Alfassi, C. Chung, CRC Press, Boca Raton, FL, p. 93 (1995).
Lombard, S.M., Isenhour, T.L., Heintz, P.H. Woodruff, G.L. and Wilson, W.E., *Int. J. Appl. Radiat. Isot.*, **19**, 15 (1968).
Lone, M.A., Leavitt, R.A. and Harrison, D.A., *Atom Data Nucl. Data Tables*, **26**, 511 (1981).
Mackey, E.A., Gordon, G.E., Lindstrom, R.M. and Anderson, D.L., *Anal. Chem.*, **63**, 288 (1991).
Mackey, E.A., Gordon, G.E., Lindstrom, R.M. and Anderson, D.L., *Anal. Chem.*, **64**, 2366 (1992).
Mackey, E.A. and Copley, J.R.D., *J. Radioanal. Nucl. Chem.*, **167**, 127 (1993).
Mackey, E.A., in *Biological Trace Element Research*, Ed. G.N. Schrauzer, Humana Press Inc., p. 103 (1994).
Magara, M. and Yonezawa, C., *Nucl. Instrum. Methods Phys. Res.*, **A411**, 130 (1998).
Matsue, H. and Yonezawa, C., *J. Radioanal. Nucl. Chem.*, **245**, 189 (2000).
Miyamoto, S., Suto, M., Shiomoto, A., Yamasaki, S., Nishimura, K., Yonezawa, C. Matsue, H. and Hoshi, M., *J. Radioanal. Nucl. Chem*, **244**, 307 (2000).
Molnár, G., Belgya, T., Dabolczi, L., Fazekas, B., Révay, Zs., Veres, Á., Bikit, I., Kiss, Z. and Östör, J., *J. Radioanal. Nucl. Chem.*, **215**, 111 (1997).
Molnár, G., Révay, Zs., Paul, R.L. and Lindstrom, R.M., *J. Radioanal. Nucl. Chem.*, **234**, 21 (1998).
Morisaki, S., Ogawa, Y., Mano, T., Yonezawa, C, Ito, Y. and Sawahata, H., *Hyomen Gijyutsu*, **47**, 456 (1996) [in Japanese].
Nakamura, A., Ebihara, M., Kobayashi, K., Ito, Y., Yonezawa, C and Hoshi, M., in *Proceedings of the Eighteenth Symposium on Antarctic Meteorites*, Ed. National Institute of Polar Research, National Institute of Polar Research, Tokyo, p. 145 (1993).
Neuwirth, W., Pietsch, W., Richter, K. and Hauser, U., *Z. Physik*, **A275**, 209 (1975).
Nielsen, F.H., in *Biological Trace Element Research*, Ed. G.N. Schrauzer, The Humana Press Inc., p. 599 (1990).
Orphan, V.J. and Rasmussen, N.C., *Nucl. Instrum. Methods*, **48**, 282 (1967).
Oura, Y., Saito, A., Sueki, K., Nakahara, H., Tomizawa, T., Nishikawa, T., Yonezawa, C., Matsue, H. and Sawahata, H., *J. Radiochem. Nucl. Chem.*, **239** 581 (1999).
Paul, R.L. and Mackey, E.A., *J. Radioanal. Nucl. Chem.*, **181**, 321 (1994).
Paul, R.L. and Lindstrom, R.M., in *Review of Progress in Quantitative Nondestructive Evaluation*, Vol. 13, Ed. D.O. Thompson, D.E. Chimenti, Plenum Press, New York, p. 1619 (1994a).
Paul, R.L. and Lindstrom, R.M., in *Diagnostic Techniques for Semiconductor Proceedings, MRS Symposium Proceedings*, Vol. 324, Eds. O.J. Glembocki, S.W. Pang, F.H. Pollak, G.M. Crean, G. Larrabee, Materials Research Society, Pittsburgh, PA, p. 403 (1994b).
Paul, R.L., *J. Radioanal. Nucl. Chem.*, **191**, 245 (1995).
Paul, R.L., Privett III, H.M., Lindstrom, R.M., Richards, W.J. and Greenberg, R.R., *Metall. Mater. Trans. A*, **27A**, 3682 (1996).
Paul, R.L., *Analyst*, **122**, 35R (1997).
Paul, R.L. and Lindstrom, R.M., in *2nd International k_0 Users Workshop Proceedings*, Ed. B. Smodis, Jozef Stefan Institute, Ljublijana, Slovenia, p. 54 (1997).
Paul, R.L., Lindstrom, R.M. and Heald, A.E., *J. Radioanal. Nucl. Chem.*, **215**, 63 (1997).
Prask, H., *Neutron News*, **5**, 10 (1994).
Riley, Jr, J.R. and Lindstrom, R.M., *J. Radioanal. Nucl. Chem.*, **109**, 109 (1987).
Ríos-Martínez, C., Ünlü, K. and Wehring, B.W., *J. Radioanal. Nucl. Chem.*, **234**, 119 (1998).
Rossbach, M., *Anal. Chem.*, **63**, 2156 (1991).
Rossbach, M. and Hiep, N.T., *Fresenius' J. Anal. Chem.*, **344**, 59 (1992).
Sakai, Y., Yonezawa, C., Magara, M., Sawahata, H. and Ito, Y., *Nucl. Instrum. Methods Phys. Res.*, **A353**, 699 (1994).
Sakai, Y., Yonezawa, C., Matsue, H., Magara, M., Sawahata, H. and Ito, Y., *Radiochim. Acta*, **72**, 45 (1996).

Sakai, Y., Ohshita, K., Kubo, M.K., Yonezawa, C. and Matsue, H., *J. Radioanal. Nucl. Chem.*, **239**, 263 (1999).
Shaw, D.M., Higgins, M.D., Hinton, R.W., Truscott, M.G. and Middleton, T.A., *Geochim. Cosmochim. Acta*, **52**, 2311 (1988).
Simsons, A. and Landsberger, S., *J. Radioanal. Nucl. Chem.*, **110**, 555 (1987).
Spychala, M., Michaelis, W. and Fanger, H.-U., *Nucl. Geophys.*, **1**, 309 (1987).
Sueki, K., Kobayashi, K. Sato, W., Nakahara, H. and Tomizawa, T., *Anal. Chem.*, **68**, 2203 (1996).
Sueki, K., Oura, Y., Sato, W., Nakahara, H. and Tomizawa, T., *J. Radioanal. Nucl. Chem.*, **234**, 27 (1998).
Tittle, C.W. and Glascock, M.D., *Nucl. Geophys.*, **2**, 171 (1988).
Trubert, D., Duplatre, G. and Abbe, J.Ch., *Appl. Radiat. Isot.*, **42**, 699 (1991).
Tuli, J.K., in *Prompt Gamma-Neutron Activation Analysis*, Ed. Z.B. Alfassi, C. Chung, CRC Press, Boca Raton, FL, p. 177 (1995).
Vogt, J.R. and Schlegel, S.C., *J. Radioanal. Nucl. Chem.*, **88**, 379 (1985).
Vossler, T., Anderson, D.L., Aras, N.K., Phelan, J.M. and Zoller, W.H., *Science*, **211**, 827 (1981).
Ward, N.I., *J. Radioanal. Nucl. Chem.*, **110**, 633 (1987).
Ward, N.I. and Mason, J.A., *J. Radioanal. Nucl. Chem.*, **113**, 515 (1987).
Ward, N.I., MacMahon, T.D. and Mason, J.A., *J. Radioanal. Nucl. Chem.*, **113**, 501 (1987).
Yonezawa, C., Tojyo, T. and Komori, T., *Bunseki Kagaku*, **35**, 782 (1986) [in Japanese].
Yonezawa, C., Haji Wood, A.K., Hoshi, M., Ito, Y. and Tachikawa, E., *Nucl. Instrum. Methods Phys. Res.*, **A329**, 207 (1993).
Yonezawa, C. and Haji Wood, A.K., *Anal. Chem.*, **67**, 4466 (1995).
Yonezawa, C., Magara, M., Sawhata, H., Hoshi, M., Ito, Y. and Tachikawa, E., *J. Radioanal. Nucl. Chem.*, **193**, 171 (1995).
Yonezawa, C., *Anal. Sci.*, **12**, 605 (1996).
Yonezawa, C., Matsue, H., Sawahata, H., Kurosawa, T., Hoshi, M. and Ito, Y., in *Cancer Neutron Capture Therapy, Proceedings of the 6th International Symposium on Neutron Capture Therapy for Cancer*, Ed. Y. Mishima, Plenum, New York, p. 221 (1996).
Yonezawa, C., Development of a neutron capture prompt gamma-ray analysis system and basic studies of element analysis using this system, Dissertation for Doctor of Science, Faculty of Science, Tokyo Metropolitan University (1997).
Yonezawa, C., Ruska, P.P., Matsue, H, Magara, M. and Adachi, T., *J. Radioanal. Nucl. Chem.*, **239**, 571 (1999a).
Yonezawa, C., Matsue, H., Adachi, T., Hoshi, M., Tachikawa, E., Povinec, P.P., Fowler, S.W. and Baxter, M.S., in *Marine Pollution*, IAEA-TECDOC-1094, p. 344 (1999b).
Yoshikawa, H., Yonezawa, C., Kurosawa, T., Hoshi, M. and Ishikawa, H., *J. Radioanal. Nucl. Chem.*, **215**, 95 (1997).
Xiao, Q.F., Chen, H., Sharov, V.A., Mildner, D.F.R., Downing, R.G., Gao, N. and Gibson, D.M., *Rev. Sci. Instrum.*, **65**, 3399 (1994).
Zeisler, R., Stone, S.F. and Sanders, R.W., *Anal. Chem.*, **60**, 2760 (1988).

3 Elemental analysis by laser ablation ICP-MS

T. E. JEFFRIES

3.1 Introduction

It may seem slightly curious to include a chapter on laser ablation inductively coupled plasma mass spectrometry (LA-ICP-MS) in a book primarily concerned with non-destructive elemental analysis. The laser ablation process is one which involves sample destruction (Fig. 3.1), albeit on a microscopic scale and local to the site of interest. Nevertheless, LA-ICP-MS has become established as a microbeam technique, complementing others such as ion microprobe analysis and electron microprobe analysis, so that it seems natural to include it here.

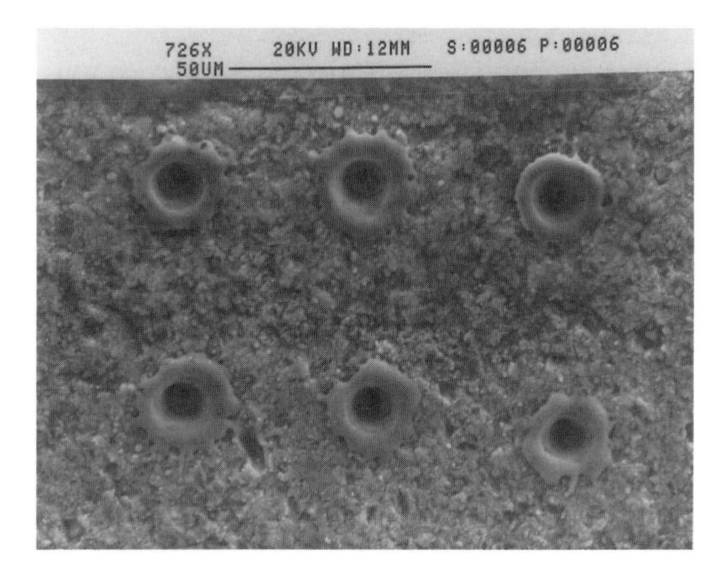

Fig. 3.1 Scanning electron microscope image of laser ablation craters in olivine. Note the rim of melted material surrounding each crater.

The intention of this chapter is to give a practical description of the LA-ICP-MS analytical technique, highlighting important instrumental characteristics and other factors affecting analysis. The text relies greatly on my own eight-year experience of the technique in two internationally renowned ICP-MS laboratories (Aberystwyth, Wales and Newfoundland, Canada) as well as more recently in my own laboratory at The Natural History Museum in London. Naturally, this chapter reflects my background as a geologist working mainly with earth science applications and developmental needs, although I have been fortunate to work with a large variety of natural and man-made materials from dinosaurs to wasps and from plastics to silicon wafers.

The chapter begins with a discussion of the background to the development of LA-ICP-MS and leads on to a description of the basic components found in a 'typical' instrument. A large part of the chapter is devoted to aspects of the analytical technique including sample preparation, instrument optimisation, data handling and factors which affect analysis. A short section on applications is given to highlight the potential of the technique across a wide range of scientific disciplines. Such is the diversity of LA-ICP-MS applications it is difficult to keep apace of recent developments. There is no intention here to provide an exhaustive literature review, but the reader is encouraged to look within and beyond the traditional literature of physics and chemistry.

By the end of the chapter it is hoped that the reader will be familiar with many of the factors to be considered prior to analysis, so that an informed choice can be made. To this end, a set of questions is posed to help the reader pull together the information presented and to facilitate further thought about the problems arising from their own applications.

3.2 Background

3.2.1 *Laser ablation*

To many, the mention of the acronym LASER (light amplification by stimulated emission of radiation) conjures to mind science-fiction movies and scenes of weapons able to wipe out cities with a single blast. For the chemist, who has many a recipe for mass destruction, the laser is seen simply as a tool to assist in the analysis of solids. Since laser action in ruby was discovered in the 1960s lasers have been put to a multitude of uses, from cleaning monuments to guided-weapons delivery – perhaps here the realisation of those science-fiction movies. Few in the developed world will have escaped the impact of lasers in, for example, music reproduction, supermarket bar code readers and in surgery.

The use of lasers to introduce solids by ablation to inductively coupled plasma mass spectrometry (LA-ICP-MS), is an area which has seen rapid development since its inception in the 1980s (Gray 1985; Arrowsmith 1987). These early studies employed ruby lasers which were found to be unstable and Nd:YAG (neodymium yttrium aluminium garnet) (1064 nm) infrared (IR) lasers. At first these provided only a bulk sampling device giving a resolution of > 100 μm (Perkins *et al.* 1991, 1993) and were unsuited to the micro-sampling required by many applications. Pearce *et al.* (1992a) and Jackson *et al.* (1992) modified the bulk sampling Nd:YAG laser using beam and energy attenuation to produce a laser beam of $\sim$40 μm. The attenuated IR laser proved to have a much wider application to micro-sampling (e.g. Jeffries *et al.* 1995a), but although it gave a much better resolution than previously, a number of other problems became apparent. Significant among them were poor ablation characteristics (Jeffries *et al.* 1995b) and laser-induced chemical fractionation (Chenery *et al.* 1992; Fryer *et al.* 1995; Jeffries *et al.* 1996; Longerich *et al.* 1996a; Figg *et al.* 1998).

Ablation of many materials by IR lasers resulted in a catastrophic ablation where a large fragment of the sample, too large to be transported to the plasma, was removed in a single shot. In addition, IR laser ablation led to chemical fractionation, which, during the course of an analysis, caused some elements apparently to increase or decrease their measured concentration relative to others. Consequently analytical precision was poor.

Because of the difficulties of the IR system a number of different lasers was investigated for use with ICP-MS. Shibata *et al.* (1993) investigated the use of a frequency-doubled Nd:YAG laser operating in the green at 532 nm, but as this did not offer any advantages over the infrared lasers it did not find popularity amongst users. It was felt that ultraviolet (UV) lasers would provide much better ablation characteristics and have reduced fractionation effects. Recently the UV excimer laser operating at 193 nm was introduced to ICP-MS analysis (Günther *et al.* 1997a; Günther & Heinrich 1999). This laser proved to have far superior ablation characteristics to the IR lasers and reduced fractionation effects and a few excimer lasers are now used very successfully. Despite the success, because the 193 nm excimer laser is expensive to purchase and run and also requires a nitrogen atmosphere as light at less than 200 nm is absorbed in air, they have not found widespread use. An alternative, relatively inexpensive ultraviolet laser source was found in the frequency-quadrupled Nd:YAG laser at 266 nm in the UV (Chenery & Cook 1993). It has been demonstrated that this laser has better ablation characteristics and the fractionation problem is reduced (Jeffries 1996). The 266 nm laser is now used extensively and has become the industry standard.

Although the 266 nm laser is the most extensively used laser in LA-ICP-MS, it does not match the superior ablation characteristics of the 193 nm excimer laser. For this reason the frequency-quintupled Nd:YAG laser operating at 213 nm has been introduced recently (Jeffries *et al.* 1998). This laser is still relatively inexpensive, has superior ablation characteristics to the 266 nm laser and, unlike the 193 nm excimer laser, can be run in air. It is too early to say whether this laser will find favour and become an industry standard.

3.2.2 *ICP-MS*

Just as lasers were being developed in the 1960s, so too the ICP was being developed as an excitation source in optical emission spectrometry (ICP-OES) (Greenfield *et al.* 1964; Wendt & Fassel 1965). Whilst ICP-OES had grown rapidly since the introduction of the first commercial instruments in the mid-1970s and proved to have a number of analytical advantages, including large dynamic range and high precision, it also proved to have a number of drawbacks, in particular the complexity of optical emission spectra. Thus the need arose for a detection system that was free from complex spectral interferences and this led to the investigation of mass spectrometers to detect ions from plasma sources.

Two major difficulties exist in coupling a plasma source to the high vacuum required in a mass spectrometric detection system: the high gas temperature ($\sim$7000 K) and pressure (10^5 Pa) of the ICP. Gray (1974) was the first to demonstrate that to couple a plasma source with a mass spectrometer was feasible by coupling a direct current plasma (DCP) to a quadrupole-based mass spectrometer. In this work the plasma was allowed to impinge on to an aperture in the tip of a cone mounted in the wall of a first vacuum stage (<0.1 Pa). Ions were focused into the second higher vacuum stage containing the quadrupole. Further studies were soon initiated using alternative plasma–mass spectrometer combinations, including microwave-induced plasmas (MIPs) and of course the inductively coupled plasmas (ICPs) of the ICP-OES systems.

The earlier DCP work provided the basis for the development of ICP-MS by Houk *et al.* (1980) and Date and Gray (1981), but in this work problems were encountered with the small cone aperture at the interface between the plasma and first vacuum stage. At the same time Douglas and French (1981) were having some success with a new sampling interface and larger aperture, in conjunction with a MIP. Shortly afterwards the new sampling interface was used in conjunction with an ICP and reported simultaneously by Date and Gray (1983) and by Douglas *et al.* (1983). By 1984, two companies had delivered the first commercial ICP-MS instruments.

Vast improvements in ICP-MS sensitivity have been made since those early days. Most manufacturers now use turbo-molecular pumps in place of diffusion pumps and the plasma–sample interface has undergone many a re-design. Instruments are now capable of delivering parts per quadrillion (ppq) (10^{-15}) detection limits for solution analysis.

In addition to quadrupole-based mass spectrometers, it is worth noting here that ICPs are used in conjunction with a number of other mass spectrometers. Several companies have introduced ICP magnetic sector mass spectrometers. These instruments operate at much greater mass resolution than quadrupole-based instruments and maintain low backgrounds. High resolution allows separation of interfering polyatomic ions from mono-atomic ions at the same nominal mass. The disadvantages are a reduced sensitivity in high-resolution mode and that it is still difficult to separate pairs of mono-atomic ions sharing the same mass. Furthermore, current magnet switching technology does not provide sufficient peak jumping speed over a large mass region for measurement of the transient unstable signals produced by laser ablation. A more recent introduction is the ICP multicollector magnetic sector mass spectrometer. Multicollector instruments allow simultaneous detection using eight or nine Faraday analogue detectors and provide impressive isotope ratio determinations. The community is beginning to use these instruments in conjunction with laser ablation (e.g. Halliday *et al.* 1998; Hirata *et al.* 1998; Hirata & Yamaguchi 1999) and it is expected that these will soon incorporate ion detectors which will aid in the measurement of transient signals.

ICP time-of-flight mass spectrometers have recently been introduced (e.g. Emteborg *et al.* 1999). These offer a greater resolution than quadrupole-based instruments and are able to cope with transient signals. The downside to these instruments is that they lack sensitivity and thus far there are few data to support their use in LA-ICP-MS.

3.3 Instrumentation

In order to appreciate the significance of LA-ICP-MS as a trace element analytical technique, it is necessary first to understand the basic operating principles of the ICP-

MS and laser ablation systems. There is now considerable diversity in both ICP-MS instrumentation and laser ablation systems in laboratories and the market-place. For the purposes of clarity, therefore, I shall confine my discussion to a description of an 'ideal' 266 nm laser ablation system (the most popular amongst users) and the major components of a basic quadrupole ICP-MS instrument. The reader must be mindful that no manufacturer could hope to provide an ideal laser for any given user, so that in-house modifications are common. It should also be remembered that the market for quadrupole ICP-MS instrumentation is highly competitive and each manufacturer will have his own slant on the basic components and may offer an array of additional features such as high-sensitivity interfaces, or cool plasmas and collision cells to reduce interferences.

3.3.1 *ICP-MS*

Figure 3.2 is a schematic diagram of a basic ICP-MS system, the operation of which may be summarised quite simply: a sample (solid or liquid) is dispersed into a stream of gas and injected into the core of an inductively coupled plasma. The sample is ionised in the plasma and is extracted through a small aperture in the tip of a cone, on which the plasma impinges, into a reduced-pressure region. Some of this extracted material passes through a further aperture into a region of high vacuum. Electrostatic ion lenses extract the positively charged ions and direct them to a quadrupole mass analyser supplied with d.c. and r.f. voltages. For a given d.c. and r.f. voltage, only ions of one mass/charge (m/z) value may be transmitted. An ion detector registers the transmitted ions. Each naturally occurring element has a unique pattern of near-integer m/z values corresponding to its (stable) isotopes, so identification of elements is straightforward. The number of ions registered at the detector will depend on the concentration of the element in the sample, and this provides a system of quantification.

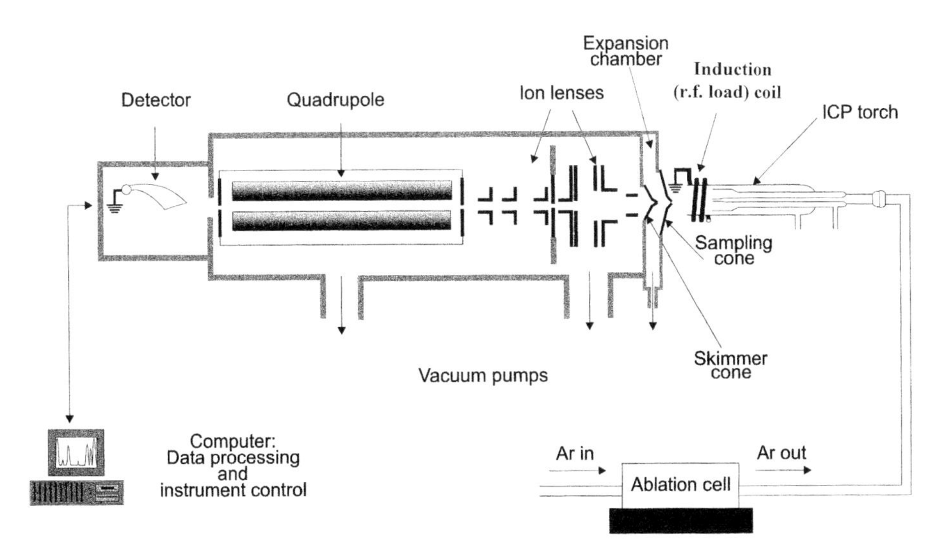

Fig. 3.2 Schematic of an inductively coupled plasma mass spectrometer. Drawing not to scale.

3.3.1.1 *The inductively coupled plasma (ICP)* An inductively coupled plasma is generated by coupling an intense radio frequency (r.f.) field to a suitable gas. In the case of ICP-MS, Ar gas at better than 99.996% purity is used. The ICP is propagated at the end of a plasma torch (Fig. 3.3) composed of three concentric fused silica tubes and positioned axially in a load coil (used to generate an intense r.f. field). To generate and sustain the plasma, Ar gas is introduced tangentially into the two outer tubes of the torch. An intense electromagnetic field is induced within the torch by supplying r.f. to the load coil. The Ar gas is seeded with free electrons by passing a high voltage spark along the inner wall of the outer tube. The free electrons accelerate in the electromagnetic field, rapidly reaching ionising energy. A plasma, which reaches $\sim$10 000 K in the core, arises when a significant proportion of the Ar atoms become ionised by non-elastic collisions with the electrons.

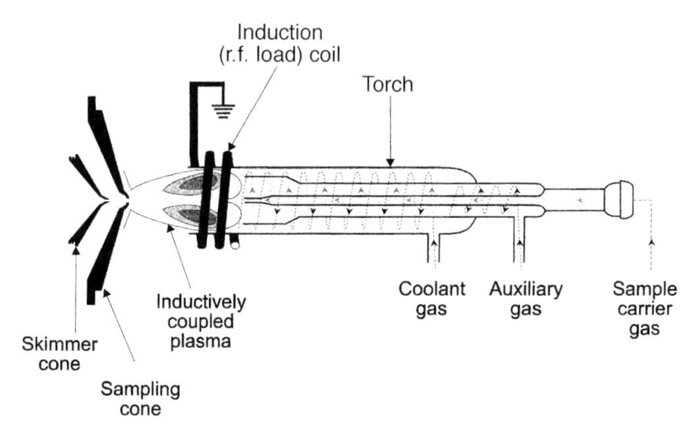

Fig. 3.3 Arrangement of torch, gas flows, r.f. induction coil, plasma and cones.

As well as sustaining the plasma, the Ar gas flowing in the outer tube of the torch carries away the heat dissipated by the plasma to the inside walls of the torch. For this reason, the gas flow in this tube (usually between 10 and 15 l min^{-1}) is referred to as the coolant gas or 'cool' gas. The gas flowing through the second tube is referred to as the auxiliary gas, 'aux' gas or sometimes, rather confusingly, the plasma gas. This gas flow is not essential to maintain the plasma, but it can be used to control the position of the plasma inside the torch, particularly to keep the plasma from the tip of the inner tube. In normal operation the auxiliary gas flow rate ranges from 0 to 1.5 l min^{-1}. The Ar gas flow to the innermost tube of the torch is introduced linearly and is used to carry the sample to the plasma. For this reason it is referred to as the sample carrier gas, or when solution analysis is implied, nebuliser gas. Flow rate for this gas depends on the sample introduction method and can have a large effect on oxide formation. It is usually in the range 0.7–1.5 l min^{-1}. The gas from this tube punches a hole through the centre of the plasma which, as a result, becomes torroidal. The temperature of the central channel at the mouth of the torch is $\sim$6000 K.

Samples may be introduced to the ICP as liquid droplets, particles or vapours in the stream of Ar gas flowing through the centre of the plasma torch. In the plasma the sample is vaporised, dissociated, atomised and ionised. For any given point in the plasma, an approximation of the extent of ionisation can be made even though the dynamic environment of the plasma is not at thermal equilibrium. Use is made of the Saha expression, which assumes local thermodynamic equilibrium and Maxwell–Boltzmann distributions (Boumans 1966):

$$\frac{n_i \; n_e}{n_a} = \frac{(2\pi m_e kT)^{3/2}}{h} \frac{2Z_i}{Z_a} e^{-E_i/kT} \quad (3.1)$$

where n_i is the ion concentration, n_e the free electron concentration, n_a the atom concentration, m_e the mass of an electron, k Boltzmann's constant, T the temperature in Kelvin, h Planck's constant, E_i the ionisation potential, Z_i the partition function of the ion, and Z_a the partition function of the atom. Saha-based percentage ion populations under typical plasma conditions for the naturally occurring elements can be found in the literature [e.g. Horlick *et al.* (1987); Houk (1986)].

As the electron population is buffered by the amount of Ar ionisation, it is governed by the Ar ionisation potential (15.8 eV) and the result is an efficient ($>70\%$) single ionisation of most naturally occurring elements. It follows that elements whose ionisation potentials are greater than Ar (e.g. F, Cl) are not efficiently ionised and those elements with a second ionisation potential less than the ionisation potential of Ar can be doubly ionised. In practice, even for the most susceptible elements, Ba and La, the doubly ionised species form only a few per cent of their ion populations in the central zone of the plasma.

3.3.1.2 *The plasma sampling interface* In ICP-MS, materials are sampled with an aperture ($\sim$1–1.2 mm diameter) in the centre of a cooled cone, referred to as the sampling cone (Fig. 3.3). This cone is positioned approximately 15–17 mm from the load coil and the large aperture punctures the plasma and allows for continuum sampling. [Because the mean free ion path ($\sim 10^{-6}$ m) in the plasma is much smaller than the diameter of the sampling aperture, a continuum flow is obtained, minimising oxide formation and ion–electron recombination.] The sampled plasma expands supersonically into a low vacuum area behind the cone referred to as the expansion chamber. Ions should reach the terminal velocity of Ar during expansion and their kinetic energies should be proportional to mass, in the absence of a plasma potential. Unfortunately a plasma potential can develop both capacitively and conductively (Houk & Thompson 1988) and because this imparts additional kinetic energy and a greater range of energies to the ions, it is undesirable. To mitigate this effect, the load coil is grounded.

A second cone, referred to as the skimmer cone, is positioned in the expansion chamber, approximately 5–10 mm behind the aperture of the sampling cone and is used to sub-sample the expanding plasma. The aperture of this cone is $\sim$0.8 mm in diameter and through this a small proportion of the plasma gas and ions is extracted into a high vacuum area containing the ion optics and quadrupole mass analyser. Molecular interferences can result from recombination of ions at the edge of the expanding plasma and in a turbulent downstream region that is impacted by shock waves (Olivares & Houk 1985; Gray 1989). To avoid this the sampling and skimmer cones are closely spaced. Most ICP-MS instruments have cones composed of high-purity Ni for general use, although Cu, Al, ceramic and solid Pt or Pttipped cones may also be used to assist with specific background and interference reduction.

3.3.1.3 *Ion focusing* The main function of the ion lenses is to separate the ions from the neutral species, which are pumped away, and transmit them to the quadrupole mass analyser. The ion lenses are housed in a region immediately behind the skimmer cone in a high vacuum usually maintained by turbo-molecular pumps, or in older instruments by oil vapour diffusion pumps. The lenses comprise cylindrical or disc-shaped electrodes to which a voltage is applied. The voltages applied to the lens array may be tuned to maximise or minimise the transmission of specific mass/charge values. It is usual to tune the lenses to maximise the transmission of an ion in the middle of the mass range of interest, whilst trying to minimise oxides or other molecular species.

One problem that can arise is the transmission of stray photons from the plasma through to the mass analyser and detector. Photons reaching the detector not only decrease its life, but are a source of background noise. The various instrument manufacturers each have a different method of minimising photon transmission and these include offsetting the ion lenses from a straight path or including a barrier or photon stop in the central channel and bending the ions around it.

3.3.1.4 *The quadrupole mass analyser* The quadrupole is a mass filter and consists of four molybdenum rods (typically 12 mm diameter, 230 mm long). These are held in a square array such that the inscribed radius r_0 produces a field between the rods which is a good approximation to the ideal hyperbolic quadrupole field (Fig. 3.4). A degree of approximation is necessary, as true hyperbolic surfaces would extend to infinity. The pair of rods positioned in the x plane are electrically connected and a (superimposed) positive d.c. voltage is applied. The other pair of rods positioned in

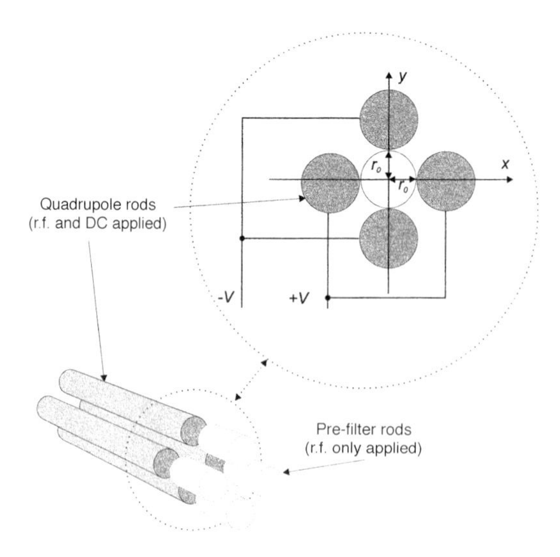

Fig. 3.4 Quadrupole rod geometry.

the *y* plane is also electrically connected, but a negative d.c. voltage is applied to this pair. Radio frequency voltages are applied to each pair of rods and for each pair the voltage applied is equal in amplitude, but 180° out of phase. [For an in-depth discussion of quadrupole technology, see Dawson (1976).]

Ions are directed from the ion lenses to the quadrupole where they interact with the applied fields and oscillate. For any given combination of r.f. and d.c. voltage applied to the rods, only ions of one particular m/z value will follow a stable trajectory all the way through the centre of the quadrupole; all others will spiral into the rods where they are neutralised. For analysis it is necessary to determine the required digital-to-analogue conversion setting of the quadrupole control system to set it at any given mass. This process is known as mass calibration and in practice it is performed using approximately ten points spanning the mass range.

Changing the m/z value transmitted by the quadrupole is achieved by controlling the d.c. and r.f. voltage applied to the rods and this allows data to be collected in several modes of operation.

Single ion monitoring. In this mode d.c. and r.f. voltages are fixed to transmit only the m/z value of the ion of interest. This mode of operation is useful for tuning the instrument and in this mode the lowest detection limits will be obtained.

Scanning or surveying mode. In this mode of operation a pre-determined mass range(s) is scanned rapidly (up to ~2500 amu s^{-1}) many times (typically 50–400). This is achieved when the d.c. and r.f. voltages are changed continuously under computer control. For LA-ICP-MS this mode of operation is particularly useful to ascertain the presence of analytes of interest in the sample before analysis proper.

Peak jumping or hopping mode. This mode is sometimes referred to as 'multiple ion detection'. In this mode several (up to 40 or so) m/z values are pre-selected and under computer control the d.c. and r.f. voltages are rapidly changed to allow transmission of each of these in turn. This is a rapid and versatile mode of operation, provides low detection limits and is especially useful where the volume of material is small.

Time-resolved analysis. In time-resolved analysis the quadrupole may be operated in any of the modes described above, but the data are stored in user-defined 'time slices' to give a real-time component to the data. In combination with the peak jumping mode, this is the preferred method of data collection during LA-ICP-MS analysis.

3.3.1.5 *Ion detection and signal handling* The different ICP-MS manufacturers employ a variety of ion detection systems in their instruments – single detectors; dual detectors combining pulse and analogue detection systems working in simultaneous or sequential modes. Most are based on the electron multiplier which consists of a flared, curved glass tube, approximately 1 mm i.d., 70 mm long, with a resistive coating of some 10^8 Ω. This is often mounted off-axis to reduce the chance of detecting stray photons. Positive ions leaving the quadrupole mass analyser are attracted to the mouth (flared end) of the tube which is held at high negative potential. The other end of the tube is grounded to provide a potential gradient along the length of the tube.

The electron multiplier is biased to operate in a highly sensitive pulse mode. A positive ion striking the flared end of the multiplier causes the ejection of one or more secondary electrons from the surface, which are accelerated down the tube. Subsequent collisions with the tube walls free further electrons and as the gain from this process is greater than unity, the electron population increases exponentially, reaching a multiplication factor of about 10^8. The space charge of electrons that builds up during this process is sufficient to inhibit further multiplication, and thus the output pulses tend to be grouped around a standard size of ~15 pC. The cloud of electrons is passed to an amplifier and then to a multichannel analyser.

A mass spectrum is produced by synchronising the accumulation of ion counts in discrete channels of the analyser. As the multiplier is operated in a pulse mode, individual ion arrivals are recorded and thus theoretical detection limits are determined only by the counting statistics of the random background counts. Background signals vary considerably from mass to mass, but at the heavy mass end of the spectrum are of the order of 10 Hz or less.

3.3.2 *Laser*

A variety of methods may be used for sample introduction to the ICP-MS including electrothermal vaporisation, slurry nebulisation, direct insertion, flow injection, ion chromatography, solution nebulisation, arc/spark ablation and laser ablation. By far the commonest method of sample introduction is the solution nebulisation method, but as with all other methods of sample introduction, excluding laser and arc/spark ablation, spatially resolved microanalysis is not achieved.

The laser ablation process, like that of ICP-MS, can be summarised simply. A pulsed laser beam is directed, focused and fired at a sample contained in an airtight cell, through which flows a suitable carrier or transport gas (normally Ar). Energy from the laser is transferred to the sample and, through various processes including disaggregation and thermal erosion from a laser-induced plasma, particles are removed, some of which are picked up in the flow of gas and transported to the inductively coupled plasma where they are ionised. Thus a basic laser ablation system consists of a laser, together with a means of controlling its energy output. In addition the system requires some optics for steering and focusing the beam, a sample cell, mounting and movement system, a sample transfer system and for convenience a means of viewing the sample and ablation event.

The frequency-quadrupled Nd: YAG laser ($\lambda = 266$ nm) is the most commonly used in LA-ICP-MS. Whilst the major components of any 266 nm Nd:YAG system have broadly similar function, there is a large variation in lasing-rod geometries, ancillary optics and in system layout (Fig. 3.5). Therefore, in this discussion, function, rather than detailed description of the components and layout must prevail.

3.3.2.1 *The Nd:YAG laser and flashlamp* The Nd:YAG laser consists of a rod of yttrium aluminium garnet ($Y_3Al_5O_{15}$) doped with approximately 3 wt% Nd_2O_3. It is held in a rod housing and a flashlamp (e.g. linear xenon flashlamp) is used as an optical pump (see below). The laser housing may also contain a filter to protect the lasing rod from UV damage arising from the flashlamp. This area of the laser is cooled, typically by a pumped water supply.

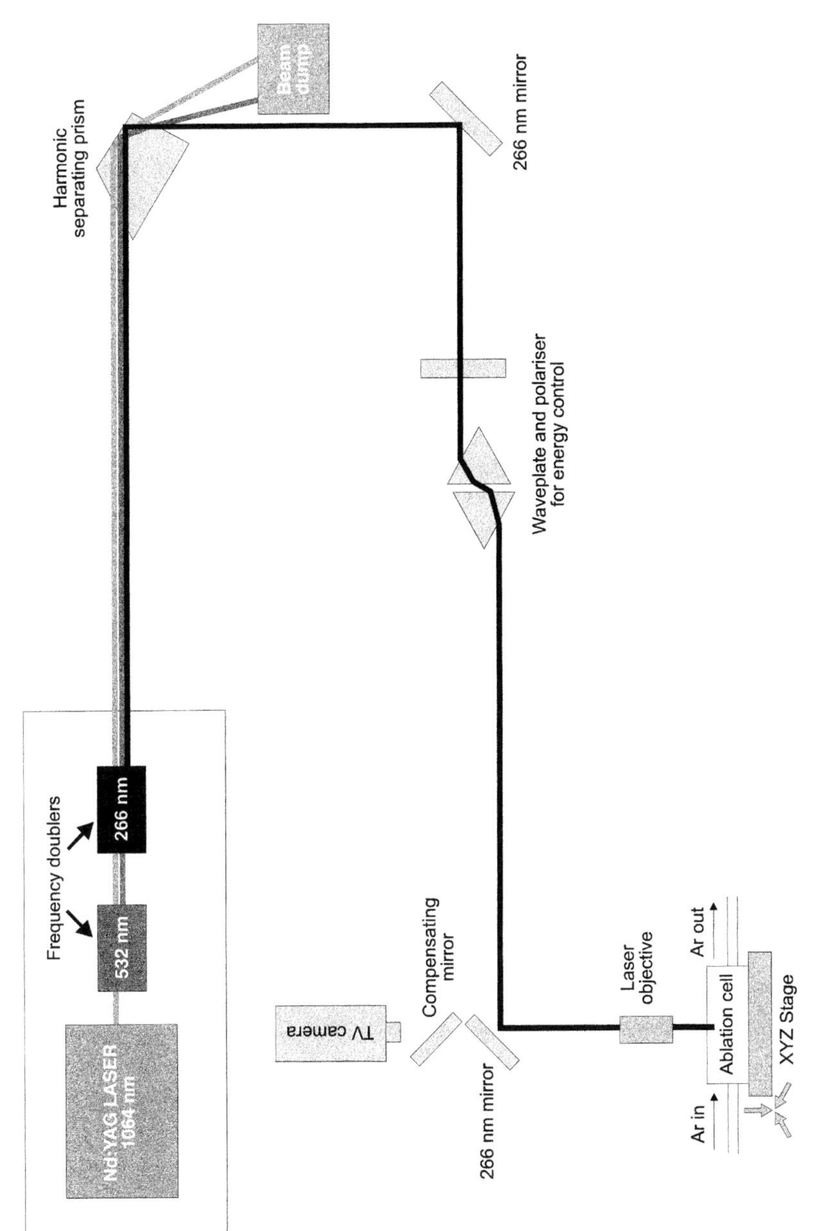

Fig. 3.5 Schematic of an 'ideal' frequency-quadrupled Nd:YAG laser ablation system.

The flashlamp radiation must be coupled into the lasing rod and this can be achieved using close-coupled ceramic reflectors. To ensure that the gas in the flashlamp is in conduction when it is triggered, it is usual to keep this constantly ionised by an applied a.c. voltage, often referred to as the simmer voltage.

3.3.2.2 *The laser cavity and lasing event* The laser cavity (Fig. 3.6) consists of the lasing rod and two mirrors in a light-tight resonant cavity. The mirrors are at either end of the rod and their separation (d) satisfies the following equation:

$$d = \frac{n\lambda}{2} \tag{3.2}$$

where λ is the characteristic wavelength and n is an integer. The first mirror, often referred to as the rear mirror, has 100% reflectivity whilst the second mirror, referred to as the optical coupler or front mirror, has a reduced reflectivity to the laser wavelength of 1064 nm.

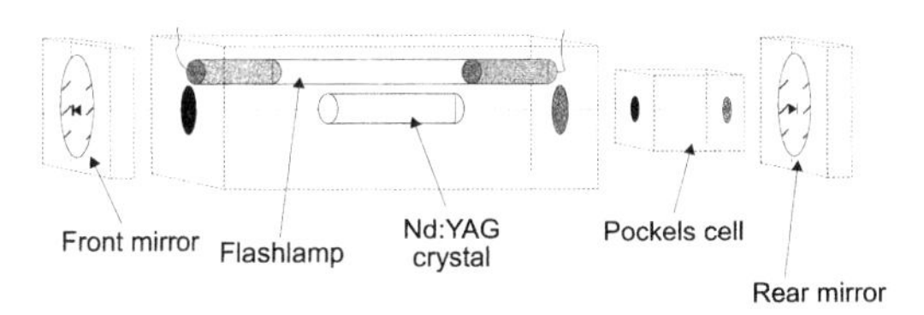

Fig. 3.6 Laser cavity showing arrangement of Nd:YAG crystal, flashlamp, optical switch (Pockels cell) and mirrors.

When the flashlamp is triggered it generates a short pulse of broadband electromagnetic radiation. This light travels through the lasing rod and specific wavelengths cause electronic transitions to occur in the Nd atoms of the rod, such that some of these assume an excited state (a process known as optical pumping) and the light is reflected back and forth by the two end mirrors. The first three excited states of Nd are shown schematically in Fig. 3.7. Optical pumping raises the atoms from the ground state E_0 to the pumping level E_3, and laser transition occurs between the metastable levels E_2 and E_1. During optical pumping an over-production of excited Nd is developed as a precursor to laser emission, a process referred to as population

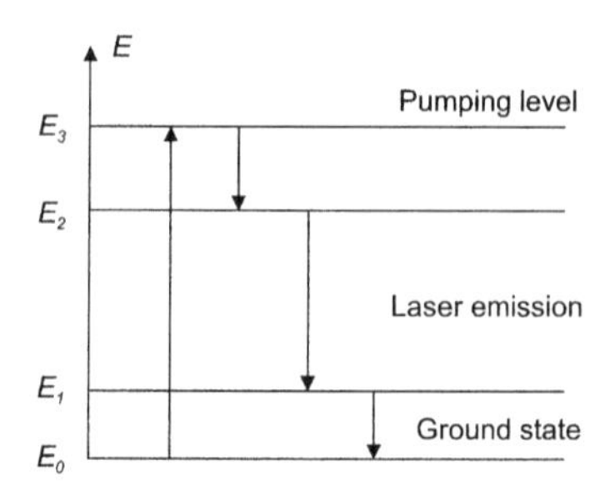

Fig. 3.7 Energy-level diagram for Nd.

inversion. The production of an over-population of excited Nd results in instability and fluorescence-induced photon production will trigger the emission of other photons. Light produced by such transitions is amplified as a result of the design of the resonant cavity. For the duration of the flashlamp discharge this process is repeated and a pulse length of some 150 μs, consisting of many spikes (referred to as relaxation oscillations), each a few microseconds in duration, is generated. A laser operated in this mode is referred to as free running or in fixed-Q operation ('Q' refers to the 'quality' of the resonator cavity).

In an alternative mode of laser operation, referred to as Q-switched, an optical switch, known as the Pockels cell or Q-switch is used to alter the quality of the laser cavity. Q-switching concentrates the output energy of the laser into giant pulses of a few nanoseconds duration. As this improves laser coupling to the sample and is necessary for frequency quadrupling, it is the preferred mode of operation. The Pockels cell comprises two polarisers and a compensating block. As the flashlamp is triggered a voltage is applied to the Pockels cell causing the two polarisers to be in opposition (planes of polarisation at 90°). This splits the polarised light into two different vibration directions and velocities and the effect is a destructive interference of the laser pulse, such that it does not reach the rear mirror of the laser cavity. As laser operation is suppressed, a larger over-population of excited-state Nd is formed. After a pre-set delay, the voltage is removed and one of the polarisers is rotated through 90°. This allows oscillation between the two mirrors and produces a single, giant pulse of a few nanoseconds in length.

3.3.2.3 *Harmonic generation (frequency quadrupling)* To produce an ultraviolet laser from the fundamental wavelength of a Nd:YAG laser at 1064 nm (infrared) the laser beam must go through a process known as harmonic generation. Nd:YAG lasers can be frequency doubled [λ = 532 nm (green)]; tripled [λ = 355 nm (ultraviolet)]; quadrupled [λ = 266 nm (ultraviolet)]; or even quintupled [λ = 213 nm (ultraviolet)] by propagation through suitable non-linear harmonic generating crystals. To generate a 266 nm frequency-quadrupled Nd:YAG laser beam first the primary or fundamental beam is passed through a frequency-doubling harmonic generator to output at 532 nm and then this second harmonic is passed through another frequency-doubling harmonic generator to give the fourth harmonic at 266 nm (Fig. 3.5). Various non-linear harmonic generating crystals may be used. Those in common use include potassium di-deuterium phosphate (KD*P) crystals and β-barium borate (BBO) crystals. Whichever crystal is used, the input beam must propagate through the crystal along a unique axis with respect to a crystallographic axis, to maximise the conversion efficiency of harmonic generation. This process is referred to as phase matching. Crystals are cut so that phase matching can be achieved by tuning the crystal angle with respect to the incoming beam. As this angle is dependent on the temperature of the crystal and heat transference occurs when the beam passes through the crystal, they are normally housed in temperature-stabilised ovens. Even so, the practice of 'warming' the crystals by passing the laser beam through them prior to laser ablation analysis, is common.

Harmonic generation conversion efficiency depends also on a number of fundamental properties of the input laser beam. Crystals are normally tailored to meet the beam characteristics of each laser, but conversion efficiency is seldom more than 5–10% for frequency-quadrupled lasers (i.e. maximum output energy at 266 nm will be 5–10% of the original 1064 nm input energy). Despite the poor efficiency of wavelength conversion, almost all frequency-quadrupled Nd:YAG laser ablation systems deliver far more than the 0.1–1 mJ (high-resolution sampling) or 1–4 mJ (bulk sampling) pulse energy required.

3.3.2.4 *Harmonic separation* Having created a laser beam containing the harmonic required, it is necessary to separate it from the other residual harmonics in the beam (the unwanted frequency-doubled beam and the remaining fundamental IR beam). Two methods are commonly used to achieve this. One approach is to use a harmonic separating prism (e.g. Pellin Broca prism) which, when placed in the beam path, causes an angle separation of the harmonics. The advantages of this system are that the resulting beam is harmonically pure and a prism can be used for a range of wavelengths. The disadvantage is that as the angle of separation is usually quite small, particularly between UV wavelengths, the separated harmonics have to travel over some distance before the required one can be 'picked off' using a mirror or similar. An alternative method employs dichroic mirrors which are coated to reflect only the wavelength of interest. As each mirror provides only a 95–99% harmonic purity, it is usual to have 3–4 of them to clean-up the beam sufficiently for laser ablation analysis. The advantage of the mirror system over the prism is that this process requires much less space. The disadvantage is that the resultant beam is never harmonically pure and as the mirrors are wavelength specific, a new set is required for each wavelength of interest.

3.3.2.5 *Energy control and focusing* The response of materials under laser ablation is extremely variable and dependent on the physical properties of the material and those of the incoming laser beam. To achieve a controlled ablation in any material, i.e. that which forms well-shaped craters at the resolution required, the laser energy must be adjusted appropriately. Arguably then, the most important feature of any laser ablation system is its means of attenuating and controlling laser pulse energy.

Consistent with the range of lasers used in laser ablation, a range of methods used to control laser energy is described in the literature. Some groups alter the Q-switch delay (above) and/or the voltage applied to the flashlamp (Jackson *et al.* 1992; Pearce *et al.* 1992a); others report the use of filters (Huang *et al.* 1993; Cromwell & Arrowsmith 1995a) or apertures (Pearce *et al.* 1992a). In my opinion, none of these methods is entirely appropriate. Filters and apertures are unlikely to provide the degree of flexibility required for many applications and, whilst making adjustments to the fundamental condition of the laser may be simple if slow, these adjustments also affect pulse duration, which has serious implications both for calibration and laser–sample interaction.

A more appropriate method of energy control makes use of the laser beam's natural linear polarisation together with standard polarising optics (Fryer *et al.* 1995). For example, a variable optical attenuator could consist of a rotatable half-wave plate and magnesium fluoride polariser. The waveplate, when turned, rotates the plane of polarisation of the laser beam entering the polariser, allowing more or less of the beam to be transmitted to the sample or directed to a beam dump. This sensitive technique does not alter beam quality, is simple to operate and is continuously variable. Manufacturers are now incorporating this and similar techniques into their new instrumentation.

An extremely useful addition to any laser ablation system is a means of measuring the laser energy such as a power meter or energy meter. These are essential when trying to standardise conditions from one sample to the next and they are invaluable aids to the diagnosis of misalignment or other beam problems.

More often than not the laser beam will require steering to fire vertically down on the sample. Beam steering is a simple process and is usually achieved with mirrors, steering prisms or combinations of these. In a laser ablation system using mirrors for harmonic separation (above), it is usual practice to combine beam steering with some last-stage harmonic separation.

The beam must also be focused onto the area of interest on the sample surface. For bulk sampling a simple plano-convex lens is all that is necessary, but for high-resolution work a higher powered reflecting or refracting objective is normally used. A variety of these objectives is available and choice will be determined by cost, damage threshold and visual image quality obtained. Briefly, whether refracting or reflecting, the objectives must be composed of UV transmitting materials (e.g. UV-grade synthetic fused silica). Good image and focus quality can be achieved with compound air-spaced refracting objectives and these are less prone to damage than those cemented with organic material. In general refracting objectives have a lower damage threshold than reflecting objectives. Dielectric-coated reflecting objectives do not have the chromatic aberrations of refracting objectives and so give excellent imaging quality. They also provide high magnification and long working distances. The disadvantages of these are their higher cost and the restriction of use to only one wavelength.

Whichever objective is used it should be remembered that ablation pits of theoretical minimum focal spot sizes will not be achieved because of thermal effects and objective aberrations.

3.3.2.6 *Sample viewing and movement* All laser ablation systems incorporate some means of sample viewing prior to and during ablation. Systems are usually set up so that viewing and ablation are co-focal. In the simplest system a monocular or binocular microscope head is used together with appropriate filters if the ablation is to be viewed. In addition, a range of higher and lower powered viewing only objectives may be added to the system to aid in sample site location. In more complex systems a microscope may be incorporated into the system (Jackson *et al.* 1992; Chenery & Cook 1993) to expand the viewing capabilities with the addition of different light sources. Many, more sophisticated systems, now use a TV camera for imaging with the output going to a TV monitor or computer screen. These systems may include zoom lenses and various light sources. Often they are almost entirely computer driven and incorporate software that can perform sample mapping, memorise site locations, save images as well as control ablation parameters.

In addition to a means of viewing the sample a means of moving it is required. This is almost always accomplished by an *X–Y–Z* stage, either a manual stage or a motorised computer-controlled stage.

3.3.2.7 *Ablation cell* The ablation cell fulfils the seemingly simple function of housing the sample during ablation, yet there is probably more variety in ablation cell design than any other component of laser ablation systems currently in use. Part of the reason for this is the change in use of laser ablation systems from bulk sampling to high-resolution work and the wide range of applications to which the instrument is put. With the decrease in volume of material removed from the sample during high-resolution micro-sampling and a desire for enhanced sensitivity, there is a greater need to maximise the amount of ablated material reaching the inductively coupled plasma. Thus, new cells have been designed to enable good gas flow dynamics and fast sample transfer. Small ablation cells with curved walls, flat sample mounting surfaces and high gas velocities would seem ideal. As quadrupole mass spectrometers are sequential instruments, the volume of the cell should be sufficient to allow gas dilution and mixing of the sample to give a steady sample input to the plasma. An ablation cell should also be of a sufficient size to hold both sample(s) and calibration materials together, because opening a cell mid-run disrupts the ICP and it may take some 40 minutes for re-stabilisation and for background signals to fall. Considering the factors above and the range of applications to which the cells are put, it is perhaps not surprising that no single design has been adopted by all. Despite this, it is fair to say that in my experience with ablation cells of widely different designs, all have given more or less similar results.

The other basic parts of an ablation cell include of course a means of entry and exit for the Ar or other sample carrier gas. On some systems gas flow into and/or out of the ablation cell is controlled by solenoid valves. This is to allow the continuation of Ar flow along the transport tubing (see below) during sample change-over and thus cause minimum disruption to the plasma. Additionally the ablation cell can be purged with Ar after sample change-over to prevent the plasma from being extinguished with atmospheric gases. The difficulties with a valve system are increased risk of leakage through the extra joints and the disruption they cause to the gas and particle flow. As the back-pressure from the sampler cone causes a back flow of Ar when the cell is open, and thus prevents entrainment of atmospheric gases, these valves are unnecessary. It is sufficient just to flush the cell with Ar by coupling the transport tubing loosely for a few moments before closing fully. Cells are normally designed to be airtight to prevent entrainment of atmospheric gases during ablation and loss of ablated material. Entrainment of atmospheric gases can increase normal background signals and compromise detection limits. Furthermore, if large enough it will extinguish the plasma.

The ablation cell also needs a window through which the laser beam can pass. For UV lasers typically this is composed of UV-grade synthetic fused silica. Whether the sample is viewed directly through a microscope or imaged by a video camera a light source will also be required. For many applications, particularly in the earth sciences, it is helpful to view the sample in transmitted light, so a window on the bottom surface of the ablation cell, through which light can pass from underneath, is useful. Other useful light sources include general illumination/incident light to show surface details and reflected light, again particularly useful in earth science applications.

It is worth noting here the development of a couple of specialist ablation cells. The first of these was developed in Australia and has since become known as the 'Australian cell'. It is used particularly in conjunction with isotope ratio measurements. The Australian cell differs from other cells in that only the sample moves relative to the point of ablation whilst the top window position is fixed. This is to enable a needle, through which the input gas flows, to be passed through the top window of the cell and fixed in a position such that the gas exits at high velocity very close to the point of ablation on the sample. The effect of this is to reduce laser-induced inter-element fractionation and thus improve isotope ratio precision. The second specialist cell is used to change the local temperature of the sample during ablation and is modelled on the heating–freezing stages used by geologists. The cell incorporates a special sample mount containing a heating coil and/or a sub-chamber of liquid nitrogen to give localised heating or cooling to the sample. Sample cooling has opened up applications in the life sciences where it is found that soft tissue can be frozen and then ablated. Cooling and heating have been employed in earth science applications, particularly for the analysis of fluid inclusions (Shepherd & Chenery 1995) where heating–freezing stages have been used for many years on standard optical microscopes. In future, it is possible that sample freezing may also assist in other applications such as the direct analysis of ice cores and the dust contained therein.

3.3.2.8 *Sample transport* Connecting a laser ablation system to an ICP-MS could not be simpler. Indeed, the simplicity of the arrangement is often difficult to grasp for the unfamiliar. All that is required is a length of suitable tubing connecting the ablation cell to the ICP-MS torch, through which the ablated material and sample carrier gas can flow. Transport efficiency is very difficult to assess accurately and suggestions range from 10 to 40% (Arrowsmith & Hughes 1988; Darke *et al.* 1989; Huang *et al.* 1993). It has been suggested that Cu tubing provides a higher transport efficiency but most groups use some form of plastic tubing because it is flexible and fairly inexpensive. The only other important tubing considerations are length and internal diameter. A small internal diameter has the advantage of minimising transport time by increasing gas velocity, with the expectation that this will reduce gravitational settling. On the other hand, if you wish to reduce the number of sample particles making contact with the transport tubing walls, a larger internal diameter is needed. Tubing lengths can vary greatly, but are typically between 1 and 2 m. Much shorter than this and a short, spiky, pulsing signal can arise from poor gas and sample mixing. At the other extreme, lengths up to 10 m have been used to smooth the signal (Broadhead *et al.* 1990; Broadhead 1991). These geometries have been the subject of a previous study (e.g. Moenke-Blankenburg *et al.* 1990) but in practice a compromise situation will give a reasonable sample transfer. My preference is for tubing of 6 mm outside diameter and 3 mm internal diameter and about 1.5 m length formed of polyurethane or PVC, but in my experience of several different plastics, there is little to choose between them. In my laboratory, sample tubing is regarded as disposable although some users clean and reuse their tubing. In either case, to avoid memory effects, tubing should be changed/cleaned on a fairly frequent basis, perhaps daily to weekly depending on the application.

3.4 Analytical technique

3.4.1 *Sample preparation*

Successful analysis begins with good sample preparation. For LA-ICP-MS, as with all highly sensitive analytical techniques, care must be exercised during sample preparation and handling to avoid contamination. One of the advantages of the LA-ICP-MS technique is that in fact very little sample preparation is really required except perhaps that of a fairly flat surface, but in most laboratories sample preparation usually goes beyond this. Laser ablation is capable of both bulk and micro-analysis, so sample preparation is likely to reflect the type of analysis being performed.

Although much less often used for this purpose now, laser ablation can be usefully used as a bulk sampling device. The advantages of bulk analysis of solids are clear: no need for lengthy digestion procedures, particularly where difficult-to-digest materials (e.g. chromites and zircons in geological samples) are present, and no concerns about those elements which are difficult to maintain in solution (e.g. Nb and Ta; see Fedorowich *et al.* 1993). The problem with laser ablation is that it is difficult to be sure that the sub-sample taken by the laser and analysed is representative of the whole. Often with geological materials rare minerals, termed accessory minerals, concentrate some elements preferentially (the nugget effect). Unless many analyses are to be performed and averaged, then the sample must be aggregated and homogenised in some way. This is usually done by crushing, pulverising and powdering to pass a fine mesh. A variety of sample preparation methods are then used on this bulked powdered material.

Pellets can be prepared by pressing the powdered sample in a die under high pressure, often having mixed it with a binding agent such a a plastic resin or alcohol (Broadhead *et al.* 1990; Denoyer 1991; Broadhead 1991; Mochizuki *et al.* 1991; van Heuzen & Morsink 1991; Durrant 1992; Magyar *et al.* 1992; Perkins *et al.* 1993; Cousin & Magyar 1994). Usefully, a spike can be added to the powder during the binding stage to act as an internal standard. The difficulty with this method has been the poor homogeneity observed. It has often proved necessary to raster the sample during ablation to even-out the signal.

Another commonly used approach involves fusing the powdered material (and spike) into a glass bead or disc, often with a flux such as those used in X-ray fluorescence (XRF) analysis (van Heuzen 1991; Perkins *et al.* 1993). In this case sample homogeneity is better than in pellets, but the disadvantages are that if a flux is used the sample becomes diluted, and the fluxes used (typically Li or Na tetraborate) leave a memory effect that is extremely difficult to remove. Fusing without a flux is also possible, but the higher temperatures required increase the risk of volatile loss.

LA-ICP-MS owes its origins to bulk analysis, but more exciting is its potential in microanalytical applications. For laser ablation micro-sampling, often the more important consideration is the preparation required for other analytical techniques that are to be applied to the sample beforehand. Typically this will include light microscopy and electron probe microanalysis. The data from the latter are frequently used for internal standardisation in LA-ICP-MS. Thus, polished samples such as thin sections and blocks are often preferred.

Polished blocks are prepared by placing the sample in the base of a mould and then covering it with a resin. Once the resin is cured the sample is exposed by polishing down the surface. It is convenient to use a mould size that will fit the sample holders of each technique applied, but samples can be made to almost any shape or size, limited only by the moulds available. The great advantage of polished blocks, particularly for calibration and reference materials, is that once the surface is covered with ablation pits it can simply be polished down and used afresh. With a little practice polished blocks can be prepared quite rapidly by most, although if they are to be used with electron probe microanalysis then a good polish will be needed.

Polished thin sections will require the skilled hands of a lapidary workshop technician for their preparation. Thin sections are frequently used in earth science applications. By viewing them in transmitted and reflected light a geologist can often identify the mineral phases present, their relationship to each other, the presence of zonation or inclusions and in some cases, the major element chemistry can also be deduced. Sections are a thin slice of material mounted, using a resin adhesive, onto a supporting glass slide. The surface is polished down mechanically and/or by hand to a known thickness. Traditionally for optical examination, rock and mineral thin sections are prepared to 30 μm thickness, but for the purposes of laser ablation it is preferable to have them thicker, up to 100 μm thick, to increase the time available for analysis before the section is drilled through by the laser. There is a huge variability in standard thin section dimensions from one country to the next. So, as with polished blocks, it is convenient to use a standardised thin section size that fits the holders of each instrument to be used.

During the analysis of thin sections, once the laser has penetrated through the sample it will start to ablate the adhesive and supporting glass slide, thus adding a contaminant to the analysis. Knowing when the laser has reached this point is important, but it is not always easy to determine. One method developed to assist here is the inclusion of an elemental spike in the resin adhesive (Longerich *et al.* 1996b). Typically, bismuth is used for this purpose. During the analysis the signal due to the spike element is monitored. Its appearance signifies the onset of ablation through to the resin and glass slide and the analysis can be stopped.

In my experience, accuracy is improved when samples and calibration standards are prepared and presented to the laser in the same manner. Preparation of blocks and thin sections allows for flat mounting in the ablation cell and polishing can improve sample viewing, so wherever possible I prefer that all samples for micro-analysis are prepared in this way. It should be remembered though, that prior to analysis polished samples should be cleaned ultrasonically to remove polishing materials from surface imperfections and if the samples have been used for electron probe microanalysis or electron microscopy the carbon coating must be removed also.

Of course, some samples presented for analysis cannot be prepared or mounted, perhaps because of their rarity or fragility or because a swift, qualitative determination is all that is required. In these circumstances the sample may be held, as is, in the ablation chamber, with the support of some pliable modelling or adhesive material. If the surface has been contaminated through handling or storage, it will be cleared in one or two laser shots and the first few moments of the analysis can be disregarded.

3.4.2 *Factors affecting analysis*

As one might expect the factors affecting LA-ICP-MS analysis fall into one of two categories: those arising from the ICP-MS and those arising from the laser.

3.4.2.1 *ICP-MS parameters* By contrast to liquid nebulisation, laser ablation generates a noisy, unstable, transient and, particularly in the case of heterogeneous samples, time-dependent signal. This signal requires a fast (or preferably simultaneous) acquisition and it is extremely important that data are recorded and can be displayed as a function of time. Observation of the signal using a real-time display during acquisition is an invaluable asset to any acquisition routine. The user can assess the signal produced from the laser, allowing the necessary changes to be made to the ablation parameters, or cease the acquisition when the laser has penetrated through the sample or starts to ablate a second phase.

Laser ablation data acquisition is a complex procedure, so it is unfortunate, but not surprising, that the software currently issued by instrument manufacturers lacks the intelligence to provide optimal operating and acquisition conditions whilst maintaining flexibility. There are several ICP-MS instrument parameters, each dependent on others, which affect the outcome of analysis. Indeed, the number of parameters is so large that users can easily run into difficulty. In addition to optimising the instrument through tuning (see below) it is necessary to define appropriate settings for quadrupole settling time, dwell time, points per peak, number of isotopes, time per sweep, sweeps per data point and number of data points. This situation is not helped by inappropriate default settings left by programmers or the lack of standard terminology amongst the different manufacturers for each of the parameters. It is true that sample introduction to ICP-MS is more often as liquids via a nebuliser than as solids introduced through laser ablation. Part of the difficulty with finding appropriate instrument settings is that the needs of these two introduction systems are vastly different.

The first parameter the user must define is the time allowed for the detector to stabilise following a jump from one selected mass to the next. This is referred to as the quadrupole settling time, but it should be noted that on many instruments quadrupole settling time may not be adjusted for individual jumps. Depending on the size of the jump the setting time required is usually between 0.1 and 10 ms, with 1 ms generally being sufficient. Ideally, quadrupole settling time should be kept to a minimum to allow stabilisation, otherwise it simply adds to overhead time. Normally, ICP-MS software acquires peak jumping data sequentially, taking one selected mass after the next in increasing order. Thus the largest jump the quadrupole is likely to make between naturally occurring masses is between ^{238}U and ^{6}Li, i.e. between the last mass selected to the first mass when starting another sweep across the range. In systems that do not allow individual settling times to be set for each jump, a settling time of 1 ms may not be sufficient for this. So rather than allowing the quadrupole to settle for longer on each mass just to accommodate this jump, it is usual to add an extra mass(es) (e.g. 5) in the sequence for which data are not required. The quadrupole may then settle (unstably) before making the next, short jump, to a mass of interest.

When dealing with transient data, such as those coming from laser ablation, the time spent measuring at each mass selected should be as short as possible, but long enough to maintain an acceptable count rate. This time is referred to as the dwell time. Laser ablation dwell times are usually set at between 8 and 10 ms so that with a 1 ms quadrupole settling time, approximately 90% of the time spent at each mass is spent in counting. This gives a happy compromise situation where dwell times are kept short and where if longer dwell times were used they would not significantly increase the proportion of the time spent counting. Some users recommend using dwell times which are multiples of the noise arising from the mains frequencies. So, in countries with 50 Hz mains, dwell times of 10 ms, 20 ms, etc. are used and for those countries which have 60 Hz mains (in particular North America), dwell times of 8.3 ms, 16.7 ms, etc. are thought better.

In laser ablation, data are normally acquired in peak jumping mode (above) and this process requires that an accurate mass calibration is maintained. The more points per peak measured, the less accurate the mass calibration needs to be, but when measuring a greater number of peaks, sensitivity is reduced and along with it, precision. For this reason it can be difficult to decide how many points per peak to measure. I take the view that in systems where the same quadrupole settling time is used for every jump even when it is less than a mass unit, the benefits of increased mean sensitivity and precision in using just one point per peak far outweigh the need to check mass calibration periodically.

The number of sweeps per analysis is defined by the number of sweeps that can be accommodated in the length of analysis required, given the dwell and settling times etc. that have been set. Often in addition to ablation time, this will include time to measure background signal prior to ablation. A time of 120 s is usually more than sufficient for measurement of both background and ablation signals. Some manufacturers offer software which can smooth the data by combining a number of sweeps together into a single data point before storing it. Although this may help detect low-intensity signals and reduce the size of data files, it may also result in the loss of short signals arising from discrete heterogeneities.

3.4.2.2 *Laser parameters* The interaction between laser beam and sample is poorly understood. Under different laser operating conditions, the ablation characteristics resulting in a given material can vary widely. Between different materials a similar diversity of ablation characteristics is found, so that it is very difficult either to predict the outcome of ablation in a given material or find the controlling influence. Despite this, it is agreed that the absorbance of the material at the laser wavelength is an important variable (Jackson *et al.* 1992; Geertson *et al.* 1994; Jeffries *et al.* 1995b) together with the energy and focus of the laser and the rate at which the laser fires per second (repetition rate).

Early in the development of the technique, laser wavelength was found to be an important factor affecting analysis. Indeed, it was largely a change in output wavelength from the IR to UV that allowed LA-ICP-MS to develop from a bulk sampling device to one capable of high-resolution micro-sampling (above). Most materials analysed are found to have poor absorbance in the IR. This means that there is a high transmission of the laser radiation and a greater chance that the energy is absorbed at an imperfection such as an inclusion, cleavage plane or in the case of thin sections, at the sample–glass-supporting-slide interface. This often results in an uncontrolled and catastrophic ablation, particularly for thin sections where laser radiation rapidly heats the adhesive resin and expansion of this causes a large sample fragment to flake off.

As mentioned above, to achieve a controlled ablation, particularly at high resolution, it is imperative that laser pulse energy and focus are controlled. Both laser energy and focus can have a large influence on the ablation pit size. A focused beam will give the smallest pit size, but even in strongly absorbing materials, too much energy will cause a catastrophic ablation. Defocusing the beam, or alternatively, spreading the beam with an expander or telescope, will increase the diameter of the pits and can increase the signal, but a greater energy may be required to achieve ablation. Increasing pit size in this way is particularly useful for bulk analysis.

For a given pulse energy and focus, the rate at which sample is removed and thus the peak signal-to-noise ratio (measured in counts per second or cps), is controlled by altering the laser pulse repetition rate (measured in Hz). It is usual to use the maximum repetition rate of the laser (for most systems 10–20 Hz) because quadrupole-based ICP-MS instruments are capable of very fast peak jumping and thus short sample ablation times (a few seconds) can provide good data. Of course if a long, stable signal is required such as when tuning the ICP-MS, or if a lower count rate is desirable then a slower repetition rate should be used.

3.4.3 *Laser-induced elemental fractionation*

Laser-induced elemental fractionation has long been recognised as a characteristic of laser ablation, yet it remains a poorly understood phenomenon. During laser ablation and transport, not all elements behave in the same way and fractionation between them can occur. Difficulties can arise with calibration of some elements (below) and determining values for isotopic ratios. Many factors appear to influence the degree of fractionation that occurs, including instrument parameters such as laser focus, pulse energy and width and wavelength, as well as ablation time, sample matrix, sample preparation and transport efficiency. It is especially severe for IR lasers and for high-resolution sampling at low energy (Cromwell and Arrowsmith 1995b). Fractionation behaviour has been characterised by, for example, Fryer *et al.* (1995), Jeffries *et al.* (1996) and Longerich *et al.* (1996a) for a large group of elements in different matrices. Depending on the element chosen to reference the data to, various patterns of behaviour have been observed and associated with different chemical and physical properties of the elements. For example, Jeffries *et al.* (1996) demonstrated that when referencing to silicon, fractionation could be explained on the basis of element melting temperature, ionic radius and field strength. The difficulty with this system is that silicon is in itself considered to fractionate strongly. Fryer *et al.* (1995) referenced their data to calcium, which is thought to exhibit less fractionation and is more often used as an internal standard. They found that the degree of fractionation divided the elements into a pattern similar to Goldschmidt's geochemical classification, i.e. lithophile, siderophile and chalcophile; the elements in each group behaving in a similar way.

Setting appropriate laser parameters can reduce fractionation. In addition, similar preparation of sample and reference material helps, as does the use of specialist ablation cells such as the 'Australian cell' (above). The most important consideration is that for an accurate calibration it is essential to use an internal standard whose behaviour is similar to the elements of interest, particularly when matrix and ablation conditions vary between sample and reference material.

3.4.4 *Instrumental optimisation and calibration*

Once laser and ICP-MS parameters have been set it remains a fairly simple procedure to acquire data, but before acquisition begins it is necessary first to optimise and calibrate the ICP-MS for the type of sample being analysed.

3.4.4.1 *Optimisation (tuning)* To those new to the technique, the process of optimising ICP-MS operating conditions for laser ablation can at first seem impossible. Certainly, in comparison to the steady signal obtained with solution optimisation, laser ablation poses a much greater problem. Without any fixed points of reference it can take 3–4 h to optimise an instrument that is completely de-tuned.

The process of optimisation is normally performed using the signal arising from ablation of a material with known concentrations of the elements of interest. Typically for the earth sciences the tuning medium used is a homogeneous glass such as NIST SRM 612 because its behaviour is similar to that of many silicate minerals. For other applications alternative tuning media may be preferred as optimal tuning conditions are, in part, dependent on matrix. Various strategies can be used to assist with tuning. For example, using higher than usual pulse energy (0.7–1.5 mJ) at a slow repetition rate (3–5 Hz) gives a reasonably stable signal in NIST SRM 612 for about 5 min after the first minute of ablation. Optimisation is an iterative process and the operator may go back to adjust various parameters again several times before being satisfied with the result. Knowing what to adjust is based partly on intuition, partly on 'knowing the instrument' and partly on understanding the effect of each parameter. As with many other techniques, the skill required to optimise an ICP-MS comes with a good deal of practice.

Instrument optimisation would become a simpler matter if the instrument could be optimised for solution introduction and then switched to laser work. Unfortunately the tuning conditions required by each of these two introduction techniques are vastly different, particularly with respect to lens settings, torch position and sample gas flow. Unless a mixed solution and laser ablation introduction method are being used Thompson *et al.* 1989; Moenke-Blankenburg *et al.* 1992; Chenery & Cook 1993; Masters & Sharp 1997; Günther *et al.* 1997b (and below), there is little point in taking this approach. Some instrument manufacturers have now built automatic tuning routines into their software. Again, whilst these work with some success for solution ICP-MS optimisation, typically these routines are iterative and do not cope well with the unstable transient signal generated by the laser ablation process.

3.4.4.2 *Calibration* Probably the greatest obstacle overcome in the development of LA-ICP-MS has been that of calibration. In the early days of the technique it was thought that finding a suitable calibration strategy for solids, whose ablation characteristics varied greatly, would be difficult, particularly in view of a lack of homogeneous, well-characterised trace element reference materials. In the end and to the surprise of everyone working on this, calibration was found to be quite straightforward for many materials.

Three basic methods of calibration are in current use for LA-ICP-MS. All require that the ablation signal of interest from the sample, with background subtracted, is referenced to the background-subtracted ablation signal from a standard reference material. Most users obtain background signals from an analysis of the sample carrier gas alone, either just prior to ablation of the sample or standard or as a separate analysis. Other sources of background can be used, especially with bulk analysis. For example, blank fusions (flux only) are made alongside rock fusions and their analyses are used as background. Clearly, in this case there is a need for a procedural blank. Similarly blank-pressed powder discs are made of 'pure' compounds (e.g. SiO_2 or $CaCO_3$) so that they have a similar matrix to the samples. These are analysed and the signal obtained is subtracted from the sample analysis. Whilst it is true that background signals during ablation of even the purest material will differ from those of a gas blank (e.g. because of plasma suppression), the difficulty is in ensuring that the signal from the blank pellet is not due to contaminants in the starting 'pure' compounds.

The simplest of the three calibration methods performed involves external calibration from a solid reference material (Gray 1985). Data are background-subtracted but otherwise untreated. The problem with this approach is that small fluctuations in laser energy can cause the ablation volume (yield) to be extremely variable from one sample to the next, even when there is little difference in their matrices (major element composition). To give any approximation to accuracy, samples and reference materials need to be closely matrix-matched. This technique, although seldom used now, does provide a useful and rapid sample reconnaissance without the need for internal standard analysis or addition.

Whilst external calibration alone is the simplest method of calibration it lacks accuracy. The most commonly used approach to calibration involves the use of internal standards, to give a much greater accuracy than external calibration alone. This technique corrects for differences in ablation yield between samples and reference materials as well as at least partially correcting for matrix effects and drift (Broadhead *et al.* 1990).

To take this approach to calibration of solid materials in their natural state (i.e. not homogenised), use must be made of a naturally occurring internal standard (Jackson *et al.* 1992; Pearce *et al.* 1992a). Typically, the internal standard used is a major component of the sample, whose concentration has been determined by an alternative technique such as electron microprobe analysis, or, for example in some minerals, is calculated from stoichiometry (Jeffries *et al.* 1995a). For many materials the major components are elements of comparatively low atomic mass relative to the elements of interest, yet matrix effects and drift are known to exhibit mass dependency. Fortunately, the majority of applications using major-component internal standardisation are of the high-resolution micro-sampling type, where the ablation yield is so small that these effects are negligible. Using internal standards and analytes of similar ablation behaviour, even in the absence of matrix-matched standards and samples, the accuracy obtained with this method of calibration is remarkable. The most widely used calibration materials for this purpose are NIST SRMs 612 and 610 (e.g. Jackson *et al.* 1992; Feng 1994; Jeffries *et al.* 1995a; Perkins *et al.* 1997; van der Putten *et al.* 1999), so much so that this series of standard reference materials has been the subject of much recent study (Pearce *et al.* 1997; Horn *et al.* 1997; Lahaye *et al.* 1997, and references therein).

Of course, in the case of bulk analysis the internal standards used can be either naturally occurring or added to the samples (Perkins *et al.* 1991, 1992, 1993; Pearce *et al.* 1992b). During bulk analysis a much greater load is placed on the ICP so that drift and matrix effects can be problematic (Broadhead *et al.* 1990). Thus, adding an internal standard of similar atomic mass to the analytes or in the middle of the mass range of interest is a commonly used approach (e.g. Jarvis & Williams 1993; Perkins *et al.* 1993). Again it is important to use an internal standard(s) whose ablation behaviour is similar to the analytes of interest.

In the final of the three basic calibration methods internal standardisation is essential, but in addition the approach requires introduction of reference and other materials using solution nebulisation along with the ablated material (e.g. Thompson *et al.* 1989; Moenke-Blankenburg *et al.* 1992; Chenery & Cook, 1993; Masters & Sharp, 1997). In this approach the sample gas flow from the ablation cell is mixed with aspirated solution from a nebuliser, prior to the ICP torch. During tuning and calibration a reference material is aspirated, but during sample ablation a blank solution is used to maintain similar plasma conditions. The introduction of reference materials in this way facilitates tuning and allows flexibility in the choice of reference material. The drawback of this technique is that by introducing water to an ICP the advantages of a 'dry' plasma (e.g. low interferences and polyatomic ion backgrounds) are lost, although some advantage can be regained by desolvating the wet material (Günther *et al.* 1997b). The greater problem with this approach is that the difference in the way reference materials and samples enter the ICP will worsen matrix effects, which in turn make correction of elemental fractionation extremely difficult. In addition, mixing the ablated volume with the gas stream coming from the nebuliser can dilute the sample and result in a loss of sensitivity. For these reasons and because highly accurate results can be obtained simply with solid reference materials, this approach to calibration has not been adopted widely.

3.4.5 *Data collection and handling*

3.4.5.1 *Data collection* The importance of collecting LA-ICP-MS data using a time-resolved technique cannot be overstated. Even for homogeneous samples it is often necessary to examine the data and exclude any heavily fractionated part. For heterogeneous samples such as zoned minerals or thin sections, time-resolved acquisition and examination is essential. To illustrate this point Fig. 3.8(a) is a time-resolved plot of the analysis of an opaque mineral in thin section. The section is 40 μm thick and mounted using resin spiked with bismuth. Ablation starts after approximately 50 s background signal collection. After 10 s ablation a second phase is encountered. This second phase (calcite) is underlying the opaque mineral but its presence could not be determined from visual examination [Fig. 3.8(b)]. After a

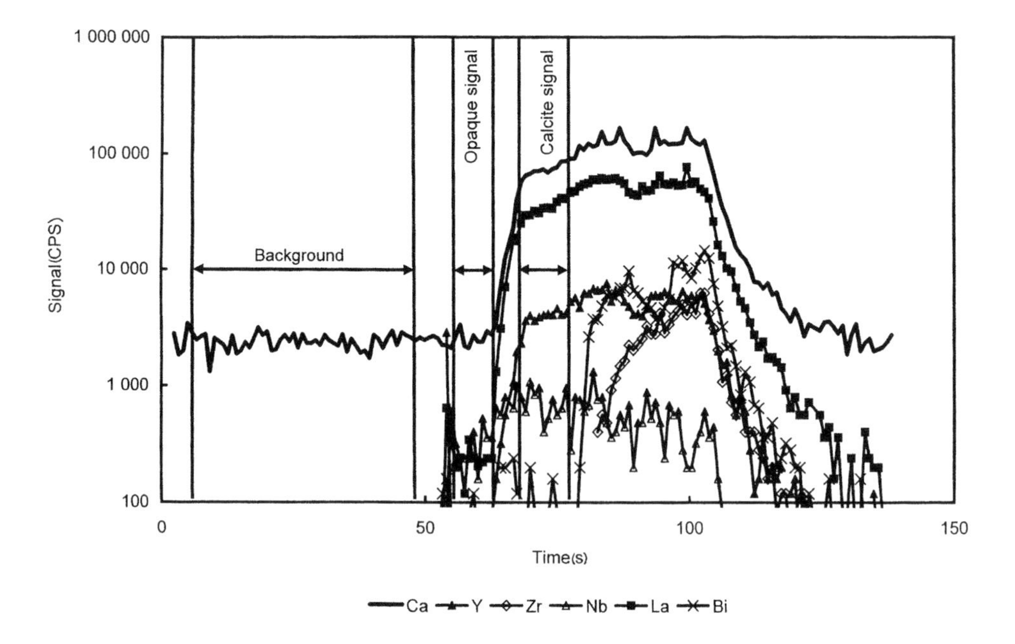

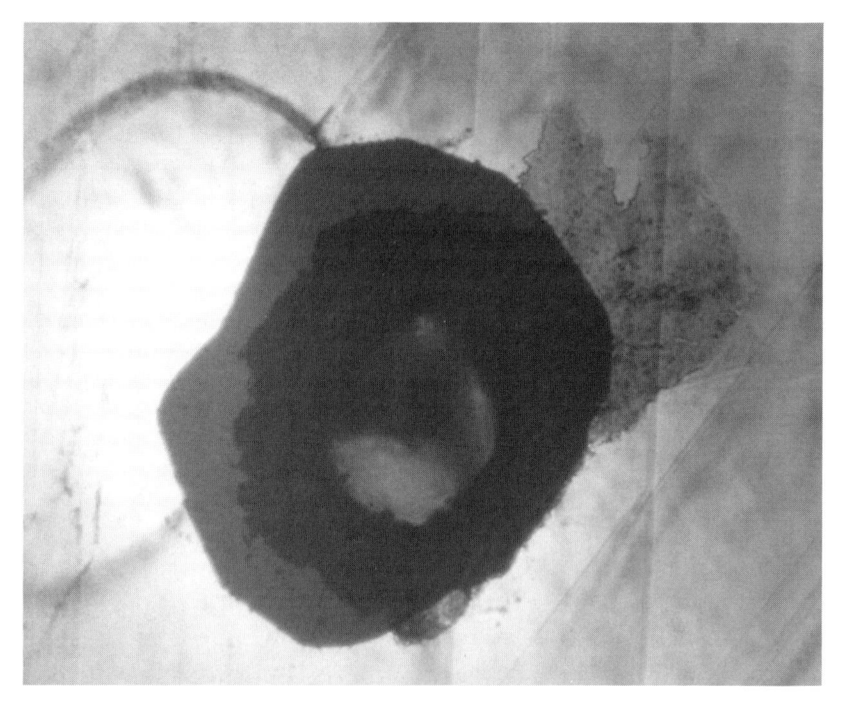

Fig. 3.8 (a) Time-resolved spectra obtained during the analysis of an opaque mineral in thin section. (b) Photomicrograph of the ablated opaque mineral and thin section from which data were obtained for (a). The rock is a carbonatite from the Igaliko dyke swarm, south-west Greenland. Note that there is an apatite crystal underlying the opaque mineral to the upper left-hand side of this image. The apatite was not ablated, but the calcite, which could not be seen, but in which the opaque was included, also underlays the mineral and was ablated late on in the analysis. Field of view 100 μm.

further 15 s, a rise in the signal for Bi, closely followed by a rise in the Zr signal (an element sometimes present in supporting glass slides) indicates that the section has been sampled right through. Note that the continued high signals from Ca, Ce and La indicate that the crater walls continue to be ablated long after the section has been penetrated. Clearly, with such a vast difference in chemistry between the opaque mineral and calcite, had the analyst relied on visual appearances and had the data been integrated together a large error in the calculated composition of the opaque mineral would have arisen.

Most instrument manufacturers now offer software with their systems which allows collection and in some cases interrogation of data using a time-resolved technique. It is unfortunate that they have been slow to develop this software, much of which remains cumbersome to use and is lacking in routines for exhaustive examination of signals and concentration calculation. Most users perform these calculations off-line using spreadsheets and in-house software.

3.4.5.2 *Data handling* Along with a table of concentrations it is usual for an analyst to quote a number of other statistics and figures of merit pertaining to the analysis (see Chapter 11). These statistics and figures are variable and depend on a number of instrumental, operating and elemental parameters. For example, precision and detection limits are affected by ablation yield and transfer efficiency, number of elements, their isotopic abundance and ionisation potential, counting time, instrument sensitivity for each element and background noise. Indeed, in LA-ICP-MS there are so many variables that quoting typical statistics and figures of merit is unreasonable. The array of laser settings used for different applications gives rise to an enormous range of ablation yields and the sensitivity of ICP-MS instruments varies with optimisation settings and between manufacturers. Nevertheless, the following, based on those commonly encountered in my experience are offered for guidance.

Detection limits. The detection limits of the system are probably the most often asked for figures, but are extremely difficult to predict. Detection limits vary by orders of magnitude from one set of instrumental parameters to the next, and one material to the next. They are largely a function of the amount of material ablated (or rather the amount ionised in the plasma, which subsequently gets through to the detector – in itself a function of tuning) and the amount of time the detector has to count each element in the ablated mass. Clearly, the more elements determined or the less material reaching the detection system, which normally equates to the less material ablated, the higher (worse) the detection limit. Users often have to reach a compromise by trading resolution and number of elements determined, to gain the detection limit required. Experience has shown that a small change in resolution can effect a large change in ablated volume and therefore detection limit. Indeed, it was once thought that vertical resolution was equal to lateral resolution and thus ablation pit volumes were a cubic function of pit diameter. It has since been shown (Jeffries 1996) that vertical resolution is extremely variable and depending on a number of laser parameters and the sample material, may extend from 0.5 to 10 times the lateral resolution. To complicate matters further, there are several methods to calculate detection limits reported in the literature, making it difficult and confusing to make formal comparisons. In general detection limits for elements most suited to ICP-MS (e.g. REEs) at low resolution (> 70 μm) are ppb to sub-ppb; at medium resolution (30–50 μm) tens of ppb; and at high resolution (< 15 μm) in the ppm range.

Precision. As with detection limit, precision is a function of many parameters, but largely a function of counting statistics. In the early days of LA-ICP-MS, the technique was often reported to have very poor precision, but as instruments have increased in sensitivity and since the introduction of UV lasers, precision has improved immensely. Precision can be improved still further by maintaining stable laboratory conditions (temperature and humidity) and during analysis by the use of appropriate internal standards. Under these conditions with modern instruments, precision is typically < 5% (relative standard deviation), with that reported from older instruments in various studies < 15% (e.g. Jackson *et al.* 1992; Chenery & Cook 1993; Jarvis & Williams 1993; Jeffries *et al.* 1995a).

Accuracy. LA-ICP-MS is a highly accurate technique. This is particularly the case for elements whose behaviour is properly matched to that of the internal standards, with or without matrix matching. Conversely, accuracy is compromised for those elements which fractionate relative to the internal standards.

3.5 Applications

3.5.1 *Literature reviews*

In recent years there has been an explosion in the application of LA-ICP-MS (Fig. 3.9). Indeed, it is often remarked that there are very few solids to which the technique could not be applied. Whilst it is beyond the scope of this chapter to provide a comprehensive review of applications to date, several excellent reviews have been published and the reader is encouraged to refer to these (Koppenaal 1992; Thiem *et al.* 1992; Moenke-Blankenburg 1993; Darke & Tyson 1993, 1994; Fryer *et al.* 1995; Ludden *et al.* 1995; Neal *et al.* 1995; Perkins *et al.* 1997; Günther *et al.* 1999). In addition, the reader is referred to the annual Atomic Spectrometry Update, published in *The Journal of Analytical Atomic Spectrometry* (various authors) which provides an up-to-date and comprehensive round-up of published work.

3.5.2 *Geological, earth and planetary sciences applications*

Much of the early driving force in the development of LA-ICP-MS came from the geological community and thus it is not surprising that applications in the geological and earth sciences are many and varied. Perhaps the commonest of these is to the analysis of minerals both to determine trace element concentration and zonation, but more especially to determine partitioning of elements between phases (e.g. Jackson *et al.* 1992; Pearce *et al.* 1992a; Chenery & Cook 1993; Jenner *et al.* 1993; Feng 1994; Günther *et al.* 1994; Chenery *et al.* 1995; Decarlo & Pruszkowski 1995; Jeffries *et al.* 1995a; Watling *et al.* 1995a; Perkins *et al.* 1997; Norman *et al.* 1998; Butler & Nesbitt

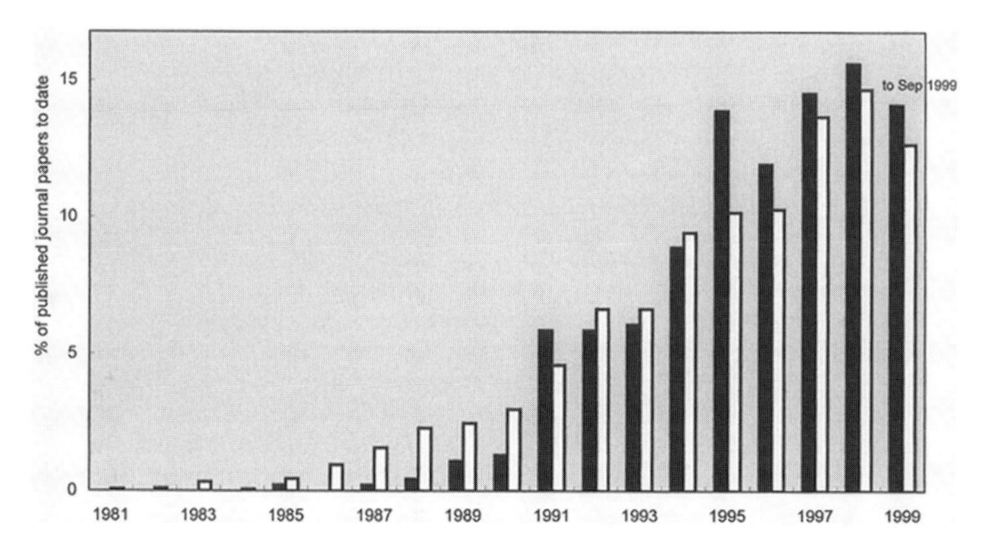

Fig. 3.9 Percentage of journal publications relating to ICP-MS and laser ablation ICP-MS published between 1981 and 1999. ■ LA-ICP-MS Papers: n = 461; □ All ICP-MS Papers: n = 3372.

1999; Dobosi & Jenner 1999; Ghaderi *et al.* 1999; Yang *et al.* 1999). With the introduction of UV lasers and their reduced fractionation, interest has grown in applications to isotopic ratio analysis, in particular U–Pb geochronological studies (e.g. Feng *et al.* 1993; Fryer *et al.* 1993; Hirata & Nesbitt 1995; Machado & Gauthier 1996; Poitrasson *et al.* 1996; Fernández Suárez *et al.* 1998, 1999; Liang *et al.* 1999). Applications development in earth and planetary sciences is ever increasing. New applications include analysis of meteorites (Hirata & Nesbitt 1997; Campbell & Humayun 1999), tephras and volcanic glasses (Westgate *et al.* 1994; Pearce *et al.* 1996, 1999), soils and coals (Durrant & Ward 1993; Guo & Lichte 1995; Lichte 1995; Querol & Chenery 1995), and fluid inclusions (Shepherd & Chenery 1995; Audetat *et al.* 1998; Günther *et al.* 1998; Shepherd *et al.* 1998; Heinrich *et al.* 1999; Loucks & Mavrogenes 1999).

3.5.3 *Life and environmental sciences applications*

With the recent upsurge of interest in environmental science and a move towards interdisciplinary research, the once clear boundary between earth and life sciences is now rather less distinct. This trend is reflected in the multitude of LA-ICP-MS applications which are generated by life and environmental scientists. Of particular interest has been trace element distribution in growth layers or structures, in anything from walrus teeth (Evans *et al.* 1995; Outridge & Evans 1995) and shells (Perkins *et al.* 1991; Imai 1992; Fuge *et al.* 1993; Imai & Shimokawa 1993; Wu & Hillairemarcel 1995; Schettler & Pearce 1996; Stecher *et al.* 1996; Price & Pearce, 1997; van der Putten *et al.* 1999), to tree rings (Watmough *et al.* 1997, 1998; Prohaska *et al.* 1998), and bones of all sorts including fish otoliths (ear bones) (Campana *et al.* 1994, 1997; Wang *et al.* 1994; Fowler *et al.* 1995; Blaber *et al.* 1996; Chenery *et al.* 1996; Gillanders & Kingsford 1996; Milton *et al.* 1997; Thorrold *et al.* 1997; Geffen *et al.* 1998; Milton & Chenery 1998; Secor & Zdanowicz 1998; Veinott *et al.* 1999). Application has also been made to the analysis of soft tissues (Wang *et al.* 1994) and most recently (and bizarrely) I have been approached to analyse wasp jaws!

3.5.4 *Archaeological and palaeontological sciences applications*

Although LA-ICP-MS is sample destructive, in high-resolution micro-sampling mode, the damage is almost invisible to the naked eye. Thus, even very old and precious palaeontological and archaeological material is frequently presented for analysis. Applications in these fields include the analysis of metal artefacts and jewellery, ancient marble, ceramics and glasses, normally with a view to determining origin (Ulens *et al.* 1994; Devos *et al.* 1999). In addition the analysis of teeth and bones is commonly required in both archaeological and palaeontological study to determine, for example, the onset of weaning and the environment during deposition (Cox *et al.* 1996; Lee *et al.* 1999).

3.5.5 *Forensic science applications*

The role of LA-ICP-MS in forensic science has been confined largely to finger-printing techniques. Here, the spectra obtained are compared to spectra from a database of analyses of the similar matrices and to determine similarities in the trace element associations. This technique has been applied to several materials including glasses, gold, cannabis and diamonds (Watling *et al.* 1994, 1995b, 1997; Walting 1998, 1999).

3.5.6 *Materials science applications*

LA-ICP-MS has been applied to a diverse array of materials including plastics, silicon wafers, metals, alloys and ceramics (Marshall *et al.* 1991; Kogan *et al.* 1994; Raith *et al.* 1995; Becker *et al.* 1996, 1997; Westheide *et al.* 1996; Baker *et al.* 1997; Gastel *et al.* 1997; Strubel *et al.* 1999). These applications are mainly to determine impurities in the materials and to check for homogeneity.

3.6 Continuing developments and final comments

3.6.1 *Continuing developments*

The LA-ICP-MS technique, though well established, is not free from problems. Its continuing development will include closer study of fundamental problems including laser-induced elemental fractionation and calibration issues such as matrix effects. It is not clear yet whether these issues will be resolved by employing different lasers (e.g. the excimer laser operating at 193 nm or the frequency-quintupled Nd:YAG laser operating at 213 nm), by improvements to the analytical technique (e.g. use of specialist ablation cells), by arithmetical means or a combination of these. Work will also continue in coupling laser ablation systems to the newly established multicollector ICP-MS and ICP-TOF-MS instruments and with it the development of new applications, particularly in the field of isotope ratio measurements.

3.6.2 *Final comments*

LA-ICP-MS has undergone rapid development since its introduction in 1985 and it is now established as a powerful trace element microprobe tool. The advantages of the technique are clear: its speed, flexibility, moderate cost, sensitivity, and diversity of application. As with all analytical techniques, LA-ICP-MS alone cannot satisfy the needs of every application, but its introduction has served to expand the arsenal of the chemist and open the door to an increasingly diverse range of scientific interest.

3.7 Questions and answers

The following problems and questions typify those with which operators are commonly concerned. There is no one 'correct' solution to the problems, but they are given here with the intention of provoking thought about your own applications.

3.7.1 *Questions and problems*

(1) You are asked to determine a range of trace elements at tens of ppm to tens of ppb levels in the mineral baddeleyite (matrix is ZrO_2). You know the concentration of Zr but at $>70\%$, analysis of even the minor isotopes causes the ICP-MS detector to trip when using the laser parameters required to obtain the detection limits asked of the analytes of interest. What can you do to obtain the data required?

(2) Calculate the volume and mass of material ablated at 25 μm resolution in a material with a density of 2.5 g cm^{-3}. Assume the ablation crater is cylindrical and its depth is equal to is diameter. Now, given an ablation rate of 10 Hz and time of 20 s, what is the mass and volume of material removed per shot? Apart from the assumptions you have been asked to make (which are less than reasonable), what else is wrong with this sort of calculation?

(3) Trace element 'fingerprinting' by LA-ICP-MS has been applied to provenance a range of materials from gold to cannabis (Watling *et al.* 1994, 1995b; Watling 1998, 1999). Calculation of analyte concentration is seldom performed and instead the user relies on element associations and viewing the spectra by eye to determine differences. What are the problems and pitfalls in this approach and what, in addition, might be done to improve the technique?

3.7.2 *Suggested answers*

(1) There are several strategies you might try to obtain these data, but it is a difficult problem likely to be resolved only by using a combination of them.

They include de-tuning the ICP-MS for Zr and reducing dwell times for this element. Measurement of minor elements (100–500 ppm) in the sample might be achieved by using a very small beam diameter and Zr as the internal standard. The sample could then be re-ablated at a larger beam diameter for determination of the trace elements, using the minor elements previously determined as internal standards. If a matrix-matched calibration standard is available then the external calibration method (not using an internal standard) could be used to obtain some preliminary results. Alternatively, you might consider using a polyatomic (interferent) as the internal standard [e.g. ZrAr (130 amu), but beware the presence of Xe in the Ar sample carrier gas].

(2) The volume of material removed is $\sim 1 \times 10^{-8}$ cm^3 and mass $\sim$30 ng. If this is achieved in 20 s with an ablation rate of 10 Hz, the volume of material removed with each shot is $\sim 6 \times 10^{-11}$ cm^3 and mass $\sim$150 pg. Apart from those already mentioned, the difficulty with this sort of calculation is the false assumption that the same amount of material is removed from a sample with each shot. There are several reasons why this is not the case. Firstly, we are assuming absolutely constant shot-to-shot laser energy output, which is never the case. We are also falsely assuming that the ablated material is all removed from the forming crater and that none of it is re-deposited inside. We are forgetting that as material is removed from the sample the point of focus changes and along with it the energy density of the laser beam. We are ignoring the effects of previous shots on the sample including deposition of material, which might be more easily ablated by the next shot, changing sample conditions such as heating and melt formation and micro-fracturing around the forming crater. Lastly, we are not taking into account the effect of the cloud of material formed during ablation on the incoming beam during subsequent shots.

It is worth noting that because we are dealing with resolution in three dimensions, a small change to the laser beam diameter will have a much larger effect on volume and mass removed. For example, doubling the beam to 50 μm diameter will give us a volume of almost 1×10^{-7} cm^3 and mass of $\sim$250 ng, nearly an order of magnitude greater.

(3) (I should make it clear that the 'fingerprinting' LA-ICP-MS technique has proved to be enormously successful in the applications to which it has been put. Nevertheless, those new to the method should be aware of the problems and pitfalls associated with it.) ICP-MS instruments and lasers can drift in sensitivity and energy. Sensitivity drifts are often non-linear over the mass range. Thus the danger in interpreting differences between one spectrum and the next by eye is that the difference may be due merely to instrument drift. A wise precaution would be to obtain a reference spectrum from a homogeneous material at regular intervals. The human eye is good at picking out gross differences, but subtle changes may be missed or overshadowed, so that the full picture is not seen. Care must be taken when defining differences on the basis of element associations because natural materials can be extremely inhomogeneous. Differences in determinations that are interpreted as arising

from two different samples may simply be a reflection of the natural variation in a single sample. Fingerprinting requires that many variables are considered simultaneously and this is very difficult to achieve meaningfully by eye or hand. The technique might benefit from the use of artificially intelligent software, such as a neural network which can 'train' on a database set and 'learn' as more samples are interpreted.

References

Arrowsmith, P. (1987) *Anal. Chem.*, **59**, 1437–1444.
Arrowsmith, P. & Hughes, S.K. (1988) *Appl. Spectrosc.*, **42**, 1231.
Audetat, A., Günther, D. & Heinrich, C.A. (1998) *Science*, **279**, 2091–2094.
Baker, S.A., Dellavecchia, M.J., Smith, B.W. & Winefordner, J.D. (1997) *Anal. Chim. Acta*, **355**, 113–119.
Becker, J.S., Breuer, U., Westheide, J., Saprykin, A.I., Holzbrecher, H., Nickel, H. & Dietze, H.J. (1996) *Fresenius J. Anal. Chem.*, **355**, 626–632.
Becker, J.S., Westheide, J., Saprykin, A.I., Holzbrecher, H., Breuer, U. & Dietze, H.J. (1997) *Mikrochim. Acta*, **125**, 153–160.
Blaber, S.J.M., Milton, D.A., Pang, J., Wong, P., Boon Teck, O., Nyigo, L. & Lubim, D. (1996) *Env. Biol. Fishes*, **46**, 225–242.
Boumans, P.W.J.M. (1996) *Theory of Spectrochemical Excitation*, Plenum, New York, pp. 156–232.
Broadhead, M. (1991) *Atom. Spectrosc.*, **12**, 45–47.
Broadhead, M., Broadhead, R. & Hager, J.W. (1990) *Atom. Spectrosc.*, **11**, 205.
Butler, I.B. & Nesbitt, R.W. (1999) *Earth Planet. Sci. Lett.*, **167**, 335–345.
Campana, S.E., Fowler, A.J. & Jones, C.M. (1994) *Can. J. Fisheries Aquatic Sci.*, **51**, 1942–1950.
Campana, S.E., Thorrold, S.R., Jones, C.M., Günther, D., Tubrett, M. & Longerich, H. (1997) *Canad. J. Fisheries Aquatic Sci.*, **54**, 2068–2079.
Campbell, A.J. & Humayun, M. (1999) *Anal. Chem.*, **71**, 939–946.
Chenery, S. & Cook, J.M. (1993) *J. Anal. Atom. Spectrom.*, **8**, 299–303.
Chenery, S., Cook, J.M., Styles, M. & Cameron, E.M. (1995) *Chem. Geol.*, **124**, 55–65.
Chenery, S., Hunt, A. & Thompson, M. (1992) *J. Anal. Atom. Spectrom.*, **7**, 647–652.
Chenery, S., Williams, T., Elliott, T.A., Forey, P.L. & Werdelin, L. (1996) *Mikrochim. Acta*, **S13**, 259–269.
Cousin, H. & Magyar, B. (1994) *Mikrochim. Acta*, **113**, 313.
Cox, A., Keenan, F., Cooke, M. & Appleton, J. (1996) *Fresenius' J. Anal. Chem.*, **354**, 254–258.
Cromwell, E.F. & Arrowsmith, P. (1995a) *Anal. Chem.*, **67**, 131–138.
Cromwell, E.F. & Arrowsmith, P. (1995b) *Appl. Spectrosc.*, **49**, 1652–1660.
Darke, S.A., Long, S.E., Pickford, C.J. & Tyson, J.F. (1989) *J. Anal. Atom. Spectrom.*, **4**, 715–719.
Darke, S.A. & Tyson, J.F. (1993) *J. Anal. Atom. Spectrom.*, **8**, 145–209.
Darke, S.A. & Tyson, J.F. (1994) *Microchem. J.*, **50**, 310–336.
Date, A.R. & Gray, A.L. (1981) *Analyst*, **106**, 1255–1267.
Date, A.R. & Gray, A.L. (1983) *Spectrochim. Acta B*, **38**, 29–37.
Dawson, P. (1976) *Quadrupole Mass Spectrometry and its Applications*, Elsevier, Amsterdam, p. 349.
Decarlo, E.H. & Pruszkowski, E. (1995) *Atom. Spectrosc.*, **16**, 65–72.
Denoyer, E.R. (1991) *Applications of Plasma Source Mass Spectrometry*, Royal Society of Chemistry, London, pp. 199–208.
Devos, W., Moor, C. & Lienemann, P. (1999) *J. Anal. Atom. Spectrom.*, **14**, 621–626.
Dobosi, G. & Jenner, G.A. (1999) *Lithos*, **46**, 731–749.
Douglas, D.J. & French, J.B. (1981) *Anal. Chem.*, **53**, 37–41.
Douglas, D.J., Quan, E.S.K. & Smith, R.G. (1983) *Spectrochim. Acta B*, **38**, 39–48.
Durrant, S.F. (1992) *Analyst*, **117**, 1585–1592.
Durrant, S.F. & Ward, N.I. (1993) *Fresenius' J. Anal. Chem.*, **345**, 512–517.
Emteborg, H., Tian, X.D. & Adams, F.C. (1999) *J. Anal. Atom. Spectrom.*, **14**, 1567–1572.
Evans, R.D., Richner, P. & Outridge, P.M. (1995) *Arch. Environ. Contamination Toxicol.*, **28**, 55–60.
Fedorowich, J.S., Richards, J.P., Jain, J.C., Kerrich, R. & Fan, J. (1993) *Chem. Geol.*, **106**, 229–249.
Feng, R. (1994) *Geochim. Cosmochim. Acta*, **58**, 1615–1623.
Feng, R., Machado, N. & Ludden, J. (1993) *Geochim. Cosmochim. Acta*, **57**, 3479–3486.
Fernández Suárez, J., Gutierrez Alonso, G., Jenner, G.A. & Jackson, S.E. (1998) *Can. J. Earth Sci.*, **35**, 1439–1453.

Fernández Suárez, J., Gutierrez Alonso, G., Jenner, G.A. & Tubrett, M.N. (1999) *J. Geol. Soc. Lond.*, **156**, 1065–1068.
Figg, D.J., Cross, J.B. & Brink, C. (1998) *Appl. Surface Sci.*, **129**, 287–291.
Fowler, A.J., Campana, S.E., Jones, C.M. & Thorrold, S.R. (1995) *Can. J. Fisheries Aquatic Sci.*, **52**, 1431–1441.
Fryer, B.J., Jackson, S.E. & Longerich, H.P. (1993) *Chem. Geol.*, **109**, 1–8.
Fryer, B.J., Jackson, S.E. & Longerich, H.P. (1995) *Can. Mineral.*, **33**, 303–312.
Fuge, R., Palmer, T.J., Pearce, N.J.G. & Perkins, W.T. (1993) *Appl. Geochem.*, **2**, 111–116.
Gastel, M., Becker, J.S., Kuppers, G. & Dietze, H.J. (1997) *Spectrochim. Acta B*, **52**, 2051–2059.
Geertsen, C., Briand, A., Chartier, F., Lacour, J.-L., Mauchien, P. & Sjöström, S. (1994) *J. Anal. Atom. Spectrom.*, **9**, 17–22.
Geffen, A.J., Pearce, N.J.G. & Perkins, W.T. (1998) *Marine Ecology Progress Series*, **165**, 235–245.
Ghaderi, M., Palin, J.M., Campbell, I.H. & Sylvester, P.J. (1999) *Econ. Geol. Bull. Soc. Econ. Geol.*, **94**, 423–437.
Gillanders, B.M. & Kingsford, M.J. (1996) *Marine Ecology Progress Series*, **141**, 13–20.
Gray, A.L. (1974) *Proc. Soc. Anal. Chem.*, **11**, 182–183.
Gray, A.L. (1985) *Analyst*, **110**, 551–556.
Gray, A.L. (1989) *J. Anal. Atom. Spectrom.*, **4**, 371–373.
Greenfield, S., Jones, I.Ll. & Berry, C.T. (1964) *Analyst*, **89**, 713–720.
Günther, D., Audetat, A., Frischknecht, R. & Heinrich, C.A. (1998) *J. Anal. Atom. Spectrom.*, **13**, 263–270.
Günther, D., Cousin, H., Magyar, B. & Leopold, I. (1997b) *J. Anal. Atom. Spectrom.*, **12**, 165–170.
Günther, D., Forsythe, L.M., Jackson, S.E. & Longerich, H.P. (1994) *Eos Trans. Am. Geophys. Union, 1994 Fall Meeting*, **75** (Suppl.).
Günther, D., Frischknecht, R., Heinrich, C.A. & Kahlert, H.J. (1997a) *J. Anal. Atom. Spectrom.*, **12**, 939–944.
Günther, D. & Heinrich, C.A. (1999) *J Anal. Atom. Spectrom.*, **14**, 1369–1374.
Günther, D., Jackson, S.E. & Longerich, H.P. (1999) *Spectrochim. Acta B.*, **54**, 381–409.
Guo, X. & Lichte, F.E. (1995) *Analyst*, **120**, 2707–2711.
Halliday, A.N., Lee, D.C., Christensen, J.N., Rehkamper, M., Yi, W., Luo, X. Z., Hall, C.M., Ballentine, C.J., Pettke, T. & Stirling, C. (1998) *Geochim. Cosmochim. Acta*, **62**, 919–940.
Heinrich, C.A., Günther, D., Audetat, A., Ulrich, T. & Frischknecht, R. (1999) *Geology*, **27**, 755–758.
van Heuzen, A.A. (1991) *Spectrochim. Acta B.*, **46**, 1803–1817.
van Heuzen, A.A. & Morsink, J.B.W. (1991) *Spectrochim. Acta B.*, **46**, 1819–1828.
Hirata, T., Hattori, M. & Tanaka, T. (1998) *Chem. Geol.*, **144**, 269–280.
Hirata, T. & Nesbitt, R.W. (1995) *Geochim. Cosmochim. Acta*, **59**, 2491–2500.
Hirata, T. & Nesbitt, R.W. (1997) *Earth Planet. Sci. Lett.*, **147**, 11–24.
Hirata, T. & Yamaguchi, T. (1999) *J. Anal. Atom. Spectrom.*, **14**, 1455–1459.
Horlick, G., Tan, S.H., Vaughan, M.A., and Shao, Y. (1987) 'Inductively coupled plasma mass spectrometry'. In: Montaser, A. and Golightly, D.W. (Eds) *Inductively Coupled Plasmas in Analytical Atomic Spectroscopy*, VCH, New York, 361–398.
Horn, I., Hinton, R.W., Jackson, S.E. & Longerich, H.P. (1997) *Geostands. Newslett.*, **21**, 191–203.
Houk, R.S. (1986) *Anal. Chem.*, **58**, 97–105.
Houk, R.S., Fassel, V.A., Flesch, G.D., Svec, H.J., Gray, A.L. & Taylor, C.E. (1980) *Anal. Chem.*, **52**, 2283–2289.
Houk, R.S. & Thompson, J.J. (1988) *Mass Spectrom. Rev.*, **7**, 425–461.
Huang, Y., Shibata, Y. & Morita, M. (1993) *Anal. Chem.*, **65**, 2999–3003.
Imai, N. (1992) *Anal. Chim. Acta*, **269**, 263–268.
Imai, N. & Shimokawa, K. (1993) *Appl. Radiat. Isotopes*, **44**, 161–165.
Jackson, S.E., Longerich, H.P., Dunning, G.R. & Fryer, B.J. (1992) *Can. Mineral.*, **30**, 1049–1064.
Jarvis, K.E. & Williams, J.G. (1993) *Chem. Geol.*, **106**, 251–262.
Jeffries, T.E. (1996) *Unpublished PhD Thesis, University of Wales.*
Jeffries, T.E., Jackson, S.E. & Longerich, H.P. (1998) *J. Anal. Atom. Spectrom.*, **13**, 935–940.
Jeffries, T.E., Pearce, N.J.G., Perkins, W.T. & Raith, A. (1996) *Anal. Comms.*, **33**, 35–39.
Jeffries, T.E., Perkins, W.T. & Pearce, N.J.G. (1995a) *Chem. Geol.*, **121**, 131–144.
Jeffries, T.E., Perkins, W.T. & Pearce, N.J.G. (1995b) *Analyst*, **120**, 1365–1371.
Jenner, G.A., Foley, S.F., Jackson, S.E., Green, T.H., Fryer, B.J. & Longerich, H.P. (1993) *Geochim. Cosmochim. Acta*, **57**, 5099–5103.
Kogan, W., Hinds, M.W. & Ramendik, G.I. (1994) *Spectrochim. Acta B*, **49**, 333–343.
Koppenaal, D.W. (1992) *Anal. Chem.*, **64**, R320–342.
Lahaye, Y., Lambert, D. & Walters, S. (1997) *Geostands. Newslett.*, **21**, 205–214.
Lee, K.M., Appleton, J., Cooke, M., Keenan, F. & Sawicka Kapusta, K. (1999) *Anal. Chim. Acta*, **395**, 179–185.

Liang, X.R., Li, X.H., Wei, G.J., Tu, X.L. & Wang, G.L. (1999) *Chinese J. Anal. Chem.*, **27**, 1121–1125.
Lichte, F.E. (1995) *Anal. Chem.*, **67**, 2479–2485.
Longerich, H.P., Günther, D. & Jackson, S.E. (1996a) *Fresenius' J. Anal. Chem.*, **355**, 538–542.
Longerich, H.P., Jackson, S.E. & Günther, D. (1996b) *J. Anal. Atom. Spectrom.*, **11**, 899–904.
Loucks, R.R. & Mavrogenes, J.A. (1999) *Science*, **284**, 2159–2163.
Ludden, J.N., Feng, R., Gauthier, G., Stix, J., Shi, L., Francis, D. & Machado, N. (1995) *Can. Mineral.*, **33**, 419–434.
Machado, N. & Gauthier, G. (1996) *Geochim. Cosmochim. Acta*, **60**, 5063–5073.
Magyar, B., Cousin, H. & Aeschlimann, B. (1992) *Anal. Procs.*, **29**, 282.
Marshall, J., Franks, J., Abell, I. & Tye, C. (1991) *J. Anal. Atom. Spectrom.*, **6**, 145–150.
Masters, B.J. & Sharp, B.L. (1997) *Anal. Comms.*, **34**, 237–239.
Milton, D.A. & Chenery, S.R. (1998) *J. Fish Biol.*, **53**, 785–794.
Milton, D.A., Chenery, S.R., Farmer, M.J. & Blaber, S.J.M. (1997) *Marine Ecology Progress Series*, **153**, 283–291.
Mochizuki, T., Sakashita, A., Iwata, H., Ishibashi, Y. & Gunji, N. (1991) *Anal. Sci.*, **7**, 151–153.
Moenke-Blankenburg, L. (1993) *Spectrochim. Acta Rev.*, **15**, 1–37.
Moenke-Blankenburg, L. Gäckle, M., Günther, D. & Kammel, J. (1990) *Special Publication of the Roy. Soc. Chem.*, **85**, 1.
Moenke-Blankenburg, L., Schumann, T., Günther, D., Kuss, H.M. & Paul, M. (1992) *J. Anal. Atom. Spectrom.*, **7**, 251–254.
Neal, C.R., Davidson, J.P. & Mckeegan, K.D. (1995) *Rev. Geophys.*, **33**, 25–32.
Norman, M.D., Griffin, W.L., Pearson, N.J., Garcia, M.O. & O'Reilly, S.Y. (1998) *J. Anal. Atom. Spectrom.*, **13**, 477–482.
Olivares, J.A. & Houk, R.S. (1985) *Anal. Chem.*, **57**, 2674–2679.
Outridge, P.M. & Evans, R.D. (1995) *J. Anal. Atom. Spectrom.*, **10**, 595–600.
Pearce, N.J.G., Perkins, W.T., Abell, I., Duller, G.A.T. & Fuge, R. (1992a) *Anal. Chem.*, **7**, 53–57.
Pearce, N.J.G., Perkins, W.T. & Fuge, R. (1992b) *J. Anal. Atom. Spectrom.*, **7**, 595–598.
Pearce, N.J.G., Perkins, W.T., Westgate, J.A., Gorton, M.P., Jackson, S.E. & Neal, C.R. (1997) *Geostands. Newslett.*, **21**, 115–144.
Pearce, N.J.G., Westgate, J.A. & Perkins, W.T. (1996) *Quaternary International*, **34–36**, 213–227.
Pearce, N.J.G., Westgate, J.A., Perkins, W.T., Eastwood, W.J. & Shane, P. (1999) *Global Planet. Change*, **21**, 151–171.
Perkins, W.T., Fuge, R. & Pearce, N.J.G. (1991) *J. Anal. Atom. Spectrom.*, **6**, 445–449.
Perkins, W.T., Pearce, N.J.G. & Fuge, R. (1992) *J. Anal. Atom. Spectrom.*, **7**, 611–616.
Perkins, W.T., Pearce, N.J.G. & Jeffries, T.E. (1993) *Geochim. Cosmochim. Acta*, **57**, 475–482.
Perkins, W.T., Pearce, N.J.G. & Westgate, J.A. (1997) *Geostands. Newslett.*, **21**, 175–190.
Poitrasson, F., Chenery, S. & Bland, D.J. (1996) *Earth Planet. Sci. Lett.*, **145**, 79–96.
Price, G.D. & Pearce, N.J.G. (1997) *Marine Pollution Bull.*, **34**, 1025–1031.
Prohaska, T., Stadlbauer, C., Wimmer, R., Stingeder, G. & Latkoczy, C. (1998) *Sci. Total Environment*, **219**, 29–39.
van der Putten, E., Dehairs, F., Andre, L. & Baeyens, W. (1999) *Anal. Chim. Acta*, **378**, 261–272.
Querol, X. & Chenery, S. (1995) *Geol. Soc. Spec. Pub. (Coal Geology)*, **82**, 147–155.
Raith, A., Hutton, R.C., Abell, I.D. & Crighton, J. (1995) *J. Anal. Atom. Spectrom.*, **10**, 591–594.
Schettler, G. & Pearce, N.J.G. (1996) *Hydrobiologia*, **317**, 1–11.
Secor, D.H. & Zdanowicz, V.S. (1998) *Fisheries Res.*, **36**, 251–256.
Shepherd, T.J., Ayora, C., Cendon, D.I., Chenery, S.R. & Moissette, A. (1998) *European J. Mineral.*, **10**, 1097–1108.
Shepherd, T.J. & Chenery, S.R. (1995) *Geochim. Cosmochim. Acta*, **59**, 3997–4007.
Shibata, Y., Yoshinaga, J. & Morita, M. (1993) *Anal. Sci.*, **9**, 129–131.
Stecher, H.A., Krantz, D.E., Lord, C.J., Luther, G.W. & Bock, K.W. (1996) *Geochim. Cosmochim. Acta*, **60**, 3445–3456.
Strubel, C., Meckel, L. & Effenberger, R. (1999) *Glastechnische Berichte*, **72**, 15–20.
Thiem, T.L., Lee, Y.I. & Sneddon, J. (1992) *Microchem. J.*, **45**, 1–35.
Thompson, M., Chenery, S. & Brett, L. (1989) *J. Anal. Atom. Spectrom.*, **4**, 11–16.
Thorrold, S.R., Jones, C.M. & Campana, S.E. (1997) *Limnology Oceanography*, **42**, 102–111.
Ulens, K., Moens, L., Dams, R., Vanwinckel, S. & Vandevelde, L. (1994) *J. Anal. Atom. Spectrom.*, **9**, 1243–1248.
Veinott, G., Northcote, T., Rosenau, M. & Evans, R.D. (1999) *Can. J. Fisheries Aquatic Sci.*, **56**, 1981–1990.
Wang, S., Brown, R. & Gray, D.J. (1994) *Appl. Spectrosc.*, **48**, 1321–1325.
Watling, R.J. (1998) *J. Anal. Atom. Spectrom.*, **13**, 917–926.

Watling, R.J. (1999) *Spectroscopy*, **14**, 16.
Watling, R.J., Herbert, H.K. & Abell, I.D. (1995a) *Chem. Geol.*, **124**, 67–81.
Watling, R.J., Herbert, H.K., Barrow, I.S. & Thomas, A.G. (1995b) *Analyst*, **120**, 1357–1364.
Watling, R.J., Herbert, H.K., Delev, D. & Abell, I.D. (1994) *Spectrochim. Acta B.*, **49**, 205–219.
Watling, R.J., Lynch, B.F. & Herring, D. (1997) *J. Anal. Atom. Spectrom.*, **12**, 195–203.
Watmough, S.A., Hutchinson, T.C. & Evans, R.D. (1997) *Environmental Sci. Technol.*, **31**, 114–118.
Watmough, S.A., Hutchinson, T.C. & Evans, R.D. (1998) *Environmental Sci. Technol.*, **32**, 2185–2190.
Wendt, R.H. & Fassel, V.A. (1965) *Anal. Chem.*, **37**, 920–922.
Westgate, J.A., Perkins, W.T., Fuge, R., Pearce, N.J.G. & Wintle, A.G. (1994) *Appl. Geochem.*, **9**, 323–335.
Westheide, J.T., Becker, J.S., Jager, R., Dietze, H.J. & Broekaert, J.A.C. (1996) *J. Anal. Atom. Spectrom.*, **11**, 661–666.
Wu, G.P. & Hillairemarcel, C. (1995) *Geochim. Cosmochim. Acta*, **59**, 409–414.
Yang, P., Rivers, T. & Jackson, S. (1999) *Can. Mineral.*, **37**, 443–468.

4 X-ray fluorescence

J. INJUK and R. VAN GRIEKEN

4.1 Introduction

One century after the discovery of X-rays, X-ray fluorescence (XRF) spectrometry has developed into a well-established and mature multielement analytical technique. There are several key reasons for this: XRF is a universal technique for solid, powder and liquid samples; it is non-destructive; it is reliable; it can yield qualitative and quantitative results; it involves easy sample preparation and handling; it has high dynamic range, from the sub-ppm level to 100% and it can cover most of the elements from fluorine to uranium. However, XRF generally provides only elemental analysis and the total concentration of the element is obtained.

X-rays are a form of electromagnetic radiation with wavelength between 10^{-2} and 10 nm. Yet, these boundaries are not clearly defined. They are produced when high-energy electrons decelerate or when electron transitions occur in the inner shells of atoms. Conventional X-ray spectroscopy is restricted to the region from 10^{-2} nm (U K_{α}) to 2 nm (F K_{α}) or in the range of 0.1–100 keV. The energy (E) and wavelength (λ) of X-ray photons are related by the equation:

$$E\,(\mathrm{keV}) = \frac{1.24}{\lambda\,(\mathrm{nm})} \tag{4.1}$$

4.2 History of X-rays

X-rays were discovered in 1895 by W.C. Röntgen, a discovery for which he was awarded the Nobel Prize in 1901. The obvious similarities of this radiation with light led to it being subjected to the crucial tests of wave optics: polarization, diffraction, reflection and refraction. But, with limited experimental facilities, Röntgen and his contemporaries could not find any evidence of these phenomena; hence the designation 'X' (unknown). In the period 1906–1911, Barkla found that X-rays could be polarized (and must therefore be waves) and that there was evidence of element-specific absorption edges and emission-line series, which he labelled K, L, M, etc.

The essential wave nature of X-rays was established in 1912 by Laue, Friedrich and Knipping who showed that X-rays could be diffracted by a crystal. The next year, Coolidge introduced the hot-filament, high-vacuum X-ray tube. The experiments of Bragg had shown that the X-ray spectrum emerging from such a device consists of a continuum, on which the line spectrum of the anode material is superimposed. In 1913, Moseley proved that the wavelengths of the lines were characteristic of the element of which the target was made and further showed that they had the same sequence as the atomic numbers. In the same period, Chadwick observed that X-rays can also be produced by bombarding a material with α particles, laying the foundation of particle- or ion-induced X-ray emission analysis (PIXE). The characteristic K absorption was first observed by de Broglie and interpreted by Bragg and Siegbahn. In 1920, Bergengren observed the effect of the chemical state of the absorber on the X-ray absorption spectra of materials.

In the period 1923–1932, von Hevesy, Coster, Nishina, Glocker and Schreiber investigated in detail the possibilities of fluorescent X-ray spectroscopy for qualitative and quantitative elemental analysis. The earliest commercially available X-ray spectrometer appeared on the market in 1938 and 10 years later, Friedman and Birks built the prototype of the first commercial X-ray secondary emission instrument. One year later, Castaing and Guinier built the first electron probe microanalyser in which a focused electron beam was used to induce X-ray emission in microscopic samples. The excitation of X-ray spectra by means of heavy ions (protons) for chemical analysis was first applied by Sterk in 1964. From the 1960s until the present day, the use of large particle accelerators (synchrotron rings) has resulted in a dramatic increase in the variety, scope and applicability of X-ray-based methods.

4.3 Properties of X-rays

When a beam of electromagnetic radiation traverses matter, it may pass through unaffected or may undergo reflection, refraction, diffraction (i.e. by slits, gratings or crystal structures), polarization, elastic and inelastic scattering, photoelectric absorption and, above 1.02 MeV, electron–positron pair production. Diffraction forms the basis of the structural investigation of crystalline materials. It is also used in wavelength dispersive (WD) modes of X-ray detection. Reflective and refractive properties of X-rays are employed in analytical methods like total-reflection X-ray fluorescence (TXRF) and in X-ray optics. Photoelectric absorption of X-ray photons is utilized in methods such as X-ray fluorescence (XRF), X-ray absorption spectrometry (XAS) and X-ray photoelectron spectroscopy (XPS). Scatter phenomena are usually indirectly employed to estimate additional properties of materials under investigation like thickness and mean atomic number.

4.3.1 *Characteristic radiation*

When matter is irradiated with either electrons of sufficiently high energy, heavy charged particles (protons or α particles) or photons (like those in XRF spectrometry) the ejection of an electron from an inner shell (usually K or L) of the atoms present may take place. As a result of the creation of these inner-shell vacancies, the atom is left in an electronically excited ionized state. The vacancy is almost immediately (within less than 10^{-16} s) filled by an electron from a higher energy level, and the difference in energy between the two levels is released in the form of an X-ray photon. These photons are specific for the particular electron transition and have a specific wavelength/energy characteristic for the atomic species involved. They are called characteristic X-ray lines. An outline of some transitions giving X-rays is displayed in Fig. 4.1. The excess energy can, however, be employed in an internal conversion

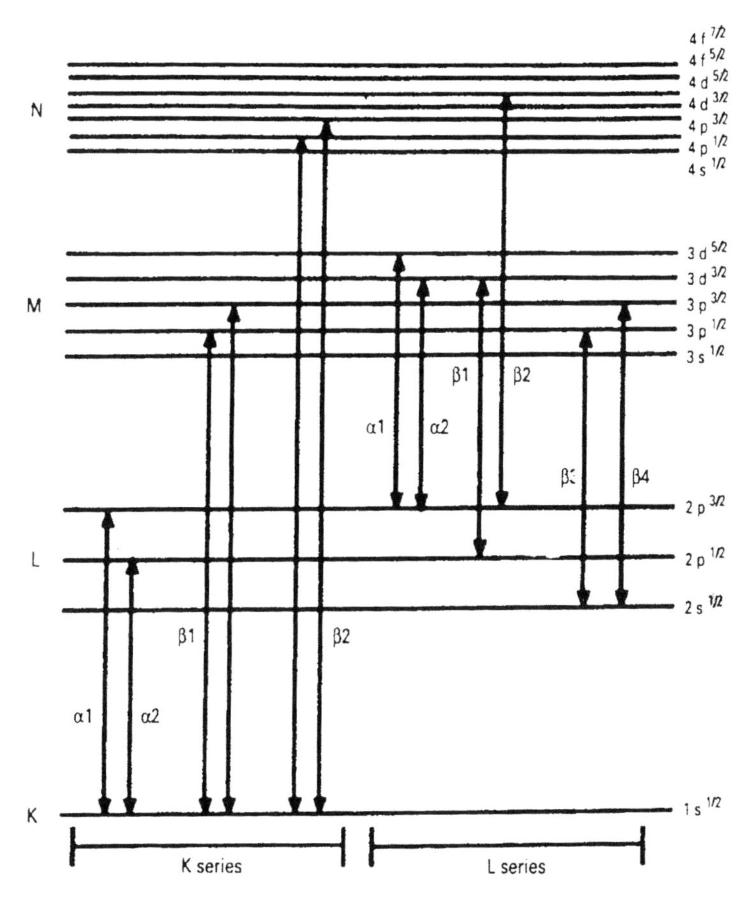

Fig. 4.1 Transitions giving X-rays. Electronic levels in the atom are denoted K, L_I, L_{II}, L_{III}, M_I, etc. and correspond to the ejection of electrons from the $1s^{1/2}$, $2s^{1/2}$, $2p^{1/2}$, $2p^{3/2}$, $3s^{1/2}$, ... $3d^{5/2}$ orbitals. The terminology adopted by Siegbahn for the characteristic X-ray lines is used. When a vacancy is created in the K shell and filled by an electron from the L or M shell, a K_α or K_β X-ray will be released. In, for example, an iron atom, the radiative $L_{III} \rightarrow K$ transition corresponds to the emission of a characteristic Fe K_α X-ray photon. The energy difference between the L and K levels is much larger than between the M and L levels, so that K lines appear at shorter wavelengths. The energy difference between, for instance, the transitions α_1 and α_2 is so small that only a single line can be observed.

process by emission of a second electron from an outer shell of the atom, leaving it in a double-ionized state. Such electrons are called Auger electrons and are used in Auger emission spectroscopy. Similar to the characteristic X-rays, Auger electrons also hold discrete energy levels, specific to the atomic species from which they were ejected.

The frequency of the characteristic photons within any series (e.g. K or L) is described by Moseley's law

$$\nu = k(Z - 1)^2 \tag{4.2}$$

where Z is the atomic number of the target element and k is a constant. As Moseley's law is not very stringent, the exact positions of characteristic X-ray lines are not calculated in practice by using Equation (4.2) but instead obtained from tables or computer databases. These sources normally give the energies and wavelengths of the peaks and additionally their relative intensities within the defined K, L, or M series. The relative intensity of a certain peak in its series is determined by the probability of the electron transition causing this particular peak. The related quantity is called the emission rate (g_j) and can be calculated from quantum mechanics. In general, the relative intensities are rather similar for most elements; e.g., for K peaks, K_α: K_β is about 100:15. A representation of Moseley's law for the K and L spectral series is given in Fig. 4.2.

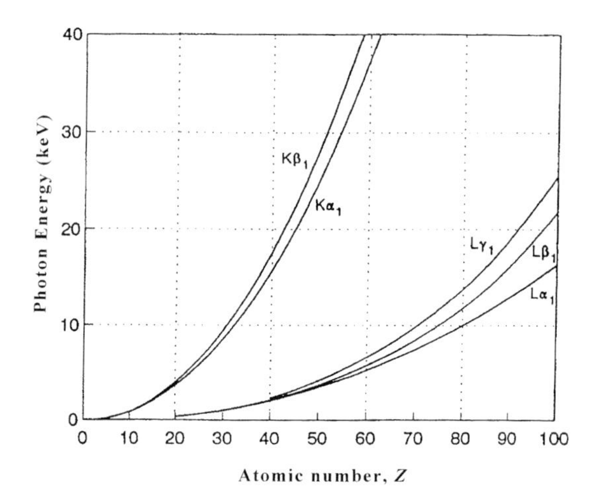

Fig. 4.2 Representation of Moseley's law: photon energies (keV) of the principal X-ray emission peaks as a function of the atomic number, Z, of the elements.

The intensity of the total K, L, and M series is a function of the fluorescence yield, ω. This gives the probability that an X-ray photon is emitted after excitation of an atom. The relationship can be described approximately by

$$\omega = \frac{Z^4}{A + Z^4} \tag{4.3}$$

where A is a constant which equals approximately 9×10^5 for the K series, 7×10^7 for the L series and 1×10^9 for the M series. In principle, the fluorescence yield, ω, can be calculated theoretically, but in practice experimental data are normally applied.

4.3.2 *Continuum*

This part of the X-ray spectrum is produced by high-energy electrons when they strike matter and collide with the atoms followed by their deceleration and X-ray photon production. This leads to a continuous band of X-ray wavelengths, known as the continuum or Bremsstrahlung radiation. The continuum has an intensity distributed continuously over a broad range of energies and a well-defined short-wavelength limit (λ_{min}) which is dependent on the accelerating voltage (but independent of the target material). The short-wavelength limit corresponds to the maximum energy (eV_0) of the electrons and is given, following Equation (4.1), by:

$$\gamma_{min}(\text{nm}) = \frac{1.24}{eV_0\,(\text{keV})} \tag{4.4}$$

The intensity is a maximum at approximately 1.5 λ_{min} and gradually falls off at longer wavelengths. The integrated intensity, I, of the continuum is related to the current i, the voltage V and the atomic number Z of the target by:

$$I = k'iZV^2 \tag{4.5}$$

As the intensity is proportional to the atomic number Z, the high atomic number elements like W, Pt and Rh are often used as targets in X-ray-producing tubes. Figure 4.3 shows the general features of a typical X-ray spectrum plotted as a function of the photon energy.

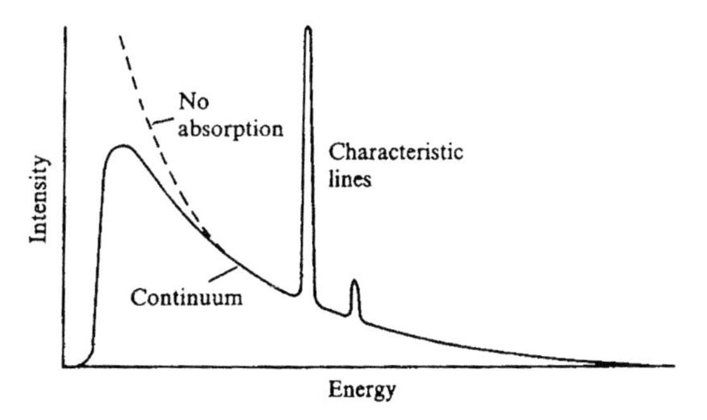

Fig. 4.3 An X-ray spectrum as a function of photon energy. The observed continuum intensity is reduced at low energies by absorption of the radiation emerging from the target.

4.3.3 *Attenuation of X-rays*

When a beam of X-ray photons of energy E is transmitted through an absorber foil of thickness d, the beam's original intensity I_0 is attenuated to a value $I(d)$. The attenuation is the result of the interaction of the X-ray photons with matter and varies with absorber thickness according to

$$I(d) = I_0\,\mathrm{e}^{-\mu_l d} = I_0\,\mathrm{e}^{-\mu\rho d} \tag{4.6}$$

where μ_l and μ are the linear and mass absorption coefficients of the material and depend on the photon energy and on the composition of the foil ($\mu = \mu_l/\rho$, where ρ is the density of the material). For a material consisting of n elements, each one present with a weight fraction w_i, the mass absorption coefficient can be approximated as a weighted sum of atomic components

$$\mu(E) = \sum_i w_i\mu_i(E) \tag{4.7}$$

where μ_i represents the mass absorption coefficient of the ith atomic species. The term $\mu(E)$ can be written as a sum of three contributions:

$$\mu_i(E) = \tau_i(E) + \sigma_i(E) + \pi_i(E) \tag{4.8}$$

The above expression reflects the three processes through which X-rays interact with matter: τ, σ and π (usually expressed in $cm^2\ g^{-1}$) represent the total cross-section for photoelectric absorption, for scattering and for pair production, respectively.

In a photoelectric interaction, X-ray photons are absorbed, as described in the previous paragraph. With respect to photon scattering, two types of absorption can be distinguished depending on whether the scattered photon loses energy during the process (Compton or inelastic scattering) or not (Rayleigh or elastic scattering). In terms of their total cross-sections (σ_C and σ_R), this can be expressed as:

$$\sigma(E) = \sigma_C(E) + \sigma_R(E) \tag{4.9}$$

The principle of Compton scattering involves the interaction of a photon with a weakly bound atomic electron. Rayleigh scattering is a process by which photons are scattered by bound atomic electrons and in which the atom is neither ionized nor excited. The energy of the photon remains unchanged. The pair production process occurs for X-rays with an energy above 1.022 MeV. However, in the 0–100 keV region, the photoelectric absorption is by far the most important for various analytical techniques. Since the photoelectric absorption takes place at various atomic levels, it is a function of atomic number. A plot of the mass attenuation as a function of a wavelength contains a number of discontinuities (so-called absorption edges) at the wavelengths corresponding to the binding energies of the electrons in the various sub-shells (Fig. 4.4). These absorption discontinuities are a major source of non-linearity between X-ray intensities and composition. As primary X-rays are absorbed and scattered, secondary fluorescence or enhancement may occur. These effects

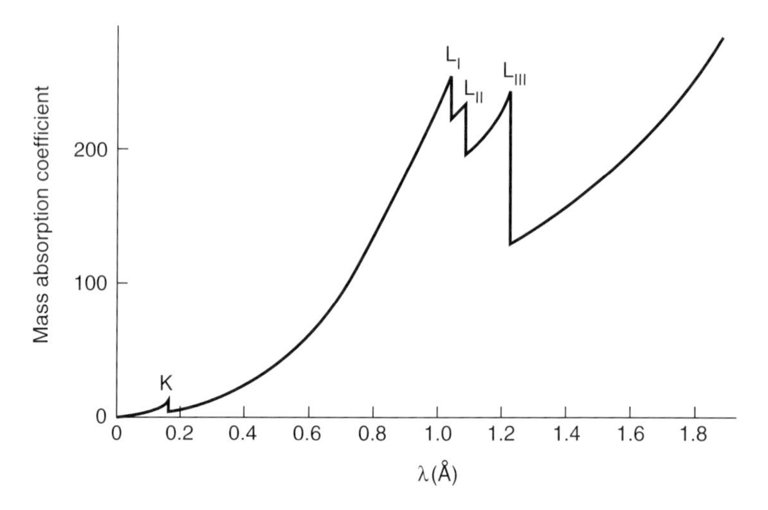

Fig. 4.4 Mass attenuation coefficient as a function of wavelength for tungsten. With the longer wavelengths of incident X-ray photons, the absorption increases.

depend on the composition of the matrix elements and may introduce large systematic errors when they are not properly accounted for. The effect is called the 'matrix effect' and it is very important in quantitative XRF.

4.4 X-ray sources

XRF spectrometry can be classified by excitation mode into three main categories:

- X-ray tube excitation
- Radioisotope excitation
- Excitation by synchrotron radiation

The most widely applied source of X-rays is the X-ray tube. In Fig. 4.5, a conventional X-ray tube is shown. Electrons, generated by heating a tungsten filament, are accelerated towards a cooled anode by means of a 10–100 kV potential difference, and when impinging on the anode the electrons produce X-radiation. A significant portion of the radiation passes through the exit window. The choice of anode material (usually a high-Z metal such as Cu, Mo, Rh, W or Ag, with good thermal conductivity and mechanical strength) and window characteristics can be critical. The conversion of electrons into X-rays is a very inefficient process, with only 1% of the total power emerging as useful radiation while the remainder appears as heat and must be dissipated by cooling the anode. In order to supply high transmission of longer wavelengths, the window is normally constructed of beryllium with a typical thickness of 0.2–1 mm.

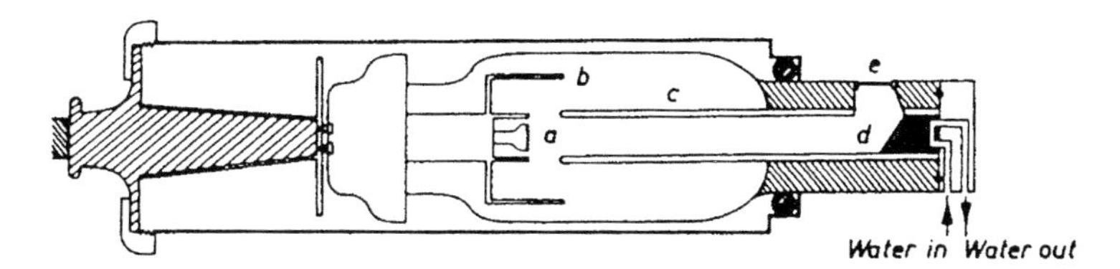

Fig. 4.5 Principle of the X-ray tube. The tungsten filament (a) is heated by the filament current, producing a cloud of electrons. The electrons are accelerated along the focusing tube (c) by the potential difference between the filament and the anode (d). Generated X-rays pass through the window (e). The purpose of the cathode cap (b) is to absorb the unused and scattered electrons and to reduce the spread of tungsten, which vaporizes from the filament (courtesy of Philips Analytical).

X-rays can also be induced by various types of radioactive decay processes. In such processes more energetic radiation (γ lines) is usually produced as well. The most commonly used radioisotopes are ^{55}Fe, ^{238}Pu, ^{244}Cm, ^{109}Cd, ^{125}I, ^{241}Am, ^{57}Co, ^{133}Ba and ^{137}Cs. Radioactive sources have the advantages of producing monochromatic radiation, being very compact and easy to use, and are portable requiring no high voltage and cooling. Costs also may be an important consideration. A disadvantage of radioactive sources is their limited life-time and the fact that they generate a relatively low flux of primary radiation (resulting in low analytical sensitivity). The selection of the radioisotope source to analyse different elements depends on many

Table 4.1 Properties of some radioisotope sources frequently used in XRF analysis

Radioisotope	Half-life (years)	X- or γ-ray energy (keV)	Photons per disintegration
^{55}Fe	2.7	Mn K (5.9, 6.5)	0.28
^{238}Pu	88	U L (13–20)	0.13
^{244}Cm	17.8	Pu L (14–21)	0.08
^{109}Cd	1.3	Ag K (22, 25)	1.07
		88	0.04
^{125}I	0.16	35	0.07
		Te K (27–32)	1.38
^{241}Am	433	59.5	0.36
^{57}Co	0.74	122	0.86

factors, including whether the energy of the radioisotope X- or γ-rays is sufficient to excite the element, and also on the energy resolution of the detector. Some properties of radioisotope sources used frequently for XRF analysis are listed in Table 4.1.

As a source of X-rays, synchrotron radiation has been employed, on a routine basis, only since 1970s. Synchrotron radiation is emitted when charged particles, with kinetic energies of the order of several billion eV (typically 3 GeV at present) are accelerated by means of a magnetic field perpendicular to their direction of motion. At the heart of a synchrotron X-ray source is an electron (or positron) storage ring, i.e. an evacuated tube through which a densely packed beam of electrons circulates. The synchrotron-produced X-ray beams have unique properties. They have a continuous energy distribution so that with the use of suitable monochromator devices (such as crystal or mirror), monoenergetic beams can be produced over a wide range of energies. The photons are highly polarized in the plane of the electron beam orbit (the storage ring), a property which reduces the scatter background in synchrotron radiation-induced XRF (SR-XRF). In addition, the most important feature of synchrotron sources, for use in XRF, is their intensity, which may be a factor of 10^4–10^{10} higher than the output of conventional X-ray tubes. This, in combination with the highly collimated and polarized character of the synchrotron radiation, has a major impact on microprobe-type methods with a high spatial resolution (about 10 μm). A disadvantage of the SR-XRF is that the source intensity decreases with time, but this can be overcome by, e.g. continuously monitoring the primary beam.

4.5 Dispersing systems and detection of X-rays

When X-rays are emitted by a sample, they must be separated into a spectrum so that the characteristic X-rays of each element can be measured individually. Two modes can be distinguished in XRF: wavelength dispersion (WD) and energy dispersion (ED). In wavelength dispersive X-ray fluorescence spectrometry (WDXRF) the emitted X-rays are dispersed spatially on the basis of their wavelengths by crystal diffraction. The angle, θ, for which the diffraction phenomenon occurs at a given wavelength λ and crystal spacing d, is given by Bragg's law:

$$n\lambda = 2d \sin \theta \tag{4.10}$$

For $n = 1$, the path difference between two rays is one wavelength and the diffraction is said to be of the first order. Second-order diffraction corresponds to $n = 2$, and so on. Thus, by measuring the diffraction angle θ, knowledge of the spacing d of the analysing crystal allows the determination of a specific wavelength, λ, with a resolution given by:

$$\frac{d\theta}{d\lambda} = \frac{n}{2d \cos \theta} \tag{4.11}$$

Commonly used dispersive crystals are LiF (420) with $2d = 0.180$ nm for the elemental range Ni–U or Si (111) with $2d = 0.626$ nm used in the elemental determination of P, S and Cl.

In energy-dispersive X-ray fluorescence (EDXRF) spectrometry, the detector receives the X-ray photons of various energies from the sample all at once and subsequently acts on them simultaneously. After photons possessing a range of energies are incident upon the detector, the detector generates a pulse of electric current with an amplitude proportional to the photon energy. The output, due to each photon, is amplified and analysed by a multichannel analyser (MCA). ED X-ray detectors are popularly associated with high-purity lithium-drifted silicon semiconductor crystals, Si(Li), although in more specialized applications, materials such as Ge and HgI_2 can be used. The basic principle of all these devices is the parallel mode of data acquisition so that X-rays of all energies are measured simultaneously. The resolution of the X-ray detector is usually defined as the full width at half maximum (FWHM) of a spectral peak and is typically less than 145 eV at 5.9 keV for a Si(Li) detector. The high-purity Ge detector is used when X-rays above 30 keV may be produced and its resolution is about 180 eV at 5.9 keV and 400 eV at 122 keV.

The fact that X-rays cause ionization in gases is the basis of gas-filled detectors such as ionization chambers, Geiger-Müller counters and proportional counters. The basic principle behind each of these radiation detectors is the photoionization of gas molecules and/or atoms followed by the separation and collection of the resulting ions by means of an electric field.

Figure 4.6 shows the energy resolution of a typical Si(Li) detector compared with a proportional counter and a wavelength spectrometer using various crystals. The proportional counter has much worse resolution except at very low energies, owing to ionization statistics, and cannot be considered seriously for ED spectrometry. The wavelength spectrometer has a resolution that varies considerably with Bragg angle, but is always better than the Si(Li) detector by a large margin. Only at high energies, above 20 keV, does the Si(Li) detector approach the wavelength spectrometer in resolution.

The most important properties that an ideal detector should possess are: sensitivity to the appropriate photon energies, proportionality and linearity. Since each X-ray photon entering the detector produces a voltage pulse, it is possible to assume that the size of the voltage pulse is proportional to the photon energy. Proportionality is required where the technique of pulse-height selection is to be used (pulse-height selection means electronically rejecting pulses of voltage levels other than those

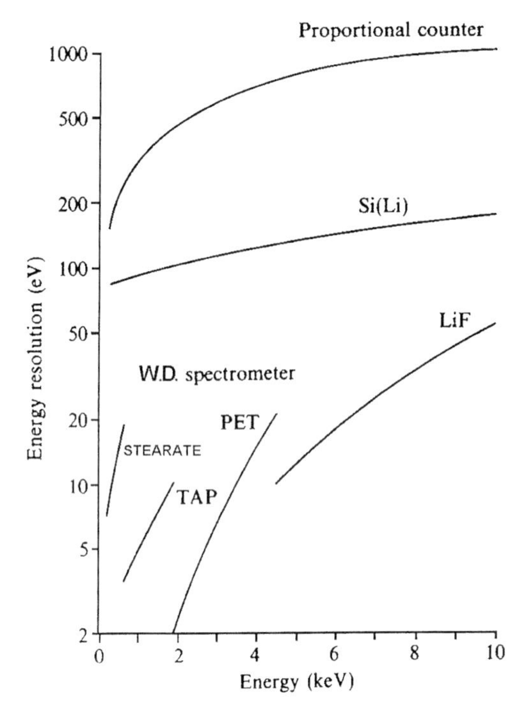

Fig. 4.6 Energy resolution of the Si(Li) detector compared to the proportional counter and wavelength spectrometer using different crystals (LiF: lithium fluoride; PET: pentaerythritol stearate; STEARATE: lead stearate; TAP: thallium hydrogen phosphate; W.D.: wave-dispersive).

corresponding to the characteristic line being measured). Alternatively, X-ray photons enter the detector at a certain rate and if the output pulses are produced at the same rate, the detector is considered to be linear. Linearity is important where the various count rates, produced by the detector, are to be used as measures of the photon intensity for each measured line.

X-ray intensities and spectra can also be recorded by photographic materials. This feature is explored in film-badge dosimetry. Furthermore, the fact that X-rays can induce visible and ultraviolet luminescence in certain materials is adapted for scintillation counters and fluorescent screens.

In the case of WDXRF, both the gas-flow proportional counter (for the measurement of longer wavelengths) and the scintillation counter (e.g. the NaI(Tl) scintillation detector for the measurement of shorter wavelengths) are generally employed. In the case of EDXRF a detector of high resolution (normally Si(Li)) must be used.

4.6 Sample preparation methods

In XRF the sample preparation procedure is very important and strongly influences the final quantitative result. If improper selection of sample from the bulk or improper sample preparation takes place, serious systematic errors can be introduced. Therefore, the following principles should as a rule be considered:

- a method should be selected which is cheap, rapid and does not introduce additional errors or contamination
- specimen preparation should be avoided as much as possible
- if possible preference should be given to simple physical sample preparation using methods such as drying, freeze-drying, homogenizing, pulverizing, cutting
- an adequately prepared specimen must be representative of the material, homogeneous and with a flat, smooth surface
- all standards and unknown samples must be presented to the analysis in a reproducible and identical manner
- standards must have similar physical properties, such as mass attenuation coefficient, density, particle size, homogeneity, etc.
- contamination control and blank specimen preparation should be available
- losses of analyte elements (particularly halogens) should be considered

A wide variety of sample types may be analysed by XRF, including solids, liquids and even gases as is shown in Table 4.2. For solid samples that are homogeneous in all three dimensions and in possession of a flat surface, sample preparation is not necessary.

Metallic specimens are usually prepared for X-ray analysis as solid disks by cutting, milling, grinding, polishing and casting. XRF analysis of metal samples involves mostly a relatively thin surface layer of the investigated specimen. Therefore, it must be representative of the bulk sample structure. The major surface preparation methods are machining (milling, turning), mechanical grinding, and polishing and etching. In any surface preparation technique employed, special care must be taken to ensure that no additional surface contamination is introduced. This care is particularly important in the case of a relatively soft metal such as aluminum, where particles of the grinding agent may penetrate to the sample surface during surface finishing. The selection of a suitable abrasive is sometimes very difficult, particularly when elements like Si, Al or Fe need to be determined. For instance, SiC and Al_2O_3 are very effective abrasives, but both contain elements of interest. In such cases electrolytic polishing or etching is recommended. For the analysis of steels, 60–240 grit alumina, zircon–alumina, or silicon carbide are found to be adequate. For the determination of calcium, calcium-free 240 grit silicon carbide has demonstrated the best results. After the preparation procedures, it is important to ensure that the surfaces are kept clean. Hence, all traces of lubricant, cutting fluid or finger marks must be removed by cleaning with isopropyl alcohol prior to examination. Occasionally, the surface of certain susceptible metals is affected by corrosion, which progresses with time. Therefore, optimum results are invariably obtained when analysing specimens without much delay after the surface preparation.

When the surface of the specimen is too rough and common methods like polishing and milling are not effective, samples can be prepared by pressing them into pellets under loads of several hundred megapascal. With such a procedure most of the specimen is compacted into a pellet with satisfactory smooth surfaces.

A number of metals and alloys that are otherwise very difficult to deal with can be

Table 4.2 Overview of the various types of materials analysed by XRF

Material sciences and industry	alloys
	superconductor materials
	semiconductors
	electrodes
	bath electrolyte
	rubber
	float glass
	graphite powder
	oil-refining products
	cement and related products
	radioactive samples
	acryl paint
	polyethylene
	carpet fibres
	cosmetics
	metals
	thermit smeltings
	coal
	asphalt and pitch
	glass sand
	natural and by-product gypsum
	thin coatings
	chemical state analysis
	silk
	reference materials
Archaeology	jewellery
	coins
	glass
	pottery
	porcelain
	mural paintings
	mummy hair
	fossil shells
	pigments
	human skeletons
	tin slags
Environmental	aqueous solutions
	aerosols
	fly ash
	biofilms
	humic substrain
	snow
	ice cores
	waste
	land and soil
Biomedical and biological	food
	wine
	hair
	bone
	cancerous tissue
	skin
	kidney
	liver
	blood
	urine
	tree rings
	leaves and vegetation
	natural drugs
	medicines
	marine organisms
	algae
	bacteria
Geological	corals
	ores and minerals
	rocks
	sediments
	clay
	bauxite
	coal
	earth crust
	interplanetary dust
	petroleum products
	chrysotile

dissolved in aqua regia (a mixture of one part concentrated HNO_3 and three parts concentrated HCl) with gentle heating to 60°C. Aqua regia is used to dissolve a number of metals (e.g. Pd, Pt, Sn, Ge) and alloys including steels, Mo, Rh, Ir, Ru, and high-temperature alloys. The strength of this acid combination is in the formation of chloro-complexes during the dissolution reaction as well as the catalytic effects of Cl_2 and NOCl. The effectiveness can additionally be improved if the solution is allowed to stand for 10–20 min before heating. Fe and V ores and crude phosphates can be dissolved using this mixture, as well as some sulfides (e.g. pyrites and copper sulfide ores), although some sulfur is lost as H_2S. HNO_3 and HCl are often used together in other proportions. So-called inverted aqua regia (mixture 3 : 1 HNO_3 : HCl) is used to oxidize sulfur and pyrites. Another aggressive decomposition reagent is the combination of HNO_3/HF. Here the complexing effect of the fluoride ion is utilized and with this acid combination a number of metals (e.g. Si, Nb, Ta, Zr, Hf, W, Ti,) and alloys (e.g. Nb–Sn, Al–Cr, Cu–Si, Ca–Si) can be dissolved. However, some elements may be lost during the digestion step, especially elements like Se, Hg and Sn, which form volatile components with various kinds of acids or their combinations

Irregularly shaped metallic specimens are prepared for analysis by embedding the them in a special wax resin (e.g. acrylic resin and methyl acrylic resin). The resulting block can be polished to an appropriate surface smoothness using abrasives such as SiC and diamond paste prior to analysis.

A common technique for geological, industrial and biological materials is the preparation of powders and pellets. Powdered specimens are prepared when the original sample is too heterogeneous for direct analysis or too brittle to form self-supporting disks or when a suitable surface finish is not possible. The XRF analysis of powder samples requires particles of uniform size range with sufficiently small grain size to yield chemical homogeneity without absorption problems. For a routine trace XRF analysis a particle size of less than 62 μm is commonly acceptable. Crushing, pulverizing and milling procedures may be used to achieve such particle dimensions. Additives are sometimes useful for the proper grinding of powders. In general, 2–10% of additive to aid the grinding, blending and briquetting process is sufficient for most materials. Once the sample has been crushed it is necessary to separate out the coarser fraction for further treatment. Sieves are used during this step. The powder specimen has to fit into the specimen chamber of the X-ray spectrometer. It is therefore desirable to have the specimen in the shape of a round, flat disk. To achieve this, presses and dies are utilized. A further procedure is to present the sample directly to the spectrometer as a loose powder, packed in cells or spread-out on film materials (so-called 'slurry' technique). The slurry technique works for water-insoluble materials. A water slurry is prepared from a few milligrams of powder and a few millilitres of water. A turbulent suspension is made followed by rapid filtration through, e.g. a Nuclepore filter. This method results in a fairly uniform thin layer suitable for XRF analysis.

Samples that are not easily soluble or tend to remain heterogeneous after grinding and pelletizing, are often treated by the technique of flux fusion. Probably the most effective way of preparing a homogeneous powder sample is by the borax fusion method (Tertian & Claisse 1982). In principle it involves fusion of the sample with an excess of sodium or lithium tetraborate and casting into a solid bead. Chemical reaction in the melt converts the phases present in the sample into glass-like borates, giving a homogeneous bead of dimensions ideal for direct placement in the spectrometer. Manual application of the technique is rather time consuming, while a number of automated and semi-automated borax bead-making machines is commercially available. The critical step of the method is the ratio of the sample-to-fusion mixture because this controls several factors like the speed and degree of chemical reaction, final mass absorption coefficient and the actual dilution factor applied to the analyte element. By using fusion aids (such as iodides and peroxides) and high atomic number absorbers as part of the fusion mixture (e.g. barium lanthanum salts) these factors can somehow be controlled. The actual fusion reaction can be performed at 800–1000°C in a crucible made from platinum, nickel or silica. All of these materials suffer from the disadvantage that the melt tends to wet the sides of the dish and it is impossible to obtain complete recovery of the fused mixture. By using graphite crucibles this problem can be partially overcome, but the best alternative is a crucible made of platinum and 3% gold.

The most severe disadvantages of the fusion techniques are the time and material costs involved, and the dilution effect of the sample, which makes it often difficult or even impossible to determine trace elements in the specimen. For this reason the low-dilution fusion technique was developed where the flux-to-sample ratio is 2:1. In most analyses, $LiBO_2$ is applied as the flux, as it is more reactive and forms fluxes of higher fluidity. Many useful recipes for the fusion method are given in the book *A Practical Guide for the Preparation of Specimens for X-ray Fluorescence and X-ray Diffraction Analysis* (Buhrke *et al.* 1998).

For aqueous samples, direct analysis leads to limits of detection in the mg l^{-1} range, which is not satisfactory for most natural water applications. Consequently, it is necessary to remove the matrix to improve detection limits. For samples of low salinity or hardness, this can be easily accomplished by simple evaporation. To achieve detection limits at the μg l^{-1} level, evaporation of about 100 ml of water is necessary. A large water sample can also be freeze-dried, and the evaporation residue can be pelletized, possibly after mixing with organic binder to reduce matrix effect variations.

Freeze-drying of 250 ml of waste water on 100 mg of graphite followed by grinding and pelletizing of the residue can lead to detection limits of a few μg l^{-1}. In the evaporation residue, all non-volatile elements are collected quantitatively and the risk of contamination is minimal. Unfortunately, taking samples to dryness causes some experimental problems such as fractional crystallization, splashing, etc.

When analyte concentrations in liquids or solutions are too high or too low, other methods including dilution or pre-concentration are frequently employed to bring the analyte concentration within an acceptable range. A variety of methods can be used for the pre-concentration of aqueous samples. These include extraction, precipitation, co-precipitation, use of ion-exchange membranes and ion-exchange resins. One of the most effective methods involves the concentration of, for instance, a litre of water on a cation exchange resin and direct analysis of the resin by EDXRF. In this procedure, trace metal contamination in drinking water to a few μg l^{-1} is easily determined. Anion exchange filters may be used to concentrate or selectively remove a particular component. An example is the determination of Cr^{3+} and Cr^{6+}. In an acid medium, an anion exchange filter paper will collect Cr^{6+}, while Cr^{3+} passes through. Subsequently, the toxic Cr^{6+} can be directly analysed on the filter paper. Another useful method for multielement trace analysis of aqueous samples is co-precipitation. Various co-precipitation agents have been proposed in the literature for XRF. By far the most useful are sodium diethyldithiocarbamates (DDTC) and pyrrolidine-dithiocarbamate (APDC) or a combination of both. The carbamates are particularly attractive because of the low solubility of their metal chelates. Many pre-concentration methods for the analysis of water by X-ray spectrometric techniques have so far been proposed. A comprehensive overview of this subject, dating back to the 1980s, was given by Van Grieken (1982).

Solid materials in aqueous samples must be filtered out through, for example, Nuclepore membranes with 0.4 μm pore size (this is considered the conventional limit between 'dissolved' and 'particulate' matter in environmental waters) and the filter then analysed as a solid. In sea-water analysis the pre-treatment procedure requires elimination of the sea salt that is present. A simple procedure involves addition of HNO_3 followed by overnight heating at 80°C.

XRF is frequently invoked for trace analysis in air pollution studies. In air, trace elements are almost exclusively in the particulate phase at typical concentrations of 50–500 ng m^{-3}. By simply drawing a large volume of air through a filter, large pre-concentration factors are easily achieved. Adequately loaded filters with air-particulate matter are presented directly to the XRF unit. Unless the filters are extremely loaded, no correction for the X-ray absorption in the aerosol material is usually necessary. Alternatively aerosol-loaded filters can be dissolved prior to XRF analysis with supra-pure concentrated (70%) HNO_3 and HF in a high-pressure digestion vessel or in a microwave digestion unit. A standard is normally added prior to the digestion procedure.

When dealing with solid biological materials, drying, powdering, homogenizing and homogeneity testing might be necessary before preparing the samples for measurement. Various treatments for biological materials include drying, lyophilization, ashing and wet digestion. The digestion method should be optimized and the recovery of the elements and precision of the procedure should be thoroughly tested.

The literature on the above topic is extensive and more complete descriptions of sample preparation procedure can be found (Injuk & Van Grieken 1993; Buhrke *et al.* 1998). Here, as a summary for this section, an elementary scheme of sample preparation procedures appropriate to environmental samples is give in Fig. 4.7.

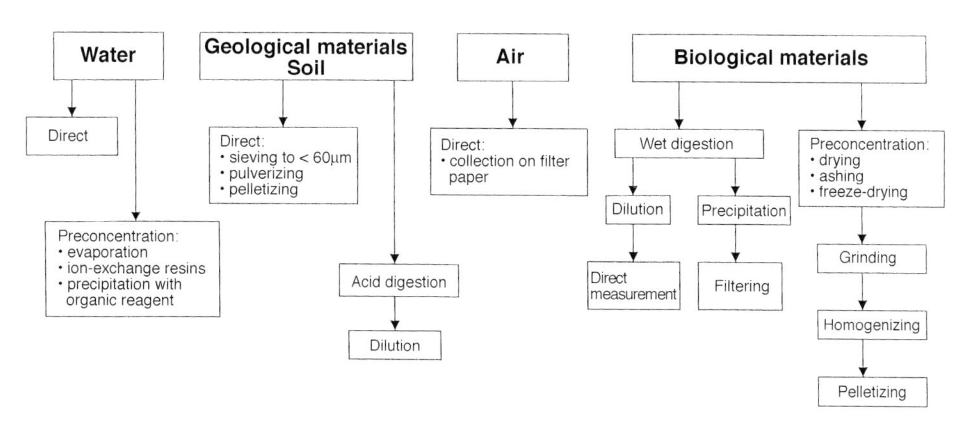

Fig. 4.7 An environmental flow-chart.

4.7 X-ray fluorescence techniques

4.7.1 *Wavelength-dispersive X-ray fluorescence (WDXRF)*

All conventional X-ray spectrometers comprise three basic parts: the primary source unit, the spectrometer itself and the measuring electronics. As was mentioned above, in WDXRF the X-rays emerging from the sample are angularly dispersed by a diffracting crystal and measured with a suitable detector (see Fig. 4.8). The prototype of

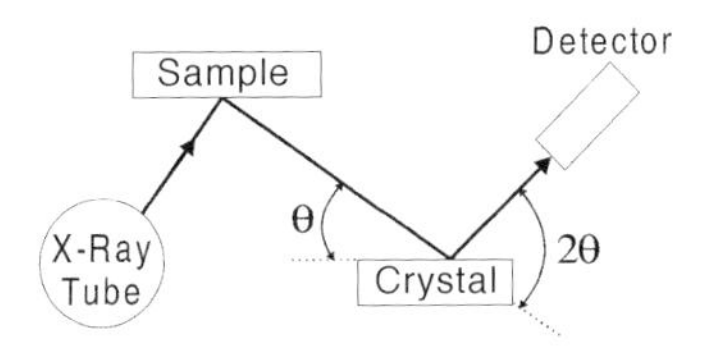

Fig. 4.8 Basic principle of the WDXRF spectrometer.

the first commercial wavelength-dispersive instrument was described by Friedman and Birks in 1948, while the first commercial instrumentation became available during 1950 and was progressively adopted by laboratories involved in the elemental analysis of a variety of sample types. Two principal categories of WDXRF instrument have been developed: single-channel or sequential and multichannel or simultaneous. Sequential instruments are designed with one or more wavelength-dispersive devices and a wide range of different elements may be determined, but for each wavelength separate counting has to be carried out with the crystal and detector set at the appropriate angles. Depending on the application, a single analysis of 15–20 elements, involving 30–40 separate spectrometer movements, may take up to 20 min. Modern spectrometers are usually provided with two X-ray tubes (e.g. one with a Cr target for long wavelengths and one with a W target for shorter wavelengths) and the interchange of dispersing crystals is possible. Analysis speed can be considerably increased by using a simultaneous X-ray spectrometer. This instrument is designed with a specified number (usually in the range 8–20) of individual fixed spectrometer channels, each optimized for the determination of a designated element. Multielemental analysis is undertaken within a few seconds by acquiring data simultaneously from all channels. Elemental determination by WDXRF is performed by measuring the background-corrected fluorescence intensity of a series of X-ray lines and comparing these data with the appropriate calibration and matrix correction functions. In Table 4.3 some characteristics of the WDXRF technique are summarized and compared to other XRF techniques.

4.7.1.1 *Applications, calibration and matrix effects* Over the past 40 years or so, WDXRF has become a well-established analytical technique for the determination of elemental compositions of raw materials, semi-finished products and final products in industrial environments. Large numbers of elemental determinations are undertaken daily, mostly in industrial practice. One of the most popular industrial uses of WDXRF is in the analysis of ores and minerals, oil and petroleum products, cement and related products, and metals like iron, aluminium and copper. Electronic and superconductor materials also present a growing interest to researchers nowadays. For example, the market demand for quality products in the microelectronics wafer industry has stimulated on-line WDXRF analysis that is able to measure both composition and layer thickness with considerable accuracy and speed. Hence, recently, a rapid growth in the use of automatic diffraction systems for industrial quality control has been seen. Reported applications and developments in the field of semiconductors involve typically phase analysis of thin films, texture analysis, strain evaluation in multilayer structures, reflectivity and topography measurements. In the

Table 4.3 Some characteristics of various X-ray techniques

Technique	Price ($)	Number world-wide	Detection limit	Spectral interferences	Matrix effects	Multielement	Preferred sample type	Vacuum
WDXRF	300000	15000	1–10 $\mu g\,g^{-1}$	Low	Medium to high[a]	Yes	Solid	Yes
EDXRF	100000	3000	1–10 $\mu g\,g^{-1}$	HIgh	Medium to high[a]	Yes	Solid	Yes
TXRF	130000 to 250000	300	$\geq$0.2 $ng\,ml^{-1}$	High	Negligible for chemical analysis[a]	Yes	Liquid (residues)	No (except for light elements)

[a] Depending on matrix.

automotive industry, WDXRF is frequently used on highly stressed material surfaces to determine their behaviour as engine or gear parts and to improve material properties. Furthermore, in the iron industry, quantitative determinations of total Fe and trace elements are necessary for optimizing various process parameters.

WDXRF has been confirmed as a valuable method in various environmental studies, e.g. monitoring environmental waters or plant evaluation (their growth, nutrient deficiency, food value, resistance to disease, etc.). In the past decade, most geochemical analyses on a routine basis were carried out using WDXRF.

Because of its high resolution, little peak overlap occurs in WDXRF ensuring that evaluation of spectra is quite straightforward. Once the concentrations of the elements of interest are deduced from the X-ray intensities, matrix effects must, in general, be considered. Frequently applied methods for the correction of matrix effects rely on the use of standards, in so-called 'compensation methods'. It is essential for a standard to have a high degree of similarity with the sample (same matrix). A standard may be a certified reference material or a sample but which is analysed by a different analytical technique. Occasionally use is made of an internal standard. Generally, the compensation methods correct for the effect of an unknown, but constant matrix. And in addition, standards are not required for the analysis of all constituents in the sample. Another approach to matrix corrections involves so-called 'mathematical methods' (i.e. fundamental parameters and methods based on the theoretical influence coefficients). Such methods can handle situations in which the matrix effect is more variable from sample to sample. In this respect, such techniques are more flexible than compensation methods but they will require more knowledge of the complete matrix. All elements contributing significantly to the matrix effect must be quantified, even if their determination is not required.

4.7.2 *Energy-dispersive X-ray fluorescence (EDXRF)*

Energy-dispersive spectrometer systems became available in the early 1970s and have traditionally been used in the field or in the laboratory for a variety of industrial, biomedical and environmental applications. The common EDXRF mode provides simultaneous determination of various elements, but EDXRF is often limited by insufficient sensitivity for low-Z elements and inter-element spectral effects. Such simultaneous detection offers a more rapid analysis than WDXRF.

Although the principles of excitation and fluorescence relevant to EDXRF are exactly the same as for WDXRF, the analytical characteristics of EDXRF are quite different owing to the differences in the response of the detector used. The X-ray spectrum emitted by an anode can be used in various excitation and detection geometries as illustrated in Fig. 4.9. A thin X-ray filter, typically of the same element as the anode material, is used to preferentially transmit the characteristic X-rays generated at the anode. This configuration provides the most efficient use of the photons generated at the anode, but is limited in choice of excitation energies. Another possibility involves the replacement of the sample by a secondary target the characteristic X-ray spectrum of which is excited by the direct output beam of the tube. These X-rays are then incident on the sample. Such a geometry has flexible choices of excitation energies but does not make efficient use of the output flux since the total system

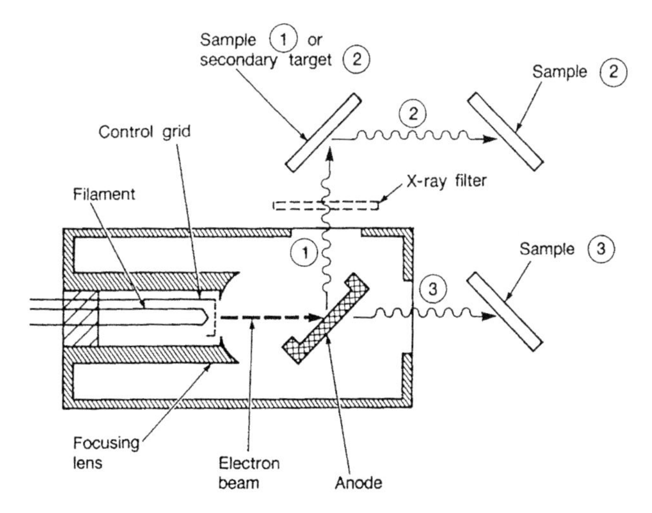

Fig. 4.9 Typical excitation geometries in tube-excited EDXRF: (a) the flux of X-rays (1) is used directly to strike the sample; (b) the primary flux (1) strikes a secondary target which generates a flux of X-rays for sample excitation; (c) the direct beam (3) is conducted through the anode towards the sample. (Reprinted with permission of Marcel Dekker, Inc.)

efficiency is proportional to the product of two solid angles. The third option is a variant of direct excitation in which the transmission of the X-rays through the anode itself is used to filter the continuum radiation. This method has certain advantages in the design of compact geometry, but is disadvantaged in having limited anode power.

Overall, the major disadvantage of the EDXRF technique is the limited energy resolution of the detector compared with a diffraction crystal. Hence, in contrast to WDXRF, line overlap is much more common, e.g. overlap of K_α and K_β lines of elements with atomic numbers $Z+1$ and Z. As there are no pure background regions between the peaks, background correction becomes more difficult and a rather sophisticated mathematical treatment of the spectra is generally required. In Table 4.3 some additional information relating to the EDXRF technique can be found.

4.7.2.1 *Applications, calibration and matrix effects* EDXRF has traditionally being used in the field or in the laboratory for a variety of industrial and environmental applications. EDXRF is ideally suited for field screening work as a result of the minimization of sample preparation and its potential portability. Such portable EDXRF instruments have experienced rapid development in recent times. Radioisotopes that have been frequently used in such instruments are now replaced with a pen-sized small tube and a tiny solid-state ED detector. With these devices, it is now possible to transport XRF spectrometers to depths of 50 m to measure, for instance, heavy metal contaminants in underground soils. Other, more frequently reported applications are *in situ* analysis of various geological, archaeological or architectural samples, artefacts and remote samples.

Monitoring and control of stack gas emission from industrial furnaces is of great importance nowadays. For this, and similar purposes, an automated EDXRF system was developed, which enables the monitoring and subsequent control of gas emissions from industrial furnaces, *in situ* measurements in the metal processing industry, or trace metals determination in the feed-water of nuclear power plants.

A very important environmental application of EDXRF is in the analysis of airborne particulate matter (see Table 4.4). It seems to be the method of choice since aerosol-loaded filters are an ideal target for EDXRF. Since the area of the sample (filter) actually analysed is only a few cm^2, however, it must be ensured that the result is representative of the entire sample. Therefore, a critical step in the procedure is the selection of an appropriate filter material. Teflon and polycarbonate (Nuclepore) filters are ideal because of their high purity and because they are surface collectors. Whatman-41 cellulose filters have been used widely because of their low cost, but they are rather thick (around 9 mg cm^{-2} versus 1.1 mg cm^{-2} for Nuclepore), which leads to more X-ray scatter and higher detection limits. They also partially collect particulate matter within their depth, so that X-ray absorption corrections become more complicated. Glass-fibre filters should be avoided because of their high inorganic impurity. High-purity quartz-fibre filters have recently been proposed, however. Before selecting a filter for a particular application, the blank count of the filter or background level of the material to be analysed must be determined. All filters contain various elements as major, minor and trace constituents as is seen from Table 4.5.

Table 4.4 Elemental concentrations and standard deviations (ng m^{-3}) measured by EDXRF in airborne particulate matter sampled in an urban area of north-western Europe

Element	Concentration
Al	350 ± 40
Si	660 ± 30
S	1400 ± 70
Cl	3500 ± 180
K	170 ± 6
Ca	260 ± 10
Ti	19 ± 1
V	5.3 ± 0.6
Cr	1.7 ± 0.2
Mn	4.8 ± 0.4
Fe	440 ± 13
Ni	2.3 ± 0.6
Cu	3.6 ± 0.6
Zn	33 ± 1
Br	5.4 ± 0.6
Rb	0.5 ± 0.3
Sr	4.8 ± 0.2
Pb	12 ± 1

In the case of homogeneous and thin samples, accurate calibration is most easily carried out by measurement of thin standards of the elements of interest, with the aim of deriving sensitivity factors (expressed in counts per seconds per μg cm^{-2}) for K_α or L_α lines. These standards should be sufficiently thin to make absorption corrections small or negligible and the element of interest should be homogeneously distributed

Table 4.5 Some trace element impurities ($\mu g\ cm^{-2}$) for Nuclepore and Teflon filters as determined by EDXRF

Element	Nuclepore filters	Teflon filters
Al	0.019	–
S	0.03	–
Cl	0.019	–
Ca	0.001	–
Cr	0.003	–
Fe	0.005	0.005
Ni	0.001	—
Cu	0.002	0.002
Zn	0.001	–
Br	0.025	–
Rb	0.003	0.002
Sr	0.005	0.005
Pb	–	0.033

over the analysed area. For calibration purposes, 'absolute methods' based on fundamental physical constants combined with experimental parameters may also be used. With regard to aerosol analysis, several factors affecting X-ray intensity must be considered, including attenuation of X-rays by individual particles (important only for particles larger than a few micrometres), accumulation of particles that have an adverse effect (in the case of heavily loaded filters) and the filter itself attenuating incident and emerging X-rays (if particles are not collected on the filter surface). Appropriate correction involves a knowledge of the particle size distribution while the two other effects can be easily eliminated by adequate sampling time and an appropriate choice of filter material.

With liquid samples there are two options for analysis. First, direct analysis with or without an internal standard and second, pre-concentration. With the latter approach the samples can be physically evaporated and concentrated to determine their $\mu g\ g^{-1}$ levels or chemical pre-concentration can be applied as was described in Section 4.6.

To date, many geochemical samples, soils and sediments have been analysed using EDXRF, and obtainable detection limits for elements such as As, Pb, Cd and Hg are normally a few ppm. A comparison of analytical results obtained for soil standard samples having recommended values is given in Table 4.6. In Fig. 4.10 an X-ray spectrum obtained from measurements of a standard soil sample is given.

EDXRF is further suitable for various environmental applications as is demonstrated in Table 4.7. All reported analyses were performed on dried, powdered samples and most detection limits were in the $\mu g\ g^{-1}$ range or below.

4.7.3 *Total-reflection X-ray fluorescence (TXRF)*

The idea and first demonstration of TXRF were presented in 1971 by Yoneda and Horiuchi and since then many publications have appeared describing both methodology and applications (Klockenkämper 1997). In contrast to traditional XRF, which uses angles of incidence of around 45°, TXRF operates at grazing angles in the vicinity of the critical angle, θ_C, of about 0.1°, below which X-rays undergo total

Table 4.6 Comparison of analytical results obtained by EDXRF for a standard soil sample (GSS–2) with certified values. All values are in μg g^{-1} if not otherwise indicated

Element or oxide	Certified value	Measured value
MgO (%)	1.04	1.2
Al_2O_3 (%)	10.31	10.3
SiO_2 (%)	73.35	72.9
K_2O	2.54	2.60
CaO	2.36	2.41
Fe_2O_3	3.52	3.35
Ti (%)	0.271	0.279
V	62	< 85
Cr	47	< 70
Mn	510	477
Ni	19.4	17
Cu	16.3	15
Zn	42.3	40
Ga	12	14
As	13.6	18
Rb	88	86
Sr	187	185
Y	21.7	21
Zr	219	202
Pb	20.2	18

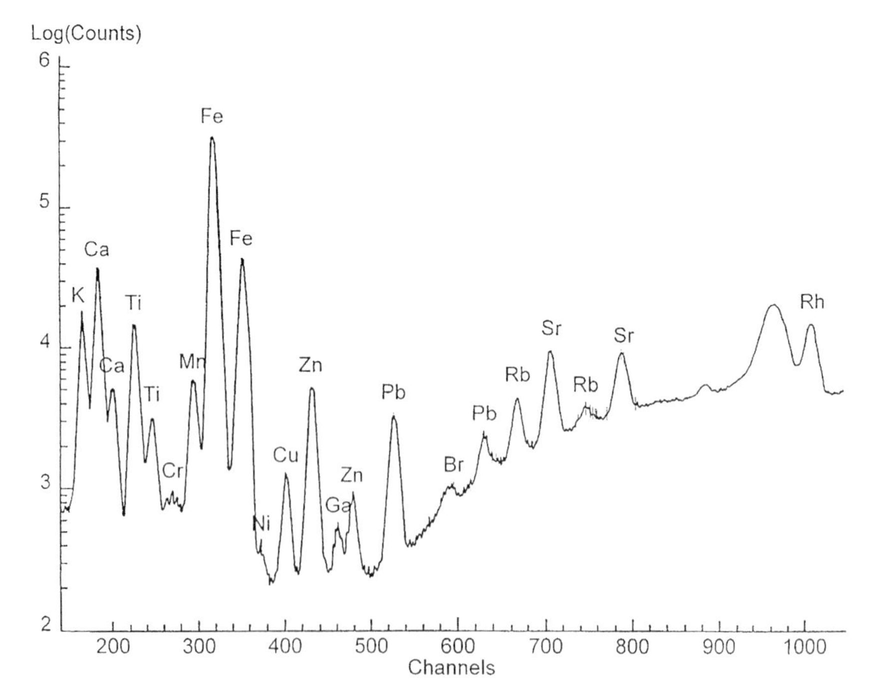

Fig. 4.10 EDXRF spectrum of a standard soil sample.

Table 4.7 Detection limits ($\mu g\ g^{-1}$) for different powdered samples analysed by EDXRF. Working conditions: Ag tube, 300 s analysis time, dry weight (Nguyen *et al.* 1998)

Element	Leaves	Fish	Hair	Coal	Sediments
S	70	74	180	620	930
Cl	20	15	20	60	90
K	8	3	7	60	80
Ca	2.5	2	5	30	50
Ti	1.5	1	1	9	13
Cr	1.2	0.9	1	8	12
Mn	1.1	0.9	1	6.5	10
Fe	1.1	0.8	1	5.6	8
Co	0.7	0.7	0.8	3.8	6
Ni	0.5	0.6	0.8	1.5	2
Cu	0.5	0.5	0.7	1.5	2
Zn	0.4	0.4	0.6	1.4	2
Se	0.3	0.3	0.2	0.4	0.6
Br	0.2	0.2	0.2	0.4	0.6
Rb	0.2	0.04	0.07	0.2	0.3
Sr	0.1	0.03	0.05	0.2	0.3
Ba	0.4	0.4	0.5	0.9	1.8
Hg	0.2	0.2	0.3	0.8	1
Pb	0.2	0.2	0.2	0.7	1

external reflection from a flat surface (Fig. 4.11). The critical angle, θ_C, varies with the photon energy and the reflector material. For Mo K_α radiation impinging on quartz, $\theta_C = 1.9$ mrad. The total reflection of X-rays is of interest as the X-rays only penetrate to very shallow depths of the reflector surface. In the above example, the penetration depth is about 74 nm. As a consequence the support material does not produce any significant interfering background radiation. This effect can be applied to a number of surface-sensitive analysis methods. The only pre-condition is that the sample must be present as a very thin layer otherwise the principle of total reflection will be lost. Fig. 4.12 shows the TXRF instrumental set-up.

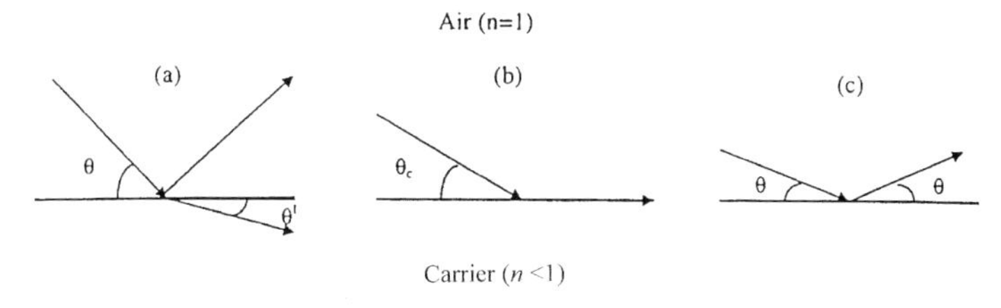

Fig. 4.11 Schematic representation of refraction and reflection phenomena. (a) $\theta > \theta_C$: when the beam penetrates into the medium and is refracted, a small portion (1%) is reflected. If the angle of incidence is reduced to smaller values one observes (b) $\theta = \theta_C$ where the beam propagates along the surface of the reflector or (c) $\theta < \theta_C$ where total reflection of X-rays occurs.

4.7.3.1 *Applications and calibration* TXRF is a well-established technique for microanalysis: less than 1 μg of a solid sample or 10 μl of a solution is required for an analysis in the parts per trillion (ppt) range. It allows the detection of extremely small amounts, i.e. below 20 pg for over 60 elements.

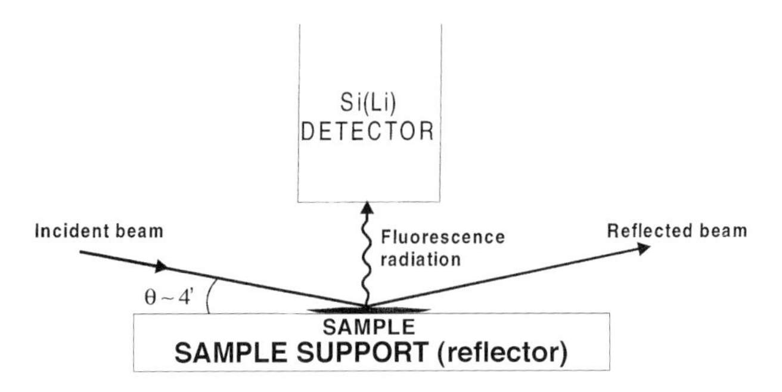

Fig. 4.12 TXRF instrumental set-up.

TXRF can be applied to the analysis of aqueous solutions, acids, airborne particulate matter, biological materials, ultra-pure metals, wafers in the semiconductor industry, etc. It is not only suitable for analysis of inorganic materials, such as ashes, sludges, sediments and soils but also organic materials like fruits, cereals, grasses and vegetables following suitable digestion and preferably matrix removal.

Because of its simplicity, TXRF is a method of choice for quality control of ultra-pure reagents, such as those needed in the electronics, cosmetics and pharmaceutical industries. Some high-grade acids, bases and solvents including high-purity water can be analysed, almost directly, down to 1 ng ml^{-1}. Further applications involve analysis of high-purity metals, such as Al or Fe and ultra-pure non-metals, e.g. Si and SiO_2. These solid products can be analysed in the sub-μg g^{-1} region after appropriate matrix separation. TXRF has been confirmed as especially useful for the determination of metal traces in light and heavy crude oils, in motor oils and in diesel fuels.

In the environmental sector especially, TXRF demonstrates particular usefulness owing to its multielemental capability for small sample amounts and its short analysis time. The analysis of water samples and air dust is routinely utilized with TXRF, as well as the analysis of organic tissue, plant material and vegetable foodstuff. In analysing rain-water or drinking water (both tap and mineral), a detection limit of ng ml^{-1} is directly accessible. For the analysis of river water, sea-water and waste waters, sample preparation is recommended to separate the suspended matter and remove the salt content. For instance, trace contaminants in samples from the Atlantic Ocean were studied at several deep-water stations, while heavy metal traces and pollutant transfers were investigated in samples from the North Sea (Injuk *et al.* 1998). In Table 4.8, some characteristics related to the analysis of aqueous samples by different XRF techniques are outlined.

Effective monitoring of air pollution assumes small sampling volumes and small collection times, as is offered by TXRF. Collection of air dust can be performed by filtration or impaction. If membrane filters are used the loaded filters (together with a standard) can be subjected to pressure digestion with HNO_3 and analysed by TXRF as usual. Another possibility is to collect air directly on glass or plexiglass carriers that are ordinarily fitted for TXRF devices. Under such conditions, various elements can be determined down to <0.1 ng m^{-3}. These detection limits are three orders of magnitude less than with conventional EDXRF. Consequently, the sampling time

Table 4.8 Some characteristics of XRF techniques for aqueous sample analysis

Technique	Sample amount	Sample preparation	Detection limits	Detectable elements
WDXRF	ml	Pre-concentration Direct analysis	$\mu g\ l^{-1}$ $mg\ l^{-1}$	$Z \geq 4$
EDXRF	e.g. 5 ml	Pre-concentration with APDC Direct analysis	$\mu g\ l^{-1}$ $mg\ l^{-1}$	$Z \geq 11$
TXRF	10–50 µl	Direct analysis Freeze-drying Extraction Pre-concentration with APDC	$\mu g\ l^{-1}$–$ng\ l^{-1}$	$Z \geq 11$

can be reduced to 1 h and the sampling volume to 0.5 m^3. Such analytical performances allow the monitoring of air pollution in the course of a day (Injuk and Van Grieken 1995).

By using SR and special X-ray tubes, which are more powerful than those available commercially, it is now possible to determine lightweight elements, such as Na, Mg and Al, on Si wafers down to the 100 fg level (Streli *et al.* 1997). Furthermore, with the SR monochromatized with a multilayer it is possible to determine 13 fg of Ni on Si wafers (Wobrauschek *et al.* 1997).

Quantitative analysis by TXRF is essentially facilitated by the use of small or minute amounts of sample. Owing to this particular operational condition, the matrix effects of conventional XRF cannot occur and a simple calibration by internal standardization can be performed in trace element analysis. Based on this easy and reliable method, calibration curves, which show a constant relationship between the atomic number and the fluorescence yield, are easily calculated.

4.7.4 *Micro-X-ray fluorescence analysis (Micro-XRF)*

Micro-XRF is a sort of microscopical variant of bulk EDXRF, where a microscopic X-ray beam is used to excite locally a small area (e.g. $10 \times 10\ \mu m^2$). Such small-dimension X-ray beams are effectively obtained using capillary optical devices. These operate through single or repeated total reflection of X-rays at the inner wall of a glass capillary tube. In principle, capillaries create a beam with size determined by the inside diameter of the tube end and not by the X-ray source or any optical aberrations. Capillaries, however, provide just one of the many possible optical elements for X-rays. Other devices, which in practice are used mostly in conjunction with SR sources, include bent crystals and multilayer X-ray mirrors. These cannot always achieve the microscopic focusing level required and need to be combined with other demagnification tools to achieve the performance of real microscopic X-ray probes such as the glass capillary.

Micro-XRF promises to be one of the best microprobe methods for inorganic analysis of various materials. It operates at ambient pressure and in contrast to particle-induced X-ray emission (PIXE) and electron probe X-ray micro-analysis (EPXMA), no charging occurs. In many instances, sample preparation is not necessary other than placing the material in the sample chamber.

During the 1980s, synchrotron facilities around the world began to implement X-ray microbeam capabilities on their beam lines for localized elemental analysis. A particular advantage of this when compared with conventional X-ray sources is the extremely high brilliance achieved with synchrotron radiation (SR) sources. A schematic illustration of an SRXRF set-up is given in Fig. 4.13.

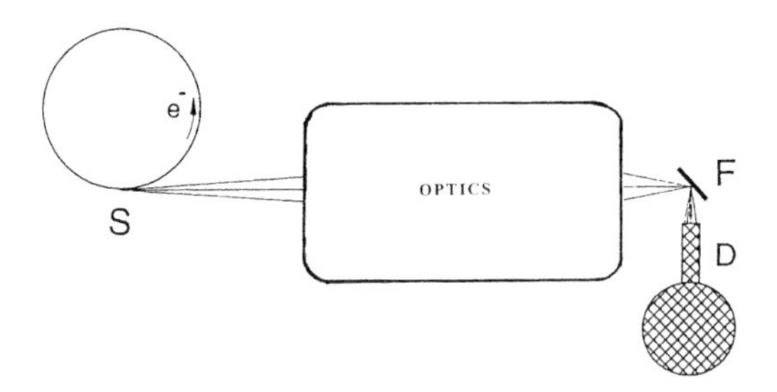

Fig. 4.13 Schematic illustration of an SRXRF set-up. The primary radiation, which originates from the synchrotron storage ring (S), is transformed into a microbeam by a suitable optical system. This microbeam is used to excite a spot on the sample (F) and the emerging fluorescence and scattered radiation is usually detected by a solid-state detector (D).

The field of micro-XRF is currently subject to significant evolution in instrumentation (lead-glass capillaries and polycapillary X-ray lenses; air-cooled micro-focus X-ray tubes; compact ED detector systems with good resolution even at high count rates; and no longer requiring liquid-nitrogen cooling). This evolution has allowed the use of micro-XRF in a wide variety of materials science investigations, in various environmental studies and in the analysis of traditional artefacts of fine structure and design.

4.7.4.1 *Applications* Presently, most applications are focused on demonstrating the spatial distribution of micro-XRF using line-scans or mapping capabilities. Among the many areas of application of micro-XRF, archaeological studies, materials science and the characterization of air-particulate matter are the most interesting.

Often, micro-XRF is employed in controlling the coating composition, thickness and homogeneity of various materials. Layer systems can be examined easily with sensitivity down to 5 nm. Furthermore, the small spot size permits measurement of coating thickness at different spots (homogeneity investigation).

In the field of environmental studies, the X-ray microprobe has been employed successfully, e.g. in an inorganic trace elements investigation of a wood sub-ring structure. The use of transmitted and scattered radiation revealed subtle changes in wood density related to changes in temperature during the growth period (Kuczumow *et al.* 1995). For air-particulate matter, a significant improvement resides in the use of microanalytical mapping. At current micro-XRF facilities, only particles of diameter above 10 μm can be readily analysed (Rindby *et al.* 1996).

As well as providing information on the concentration of certain elemental species, chemical information on sample materials can also be obtained. This is achieved by scanning the energy of a monochromatized X-ray beam over the absorption edge of the element of interest.

Based on the present situation, it is clear that progress in the field of X-ray microanalysis using X-ray sources is advancing at a rapid rate (Adams *et al.* 1998). Use of commercial laboratory instrumentation incorporating capillary optics combined with rapid scanning and compositional mapping capability is expected to grow. Already new systems from companies such as Horiba, EDAX and XCO have appeared (Pella and Lankosz 1997).

4.8 Conclusions, trends and the future

The role that the XRF method plays in modern society is confirmed by the variety of applications to which the technique is put daily in different analytical laboratories world-wide. Conventionally, XRF is able to quantify all elements in the periodic table from fluorine. Newer developments, however, yield determinations of low atomic number elements including B, C, O and N. Accuracies of 1% and better are now possible for most atomic numbers and elements are detectable in many cases to the low ppm level. Excellent data-treatment software is available which allows the rapid application of quantitative and semi-quantitative procedures.

Since XRF is one of the best understood and established techniques, fundamental research is rather limited and is directed mainly to the area of matrix correction models and the evaluation of X-ray cross-sections.

A clear trend is emerging in X-ray tube technology. Almost all manufacturers are now using end-window tubes as opposed to side-window designs. An advantage of the end-window geometry is the closer optical coupling between tube and sample. The shape of the anode and filaments has also been redesigned to withstand higher electrical currents and hence higher emission yields. To complement this, the X-ray tube window is getting thinner allowing better transmission of low-energy X-rays.

An important advance involves data treatment. This has led to powerful semi-quantitative software programs. Thanks to the increasing power of personal computers and the availability of fundamental parameter programs and highly interactive graphic programs, XRF analysis of any solid or liquid sample is easily accessible to anyone.

In WDXRF, most of the recent improvements have been in the light element region. Parameters such as the fluorescent yield, the primary X-ray spectrum, the X-ray optics, multilayer crystals and their reflectivity, the detector window and the electronics have been optimized to extend the element detectability of XRF to beryllium. As a result new application areas have been opened such as the determination of nitrogen in ceramics or beryllium in copper alloys.

Spectacular advances have also been witnessed in EDXRF. Portable EDXRF instrumentation is undergoing rapid development. Radioisotopes that have been frequently used in such instruments are now being replaced with a pen-sized X-ray tube. Tiny ED detectors and conical capillaries are used as well. Small transmission or micro-focus tubes with low-power exigency are now available from most manufacturers. With these miniature detectors and X-ray tubes it is possible to transport an

XRF spectrometer to depths of 50 m to measure, for instance, heavy metal contaminants in underground soil.

The field of micro-XRF is currently subject to significant evolution in instrumentation including use of lead-glass capillaries and polycapillary X-ray lenses, air-cooled micro-focus X-ray tubes, compact ED detector systems with good resolution even at high count rates and instruments that no longer require liquid-nitrogen cooling. This evolution has allowed the use of X-ray microfluorescence and imaging techniques to become powerful tools in, e.g. the semiconductor industry and forensic-type investigations.

References

Adams, F., Janssens, K. & Snigirev, A. (1998) *Journal of Analytical Atomic Spectrometry*, **13**, 319–331.

Buhrke, V.E., Jenkins, R. & Smith, D.K. (eds) (1998) *A Practical Guide for the Preparation of Specimens for X-Ray Fluorescence and X-Ray Diffraction Analysis*. John Wiley & Sons, New York.

Injuk, J. & Van Grieken, R. (1993) Sample preparation for XRF. In: *Handbook of X-ray Spectrometry–Methods and Techniques* (eds R.E. Van Grieken & A.A. Markowicz), pp. 453–488. Marcel Dekker, New York.

Injuk, J. & Van Grieken, R. (1995) *Spectrochimica Acta*, **B50**, 1787–1803.

Injuk, J., Van Grieken, R. & de Leeuw, G. (1998) *Atmospheric Environment*, **32**, 3011–3025.

Klockenkämper, R. (1997) *Total Reflection X-Ray Fluorescence Analysis*. John Wiley & Sons, New York.

Kuczumow, A., Rindby, A. & Larsson, S. (1995) *X-Ray Spectrometry*, **24**, 19–26.

Nguyen, T.H., Boman, J. & Leermakers, M. (1998) *X-Ray Spectrometry*, **27**, 265–276.

Pella, P.A. & Lankosz, M. (1997) *X-Ray Spectrometry*, **6**, 327–332.

Rindby, A., Voglis, P. & Attaelmanan, A. (1996) *X-Ray Spectrometry*, **25**, 39–49.

Streli, C., Wobrauschek, P., Bauer, V., Kregsamer, P., Görgl, R., Pianetta, P., Ryon, R., Pahlke, S. & Fabry, L. (1997) *Spectrochimica Acta*, **B52**, 861–872.

Tertian, R. & Claisse, F. (1982) *Principles of Quantitative X-Ray Fluorescence Analysis*. Wiley-Heyden, London.

Van Grieken, R. (1982) *Analytica Chimica Acta*, **143**, 3–34.

Van Grieken, R. & Markowicz A. (eds) (1993) *Handbook of X-Ray Spectrometry, Methods and Techniques*. Marcel Dekker, New York.

Wobrauschek, P., Görgl, R., Kregsamer, P., Streli, C., Pahlke, S., Fabry, L., Haller, M., Knöchel, A. & Radtke, M. (1997) *Spectrochimica Acta*, **B52**, 901–906.

5 Elemental analysis in scanning electron microscopy

M. TALIANKER

5.1 Generation of X-ray radiation in scanning electron microscope

When the electron beam in the scanning electron microscope (SEM) is swept across the surface of a specimen, the resultant X-rays are generated through the interaction of the incident beam electrons and the atoms within the small irradiated volume of the target. A typical example of a spectrum of X-radiation obtained from such a target is shown in Fig. 5.1 where X-ray intensity is represented as a function of the wavelength of the X-rays. Examination of this figure reveals immediately two main features. The first is the presence of a broad continuous background with a sharp cut-off on the short-wavelength side of the spectrum at λ_{min} and a maximum at approximately $1.5\lambda_{\mathrm{min}}$. The X-ray intensity then gradually falls at longer wavelengths. This background represents the so-called 'Bremsstrahlung', or 'continuum', or 'white' radiation. The continuum radiation occurs as a result of deceleration of electrons of the incident beam. When the electrons impact the specimen target they are slowed giving up their energy in numerous collisions with the atoms of the specimen. In accordance with classical electrodynamics the decrease of kinematic energy of the decelerated charge must be released in the form of electromagnetic radiation.

The short-wavelength cut-off at λ_{min} is related to the maximum voltage V_0 applied to the electron gun where the incident electrons are accelerated and can be calculated from the equation:

$$\frac{hc}{\lambda_{\mathrm{min}}} = eV_0 \tag{5.1}$$

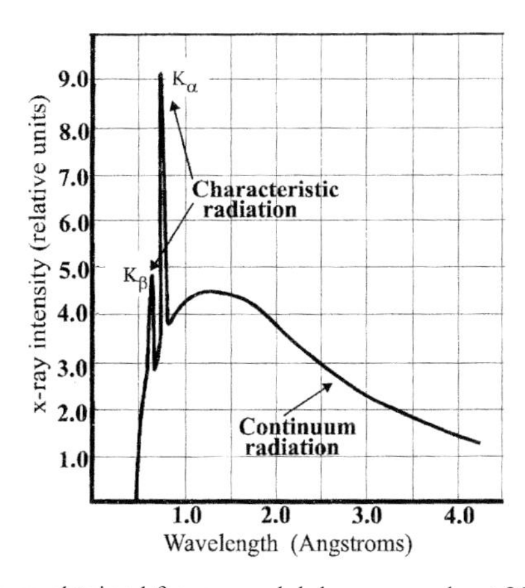

Fig. 5.1 The X-ray spectrum obtained from a molybdenum anode at 25 kV.

Here h is Plank's constant, c is the velocity of light and e is the charge of the electron.

The second feature of the X-ray spectrum is the presence of very intense but comparatively narrow peaks that represent the so-called 'characteristic lines' which are superimposed on the continuous background. These characteristic X-rays are generated in the following mechanism. When incident electrons bombard the specimen they can eject electrons from the specimen atoms thus creating unoccupied sites in their inner shells (K, L or M) leaving atoms in an excited or ionized state. An atom returns to its ground state when an electron from any one of the outer shells jumps to an unoccupied site. This event is accompanied by emission of a characteristic X-ray photon at a specific wavelength, λ, corresponding to the energy difference, ΔE, between the initial and final states of the atom:

$$\frac{hc}{\lambda} = \Delta E \qquad (5.2)$$

Figure 5.2 shows schematically the diagram of atomic energy levels and possible transitions of electrons between these levels, assuming that the zero energy level corresponds to a normal atom in its ground state. For example, ejection of an electron from the K shell raises the energy of the atom to the K level with value E_K. When the 'hole' in the K shell is filled by an electron from, say, the L_{III} level, the atom reduces its energy from E_K to E_{III}. This transition is accompanied by the emission of K_{α_1} characteristic radiation, and the arrow indicating the K_{α_1} spectrum line is drawn accordingly from the K level to the L_{III} level. The discrete energy levels and the energy differences between the levels are specific for atoms of a given atomic number, and, therefore, the wavelengths of the characteristic X-rays are unique to the atoms of a given element. Hence, measurement of the energy (or wavelength) of the X-ray

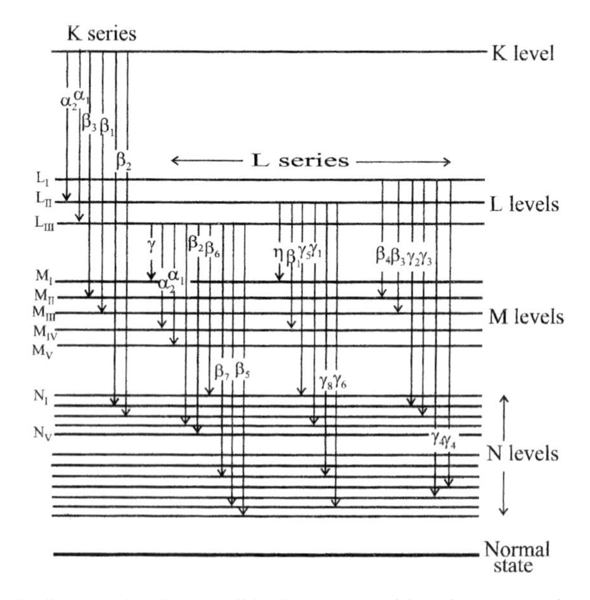

Fig. 5.2 A schematic diagram showing possible electron transitions between various energy levels in an atom.

photons emitted from the specimen's atoms allows identification of the elements in the specimen and is used for analytical purposes. This is the main principle upon which the elemental analysis in the SEM is based.

5.2 Spectral notation of X-ray emission lines

From the diagram in Fig. 5.2 it can be seen that all electron transitions in the excited atom accompanied by the emission of characteristic X-rays can be divided into several groups depending on the type of inner shell (K, L or M) from which the electron was initially ejected. Accordingly, each group of transitions gives rise to emission of K, L or M series of lines in the spectrum of characteristic X-rays. For instance, the generation of a K series of lines K_{α_1}, K_{α_2}, K_{β_3}, K_{β_1}, etc. is due, respectively, to the transitions $K \rightarrow L_{III}$, $K \rightarrow L_{II}$, $K \rightarrow M_{II}$, $K \rightarrow M_{III}$, etc.

Table 5.1 shows an example of electron transitions in the excited Mo atom and corresponding spectral notations of the X-ray emission lines. Note that emission of the L series of lines is due to transitions of the atom from three closely spaced levels L_I, L_{II} and L_{III} to M and N levels.

Table 5.1 Principal emission lines from the X-ray spectrum of molybdenum

Electron transition	Notation of X-ray line	Intensity (relative units)	Wavelength (Å)
K series			
K–L_{III}	K_{α_1}	100	0.709
K–M_{II}	K_{α_2}	50	0.713
K–M_{III}	K_{β_1}	15	0.632
K–M_{II}	K_{β_3}	15	0.633
K–N_{II},N_{III}	K_{β_2}	5	0.621
K–N_{IV},N_V	K_{β_4}	weak	0.620
L series			
L_{III}–M_V	L_{α_1}	100	5.406
L_{III}–M_{IV}	L_{α_2}	10	5.414
L_{II}–M_{IV}	L_{β_1}	80	5.176
L_{III}–N_V	L_{β_2}	60	4.923
L_I–M_{III}	L_{β_3}	30	5.013
L_I–M_{II}	L_{β_4}	20	5.048
L_{III}–N_I	L_{β_6}	weak	5.048

The ejection of an electron from the K shell requires a certain amount of energy, which is specific for the atom of a given element. This energy value is called the critical excitation potential $E_{K(cr)}$. In addition, it can be said that the energy of the incoming electrons of the primary beam must exceed the value $E_{K(cr)}$ in order to excite the K shell in the atom. Similarly, the critical excitation potentials necessary to pull an electron away from the L shell and M shell are designated $E_{L(cr)}$ and $E_{M(cr)}$, respectively. It should be noted that in practice, in order to ensure an adequate intensity of X-ray lines, the energy of the beam electrons (E_0) must exceed the critical excitation potential by approximately a factor of two (or more). In other words, the overvoltage, U, defined as the ratio $U = E_0/E_{(cr)}$, must be of value two or greater.

From Fig. 5.2 it can be concluded that if incident electrons in the beam have sufficient energy to remove an electron from the K shell, they can also expel any L or M shell electrons and, therefore, the K, L and M series of lines should appear simultaneously.

It is important to note that not all electron transitions are permitted by the selection rules of quantum mechanics. Besides, allowed transitions differ in their occurrence probability. Therefore, characteristic X-ray lines vary in intensity (see Table 5.1) and only a few important lines with the greatest intensity (K_α, K_β, ...) are used in practice for elemental analysis.

5.3 The penetration of beam electrons into a bulk specimen: spatial resolution of X-ray microanalysis

When an electron beam strikes a bulk target the incident electrons, before they are stopped, penetrate into the material to a depth of up to several microns, depending on the composition of the sample. The main processes controlling electron penetration are: (1) elastic scattering that causes the electrons to deviate from their initial direction of movement without loss of their energy; and (2) inelastic scattering resulting in a gradual loss of energy of the incident electrons, mostly with little change in the direction of their movement.

There are two main mechanisms involved in the process of elastic scattering: (a) large-angle Rutherford collisions and (b) small-angle multiple scattering events, which together may also lead to a large change in the moving direction of the incident electrons. A proportion of the incident electrons may even be scattered back to the surface and escape from the specimen. This is the so-called backscattering effect.

Turning to inelastic scattering, it can be said that although the majority of the electron energy is dissipated as heat, a fraction of the lost energy is converted to X-rays. As was described in Section 5.1, characteristic X-rays are produced as a result of ionization processes in the inner shells of target atoms caused by incident electrons, provided that the energy of the incident electron is greater than the critical excitation potential $E_{j(\mathrm{cr})}$ of the atom of the target (j = K, L, or M for excitation K, L, M X-ray series, respectively). As soon as the energy of the incident electron falls below $E_{j(\mathrm{cr})}$, no further production of characteristic X-rays occurs, although the process of electron deceleration will continue until the energy decreases to zero.

As a result of the scattering effects the electrons are eventually spread over the target within a pear-shaped micro-volume, just below the surface. The size and form of this region of electron spreading depends on both the incident beam energy and the mean atomic number of the target material. It is obvious that characteristic X-rays are generated only from this region.

The region of electron spreading in the bulk specimen is shown schematically in Fig. 5.3, where D is the diameter of the incident electron beam impacting the sample surface, and R_X is the depth of the X-ray production region, the so-called X-ray range. The volume within the dashed line represents the overall electron distribution in the specimen, while the region between the dashed and solid curves corresponds to

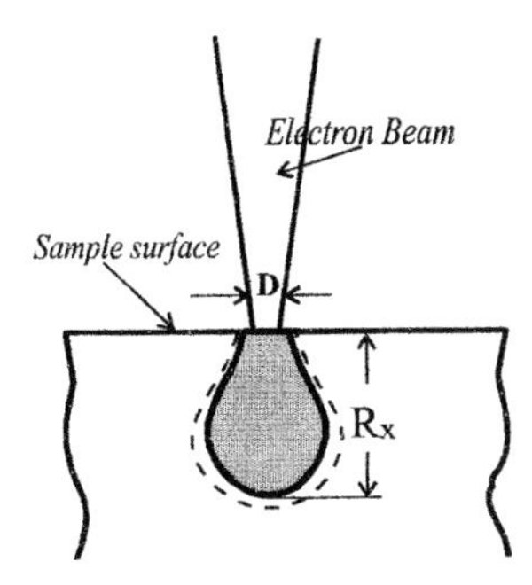

Fig. 5.3 Schematic illustration of the effect of penetration of electrons into a bulk specimen. The solid line outlines the X-ray production region.

electrons with energy below the ionization potential; this latter region, therefore, does not serve as a source of characteristic X-rays.

If we now define *the spatial resolution* of X-ray microanalysis as the minimal distance between two adjacent areas of the specimen surface from which two independent analytical results can be received, it will be easily seen that it is the size of the X-ray production region that actually determines the spatial resolution of the microanalysis. Assuming that the lateral X-ray production range is approximately the same as R_X it can be said that R_X represents the X-ray source size and, in fact, determines the spatial resolution.

The estimation of the value R_X as a measure of spatial resolution has been the subject of much theoretical and experimental work. For calculations, instead of R_X it is convenient to use the value ρR_X, the so-called 'mass range', and the general expression developed for ρR_X can be presented in the form

$$\rho R_X = K(E_0^n - E_{(cr)}^n) \tag{5.3}$$

where E_0 and $E_{(cr)}$ (in keV) are, respectively, the acceleration voltage and the critical potential for excitation of particular X-ray series in a given matrix with density ρ (in $g\,cm^{-3}$). If R_X is expressed in μm, then the constants K and n are 0.064 and 1.68, respectively (Andersen & Hasler 1965).

Equation (5.3) shows that the X-ray range R_X depends not only on the density of the matrix but also on the difference between the energy of incident electrons, E_0, and the critical excitation energy, $E_{j(cr)}$ (j = K, L or M), of the X-ray line measured. This implies that in order to obtain spatial resolution of analysis of less than 1 μm, the energy of the electron beam should be chosen correctly, depending on the material of the matrix and the element to be analysed. For example, consider calculations from Equation (5.3) of the X-ray source size for the Si particle in an Al matrix and in a Cu matrix. For Si K_α the value $E_{K(cr)}$ is 1.84 keV, and the densities are $\rho_{Al} = 2.7\ g\ cm^{-3}$ and $\rho_{Cu} = 8.93\ g\ cm^{-3}$. If an operating voltage of 20 keV is used, then for the Si particle in an Al matrix, $R_X^{Si(Al)} \cong 3.5$ μm while for the Si particle in a Cu matrix, $R_X^{Si(Cu)} \cong 1.05$ μm. It is seen that the X-ray range for Si K_α in Al is substantially greater than that for Si K_α in the Cu matrix. Therefore, in order to reduce the size of the X-ray source to a value of less than 1 μm, the voltage of the electron beam E_0 has to be set below about 11 keV. This example clearly demonstrates that the spatial resolution of analysis is actually a function of two main parameters. First, the incident beam energy E_0, i.e. the accelerating voltage in SEM, and second, the specimen composition. By adjusting the operating voltage and by proper choice of the analysed X-ray line (of type K, L or M) spatial resolution of analysis of less than 1 μm can be obtained.

5.4 Collection of X-rays in SEM: energy resolution

As was seen in previous sections, chemical analysis in the scanning electron microscope relies on the emission of characteristic X-rays generated from a small volume of specimen that is being bombarded by incident beam electrons. It is obvious that reliable detection and recording of the generated X-rays is of great importance. In the scanning electron microscope this problem is solved by attaching an X-ray spectrometer, the basic purpose of which is to acquire an X-ray spectrum, i.e. to measure the intensity of the X-rays as a function of their energy (or wavelength). Two types of spectrometers can be used in SEM:

(1) The wavelength-dispersive spectrometer (WDS), or crystal spectrometer, which employs one or more diffracting crystals in order to separate the X-rays of various wavelengths and to detect them *sequentially* using a suitable detector (usually a gas-flow proportional counter).

(2) The energy-dispersive spectrometer (EDS), which combines a special solid-state detector with processing electronics that allows *simultaneous* recording of the *whole spectrum* of X-rays; dispersing X-rays into a spectrum is performed according to their energy rather than their wavelengths.

Let us consider each type of spectrometer in detail.

5.4.1 *The wavelength-dispersive spectrometer*

The principle of a wavelength-dispersive spectrometer with Johansson geometry (Johansson 1933) is illustrated in Fig. 5.4. An electron beam strikes the specimen and a cone of X-rays generated from the point source S on the sample can be reflected by an analysing crystal according to Bragg's diffraction law

$$2d \sin\theta = n\lambda \tag{5.4}$$

where n is an integer (only the first-order reflection with $n=1$ is considered), λ is the X-ray wavelength, d is the interplanar spacing of the analysing crystal and θ is the incidence angle of the X-rays.

In Johansson geometry the crystal's reflecting planes are bent to a radius $2r$, and the surface of the crystal itself is ground to radius r. On the basis of geometrical considerations it is easy to show that if the crystal and the point source S are located on a circle of radius r (Rowland circle), then all X-rays originating from this point source will have the same incident angle θ with the reflecting plane over the whole area of the

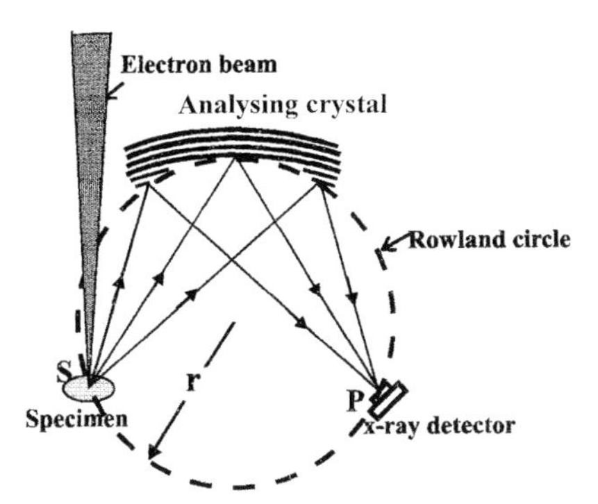

Fig. 5.4 X-ray spectrometer geometry with precise Johannson focusing conditions.

crystal. Therefore, if, for example, Bragg's law is obeyed for X-rays with a particular wavelength (λ_1), then they will be diffracted at the angle θ and will be precisely focused at the point P lying on the same Rowland circle, and the detector placed at point P will collect them. Special electronics will convert detector signals into the corresponding X-ray intensity.

It is essential that the source of X-rays, the crystal and the detector are all located on the Rowland circle. In order to collect X-rays of different wavelength, for example λ_2, the position of the source on the Rowland circle must be changed with respect to the crystal. Thus the crystal will 'see' the X-rays at a different angle of incidence, and if the diffraction conditions are satisfied for the wavelength λ_2, then the X-rays of wavelength λ_2 will be brought to a focus at another point on the Rowland circle and, correspondingly, the detector should also be moved to this new point.

In practice the position of the X-ray source (i.e. the position of the specimen) is not changed and focusing requirements are maintained only by co-ordinated moving of the crystal and the detector. In the so-called 'linear spectrometer' the analysing crystal moves along a straight line away from the specimen, thus gradually changing the angle θ, whilst the detector movement, which is rather complex, is such that the Rowland circle geometry is maintained, as shown in Fig. 5.5. It is easy to see from Fig. 5.5 that the crystal-to-source distance L is related to θ by the equation:

$$\frac{L}{2} = r \cdot \sin\theta \tag{5.5}$$

where r is the radius of the Rowland circle.

Combining Equation (5.5) with Bragg's law given by Equation (5.4) for the first-order reflection we obtain

$$\lambda = L \cdot \frac{d}{r} \tag{5.6}$$

where d is the interplanar spacing of the analysing crystal.

Thus, after calibration of the spectrometer, the position L of the crystal gives the wavelength, λ, directly and the system can be easily aligned for detection of the characteristic radiation of a given element.

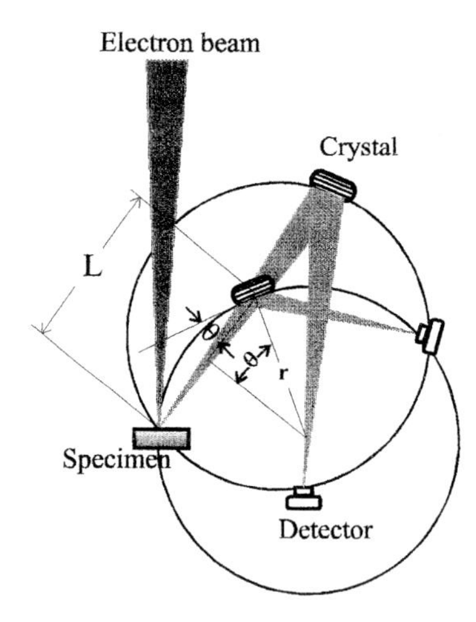

Fig. 5.5 The geometry of a linear spectrometer: the analysing crystal moves along a linear path while the detector movement is such that the X-ray point source, crystal and detector are all located on the Rowland circle.

There is, however, a substantial limitation of the system: only one particular wavelength can be collected by the spectrometer at a given moment, so the system, through its design, is unable to detect simultaneously all characteristic lines of the spectrum. In order to perform analysis the electron beam is targeted over a selected area of a specimen, and the spectrometer is set to detect only one characteristic line of a chosen element. The X-ray signal is collected over a period of 30–100 s, and only then can the spectrometer be realigned for detecting a further element.

Figure 5.6 shows a typical arrangement for the scanning electron microscope which incorporates a wave-dispersive spectrometer with linear crystal movement. In the majority of microscopes a specimen chamber is designed so that it is possible to equip the instrument with more than one crystal spectrometer, thus allowing the analysis of more than one element at a time. It is worth noting that in such spectrometers the usual range of incidence angles is approximately 12–70°. This implies that the range of wavelengths that the analysing crystal can focus is only about a few tens of Ångstroms. It is clear, therefore, that one crystal cannot cover all spectra of K (L or M) radiation for all elements in the periodic table. Therefore, a WDS spectrometer often has an option of at least two (or several) diffracting crystals with different *d* values of reflecting plane, thereby allowing optimal performance in different wavelength ranges.

Table 5.2 lists some of the most commonly used analysing crystals and the range of elements that they cover.

In order to characterize the performance of an X-ray spectrometer two important points should be considered:

(1) The *resolution of the spectrometer*, i.e. its ability to resolve two adjacent peaks in the spectrum. The resolution is usually measured in energy units (eV) and is defined as the minimal energy difference between two spectral lines at which these lines can be perceived as two separate peaks.

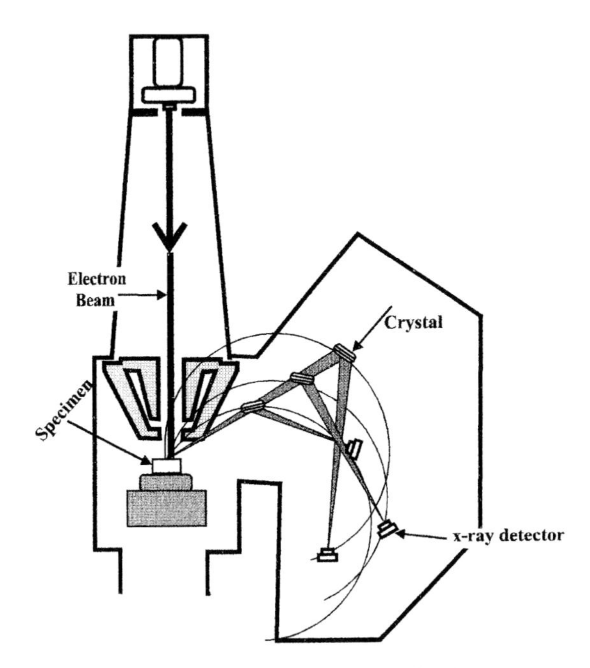

Fig. 5.6 Schematic representation of an SEM with a wave-dispersive linear spectrometer attached.

Table 5.2 Common diffracting crystals used in wave-dispersive spectrometers

Crystal	2*d* spacing for reflecting plane (Å)	Wavelength range (Å)	Range of detectable elements	
			For K radiation	For L radiation
α-quartz	6.687	1.39–6.28	S–Cu	NB–W
LiF (lithium fluoride)	4.028	0.87–3.8	Ca–Ge	Sb–U
KAP (ortho-phthalate potassium hydrogen)	26.632	5.54–25	O–P	V–Nb
RAP (ortho-phthalate rubidium hydrogen)	26.121	5.4–24.5	F–P	Cr–Mo
TAP (thallium acid phthalate)	25.75	7.31–23.79	O–Si	Cr–Rb
PET (pentaerythritol)	8.742	1.8–8.3	Si–Fe	Rb–Dy
STE (stearate)	98.0	22–85	B–O	S–V

(2) *Peak-to-background ratio* and intensity of X-ray signal, both of which define the ability of the spectrometer to achieve good counting statistics and to provide rapid and accurate measurement of the intensity of the spectral line.

With regard to the spectrometer having Johansson geometry it can be said that this fully focusing system gives, in comparison with other types of spectrometers, superior resolution (5–10 eV at 5.9 keV X-ray photons) and the highest possible peak-to-background ratio. This excellent capability of the instrument allows it to detect very small amounts of elements, less than 1000 ppm, and to achieve good detection sensitivity for light elements (minimum $Z = 4$ for Be).

Although a fully focusing spectrometer has excellent performance characteristics, the majority of commercial instruments are built on the basis of semi-focusing Johann geometry (Johann 1931). The main difference with a fully focusing spectrometer is observed in Fig. 5.7 where the analysing crystal is also bent to a radius of curvature $2r$ but not ground to be completely tangent to the Rowland circle with radius r. As a result, X-rays of wavelength λ that are reflected from the peripheral parts of crystal will not be focused on the Rowland circle, and this de-focusing effect causes some loss of resolution. This drawback is in practice compensated for by good-quality crystals and by improved design of spectrometer mechanics that permit the setting of the correct diffracting angle with great accuracy.

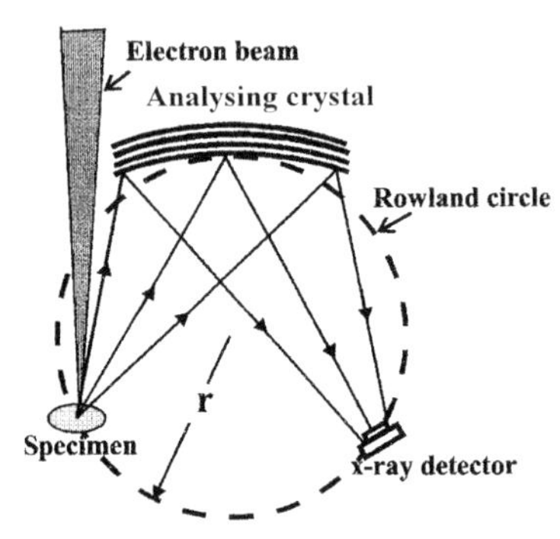

Fig. 5.7 X-ray spectrometer with the semi-focusing Johann geometry.

5.4.2 *The energy-dispersive spectrometer*

A schematic representation of an energy-dispersive spectrometer (EDS) with solid-state detector is shown in Fig. 5.8. The main part of the EDS system is a semiconductor Si(Li) detector (typically of area 10 mm^2), which is kept at liquid-nitrogen temperature in order to reduce the electronic noise. Because of the low-temperature

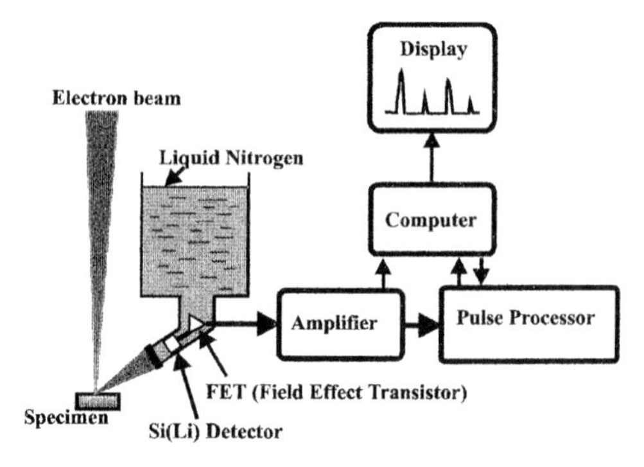

Fig. 5.8 Basic components of an energy-dispersive spectrometer with a solid-state detector.

requirement the detector must be maintained under vacuum in a special container, which is isolated from the microscope column in order to prevent condensation of water and vacuum oil vapours on the cold detector surface. Practically, the detector assembly is sealed under vacuum in a tube with a very thin 'window' through which X-rays pass before they reach the Si(Li) semiconductor. A thin sheet of beryllium, 7–12 μm thick, can be used as the window: it is transparent to X-rays but firm enough to withstand atmospheric pressure when the microscope column is vented to air. Unfortunately, the Be window absorbs soft X-rays, i.e. photons of energy less than approximately 1 keV (about the energy of Na K_α radiation). Therefore, in modern systems, beryllium is replaced with a specially prepared, very thin polymer film. In this way, ultra-soft X-rays, like B K_α (or even Be K_α) can be detected.

X-ray photons passing through the 'window' interact with the detector and create electron–hole pairs within the semiconductor. The number of these pairs is proportional to the energy of the incoming X-rays. In a Si(Li) detector the energy required to create one pair is about 3.8 eV. The integrated charge produced by a single X-ray photon is pre-amplified by a special field effect transistor (FET) and converted into a voltage pulse. This signal is then amplified. The output voltage pulse, being proportional to the energy of the X-ray photon, is then processed by a multichannel analyser where pulses are sorted by their voltage level and stored as a number of counts in the corresponding memory channels. Each memory channel is assigned to a specific, narrow range of energies of X-ray photons so that all channels together cover a sufficiently wide region of the X-ray spectrum. If the number of channels is 1024 (which is usual), and the typical energy interval is 20 eV per channel, then the range of the X-ray spectrum recorded will be 0–20 keV.

The information contained in each channel is displayed by a computer on the monitor screen in the form of a histogram illustrating the number of counts (i.e. X-ray intensity) versus the energy of the X-ray photons. Figure 5.9 is an example of an EDS

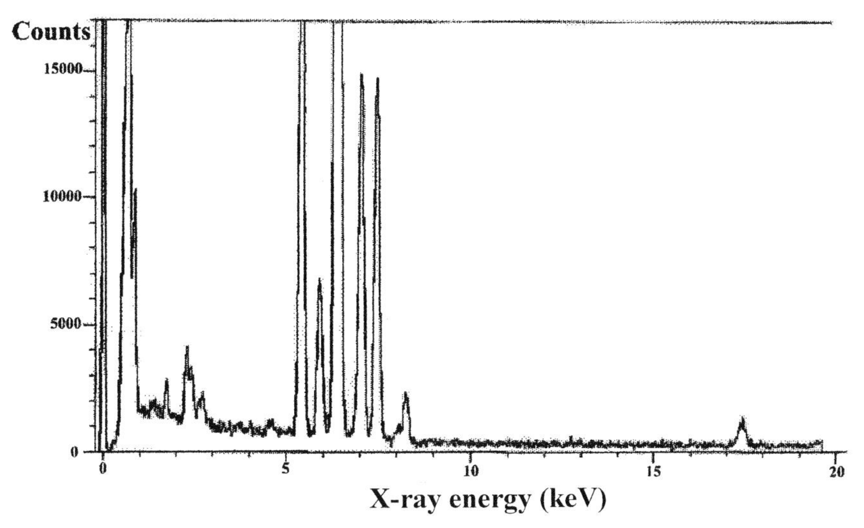

Fig. 5.9 An example of an X-ray spectrum which has been acquired with an energy-dispersive spectrometer.

spectrum. The spectrum can be easily stored in the computer database and then readily processed by appropriate computer software allowing direct identification of elements according to the position (energy) of the characteristic peaks.

An outstanding advantage of the energy-dispersive spectrometer is the speed with which the entire X-ray spectrum can be acquired. This is because the EDS system collects and processes simultaneously all X-ray photons emitted from a specimen. The spectrum of the type shown in Fig. 5.9 can be collected easily in a short time, usually between 30 s and 2 min. The WDS instrument must collect peak after peak sequentially and therefore, the time required for a full spectral scan using the wave-dispersive spectrometer will be about 1–1.5 h. Additionally, the EDS spectrometer uses only one detector that covers all spectral ranges, in contrast to the WDS system where two (or more) proportional counters must be used – each for the spectral range of a suitable diffracting crystal.

It should be emphasized that the speed of EDS analysis also reduces instrumental problems associated with the stability of the electron beam current and electronic drift, whilst in the case WDS the requirements for stability of the system are of great importance, because the intensity measurements of X-ray peaks are not performed simultaneously.

A great advantage of the solid-state detector is that its 'head' has a relatively small size and hence it can be placed very close to the specimen facilitating the collection of X-rays at a wide solid angle. This markedly increases the detection sensitivity of the system and provides improved statistical data. Therefore, the elemental analysis in an SEM coupled with the EDS analyser can be performed at substantially reduced beam currents, well below those used in an SEM equipped with the WDS system.

Despite all the advantages, the EDS has a substantial disadvantage when compared with the WDS system – its poor energy resolution. As previously discussed, the energy resolution of a spectrometer system is defined by its ability to separate two adjacent peaks in the energy spectrum. While the resolution of the WDS (5–10 eV) is determined mainly by the accuracy of focusing the X-rays, in the case of the EDS system the resolution is determined by the detector. The resolution, δ, of the detector can be calculated in accordance with (Woldseth 1973)

$$\delta = (P^2 + I^2)^{1/2} \tag{5.7}$$

where δ is defined as the full width (in eV) at half maximum (FWHM) of the observed standard X-ray peak (see Fig. 5.10), the term P is the electronic noise contribution and I is the intrinsic line width of the detector, which can be estimated from the formula

$$I = 2.35(F\varepsilon E)^{1/2} \tag{5.8}$$

where F is the Fano factor, ε (in eV) is the mean energy necessary to create an electron–hole pair in the detector and E is the energy of the X-ray photons. Assuming that the electron noise is $P = 0$, the theoretical limit for a Si(Li) detector can be estimated. Taking $F = 0.11$, $\varepsilon \approx 3.8$ eV (for Si) and $E = 5.9$ keV (for Mn K_α), then $I = 2.35$ $(0.1 \times 3.8 \times 5900)^{1/2} = 111$ eV is obtained. Practically, owing to electronic noise, the best energy resolution an Si(Li) detector can achieve is about 130 eV.

The relatively poor energy resolution of the EDS detector leads to difficulties in resolving pairs of peaks with very small energy differences. Such situations occur in

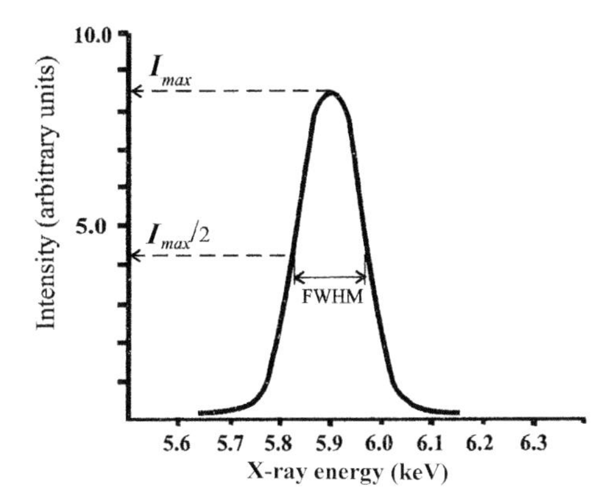

Fig. 5.10 Characteristic Mn K_α peak, which is used for the determination of the energy resolution of an EDS detector. The full width at half maximum of the peak is generally taken to be the resolving power of the detector.

multi-component materials. Here the pairs of unresolved peaks are called 'overlaps'. Although in many cases these overlaps can be handled through deconvolution procedures, in some cases the solution to the problem can be extremely difficult, in particular when one of the overlapped elements is present in only small amounts. Examples of the most difficult overlap situations are listed in Table 5.3.

5.5 Quantitative micro-analysis in SEM

As was discussed in Sections 5.1–5.3, characteristic X-rays generated from the micro-volume of a specimen in SEM carry information which can be used for identification of atoms within this volume. These X-rays can also be used for quantitative deter-

Table 5.3 Examples of X-ray peak overlaps in EDS micro-analysis

K_α and K_β lines of neighbouring transition metals, particularly pairs: Ti K_β (4.931 keV) – V K_α (4.952 keV) V K_β (5.427 keV) – Cr K_α (5.414 keV) Mn K_β (6.490 keV) – Fe K_α (6.403 keV)
Ti K_α (4.510 keV) – Ba L_α (4.467) keV)
S K_α (2.308 keV) – Mo L_α (2.293 keV) – Pb M_α (2.346 keV)
Na K_α (1.041 keV) – Zn L_α (1.009 keV)
Zr L_α (2.042 keV) – Pt M_α (2.051 keV) – P K_α (2.015 keV) – Ir M_α (1.980 keV)
Nb L_α (2.166 keV) – Hg M_α (2.196 keV)
Si K_α (1.74 keV) – W M_α (1.775 keV) – Ta M_α (1.710 keV) – Rb L_α (1.694 keV)
Al K_α (1.487 keV) – Br L_α (1.480 keV)
Y L_α (1.922 keV) – Os M_α (1.910 keV)
O K_α (0.523 keV) – V L_α (0.510 keV)
Mn L_α (0.636 keV) – Fe L_α (0.704 keV) – F K_α (0.677 keV)

mination of the element concentrations. Quantitative analysis of the bulk specimen in SEM is based on accurate measurement of the intensity of the characteristic X-rays of the analysed element in the specimen and comparing this intensity with that obtained from this element in a standard material.

In the situation when the composition of the specimen is very close to that of the standard the concentration C_i of an element i in a bulk sample can be calculated easily from the basic equation given by

$$C_i = C_i^* \frac{I_i}{I_i^*} \tag{5.9}$$

where I_i and I_i^* are the measured X-ray intensities from the sample and from the standard, respectively, and C_i^* is a weight fraction of the element of interest in a particular standard.

However, when the elemental content of the standard substantially differs from that of the specimen, Equation (5.9) must be modified and the measured intensities I_i and I_i^* have to be corrected for several effects.

5.5.1 *ZAF correction method*

In the ZAF method, used for bulk specimens, the form of the correction is:

$$C_i = C_i^* \frac{I_i}{I_i^*} \cdot \frac{1}{K} \tag{5.10}$$

The coefficient K is usually represented as the product of three major correction factors:

$$K = Z \times A \times F \tag{5.10a}$$

Terms Z, A and F denote corrections required for the effects of 'atomic number', 'X-ray absorption' and 'fluorescence', respectively, and for this reason the correction procedure is often referred to as 'ZAF' correction.

It will be shown below that the values of the factors Z, A and F are functions of elemental composition of the material analysed. While the composition of the standard is, of course, known, the composition of the specimen is, in fact, the purpose of the analysis. This situation is usually resolved by a procedure of iteration: as a first approximation the weight fraction of each element present in the sample is estimated on the basis of Equation (5.9). Then the correction coefficients Z, A and F can be calculated, and from Equation (5.10) the corrected values for the concentrations are obtained. These results are used in recalculating the coefficients Z, A and F and to obtain from Equation (5.10) new and more accurate values of C_i. By repeating these iteration steps the convergence of results can be quickly achieved.

For a better understanding of the essence of all corrections used in the ZAF method consider the physical aspects of the problem concerning the relationship between the observed characteristic X-ray intensities and concentration of elements in a target specimen.

5.5.1.1 *The atomic number effect* As was noted earlier in Section 5.3, while beam electrons are penetrating the target material they are slowed and lose their energy, part of which is converted into characteristic X-rays. Generation of characteristic X-rays occurs as a result of the ionization of atoms followed by their relaxation to the ground state. Thus, the number of emitted X-ray photons, i.e. the intensity of the characteristic X-radiation, should be proportional to the number of ionization events occurring along the path followed by the incident electrons. This number depends on three important parameters:

- the ionization cross-section, Q, of the atom being ionized
- the stopping power, S, of the material
- the backscatter factor, R

The role of these parameters is discussed below.

The ionization cross-section $Q_i(E)$ is defined as the probability per unit path length of an electron of given energy to ionize a particular electron shell (K, L or M) of the atom of element i. The ionization cross-section is a function of the electron energy (Bethe 1930) and is roughly proportional to ln U/U (where $U = E_0/E_{i(\mathrm{cr})}$ is the over-voltage; see Section 5.2):

$$Q_i(E) = \frac{\mathrm{Const}_j}{E^2_{i(\mathrm{cr})}} \cdot \frac{\ln U}{U} \tag{5.11}$$

Here, Const_j is a specific constant for the shell (j = K, L or M), and the subscript i denotes the element.

A typical curve, representing the variation of the K-shell ionization cross-section $Q_i(E)$ with electron energy E is shown in Fig. 5.11 (Green & Cosslett 1961). Here $Q_i(E)$ is replaced by the value $Q_i(E) \cdot E^2_{\mathrm{cr}}$, which is plotted as a function of U; in this representation the graph is approximately the same for all elements. It can be seen from Fig. 5.11 that the K shell of the atom cannot be ionized by electrons of energy lower than the critical excitation energy of the K shell, i.e. the ratio E/E_{cr} must be

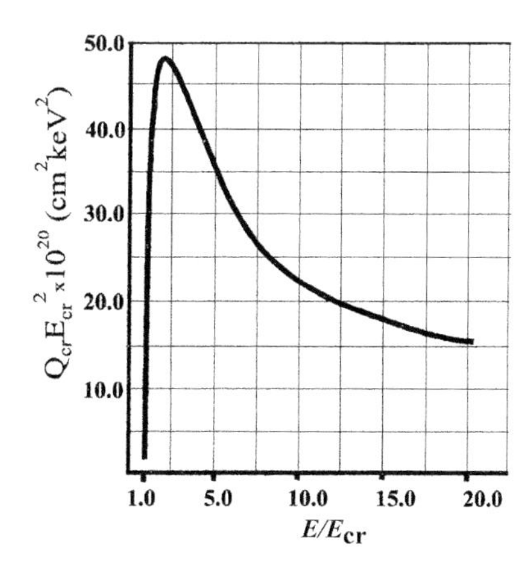

Fig. 5.11 Typical curve showing variation of ionization cross-section of the K shell with electron energy. The value $Q_i(E)\ E^2_{\mathrm{cr}}$ is plotted as a function of the over-voltage ratio E/E_{cr}.

greater than 1. From approximately $E = 2E_{cr}$ the curve gradually descends, but for high electron energies the value $Q_i(E) \cdot E_{cr}^2$ tends to become constant.

It follows from the definition of $Q_i(E)$ that the number of ionization events dn_i (for the specific shell of element i) occurring along the length dz of the electron path is

$$dn_i = N_0^{(i)} \cdot Q_i(E)\,dz \tag{5.12}$$

where $N_0^{(i)}$ is the number of i atoms per unit volume of the specimen. This value can be calculated as

$$N_0^{(i)} = C_i \cdot \frac{\rho}{\mu_i} \cdot N_A \tag{5.13}$$

where N_A is Avogadro's number, μ_i is the atomic weight of element i in a material of density ρ having concentration C_i of element i.

As can be deduced from Fig. 5.11, while each ionization event should decrease the energy of the incident electron, the likelihood of causing the next ionization gradually increases, until the kinetic energy of the electron falls below E_{cr}. It is, therefore, extremely important to know how the energy of the electron changes with depth below the specimen surface in order to estimate correctly the number of X-ray photons generated in the sample.

For a material of density ρ the energy change can be characterized by the stopping power, S, which is defined as the rate at which the electron loses its energy while traversing a small mass thickness $d(\rho z)$:

$$S = \frac{dE(z)}{d(\rho z)} \tag{5.14}$$

An expression for the stopping power in the electron energy region $1 < E < 50$ keV can be obtained from Bethe (1930) in the following form:

$$S = 2\pi e^4 N_A \frac{Z}{A} \cdot \frac{1}{E} \ln \frac{1.166E}{J} \tag{5.15}$$

Here e is the electronic charge, Z is the atomic number, A is the atomic weight of the scattering element and N_A is Avogadro's number. The parameter J, which can be taken as proportional to Z, is called the mean ionization potential of the element concerned.

For a multielement target it was found experimentally that the stopping power obeys the additive law, namely

$$S = \sum_i C_i \cdot S_i \tag{5.15a}$$

where C_i is the concentration of element i in the specimen.

It can be seen from Equation (5.15) that since J increases with increasing atomic number, the logarithmic term decreases with Z. Besides, as Z increases the ratio Z/A tends to decrease. It can be said, therefore, that the stopping power decreases with an increase in the average atomic number of the target.

The total number of ionization events caused by an electron can be obtained by integration of Equation (5.12) along the depth z_a of 'active' electron penetration, where the electron is capable of ionizing atoms of element i:

$$n_i = N_0 \int_0^{z_a} Q_i(E)\mathrm{d}z = N_A \frac{\rho}{\mu_i} C_i \int_0^{z_a} Q_i(E)\mathrm{d}z \tag{5.16}$$

By changing the variable of integration in Equation (5.16) to E and substituting $S(E)$ from Equation (5.14) we obtain:

$$n_i = N_A \frac{\rho}{\mu_i} C_i \int_{E_0}^{E_{cr}} Q_i(E) \frac{\mathrm{d}E}{\mathrm{d}E/\mathrm{d}z} = \frac{N_A \cdot C_i}{\mu_i} \int_{E_0}^{E_{cr}} \frac{Q_i(E)}{S(E)} \mathrm{d}E \tag{5.16a}$$

At this point account should be taken of some fraction of incident electrons, η, that are scattered back and the quantity of backscattered electrons that escape from the specimen (see Section 5.3). Such a 'loss' of potentially ionizing electrons results in a reduction of the number of ionization events. This is quantified by the backscatter factor, $R = 1 - \eta$:

$$n_i = R \frac{N_A \cdot C_i}{\mu_i} \int_{E_0}^{E_{cr}} \frac{Q_i(E)}{S(E)} \mathrm{d}E \tag{5.16b}$$

If a similar calculation is performed for the standard, then the ratio n_i/n_i^*, which actually represents the ratio of intensities generated by element i in the specimen and in the standard, can be obtained:

$$\frac{I_i}{I_i^*} = \frac{n_i}{n_i^*} = \frac{R \cdot C_i \int_{E_0}^{E_{cr}} \frac{Q_i(E)\mathrm{d}E}{S(E)}}{R^* \cdot C_i^* \int_{E_0}^{E_{cr}} \frac{Q_i^*(E)\mathrm{d}E}{S^*(E)}} \tag{5.17}$$

Here the superscript * denotes the quantities associated with the 'standard'.

Equation (5.17) can be rewritten in the form

$$C_i = C_i^* \cdot \frac{I_i}{I_i^*} \cdot \frac{1}{Z} \tag{5.18}$$

where

$$Z = \frac{R \cdot \int_{E_0}^{E_{cr}} \frac{Q_i(E)\mathrm{d}E}{S(E)}}{R^* \cdot \int_{E_0}^{E_{cr}} \frac{Q_i^*(E)\mathrm{d}E}{S^*(E)}} \tag{5.19}$$

Comparing Equation (5.18) with Equations (5.10) and (5.10a) it is possible to attribute the coefficient Z to the correction factor with respect to the effect of atomic number.

It is now instructive to discuss the implication of Equation (5.19). As was previously noted, an increase in the average atomic number of the matrix results in a decrease of stopping power. This implies that more ionization events will occur along the electron path and, therefore, more X-ray photons will be generated. It follows that concentration C_i of element i should generate a characteristic line with greater intensity in a higher atomic number matrix than in a lower one. In contrast, the backscattering effect, represented by Equation (5.17) through coefficient R, works in an opposite manner: electron backscattering increases with increasing average atomic number leading to a 'loss' of X-ray intensity and to the approximate compensation of the stopping power effect.

The same can be said concerning the behaviour of S and R as functions of the initial energy of the incident electrons: these two factors vary inversely with E_0 and tend to neutralize each other. The conclusion is that these two factors, namely stopping power, S, and backscattering, R, tend to cancel each other. However, the calculation of Z correction must be performed with great care, especially at low beam energies between 500 eV and 10 keV, where R varies strongly with the beam energy (Darlington 1975).

A full description of the calculation procedures, and theoretical expressions for Q, S and R, that are necessary for integration, can be found in the original studies of Castaing (1951) and Duncumb and Reed (1968).

5.5.1.2 *The absorption effect* It is important to realize that the intensity of the X-rays generated within the target differs from the intensity of emitted X-rays, i.e. the intensity as measured by the X-ray detector. This is due mainly to the absorption effect which will be considered now.

Figure 5.12 shows a schematic representation of the X-ray production region in a specimen. X-ray emission takes place from a layer $d(\rho z)$ at mass-depth (ρz) with the X-ray detector position at take-off angle θ with respect to the surface of the specimen.

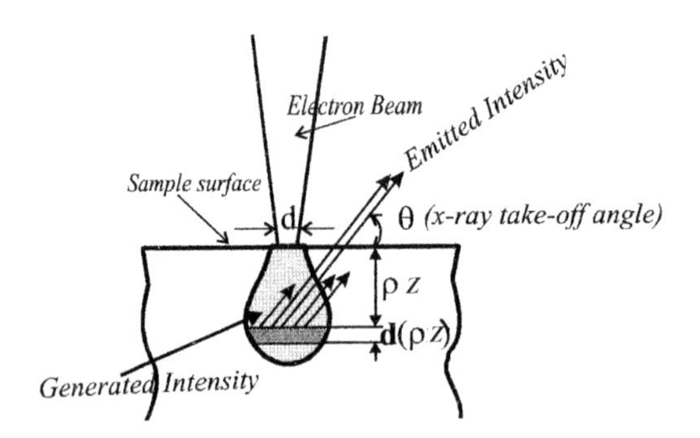

Fig. 5.12 Schematic illustration used to calculate the effect of X-ray absorption on elemental analysis. The larger shaded area represents the X-ray production region of the specimen. Also shown (darker region) is a layer $d(\rho z)$ *at a mass-depth* (ρz) emitting a portion of the total X-rays. The position of the X-ray detector is at take-off angle θ with respect to the surface of the specimen.

The function $\Phi_i(\rho z)$ is defined as the distribution of characteristic X-ray production with depth (subscript i refers to element i). Thus, the X-ray intensity generated in a layer of mass thickness $d(\rho z)$ at mass depth (ρz) would be:

$$dI^{(0)} = \Phi_i(\rho z)\, d(\rho z) \qquad (5.20)$$

It can be observed from Fig. 5.12 that before the X-rays generated in the layer $d(\rho z)$ are collected by the detector, they must pass through distance $\rho z\,(1/\sin\theta)$ where θ is the take-off angle. Therefore, owing to the absorption effect and in accordance with Beer's law, the X-ray intensity will be reduced to:

$$dI = \Phi(\rho z)\exp\left\{-\frac{\mu_i}{\rho}\left(\frac{\rho z}{\sin\theta}\right)\right\}d(\rho z) \tag{5.21}$$

Usually, the ratio $\mu_i/\rho \sin\theta$ is denoted as χ_i:

$$\chi_i = \mu_i/(\rho \cdot \sin\theta)$$

where μ_i/ρ is the mass absorption coefficient of element i for the specific characteristic radiation. For a multi-component material it follows that $\mu/\rho = \Sigma_i C_i(\mu/\rho_i)$ where C_i is the weight fraction of the element i.

The total X-ray intensity from element i, which is detected by the spectrometer, can be obtained by integration through all layers of the target:

$$I_i = \int_0^\infty \Phi(\rho z)\exp(-\chi_i \cdot \rho z)d(\rho z) \tag{5.22}$$

The fraction of the X-ray intensity detected by the spectrometer with respect to the overall intensity generated within the specimen will be given by the equation:

$$f(\chi) = \frac{\int_0^\infty \Phi(\rho z)\exp(-\chi_i \rho z)d(\rho z)}{\int_0^\infty \Phi(\rho z)d(\rho z)} \tag{5.23}$$

If, in analogous manner, the function $f^*(\chi)$ is used to define the standard, then the absorption correction factor in Equations (5.10) and (5.10a) can be given as:

$$F = \frac{f(\chi)}{f^*(\chi)} \tag{5.24}$$

It is worth noting that the range of $f(\chi)$ must be from 0 to 1 because the value of $f(\chi)$ is considered as the *fraction* of the generated intensity actually reaching the detector. It is extremely important to maximize $f(\chi)$ in order to obtain accurate analysis. In practice the condition $f(\chi) > 0.7$ should be maintained. To maximize $f(\chi)$ a low over-voltage ratio, $U = E_0/E_{cr}$, should be used (since this will decrease the region of electron spreading in the specimen), and the take-off angle θ should be as high as possible.

5.5.1.3 *The fluorescence effect* Characteristic X-rays can be generated not only when atoms in the target are excited by the electrons of the primary electron beam, but also induced by other X-rays. This effect is called X-ray fluorescence and occurs as a direct consequence of the absorption of characteristic radiation of one element (B) in the target by the atoms of another element (A), provided that the energy of the absorbed X-ray photon is sufficiently high to cause ejection of an electron from the inner shell of atoms of element A. When the atom of element A returns to its ground state the characteristic fluorescent X-rays are emitted. It should be noted that the fluorescence effect may also be induced by white radiation, in particular in those specimens of high mean atomic number. The characteristic X-radiation caused by the fluorescence is often called 'secondary' radiation to distinguish it from the 'primary' X-rays excited by the electron beam.

The effect of X-ray fluorescence should be of particular concern when the energy of the primary radiation is slightly higher than the critical excitation potential (the absorption edge) of the atom of element A, which is excited to emit fluorescence (secondary) radiation. For example, the fluorescence effect can be considerable in materials such as steel, where the K_α radiation of iron excites atoms of Cr producing Cr K_α X-rays. As a result, the intensity of the Fe K_α radiation emitted from the specimen is decreased owing to absorption, while the measured K_α radiation from the Cr will always be raised as a result of fluorescence.

It is easy to understand that the volume of material in which the fluorescence effect occurs is substantially larger than that in which the primary X-rays are generated. Therefore, for those specimens in which the fluorescence effect can be strong careful treatment of the measured intensities is necessary.

The fluorescence correction coefficient F_i can be represented by the following:

$$F_i = \frac{1 + \sum_i \frac{I_{ij}}{I_i}}{1 + \sum_j \frac{I^*_{ij}}{I^*_i}} \tag{5.25}$$

The subscript j spans all elements in the target that can excite element i by their X-radiation; the quantities relating to the 'standard' are denoted by the superscript *. I_{ij} is the intensity of secondary characteristic radiation from the element i induced by element j, and the quantity I_i is the intensity of the primary X-rays generated by the element i. The complete expression of the ratio I_{ij}/I_i is rather complex and a full derivation of the correction formula for fluorescence can be found in Reed (1965) and Heinrich (1981).

5.5.2 *The peak-to-local continuum method*

The peak-to-local continuum method minimizes the need to calculate correction factors used in the ZAF method, thus simplifying substantially the procedure of quantitative analysis.

The method is based on the experimental observation that the contribution of one electron from the incident beam to the intensity I_i of the characteristic X-ray line for element i is roughly proportional to the contribution of one incident electron to the intensity W of the so-called local continuum radiation, i.e. white X-rays in a narrow energy band centred around this characteristic line. Another observation is that the function $\Phi_i(\rho z)$, representing the distribution of X-ray production with depth, is approximately the same for both the characteristic X-rays and the continuum radiation. This implies that the fraction of absorbed characteristic X-rays, while they pass through the sample towards the detector, is the same as that for the local continuum X-rays. In turn, this means that the ratio of the two intensities, I_i/W, is actually independent of the specimen mass-thickness. Therefore, the concentration of element i in the bulk specimen can be represented by a simple ratio, i.e. $C_i = k\,(I_i/\mathrm{W})$. The value of the constant k can be readily determined if a standard is used for measurements.

5.6 Summary

In this chapter an overview of the main concepts of X-ray elemental analysis in the SEM and a description of the necessary instrumentation were presented. The emphasis was on the underlying physical principles that, hopefully, will help the reader understand how the analytical system works, which parameters should be considered and which experimental conditions must be satisfied in order to perform successful micro-analysis and to obtain reliable results.

It was emphasized that for analysis with optimum accuracy (of about 1% of content) the use of a standard is required. The standard must be as similar to the specimen being analysed as possible, and the conditions under which the X-ray spectrum is recorded should be identical for both the specimen and the standard. In practice, the operator need only measure the intensity of emitted characteristic X-rays from the specimen since the data relating to standards have been previously stored in the computer database.

The intensity data obtained and information concerning the accelerating voltage, geometry of mutual positions of the specimen and the detector in the spectrometer are together sufficient for calculation of elemental concentrations.

The results of the analysis depend crucially on the accuracy of the X-ray intensity measurements. Therefore, much attention must be paid to correcting any non-linear responses of the measuring system (e.g. dead-time losses in the detector) or any other errors. Also, it is important to ensure the high quality of the specimen surface and to take into account possible changes occurring in the specimen during its preparation and in the course of the analysis. This topic is beyond the scope of this chapter and the interested reader is referred to the literature (Goldstein & Yakowitz 1977; Belk 1979; Goldstein *et al.* 1992; Williams *et al.* 1995).

References

Andersen, C.A. & Hasler, M.F. (1965) X-ray optics and microanalysis. In: *Proc. IV International Congress on X-ray Optics and Microanalysis* (eds R. Castaing, P. Deschamps & J. Philibert). Orsay, p. 310.

Belk, J.A. (ed.) (1979) *Electron Microscopy and Microanalysis of Crystalline Materials*. Applied Science Publishers, London.

Bethe, H.A. (1930) *Ann. Phys.*, **5**, 325.

Castaing, R. (1951) Thesis, Onera Publication No. 55, University of Paris.

Darlington, E.H. (1975) *J. appl. Phys.*, **8**, 85.

Duncumb, P. & Reed, S.J.B. (1968) *Quantitative Electron Probe Microanalysis* (ed. K.F.J. Heinrich). NBS Special Publication, Vol. 298, p. 133.

Goldstein, J.I. & Yakowitz, H. (eds) (1977) *Practical Scanning Electron Microscopy*. Plenum Press, New York.

Goldstein, J.I., Newbury, D.E., Echlin, P., Joy, D.C., Romig, A.D. Jr., Lyman, C.E., Fiori, C.E. & Lifshin, E. (1992) *Scanning Electron Microscopy and X-ray Microanalysis*, 2nd edn. Plenum Press, New York.

Green, M. & Cosslett, V.E. (1961) *Proc. Phys. Soc.*, **78**, 1206.
Heinrich, K.F.J. (1981) *Electron Beam X-ray Microanalysis*. Van Nostrand Reinhold, New York, pp. 303–328.
Johann, H.H. (1931) *Z. für Physik*, **69**, 185.
Johansson, T. (1933) *Z. für Physik*, **82**, 507.
Reed, S.J.B. (1965) *Br. J. appl. Phys.*, **16**, 913.
Williams, D.B., Goldstein, J.I. & Newbury, D.E. (eds) (1995) *X-ray Spectrometry in Electron Beam Instruments*. Plenum Press, New York.
Woldseth, R. (1973) *X-ray Energy Spectrometry*. Kevex, Burlingamer.

6 Elemental analysis by AES, XPS and SIMS

H.J. MATHIEU

6.1 Introduction

Elemental analysis of surfaces is made possible by probing with electrons, photons or ions. The common parameter of surface chemical analysis is its small information depth from within the first few nanometres. In many practical cases the composition of the outermost surface layers of solids is different from the bulk. Examples are protective passive films of metals having thickness of a few nanometres.

Surface analysis covers the whole sub-micron range from the outer interface air/solid, i.e. the first monolayer, to several hundred nanometres. The methods presented in this chapter allow access to internal interfaces buried below the surface typical of, for example, semiconductors. In addition, gradients of concentration may be detected either non-destructively or by ion depth profiling. In general, the methods presented cover all solid-state materials used in modern science, i.e. devices used in physical, chemical, biological or medical applications or in processes that determine the surface chemical composition. Because of their very high surface sensitivity, these methods are subject to contamination effects at the surface. Therefore, a well-controlled preparation of the sample is required. These methods are not only capable of visualizing the surface, but in addition provide chemical information with high elemental and molecular sensitivity below the monolayer coverage. All three spectroscopies described need ultra-high vacuum (UHV) conditions for analysis, i.e. pressures below 10^{-8} mbar. The latter condition may present certain limitations for samples with a spatial surface geometry with atoms or molecules attached to the surface, as in the case of organic or biomedical applications. Last, but not least, the probing beam might not be as non-destructive as assumed and this may lead to a certain modification at the site of analysis.

The principle of chemical surface analysis is illustrated schematically in Fig. 6.1. A primary source is directed towards the surface of a solid sample placed inside the

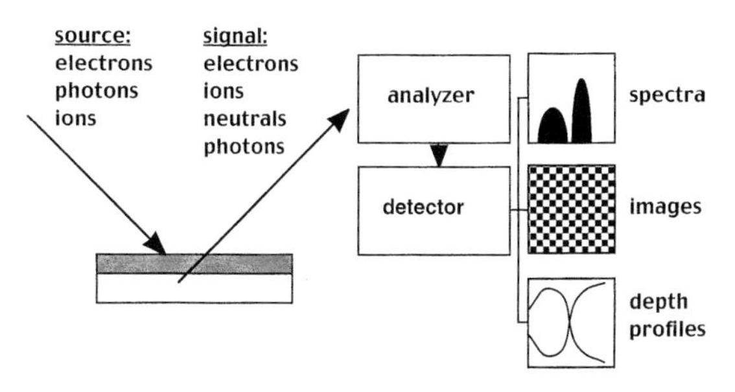

Fig. 6.1 Principle of surface analysis.

spectrometer. The primary particles interact with the atoms and/or molecules of the first few monolayers of the solid and provoke the emission of secondary particles. These secondary particles carry specific elemental and chemical information on the composition and nature of the emitting volume. Measured quantities will be typically mass, charge or energy depending on the method used. A monolayer with a thickness ranging from a few Ångströms to a few nanometres contains approximately 10^{15} atoms. The experiment may require that one probes only the first monolayer or determines the thickness of a thin film on the sample. The UHV environment is necessary to guarantee contamination-free analysis conditions. An analyser is placed close to the surface incorporating a detector counting the transmitted particles. If the primary probing beam or the analysis is raster scanned lateral information is available leading to elemental or molecular mapping or line-scans. By changing either the primary particle conditions, the angle of emission of the secondary particles or using additional ion bombardment it is possible to obtain compositional information, i.e. depth profiles displaying changes of atomic concentration as a function of the distance from the surface. This latter option may not be desirable for every sample, because of the destructive nature of the latter process. Extensive data reduction is often necessary to retrieve the required information. In many cases computer software is available to extract qualitative and quantitative information of the detected elements and their surface chemistry.

6.2 Electron spectroscopies

Electron spectroscopies for elemental and chemical surface analysis are known by the acronyms ESCA (electron spectroscopy for chemical analysis), as proposed by Siegbahn *et al.* (1967) and AES (Auger electron spectroscopy). Nowadays ESCA is often replaced by XPS (X-ray photoelectron spectroscopy), which uses soft X-rays as primary particles. If electrons are used for excitation, then AES named after Auger (1925) is considered. The principle of XPS and AES is shown schematically in Fig. 6.2 indicating that electron spectroscopies deal with the measurement of the kinetic energies of electrons emitted from the sample due to interaction with an impinging beam.

The primary X-rays and/or electrons arriving at the surface under different angles and energies penetrate to certain depths. They create secondary particles like photons as well as secondary electrons, Auger electrons or even ions. The escape depth of the emitted particles is determined, however, by their kinetic energy, their emission angle

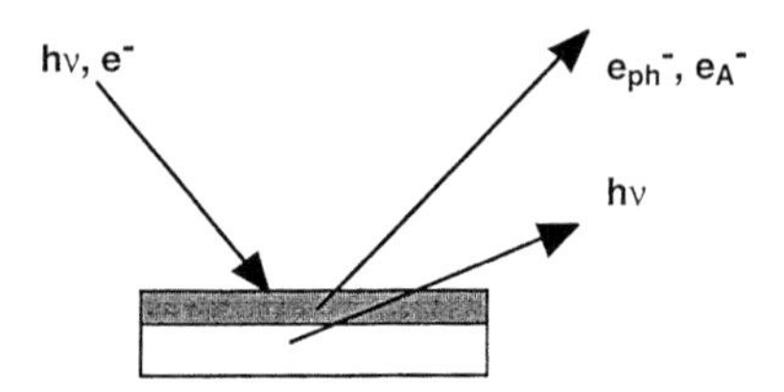

Fig. 6.2 Electron spectroscopies for chemical analysis (schematic).

as well as the mass density of the surface layers. Secondary electrons have a much smaller escape depth than secondary X-rays, because of a much smaller mean-free-path of electrons in solids. Before presenting the two spectroscopic methods involving electrons, it should be recalled that an electronic state of an atom is characterized by four quantum numbers: n (principal quantum number), l (azimuthal quantum number), s (spin quantum number) and j (total angular moment with $j=l+s$). AES and XPS use, in general, different nomenclatures, i.e. in AES letters like K, L, M, etc. and in XPS 1s, 2s, 2p, etc. are used for identifying electronic states. Table 6.1 gives examples of the two classifications of symbols used.

Table 6.1 Quantum numbers for electronic states used in AES and XPS

n	l	$j=l+s$	AES	XPS
1	0	1/2	K	1s
2	0	1/2	L_1	2s
2	1	1/2	L_2	$2p_{1/2}$
2	1	3/2	L_3	$2p_{3/2}$
3	0	1/2	M_1	3s
3	1	1/2	M_2	$3p_{1/2}$
3	1	3/2	M_3	$3p_{3/2}$
3	2	3/2	M_4	$3d_{3/2}$
3	2	5/2	M_5	$3d_{5/2}$
etc.	etc.	etc.	etc.	etc.

6.2.1 *X-ray photoelectron spectroscopy (XPS)*

Two types of sources are commonly used in XPS: either Mg $K_{\alpha_{1,2}}$ or Al $K_{\alpha_{1,2}}$ radiation with an energy of 1253.6 $\pm$ 0.7 or 1486.6 $\pm$ 0.85 eV, respectively. Both sources use a doublet ray. Al $K_{\alpha_{1,2}}$ radiation is often monochromatized by elimination of the α_2 ray to result in a single Al $K_{\alpha 1}$ with a much smaller full width at half maximum (FWHM) of the primary source allowing measurement of energy differences, ΔE, <0.5 eV of the emitted electrons. Under X-ray exposure the sample will be photoionized according to the processes illustrated in Fig. 6.3.

It is assumed that the primary X-ray creates (a) a photoelectron at the core energy level E_K. This excitation is followed by either a relaxation process (b) or (c). In (b) a secondary X-ray is created after filling the hole at the E_K level with a valence band electron. In (c) a 'tertiary' electron called the Auger electron is emitted after an electron transition from a lower level, e.g. E_V to the core level E_K. XPS mainly explores excitation processes leading to the emission of photoelectrons. All solid elements can be detected with $Z \geq 3$. A more detailed representation of the photoionization process is shown in Fig. 6.4, which illustrates how the binding energy of the emitted electron is determined experimentally.

Following excitation by a primary photon with energy $E = h\nu$, the energy conservation principle is applied to give

$$h\nu = E_b + E'_{kin} + \Phi_S \tag{6.1}$$

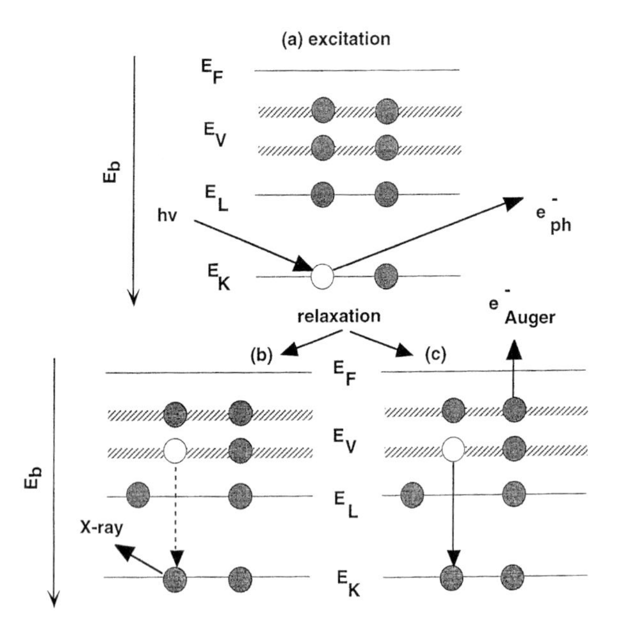

Fig. 6.3 Principle of the photoelectronic effect: the excitation and relaxation processes are shown indicating schematically the different binding states with the Fermi energy (E_F) taken as the reference level (equal to zero). The valence band energy is E_V followed by a discrete level E_L and the core level E_K.

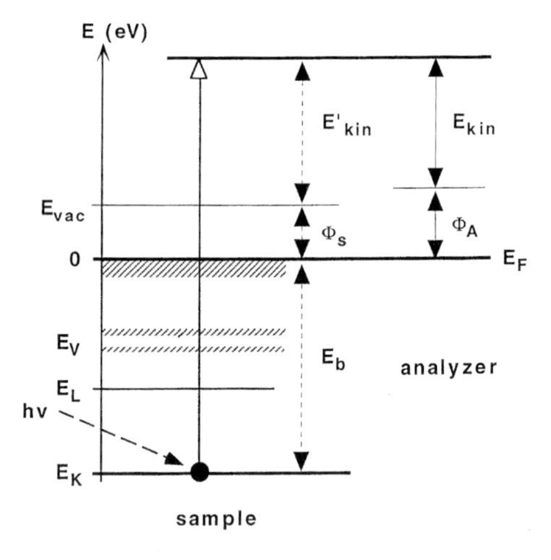

Fig. 6.4 Energy balance of the photoelectric process.

where E_b and E'_{kin} are the binding energy and apparent kinetic energy of the emitted photoelectron with respect to the Fermi level. Φ_S is the work function of the sample S. The kinetic energy, E'_{kin}, of the photoelectron within the analysis chamber is not measured; it is the kinetic energy, E_{kin}, which is actually determined by the detector. These two kinetic energies are related by

$$E_{kin} + \Phi_A = E'_{kin} + \Phi_S \tag{6.2}$$

where Φ_A is the detector work function of the analyser. The latter is a constant for a given detector. Insertion of Equation (6.2) into Equation (6.1) gives the fundamental XPS equation relating binding energy and measured kinetic energy:

$$E_b = h\nu - (E_{kin} + \Phi_A) \tag{6.3}$$

Equation (6.3) is applied to the calibration of the analyser using the following standard energies:

Au $4f_{7/2}$	84.0 eV
C 1s	285.0 eV
Cu $2p_{3/2}$	932.7 eV

Equation (6.3) assumes that the Fermi level of sample and analyser are identical, which signifies that there is good electrical contact between them. This is in general a good assumption for metals. For semiconductors or insulators, however, this may not be true. Owing to photoelectron emission and loss of electrons the sample may charge-up positively. This charging of the surface makes it more difficult for the remaining electrons to be emitted and as a consequence the measured kinetic energy appears to be reduced. The calculated binding energy shifts to higher values. This effect can be overcome by a charge compensation device built into many XPS spectrometers. This device supplies the sample with electrons of low energy (≤ 10 eV) to compensate for the positive charge-up of the sample. Negative charging due to excessive adsorption of negatively charged particles may be overcome by low positive ion beam argon bombardment. It is important to match electron emission and electron compensation to avoid any influence of charging on the analysed sample.

Figure 6.5 gives an example of a survey spectrum of an Fe–25Cr alloy showing the normalized intensity as a function of the binding energy, E_b. A typical acquisition time for a survey spectrum is a few minutes.

The intensity is determined by the number of detected photoelectrons at a given energy normalized by the binding energy. It is observed that the main 2p doublet of iron and chromium, as well as other minor 3s and 3p peaks of iron are present.

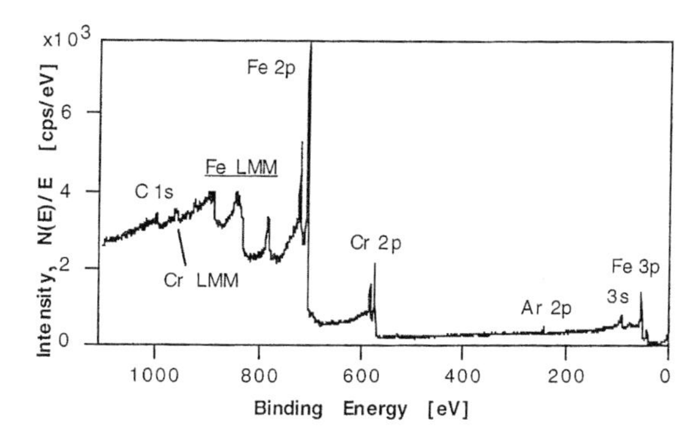

Fig. 6.5 XPS survey of an Fe–25Cr metal alloy measured after sputter cleaning with argon gas using Al K_α monochromatic radiation.

Carbon and argon are detected as well. The latter stems from the sputter cleaning effect. In addition to the photoelectron peaks a family of iron Auger peaks (labelled Fe LMM), which are peaks that are created during the relaxation process of the metal, is seen. The chromium Auger peaks (Cr LMM) are of lower intensity because there is only 25% chromium in the alloy. They are detected close to 997 eV.

A spectrum acquired with the sufficiently high-energy resolution of a few hundred meV indicates the binding energy of the emitted photoelectrons. XPS is able to indicate the chemical state of the emitting atom. For light elements the 1s levels are very specific for the binding state. For heavier elements 2p, 3d and 4f transitions are normally used for determining the chemical state. As for the transition metals, 2p levels are very sensitive for the measurement of binding energy. This is illustrated in Fig. 6.6 which shows the difference between the 2p spectra of Fe metal and Fe_2O_3.

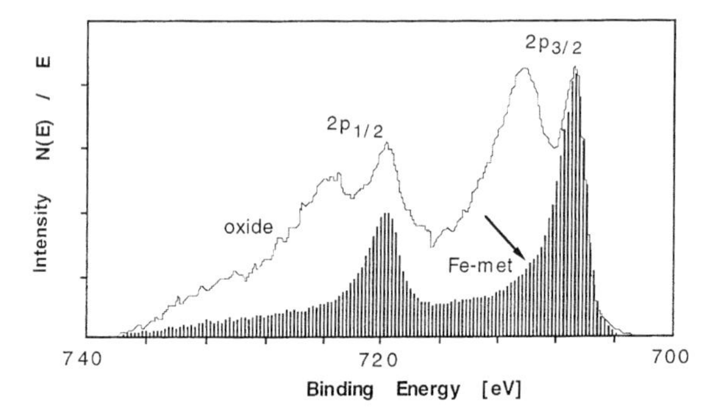

Fig. 6.6 XPS narrow scan spectra of normalized Fe 2p peaks of the metal (Fe-met) and a thin Fe_2O_3 film grown on top of an Fe substrate. The thickness of the overlayer oxide is estimated to be less than 1 nm.

In Fig. 6.6 it is noted that the 2p doublet ($2p_{3/2}$ and $2p_{1/2}$) of the metal has slightly asymmetric peaks. Their peak width is smaller than that of the oxide. The thickness of the oxide film formed in situ after sputter cleaning of the Fe–Cr alloy is estimated to be less than 1 nm. The information depth is characterized by the escape depth, which is a function of the mean-free-path of the emitted electrons [cf. Equation (6.7b) below]. Owing to an escape depth of a few nanometres, both metal and oxide peaks are detected simultaneously. XPS is able to detect different chemical binding states of solid elements with $Z \geq 3$, if the energy resolution of the spectrometer is sufficiently good. Typical spectrometer resolutions are in the range 800–400 meV, determined by both the X-ray source and the analyser. Peak-fitting routines are available on most modern instruments. Table 6.2 gives, as an example, the photoelectron energies of the Ti $2p_{3/2}$ peak for different compounds (Briggs & Seah 1990). In addition, kinetic energies of the Auger peaks, which are observed in the XPS spectrum, are given. The presence of the Auger peaks (cf. Fig. 6.3) is due to the de-excitation processes after the emission of a photoelectron.

The sum of the binding energy, E_b, of a photoelectron and the kinetic energy, E_{kin}^{AES}, of an Auger transition is called the Auger parameter, α, defined as:

$$\alpha = E_{kin}^{AES} + E_b \tag{6.4}$$

Table 6.2 Photoelectron energies and kinetic (Auger) energies (eV) of Ti $2p_{3/2}$ (reference C 1s 284.8 eV)

Compound	E_b	$L_3M_{2.3}M_{4.5}$	α
Ti	454.0	419	873.0
TiC	454.6	418.2	872.8
TiO_2	458.7	418.6	877.3
Ti_2O_3	457.4	416.8	874.2
TiO	458.4	414.8	873.2
Na_2TiF_6	462.6	409.8	872.4
K_2TiF_6	462.1	409.4	871.5

The kinetic energy of the Ti $L_3M_{2.3}M_{4.5}$ Auger peak together with the Auger parameter is also listed in Table 6.2. The difference in the binding energy, α', between the photoelectron (E_b) and Auger (E_{kin}^{AES}) peaks, respectively, can be used to identify different oxidation states of an element. This difference is calculated by:

$$\alpha' = h\nu - \alpha = E_b^{AES} - E_b \tag{6.5}$$

Tabulated values of the Auger parameter are found in the specialized literature or in databases (see Briggs & Seah 1990; Powell & Jablonski 1999). In practice, the application of the Auger parameter is useful, in particular when there is insufficient spectrometer energy resolution.

In XPS either singlet (s transitions) or doublet (p, d, f, etc. transitions) peak structures are observed. The multiple splitting is due to spin–orbit coupling. The total angular moment ($j = l \pm s$) is determined by the azimuthal and spin momentum l and s, respectively. Except for the s transition one observes doublets for all other transitions. The intensity ratio is determined by the final states with either $s = +1/2$ or $s = -1/2$ according to

$$\frac{I(l+1/2)}{I(l-1/2)} = \frac{2J_{1+1/2}+1}{2J_{1-1/2}+1} \tag{6.6}$$

where $J = L + S$ represents the sum of the individual angular moments. With the help of Table 6.1 the corresponding values of J are determined. In agreement with Equation (6.6) experiment shows intensity ratios close to 2:1, 3:2 or 4:3 for 2p, 3d or 4f doublets, respectively. This is also true of the intensity ratios for Fe and Ti $I(2p_{3/2})/I(2p_{1/2})$ shown in Figs 6.6 and 6.9, respectively.

The peak intensity allows quantifying the measurement. Assume that dI is the electronic contribution of a monolayer with thickness dz. Then the differential contribution of an element of atomic density c_A at depth z of that layer will be

$$dI_A = gc_A(z)\,e^{-z/\lambda\cos\theta}\,dz \tag{6.7a}$$

where g is an instrument- and method-dependent variable. The emitted electron with an initial kinetic energy E_{kin} is attenuated on its way towards the surface, assuming exponential decay and taking into account the mean-free-path (IMFP), $\lambda(E_{kin})$, of the electron trajectory, which is dependent on the kinetic energy. θ is the collection angle of the analyser with respect to the surface normal. Integration gives:

$$I_A = g \int_0^{\infty} c_A(z) \exp\left(-\frac{z}{\lambda \cos\theta}\right) dz \tag{6.7b}$$

The infinity limit is, in practice, often replaced by 3λ because more than 95 of all the created electrons come from this escape depth.

For XPS, g is defined by

$$g = K I_{RX} \sigma_A \tag{6.7c}$$

where K contains instrument-dependent constants (like the transmission function for electrons), I_{RX} is the primary flux density and σ_A is the photoemission cross-section for a specific orbital of element A. In order to keep the calculation simple, one neglects, in general, the surface roughness of the sample and angular asymmetry of electron emission.

Assuming a homogeneous density distribution, integration of Equation (6.7b) yields

$$I_A = K I_{RX} \sigma_A c_A \lambda \cos\theta\, \Omega_0 \tag{6.8}$$

where it is assumed that the emission of the photoelectrons is symmetric and that there is a solid angle of acceptance at the analyser, Ω_0. In the case of a thin overlayer (film) of uniform composition on a substrate (sub) of thickness $d_{film} \leq 3\lambda$, its thickness can be calculated by

$$d_{film} = \lambda_A^{film} \cos\theta \ln\left(1 + \frac{I^d_{A,film}\, I^0_{A,sub}}{I^d_{A,sub}\, I^0_{A,film}}\right) \tag{6.9}$$

where $I^d_{A,film}$ is the measured intensity of element A within the film of thickness d, while $I^0_{A,sub}$ is the intensity of a standard of element A in the substrate. In a first-order approximation the standard intensities are assumed to be equal for film and substrate, if the same element is in different oxidation states. Thicknesses can be extracted from narrow scan data (cf. Figs 6.6 and 6.9).

In general, a hemispherical analyser is used for XPS analysis. A typical schematic representation of such an XPS analyser is shown in Fig. 6.7. The source can be either achromatic ($K_{\alpha_{1,2}}$ of Mg or Al) or monochromatic (Al K_{α_1}) radiation. Depending on the manufacturer a focused or unfocused X-ray source is used. If the source is not focused, the analysed area is determined by the entrance lens of the analyser. A lateral resolution of a few microns is possible. An example for imaging XPS will be shown at the end of the AES section (Section 6.2.2). The monochromatic source reduces not only the width of the primary Al K_α line to values below 500 meV, but also reduces the background radiation considerably. It also prevents electrons from the achromatic source reaching the sample. The design of the analyser defines the collection efficiency of the spectrometer, which is responsible for the sensitivity of the analyser. A take-off angle with respect to the surface, θ_S, can also be defined.

An example of a high-energy resolution XPS narrow scan of C 1s is shown in Fig. 6.8(a) for polyethylenterephthalate (PET). The measured envelope curve has been fitted with six bands all of which are identified in the caption. It shows different binding states between carbon, hydrogen and oxygen typical for polyethylenterephthalate (PET), the chemical structure of which is shown in Fig. 6.8(b)

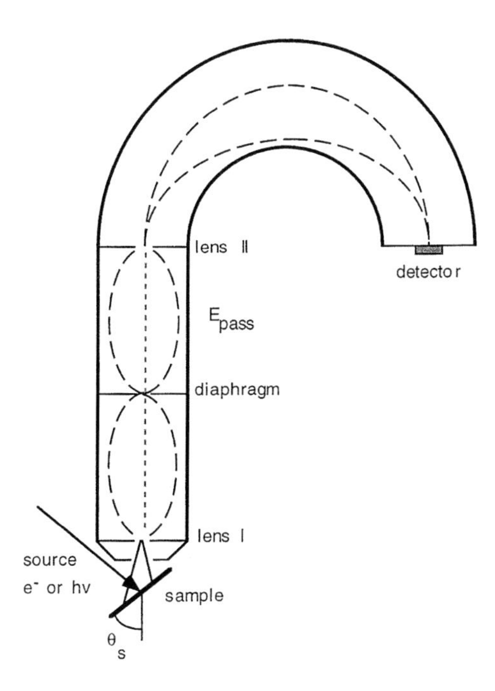

Fig. 6.7 Schematic representation of an electronic hemispherical XPS analyser with X-ray source *hv*, sample holder, and energy filter with diaphragm and lenses.

The functional groups 1–3 and 6 belong to the PET, whereas groups 4 and 5 are typical of the N_2-plasma treatment. Figure 6.8(a) illustrates the possibility of XPS identifying, qualitatively, different binding states obtained at different binding energies measured relative to the C–H binding energy at 284.7 eV. The difference in binding energy is called the chemical shift, ΔE. Their values for the data of Fig. 6.8(a) together with the relative intensity of the areas are: (1) C–H, 55.4%; (2) C–O, ΔE = 1.7 eV, 6.6%; (3) O–C=O, ΔE = 4.1 eV, 6.9%; (4) C=N, ΔE = 3.0 eV, 8.1%; (5) C–N, ΔE = 0.8 eV, 11.2%; (6) $\pi \rightarrow \pi^*$ shake-up, ΔE = 8.2 eV, 1.7%.

Non-conducting samples usually charge-up positively during photoelectron emission. This charging can be neutralized by an additional electron source providing low-energy electrons to compensate for the loss of electrons during photoelectron and Auger electron emission. Different electronic and ionic neutralizers for negative and positive charge compensation are available. Such charge effects can be a major limiting factor in XPS, because they may lead not only to a total shift of the whole spectrum but increase also the line-width of a photoelectron transition, if the charge compensation is not uniform over the analysed area. Another limiting factor in XPS may be X-ray degradation. Examples of polymer degradation during X-ray analysis are shown elsewhere (Coullerez *et al.* 1999).

A sample holder with a tilt facility allows angular resolved X-ray photoelectron spectroscopy (ARXPS) at different emission angles. This technique probes the sample composition at shallow depths non-destructively. Note that θ was previously defined with respect to the sample normal. Application of Equation (6.8) at different take-off

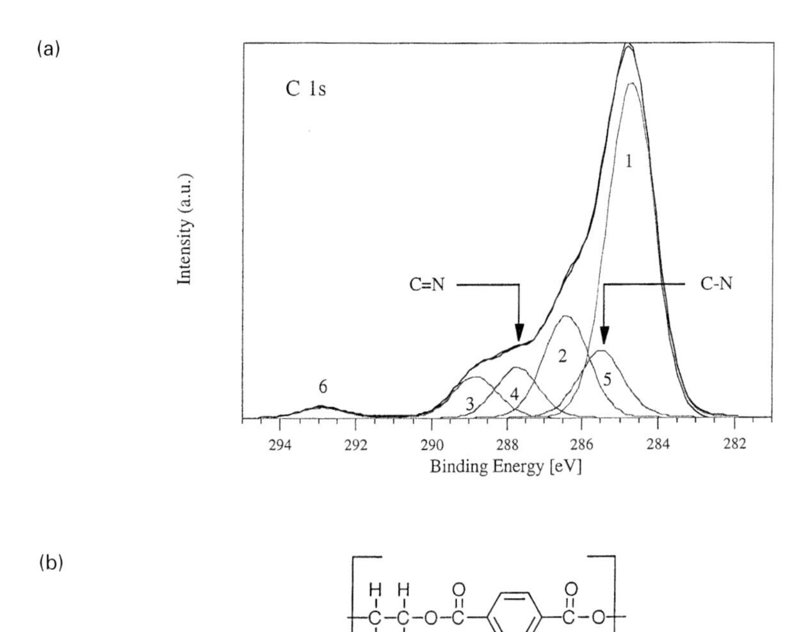

Fig. 6.8 (a) Narrow XPS scan of polyethyleneterephthalate (PET) shown with fitted bands after N_2 plasma treatment with functional groups: (1) C–H and C–C; (2) C–O; (3) O–C=O; (4) C=N; (5) C–N; (6) aromatic shake-up peak, π^*, measured with monochromatic Al K_α (Pitton 1996). (b) Chemical structure of polyethyleneterephthalate (PET) with three functional groups: C–H; C–O; O–C=O including the aromatic group.

angles, θ, is denoted *angle-resolved XPS* (ARXPS) and is able to yield a depth profile within the first 3–5 nm below the sample surface. Such a profile is obtained just by variation of the tilt of the analyser, with small angles used to probe close to the surface and larger angles for probing the bulk. The major restriction of ARXPS is surface roughness, even in the order of several nanometres.

Figure 6.9(a, b) gives an example of the use of ARXPS. It shows a naturally formed oxide of a Ti sheet metal that was sputter cleaned by argon bombardment just prior to the ARXPS measurements. At $\theta = 70°$ [Fig. 6.9(a)] from surface normal the IMFP is 0.41 nm. Deconvolution (fitting) of the envelope curve exhibits the $2p_{3/2}$ and $2p_{1/2}$ doublets of the main oxides TiO_2 as well as small amounts of the suboxides TiO (A1 and A2) and Ti_2O_3 (B1 and B2). Peaks were assumed to be asymmetric. Because of the small thickness of the oxide layer, it is observed at a larger probing depth, with $\theta = 10°$ [Fig. 6.9(b)] corresponding to an IMFP of 1.18 nm, that there is a metal doublet together with peaks of different Ti oxides (TiO_x) with varying stoichiometry, i.e. x = 2, 1.5 and 1, respectively.

The thickness of the film was determined to be 1.5 nm by applying Equation (6.9) to Fig. 6.9(b) data at $\theta = 10°$ with $\lambda = 1.2$ nm and assuming $I^0_{\mathrm{film}} = I^0_{\mathrm{sub}}$. A number of algorithms for performing a deconvolution to extract undestructive depth profiles from ARXPS data has been published (Tanuma *et al.* 1988; Tyler *et al.* 1989; Tanuma

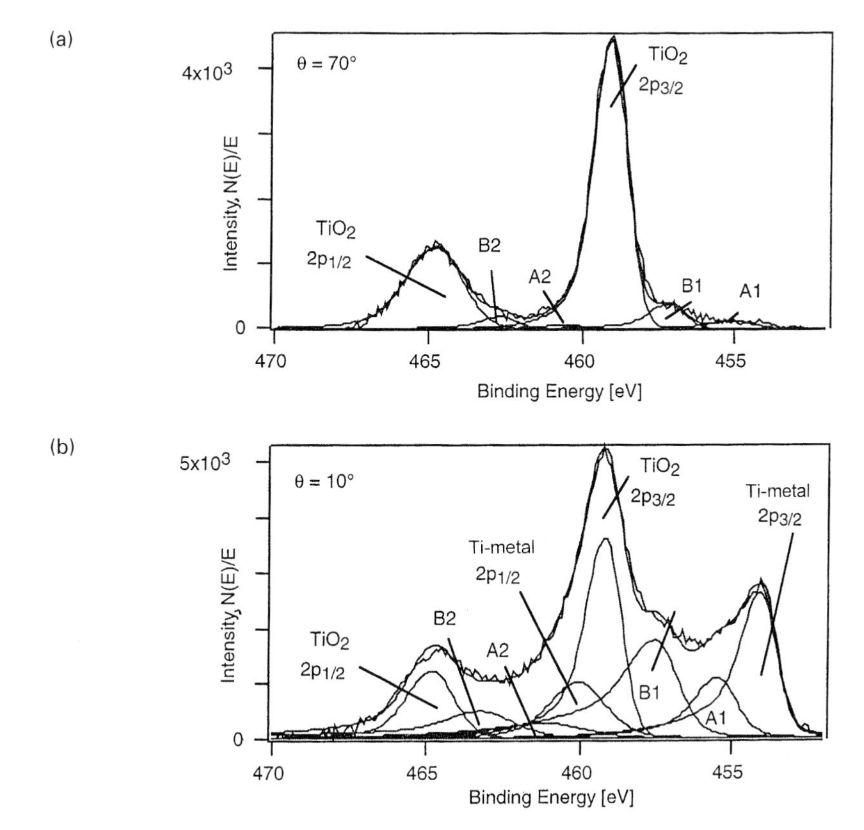

Fig. 6.9 Narrow scan data of TiO_2 on Ti sheet metal at different take-off angles (a) 70° and (b) 10° with respect to the surface normal. A1 and A2 represent TiO $2p_{3/2}$ and $2p_{1/2}$ respectively, and B1 and B2 represent Ti_2O_3 $2p_{3/2}$ and $2p_{1/2}$, respectively.

et al. 1991; Ratner & Castner 1997). The principle of ARXPS is illustrated in Fig. 6.10 for two take-off angles.

The angle of collection is defined with respect to the sample's surface normal and the analyser axis and determines the escape depth, Λ, calculated using Equation (6.10):

$$\Lambda = 3\lambda(E)\cos\theta \tag{6.10}$$

This approximation assumes that 95% of the electrons are emitted within a distance of three times the IMFP λ multiplied by the cosine of the take-off angle. λ (in nm) can be estimated from an empirical formula proposed by Seah and Dench (1979)

$$\lambda = C_{SD}a^{3/2}E_{kin}^{1/2} \tag{6.11}$$

with a constant $C_{SD} = 0.41$ for elements (0.72 for inorganic compounds and 0.11 for polymers), E_{kin} the kinetic energy of the emitted electron and a (m^3) the average monolayer thickness calculated for a cubic crystal, given by:

$$a^3 = \frac{M_A}{\rho N_A n} \tag{6.12}$$

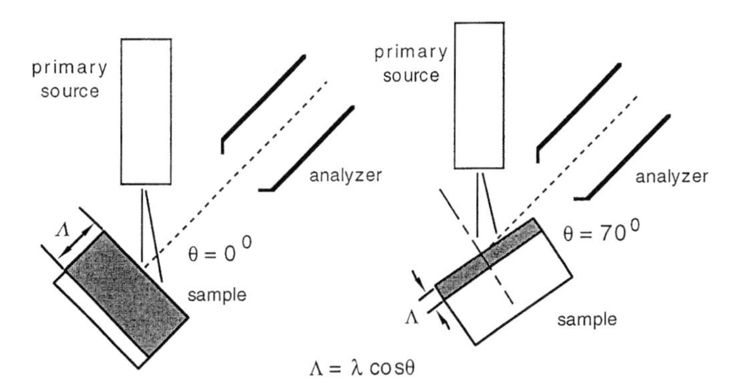

Fig. 6.10 Principle of angular-resolved measurements (schematic).

in which M_A is the atomic weight of element A, ρ the mass density, N_A Avogadro's number and n the number of atoms in the molecule (e.g. $n = 1$ for Si or 3 for SiO_2). Typical values of a lie between 0.2 and 0.3 nm for most solid elements and components. It should be emphasized that Equation (6.11) is only a rough estimate with errors of up to 50%. More precise values are obtained from a formula proposed by Tanuma *et al.* (1988, 1991)

$$\lambda = \frac{E_{kin}}{E_{pl}^2 \left[\beta \ln(\gamma E_{kin}) - (C/E_{kin}) + (D/E_{kin}^2)\right]} \tag{6.13}$$

based on experimental optical data. It is a modified form of the Bethe equation (Bethe 1930), where $E_{pl} = 28.8\ (N_V \rho/M)$ is the free-electron plasmon energy (eV), and N_V the number of valence electrons per atom. Values for parameters b, γ, C and D are obtained from fits to the computed IMFPs for all of the 30 elements available in the original paper (Tanuma *et al.* 1991). The RMS deviations of the fits vary between 0.1 and 1%, while the maximum deviation at any one energy was 2.5%. Figure 6.11 shows the IMFP calculated by Tanuma *et al.* (1991, 1994) for a polymer (polystyrene), an inorganic compound (Al_2O_3) and an element (Au). Notice that the IMFP of a polymer is significantly higher than those of elements ranging from 1.2 to 3.5 nm at a kinetic energy of 1 keV. Additional IMFP calculations have been made for

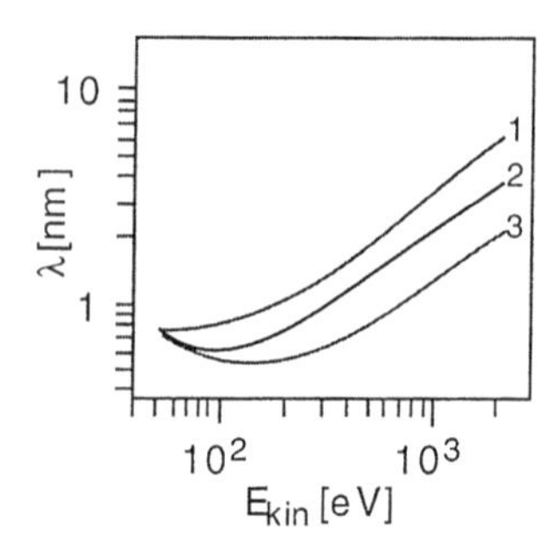

Fig. 6.11 Calculated IMFPs as a function of kinetic energy E_{kin} for (1) polystyrene, (2) Al_2O_3 and (3) Au (see Tanuma *et al.* 1991, 1994).

elements, inorganic and organic compounds and are available from NIST (1997). A review of the evaluation and calculation of IMFPs is available from Powell and Jablonski (1999).

As shown in Fig. 6.5, a photoelectron spectrum displays a large background which is formed by photoelectrons having lost energy during their escape. A formalism for background subtraction evaluation for chemical state and quantitative information has been proposed by Tougaard and is available as a commercial software (Tougaard & Jansson 1992). To apply this formalism the spectrum has to be measured in the vicinity of the peak with a window of width larger than 100 eV.

A method of relative sensitivity factors is used in XPS (as in AES) to calculate the composition of a material from spectral data. It is based on the assumption that the peak signal intensity for a chemical element is linearly proportional to the number of atoms present, but that the proportionality constant is element dependent. The method described does not make reference to any physical models of excitation, emission or detection. The only basic assumption of the *relative sensitivity method* is that the signal intensity, I_A, from each element, A, is proportional to its abundance (atomic concentration), c_A, in the analysed volume according to Equation (6.8)

$$I_A = K S_A c_A \tag{6.14}$$

where S_A represents the normalized cross-section with $S_A = \sigma_A/\sigma_F$ of a photoemission cross-section for a specific orbital of element A with respect to an arbitrarily chosen elemental orbital. In many cases the F 1s transition is used as a reference. I_A/S_A can be considered as a sensitivity-corrected intensity. Commercial instruments provide a set of dedicated sensitivity factors for a particular analyser. If K is known or kept constant for a certain measurement Equation (6.14) can be solved for c_A. More commonly, the constant K is eliminated by forming a ratio such as

$$\frac{N_A}{N_x} = \frac{I_x/S_x}{I_A/S_A} \tag{6.15}$$

or more general as the molar fraction c_A of element A:

$$c_A = \frac{I_A/S_A}{\sum_{i=1}^{n} I_i/S_i} \tag{6.16}$$

In this form only the ratio of each sensitivity factor relative to the others is important. Note that the units of concentration have not been specified. Typically in both XPS and AES the relative sensitivity factors are defined to give atomic concentrations, c_A. However, Equation (6.16) may also be used to calculate weight fractions. The sensitivity factor for weight, S_A^w, can be derived from the atomic concentration factors S_A by $S_A^w = S_A/M_A$, where M_A is the atomic mass of element A. Relative sensitivity factors for XPS and AES can be found in handbooks or can be derived from the calibration of pure standards directly on the instrument. It must be remembered, however, that the atomic density may change with composition for a given alloy or compound and that a correction for such variation may be necessary. Finally, quantification can also be performed for functional groups, i.e. the different C_xH_y groups of polymers or different oxidation states of metals.

6.2.2 *Auger electron spectroscopy (AES)*

In Fig. 6.2 it was seen that excitation with photons may result in the direct emission of a photoelectron. During the relaxation process either a secondary photon or an Auger electron is emitted. If the primary photon beam is replaced by primary electrons it is possible to create directly secondary and Auger electrons during the relaxation process (Auger 1925). The so-called Auger process involving three electron levels is illustrated schematically in Fig. 6.12.

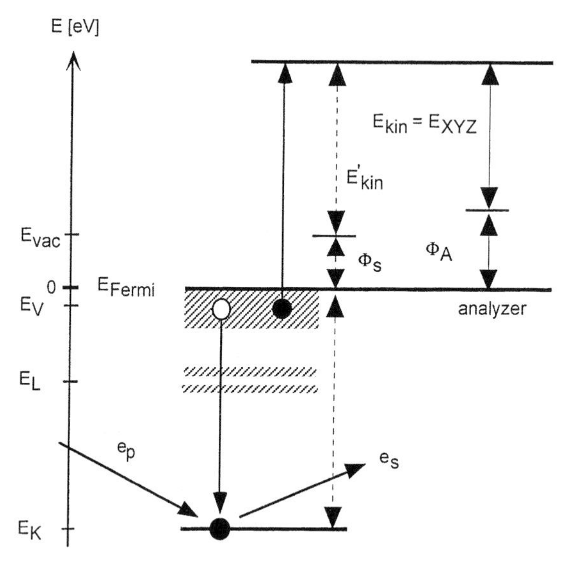

Fig. 6.12 The Auger electron process.

An atom A is assumed to be ionized by creating a secondary electron on the K level. An electron from a level closer to the Fermi level, for instance from the valence band, V, may fill this hole during an internal electron transfer process which liberates energy ΔE corresponding to the difference between those two levels, i.e. $\Delta E = E_K - E_V$. This ΔE is transferred either to a third electron or to a secondary photon. This third electron is labelled the KVV Auger electron indicating the three energy levels involved. According to quantum physics, the probability of photon emission is low, in particular for atoms with low atomic number Z. For $Z \leq 30$ the probability of a KVV or LVV Auger transition is still high. For heavier elements ($Z > 40$) MNN or MVV transitions are observed. This suggests that all solid elements with a minimum of three electrons (Li) can be detected by AES with sensitivities approximately equal to those of XPS (limit 10^{12}–10^{13} atoms cm^{-2}). The energy of the emitted Auger electron is determined basically by the difference between the three energy levels involved, i.e. levels K, V and V or more general X, Y and Z

$$E_{XYZ} = E_X(Z) - E_Y(Z + \delta) - E_Z(Z + \delta) - \Phi_A \tag{6.17}$$

where Φ_A is the work function of the analyser. All energy levels depend on the atomic number Z. However, after ionization and emission of the secondary electron the remaining electrons (their number is now $N-1$) suffer from an increased attraction by the N protons. As a consequence all levels show an increase in binding energy with respect to the Fermi level. Therefore a small correction factor, δ, is introduced to add to the atomic number Z, because of the augmented additional binding energy. Values of δ are between 0 and 1; in practice, the average binding energy between the same levels of two neighbouring atoms is used for estimating $E_Y(Z+\delta)$ and $E_Z(Z+\delta)$.

The number of electrons detected by the analyser (cf. Fig. 6.7) determines the intensity of a given elemental Auger transition. The basic Auger equation is similar to that of XPS [Equation (6.7c)] with integration limits between 0 (surface) and approximately 3λ:

$$I_A = g' \int_0^{\infty} c_A(z) \exp\left(-\frac{z}{\lambda \cos\theta}\right) dz \tag{6.7d}$$

The instrumental constant g', however, is defined differently

$$g' = K I_0 (1 + r_A) \sigma_{A,e} \tag{6.18}$$

where I_0 is the primary beam current and r_A the factor of backscattered secondary electrons. These electrons are backscattered from other atoms of the matrix and can create additional Auger electrons. The Auger cross-section, $\sigma_{A,e}$, is specific for the given element and transition. In an analogy with XPS use is made of relative sensitivity factors, S_A, normalized to an arbitrarily chosen element, like the Ag_{MNN} transition defined by

$$S_A = \frac{\sigma_{A,e}}{\sigma_{Ag_{MNN},e}} \tag{6.19}$$

These S_A factors are instrument specific, depending on the energy transmission, detection efficiency of the analyser and the backscattered electrons in the matrix.

The lateral resolution is primarily determined by the diameter of the primary beam. As is illustrated in Fig. 6.13, however, backscattered electrons contribute to an increase of the volume from which electrons are emitted.

The measured Auger intensity, I_A [Equation (6.7d)], is an integration over the excited volume of the matrix determined by the IMFP [see Equations (6.10)–(6.13)] and the emission angle, θ. In general, the IMFP of Auger peaks is lower than that of XPS peaks because of the smaller kinetic energies of the main Auger transitions. Angular resolved Auger electron spectroscopy (ARAES) is possible, if an hemispherical analyser is used in a similar way to that in ARXPS (cf. Fig. 6.10).

A typical survey spectrum is given in Fig. 6.14 illustrating an oxidized iron–chromium alloy. The intensity $N(E)$ is shown as a function of the kinetic energy of the Auger peaks. In addition, the numerically differentiated spectrum is given. LMM and KLL Auger families are labelled for the major peaks including a surface contamination of carbon and a small peak of argon after a light surface cleaning which removed approximately 1.5 nm of oxide. For quantification, the area below the peak after background subtraction is used to determine the surface composition with the

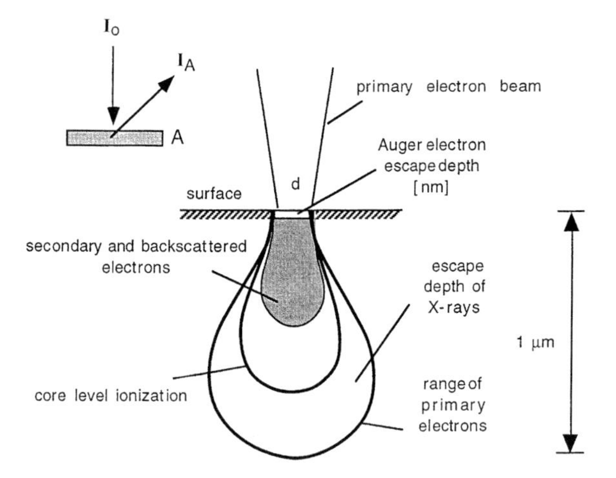

Fig. 6.13 Schematic representation of the escape depths of Auger electrons and secondary X-rays during combardment with primary electrons of intensity I_0 and beam diameter, d.

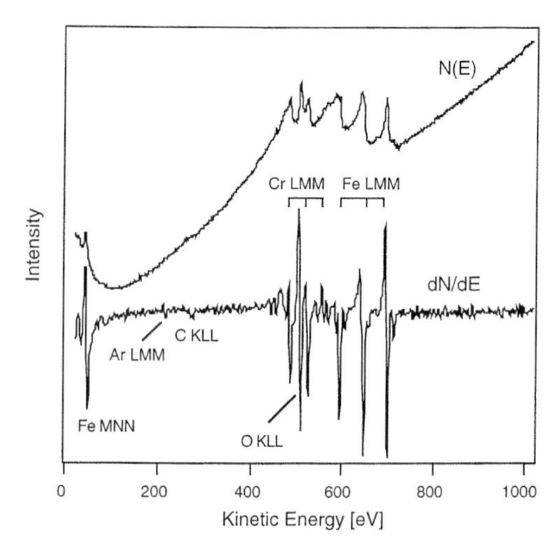

Fig. 6.14 Spectrum of an oxidized Fe–25Cr alloy after light argon sputter cleaning in the $N(E)$ or $dN(E)/dE$ mode as a function of the kinetic energy.

help of Equation (6.16) with a precision of only $\pm 30\%$ if instrumental sensitivity factors are used.

Figure 6.14 indicates that such background subtraction is not straightforward, owing to inelastic scattering of the created Auger electrons. Differentiation of $N(E)$ removes the inelastic background. However, valuable information on the scattering of emitted electron is lost. For quantification, peak-to-peak intensities of the dN/dE data are used assuming that the peak area is symmetrical (Gauss–Lorentzian), an assumption that is often not valid. These are the reasons for the $\pm 30\%$ errors in quantitation of Auger data. It should be mentioned that in the case of XPS, data quantification is easier due to the fact that photoelectron transitions involve only one electron and give narrower peak widths (0.5–2.5 eV) compared to Auger transitions where three electrons are always involved rendering peak widths of up to 15–25 eV. To improve the precision of the experiment a narrow scan window around a given peak is usually measured. In general, Auger analysis is an elemental analysis and only in a few cases do peak-shape changes permit easy distinction between different binding states, as, for example, in light elements such as carbon, aluminium, silicon or titanium.

Lateral information can be obtained in the form of a line-scan or a mapping. Such information is obtained by determining the intensity of a chosen peak as function of the scanned position of the electron beam on the sample. The intensity is measured by taking the difference between a peak maximum and the background or the difference between the maximum and the minimum for the $N(E)$ or dN/dE data, respectively. An example of an Auger map is illustrated in Fig. 6.15(a), where the secondary electron image (SEM) of gold islands on graphite is shown, which is commonly used to evaluate an instrument's secondary electron imaging capabilities. In the example illustrated a primary electron column and Schottky thermal field emission electron source were used. A finely focused electron beam with SEM imaging performances down to 10 nm is provided. A high-transmission cylindrical mirror analyser (CMA) was used allowing Auger imaging with minimal analyser shadowing. In Fig. 6.15(b), high spatial resolution Auger imaging of a scanning Auger spectrometer is demonstrated. This is particularly useful in the semiconductor industry for device characterization and failure analysis (Physical Electronics 1997).

Typical acquisition times are 5–30 min per map. Auger mapping is a way of determining semi-quantitative correlations of chosen elements, e.g. Au and C are complementary with a lateral Auger resolution of 10–50 nm depending on the surface topography. Practical analysis may be limited by the high electron beam current densities applied for high lateral resolutions. Typical sensitivity limits for Auger analysis of 0.1 or 1% of a monolayer may be reduced in mapping situations by up to a factor of 10–100.

Modern XPS systems also have imaging capabilities in the low-micron range. Figure 6.16 provides an example showing platinum pads on a silicon wafer. The Pt lines deposited by electron-beam lithography have a width of 10 μm and are separated by 15 μm (A), 10 μm (B) and 5 μm (C). A line-scan over four Pt lines (Fig. 6.17) has been produced retrospectively from such an image indicating a clear separation of the lines with a lateral resolution well below 3 μm.

6.2.3 *Sputter depth profiling with AES and XPS*

Probing depths for XPS and AES are typically $< 3\lambda$ corresponding in general to layers of less than 5–10 nm thickness. For thicker layers Auger or photoelectron emission is combined with sputter removal to access deeper layers or buried interfaces. In general argon ion bombardment is applied where typical ion energies ranging from a few hundred eV to a few keV are used. This locally destructive method allows work with sputter rates ranging typically from 0.1 to 100 nm min^{-1}. The latter can be changed by variation of the raster size or the ion beam current density. The principle of the method is illustrated in Fig. 6.18 indicating the zone analysed either

(a)

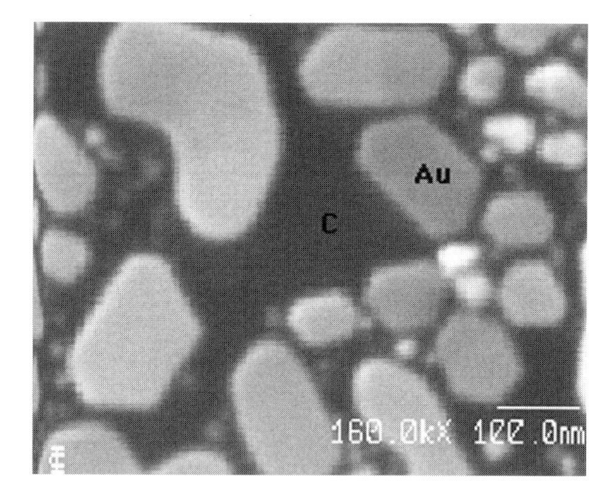

(b)

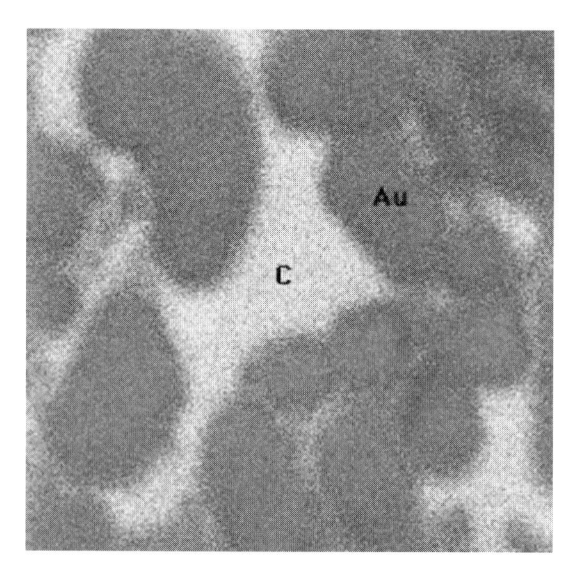

Fig. 6.15 (a) Secondary electron image and (b) 10 000 × colour overlay Auger maps of a graphite sample covered with gold islands. Images for the gold on graphite sample were obtained with a 25 keV, 2 nA primary electron beam at a magnification of 160 000 × with a CMA. (Reproduced with permission of Physical Electronic Industries.)

by Auger or photoelectrons together with the ion beam crater (xy) defined by the raster size of the ion beam.

The focused ion beam with a typical diameter of 0.1–0.3 mm at 2–3 keV is rastered across the surface alternating either electron (AES) or X-ray bombardment (XPS). Beam alignment is important to match the ion beam crater and analysed zone. The size of the latter should be small compared to the ion beam crater size in order to avoid crater edge effects. In general, owing to the size of the ion beam diameter sharp crater edges are not obtained. In addition, continued ion bombardment leads to a

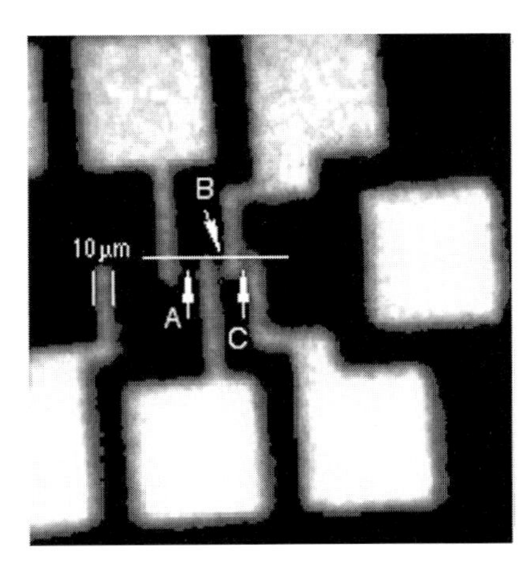

Fig. 6.16 XPS image of Pt pads on a Si wafer. The 10 µm width of the Pt line is indicated. The separation of the four lines is (A) 15 µm, (B) 10 µm and (c) 5 µm with an intensity scale shown on the right. Data were obtained with a Kratos Ultra XPS system.

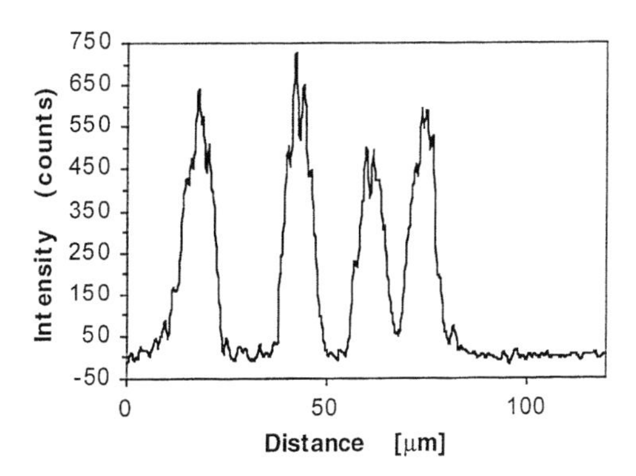

Fig. 6.17 XPS line-scan over four Pt lines as shown in Fig. 6.16. Data were obtained with a VG ESCLAB 250.

roughening of the surface with sputter time. Such roughening can be reduced by rotation of the sample, which leads to a variation of the angle of incidence of the sputter ions. This induced roughness is more pronounced for crystalline than for amorphous targets. Interaction of the primary ion beam with surface atoms and molecules may lead to atomic mixing and preferential sputtering depending on the sputter yield of the sputter material.

A detailed discussion of the depth resolution, which relates to the goodness of the detectable interface width, can be found elsewhere (Hofmann 1994, 1998). The best depth resolution values reported are in the order of 1–2 nm for amorphous materials. The sputter yield Y_M (atoms per ion), which is inversely proportional to binding energy, is related to the sputter rate (depth per time) by

$$v = \frac{J\,Y_M A}{\rho\, e\, N_A} \qquad (6.20)$$

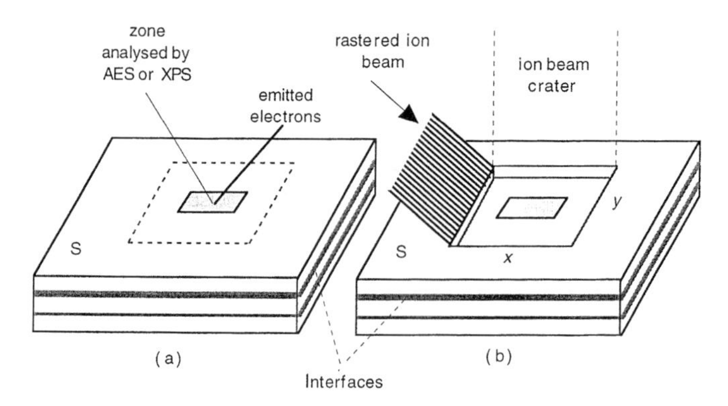

Fig. 6.18 Principle of sputter depth profiling. The analysed zone in (a) is defined by lateral resolution of AES or XPS. The surface S is bombarded with ions with an ion beam crater determined by the rastered ion beam in the x–y direction (b).

where J is the ion flux density, A is atomic number, ρ is the mass density, e is the electronic charge and N_A is Avogadro's constant. Sputter yields of the elements are tabulated elsewhere (for argon yields see Mathieu 1997). For inorganic compounds and oxides a sputter depth profiling standard was developed (Seah *et al.* 1984), which allows calibration of sputter rates and conversion of the sputter rate to extrapolate values for other materials under different experimental conditions (variation of ion beam energy, angle of incidence, etc.). A few sputter rates are given in Table 6.3. Additional sputter rates can be found elsewhere (Hofmann & Sanz 1982–83).

Table 6.3 Normalized sputter rates (relative to Ta_2O_5) of a few selected elements and oxides for 3 keV argon

Material	Normalized sputter rate
Al	0.95
Al_2O_3	0.63
Ag	4.64
Au	4.10
Fe_2O_3	0.92
Nb_2O_5	1.04
NiO	1.00
Si	0.90
SiO_2	0.85
Ta_2O_5	1.00
TiO_2	0.82

Conversion of the measured sputter time into a depth z is done by integrating the total sputter time, t, according to:

$$z = \int_0^{\infty} v \, dt \tag{6.21}$$

Figure 6.19 illustrates an example of an Auger depth profile. It shows the result of an anodically formed thin oxide film of titanium. It is observed that the oxygen and

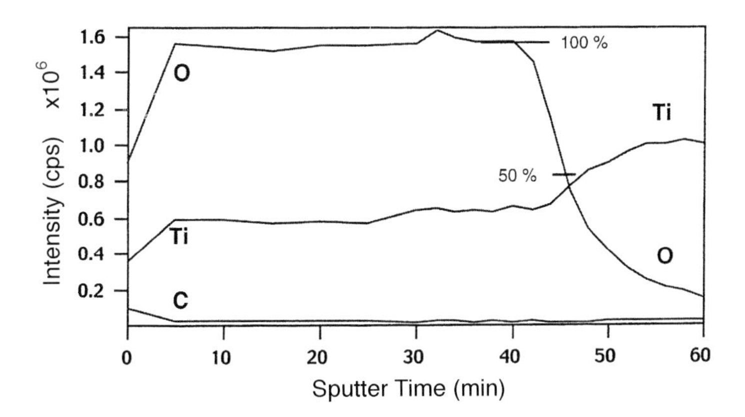

Fig. 6.19 AES sputter profile of an anodic TiO_2 film on Ti with argon at 3 keV ion beam energy. The thickness determined at 50% of the O signal at the TiO_2/Ti interface with respect to the Ta_2O_5 standard is 137 nm.

titanium signals increase during the first few minutes during which the carbon signal decreases. This is due to the removal of carbon contamination present on the outer surface. At the oxide/metal interface there is a decrease of the oxygen signal coupled with an increase of the titanium signal.

The film thickness was determined (by reference to Table 6.3) to be 36.9 nm after conversion of the time necessary to reach the interface. As a measure of the oxide thickness the value at the 50% oxygen signal relative to the plateau value (100%) is taken. Applying Equation (6.16) to calculate the respective atomic concentration of Ti and O (c_{Ti} and c_{O}) within the film gives $c_{\mathrm{Ti}} = 37\%$ and $c_{\mathrm{O}} = 63\%$. These values are close to the theoretical values of 33% and 66%. This indicates that under the applied conditions only little preferential sputtering of oxygen relative to titanium took place. If preferential sputtering of oxygen exists, reduction of oxides and a change of stoichiometry from the thermodynamically stable value to a lower oxidation state are observed. Figure 6.19 cites the case using a Ta_2O_5 standard which is reduced during sputtering to intermediate suboxides and TaO.

In materials science a multilayer structure is often employed. Such a structure is illustrated in Fig. 6.20 which shows schematically an example of packaging material comprising thin layers of non-stoichiometric SiO_x as oxygen barrier layers on polyethylenterephthalate (PET).

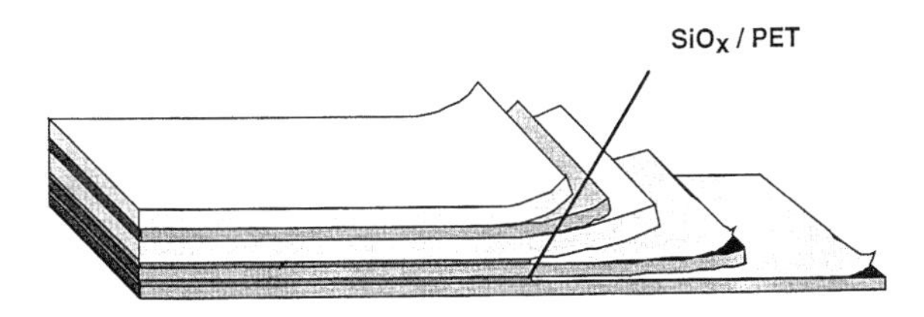

Fig. 6.20 Schematic representation of a multilayer material with SiO_x/PET internal interfaces.

The usefulness of sputter depth profiling is illustrated in Fig. 6.21 which shows the profile of a 100 nm thick $SiO_{1.85}$ film obtained by reactive sputter deposition on a PET substrate (a) before and (b) after N_2-plasma treatment (Pitton 1996). Comparison shows that the interface width measured between the silicon oxide film and the PET at 84% and 16% of the oxygen O 1s plateau value is increased, respectively, from 22.8 to 33 nm resulting in a better adhesion of the film to the substrate. These examples demonstrate the usefulness of depth profiling, even if the employed method is locally destructive and some information on the chemical nature is lost. Apart from SIMS profiling (see below) there is currently no other method available that is capable of determining with high sensitivity the stoichiometry and thickness of films or of controlling the presence of impurities at an internal interface of films with submicron thickness.

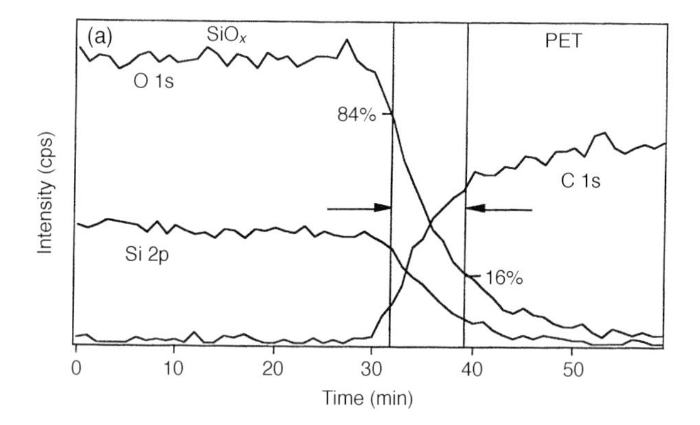

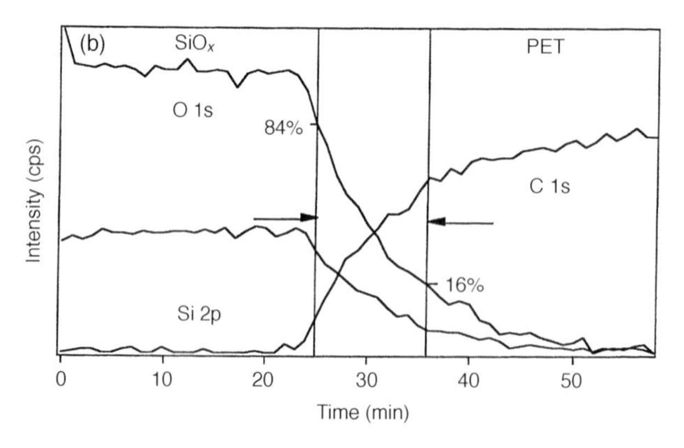

Fig. 6.21 Comparison of XPS depth profiles of 100 nm $SiO_{1.85}$ obtained by reactive PVD sputter deposition on polyethyleneterephthalate (PET) with (a) and without (b) modification by N_2 plasma treatment (Pitton 1996).

6.3 Secondary ion mass spectrometry (SIMS)

Secondary ion mass spectrometry (SIMS) (Briggs & Seah 1992) is more sensitive than XPS or AES (Briggs & Seah 1990). SIMS involves measuring secondary ions, which form only a portion of the particles (neutrals, electrons, photons) emitted during a bombardment with primary ions. Reasonable quantification of the ion intensity, however, can only be achieved if suitable standards are available. The considerable influence of the matrix controls the number of emitted charged particles (ions) (Benninghoven *et al.* 1987). It represents the severest limitation of this technique for the quantitative analysis of both inorganic and organic materials.

Under primary bombardment, the solid surface emits secondary particles. As illustrated in Fig. 6.22, the bombardment with primary ions induces the emission of neutral, positive or negative fragments and clusters. During the bombardment the surface layer is altered and the degree of perturbation is dependent on the flux of the primary ion beam.

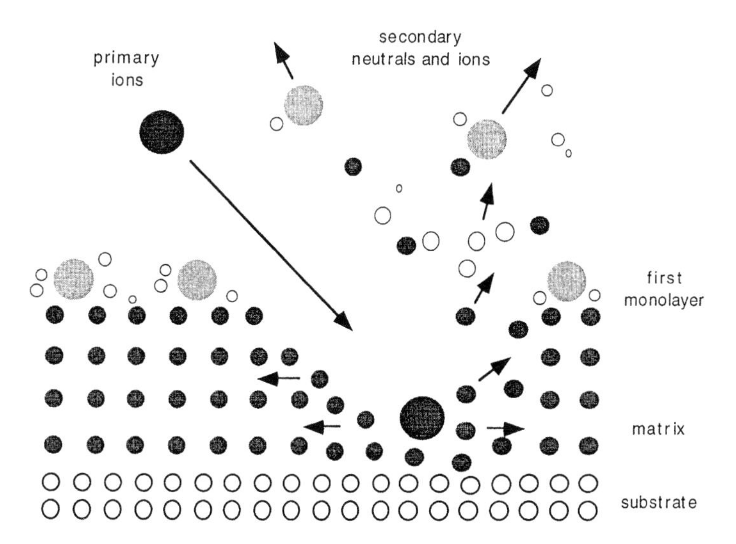

Fig. 6.22 Principle of secondary ion mass spectrometry (SIMS).

SIMS can operate in two modes: static and dynamic. The static regime is determined by a primary particle density of $\leq 10^{12}$ particles cm^{-2}. This number represents 1‰ of particles in the first monolayer (10^{15} particles cm^{-2}). Using the static mode it is postulated that less than 1‰ of the first monolayer is perturbed. Beyond this limit, however, the top surface layer is destroyed and dynamic SIMS is relevant. Figure 6.22 illustrates basically the circumstances of dynamic SIMS, because layers below the first monolayer are altered. In this mode depth profiles are obtained by using the primary ion beam for ion milling. Both SIMS modes require, in general, dedicated spectrometers because respective primary ion beam fluxes can be different by several orders of magnitude.

The objective of SIMS is to analyse the emitted ionized particles as a function of the ratio of mass m and electric charge z, m/z. As mentioned already, only a small fraction of emitted fragments is charged, a typical value being $\leq 5\%$ for oxides. However, this value can be considerably smaller, with values below 10^{-4} being observed.

The SIMS intensity measured for a given target is described by the following equations

$$I_A^{\pm} = I_p \, Y_{tot} \, c_A \, \gamma_{A(M)}^{\pm} \, \eta_A \tag{6.22}$$

and

$$Y_M^{\pm} = Y_{tot} \gamma_{A(M)}^{\pm} \tag{6.23}$$

in which $I_A^{\pm}$ is the detected ion current of species A, I_p is the applied primary ion beam current (1 nA cm^{-2} corresponds to approximately 10^{10} particles cm^{-2} s^{-1}), Y_{tot} the total sputter yield of secondary particle A particles, c_A the fractional (atomic) concentration of species A, $\gamma_{A(M)}^{\pm}$ the ionization probability for species A in a given matrix M, η_A the transmission of the analyser for species A, which also includes the detector efficiency for species A. $Y_M^{\pm}$ is defined as the secondary ion yield of species A (secondary positive or negative ions per primary ion) depending on the matrix M.

The value of $Y_M^{\pm}$ varies over several orders of magnitude and is strongly influenced by the electronegativity of the material (host matrix) from which the ion is emitted. In addition, surface contamination may influence ionization probability. The ionic nature of the primary particles plays an important role in terms of possible charge exchange with the sputtered secondary particles and the surface charge accumulation, which is particularly important for insulating matrices. It is common to use a simultaneous electron bombardment to compensate for the charge of the primary beam in order to neutralize non-conducting surfaces.

The choice of primary particle is important for optimization of the total sputter yield, Y_{tot} determined by the mass ratio of primary and secondary particles as well as the binding energy of the sputtered secondary particle.

Typical primary energies are 2–15 keV and three models have been developed for sputtering:

- ion–surface atom collision model (Biersack 1987)
- collision cascade model (Sigmund 1981)
- high mass molecular species sputter model (Benninghoven 1983; Benninghoven *et al.* 1987)

Elemental sputter yields at low ion beam energies for argon have been reported by Seah (1981). Sputter and ionization yields for inorganic and organic materials can be found elsewhere (Wilson *et al.* 1989; Briggs & Seah 1992). Each type of beam source has different characteristics in terms of yield, ease of use with insulating materials, spatial resolution, and induced beam damage with continuous or pulsed beams. The lateral resolution is determined by the choice of primary ion beam as well. For the highest lateral resolution obtainable today (approx. 0.1 μm) liquid metal Ga^{+} ions are preferred.

There are four basic mechanisms which lead to ionization (Vickerman & Swift 1997):

- electron bombardment
- plasma
- surface ionization
- field ionization

Reactive species like Cs^+ and O_2^+ are employed to influence the ionization probability. Because of its larger mass, the caesium ion renders higher yields of heavier atoms or molecular fragments. For a discussion of the ionization process the reader is referred to specialized literature (Benninghoven *et al.* 1987; Vickerman *et al.* 1989; Rivière 1990; Vickerman & Swift 1997). The basic arrangement for a SIMS experiment is shown in Fig. 6.23.

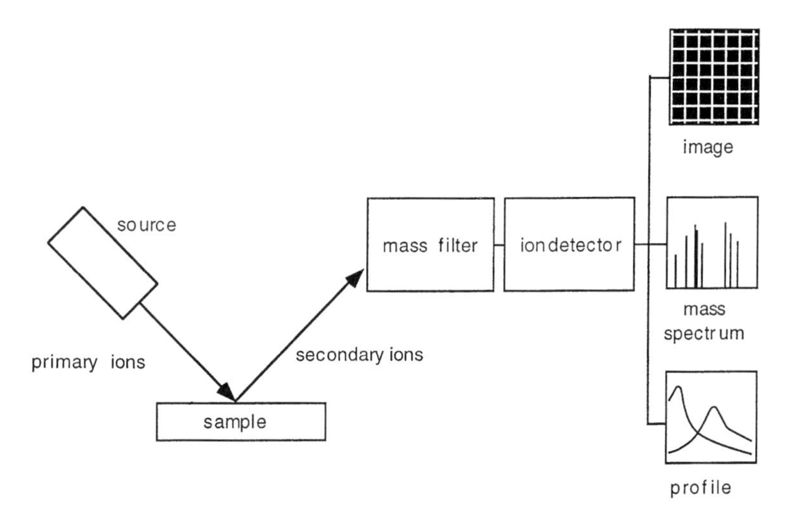

Fig. 6.23 Schematic representation of a SIMS experiment.

The main components of ultra-high vacuum (UHV) spectrometer systems are the primary ion source, the mass spectrometer including ion optics and the detector that provides the mass spectrum. The mass resolution is dependent on the type of analyser used (quadrupole, magnetic sector, time-of-flight, etc.) with mass resolution $m/\Delta m$ varying between 1 and >5000. Depending on the type of ion beam source used images can be produced with sub-micron lateral resolution, e.g. using Ga^+ or In^+ sources. Profiles are obtained in dynamic SIMS mode. Depending on the ion current densities applied, depths from a few nanometres to a few microns are probed. As previously mentioned, surface roughening may be a limiting factor for depth resolution at buried interfaces.

A typical experimental set-up for dynamic SIMS, with a duoplasmatron source and several electrostatic and magnetic analysers, is shown in Fig. 6.24. Such an ion-microprobe is usually equipped with a choice of different ion sources to optimize ionization and imaging capabilities dependent on the sample and problem to solve. These ion sources are operated in a continuous current mode.

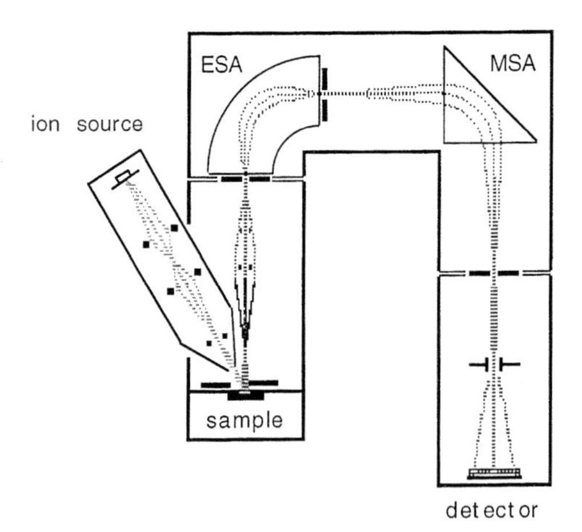

Fig. 6.24 Schematic illustration of a dynamic SIMS instrument with ion source, electrostatic (ESA) and magnetic analyser (MSA) for increased mass resolution.

A mass spectrum can be obtained with a magnetic analyser that uses mass discrimination by applying

$$\frac{m}{e} = \frac{\eta_H |B|^2 r^2}{2V} \tag{6.24}$$

where e is the electronic charge (1.607×10^{-19}C), η_H is the ratio of the electronic charge to a mass of one amu (9.58×10^7 C kg^{-1}), B is the magnetic field intensity and r is the radius of the path taken by the analysed ion.

Time-of-flight (ToF) spectrometers use pulsed ion beam sources. The relationship between acceleration voltage, V_{acc}, and kinetic energy of the secondary ions, E_{kin}, is described according to

$$eV_{acc} = \frac{m}{2}v^2 \tag{6.25}$$

where v is the speed of the accelerated ion. This leads to a simple relation between time-of-flight, t, and the square-root of the ion mass, m

$$t = \frac{L}{v} = \frac{L}{\sqrt{2eV_{acc}}} m^{1/2} \tag{6.26}$$

where L is the distance between sample and detector. Picosecond pulses of the primary ion beam increase the precision of the time measurement and hence the mass resolution. A typical ToF analyser is illustrated in Fig. 6.25 showing pulsed ion source and electronic charge compensation. Ions emitted from the sample are accelerated over a distance d (typically a few millimetres) between the sample and the immersion lens, IL. Three electrostatic analysers (ESA) serve as mass filters leading to a mass resolution $m/\Delta m > 5000$ for a mass of 100. Modern spectrometers are equipped with a fast entry lock allowing the introduction of a sample into the UHV within a few minutes.

As mentioned above, quantification in SIMS is difficult owing to the fact that the

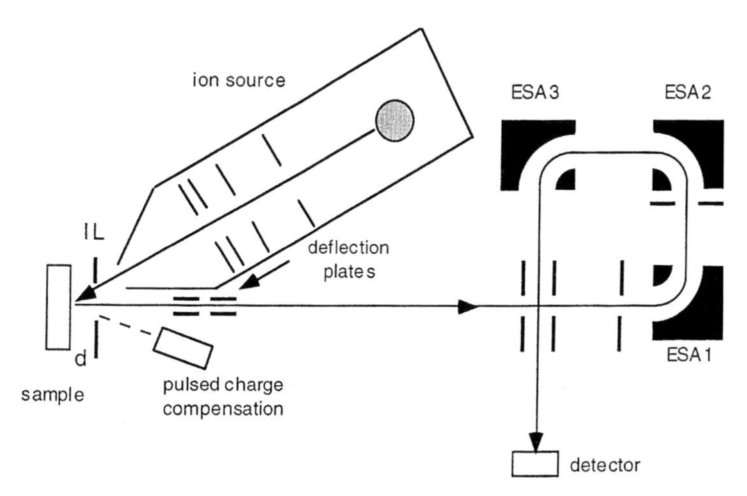

Fig. 6.25 Schematic representation of a time-of-flight (ToF) analyser with a pulsed ion source, and electronic charge compensation, electrostatic analysers (ESA) and immersion lens (IL). The distance between IL and the sample is given by *d*.

ionization probability $\gamma^{\pm}_{A(M)}$ in Equation (6.22) is a strong function of the matrix M. In the situation when the matrix is kept constant or does not vary for different elements with low atomic concentrations, i.e. small in comparison to the matrix concentration, quantification is possible by applying:

$$\frac{c_A}{c_B} = \frac{Y_B\, \gamma^{\pm}_{B(M)}}{Y_A\, \gamma^{\pm}_{A(M)}} \frac{I^{\pm}_A}{I^{\pm}_B} \tag{6.27}$$

This method allows the measured intensities $I^{\pm}_A$ and $I^{\pm}_B$ to be correlated to determine atomic concentration ratios. Useful ion yields $Y^{\pm} = Y\gamma^{\pm}$ for semiconductor matrixes, like Si or GaAs, are found in the literature (Wilson *et al.* 1989). In the absence of a dominant matrix, only semi-quantitative analysis is possible permitting the determination of relative concentration ratios after subtraction of surface impurities, and in particular hydrogen, which often makes measurement unreproducible because of its presence in the UHV (Léonard *et al.* 1998). Another way to quantify ion intensities is to control the ionization probability $\gamma^{\pm}$. To a certain extent this is possible by a post-ionization technique using electron gas, plasma or laser ionization. These techniques are usually known as sputtered neutral mass spectrometry (SNMS) (Oechsner 1995). The most quantitative method is resonant laser ionization which gives ionization probabilities close to 100% (Arlinghaus & Joyner 1996). The reader is referred to a review of post-ionization techniques (Mathieu and Léonard 1998).

A typical mass spectrum derived from a silicon wafer at low mass resolution is shown in Fig. 6.26(a) illustrating the possibility of measuring the different isotopes of Si as well as hydrogen. The same measurement in the high mass resolution mode in the vicinity of $m = 56$ Da is shown in Fig. 6.26(b). The spectrum indicates that one can easily discriminate between positive ions of $^{56}Fe^{+}$ and $^{28}Si^{+}_{2}$ as well as between other Si-containing fragments and oxygen, nitrogen and hydrocarbons.

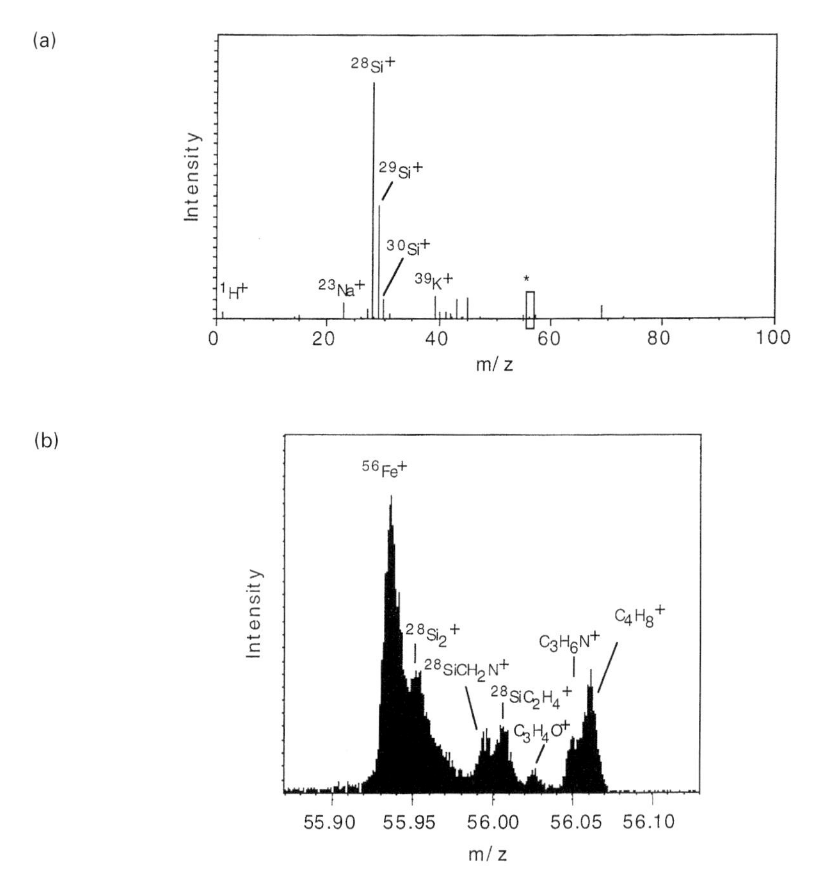

Fig. 6.26 (a) Low mass resolution spectrum of an Si wafer. The mass around $m = 56$ (*) is expanded in (b) which shows the high resolution mass spectrum of an Si wafer indicating the different impurities.

Simpler and cheaper analysers like the quadrupole mass analyser with a mass resolution ($m/\Delta m$) between 0.1 and 1 do not allow discrimination between such species. Another important feature of ToF analysers is that the entire mass spectrum is obtained within a fraction of a second, because typical flight times are of the order of microseconds. However, longer acquisition times will lead to better signal-to-noise ratios (the signal background can be practically zero). The mass range of ToF analysers is typically 0–1000 amu, although larger detected masses are possible (Schueler 1992), in particular when ionizing metal substrates like silver are used (Benninghoven *et al.* 1993). A library of static SIMS spectra has been developed and is continuously being extended (Vickerman *et al.* 1996).

Figure 6.27 shows a depth profile of an Fe–Cr–Nb alloy, in which positive ion fragments are measured as a function of sputter time. It was obtained by applying a 15 keV Ga^+ beam in the dynamic mode to a naturally formed passive film. The sputter rate was approximately 0.1 nm min^{-1}. Owing to ionization probabilities for the Fe and Nb fragments that are much higher in the oxide than in the underlying metal alloy, all detected ion signals decrease towards the metallic matrix. This

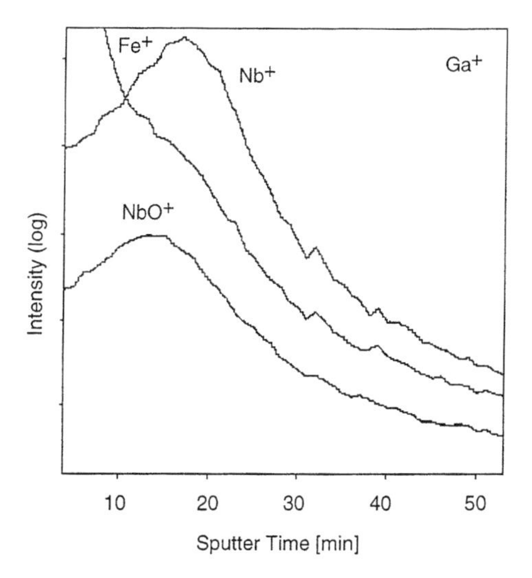

Fig. 6.27 Depth profile of Fe–Cr–Nb alloy (see Landolt *et al.* 1994).

phenomenon is typical for metals bombarded with inert ions like argon or non-reactive ions like Ga^+.

In general, when imaging, operation in the spectroscopic high-mass resolution mode does not take place. Rather the imaging mode is employed with lower mass resolution in order to use higher intensities and improve the quality of the image. Figure 6.28 illustrates the possibility of measuring, with submicron resolution, the lateral distribution of a given ion fragment. The example shows the image of a photochemically treated diamond surface, onto which an organic molecule was linked. The organic MAD molecule has two active ends, one of them containing a fluor (Léonard *et al.* 1998). The pattern was obtained by masking the diamond sample covered with the MAD during illumination. Thus, only areas exposed to light reacted with the MAD. This image represents a scanned area of 128 × 128 pixels. In each pixel

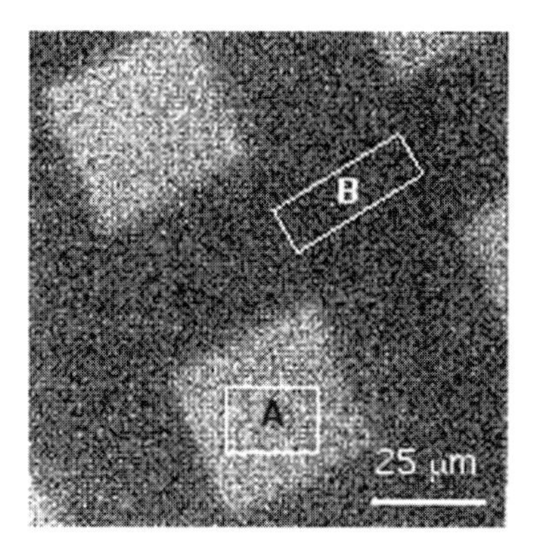

Fig. 6.28 ToF-SIMS image of $^{19}F^-$ distribution. A mass spectrum of the regions of interest A and/or B may be obtained.

the total mass spectrum was stored. Retrospective analysis allows either the display of mass spectra of certain regions of interest (labelled area A or B in Fig. 6.28) or the intensity along a chosen line. This illustrates full control of the mass analysis of the sample by the operator. The lateral resolution of SIMS can be in the submicron range – not as good as in AES but better than in XPS (cf. Figs 6.15–6.17). The choice of correct method is not simple. It will be determined not only by the physical properties of the sample, but also by the questions to be answered.

6.4 Summary

Electron and ion spectroscopies for elemental and molecular surface-sensitive analysis have been briefly described. All three techniques use spectrometers in an ultra-high vacuum environment and provide elemental information on the chemical composition from within the first few monolayers (5–10 nm) of most solid materials with high sensitivity. Auger electron spectroscopy (AES) is useful mainly for conducting samples, whereas X-ray induced photoelectron spectroscopy (XPS) and secondary ion mass spectrometry (SIMS) are able to perform analysis on insulators like ceramics, polymers and biomaterials. AES and XPS are quantitative, whereas SIMS is only semi-quantitative usually when applied to analysis of semiconductors. SIMS is the most sensitive of all three methods, although this renders the technique in many cases vulnerable to unwanted surface contamination acquired through sample

Table 6.4 Compared performances of XPS, AES and SIMS

Technique	XPS	AES	SIMS
Principle			
excitation	$h\nu$	e^-	ions
emission	e^-	e^-	ions
analysis	E_b	E_{kin}	m/z
Information	elemental binding energy	elemental micrographs	elemental molecules isotopes
Atomic number	$Z \geq 3$	$Z \geq 3$	all
Insulators	yes	no	yes
Modes			
spectroscopic	yes	yes	yes
imaging	yes	yes	yes
profiling	yes	yes	dynamic SIMS
Resolution			
depth resolution	1–5 nm	1–5 nm	1–5 nm
lateral resolution	< 5 μm	30–100 nm	80–200 nm
mass resolution $m/\Delta m$			1–8000
Sensitivity			
volume (weight cm^{-3})	mg	mg	μg
surface (weight cm^{-2})	0.01 ng	0.01 ng	10^{-12}g–10^{-15}g
Precision (%)	±3–20	±5–30	±10–1000
Quantification	yes	yes	semi

handling and transfer. Data reduction is straightforward based on the available literature and handbooks. For static SIMS a database is currently being developed (Vickerman *et al.* 1996). Access to buried interfaces below 5 nm is available through destructive depth profiling. Imaging is possible for all three methods ranging from a few micrometres (XPS) to sub-micron (SIMS) and nanometre analysis (AES) with state-of-the-art instruments. Table 6.4 summarizes the principles and main characteristics of each technique.

6.5 Acknowledgements

I would like to thank D. Léonard and N. Xanthopoulos for discussions and assistance with some of the figures. I would also like to express my gratitude to C. Blomfield (Kratos Ltd), J. Wolstenholme (VG Scientific Ltd) and Dennis Paul (Physical Electronic Industries) for some of the images. Financial support from the Swiss Priority Program on Materials Research is acknowledged.

Dedication

This chapter is dedicated to Pascal A. Mathieu who died much too early at the age of 27. Pascal worked on his PhD thesis in Organic Chemistry at the ETH Zürich, in Professor D. Seebach's group. He had just published his first paper in *Helvetica Chimica Acta*.

References

Arlinghaus, H.F. and Joyner, C.F., *J. Vac. Sci. Technol. B*, **14**, 1 (1996).
Auger, P., *J. Phys. Radium*, **6**, 205 (1925).
Benninghoven, A., *Ion Formation from Organic Solids*. Springer, Berlin (1983).
Benninghoven, A., Rüdenauer, F.G. and Werner, H.W., *Secondary Ion Mass Spectrometry Basic Concepts, Instrumental Aspects, Applications and Trends*. John Wiley, New York (1987).
Benninghoven, A., Hagenhoff, B. and Niehuis, E., *Anal. Chem.*, **65**, 630A (1993).
Bethe, H., *Ann. Physik*, **5**, 325 (1930).
Biersack, J.P., *Ion Beam Modification of Insulators*. Elsevier Science, Amsterdam, p. 648 (1987).
Briggs, D. and Seah, M.P., *Practical Surface Analysis: Auger and X-ray Photoelectron Spectroscopy*, 2nd edn, Vol. 1. John Wiley, Chichester, 1990.
Briggs, D. and Seah, M.P., *Practical Surface Analysis: Ion and Neutral Mass Spectroscopy*, Vol. 2. John Wiley, Chichester (1992).
Coullerez, G., Chevolot, A., Léonard, D., Xanthopoulos, N. and Mathieu, H.J., *J. Surf. Anal.*, **5**, 235 (1999).
Hofman, S. and Sanz, J.M., *J. Trace Microprobe Techniques*, **1**, 213 (1982–83).
Hofmann, S., *Surf. Interface Anal.*, **21**, 673 (1994).
Hofman, S., *High Temp. Mater. Proces.*, **17**, 13 (1998).
Landolt, D., Schmutz, P., Mathieu, H.J., *Mater. Sci. Forum*, **185–188**, 313 (1994).
Léonard, D., Chevolot, Y., Bucher, O., Sigrist, H. and Mathieu, H.J., *Surf. Interface Anal.*, **26**, 783 (1998).
Mathieu, H.J. and Léonard, D., *High Temperature Materials and Processes*, **17**, 29–44 (1998).
Mathieu, H.J., Auger electron spectroscopy. In: *Surface Analysis – The Principal Techniques*, J.C. Vickerman (ed.). John Wiley, Chichester, p. 99 (1997).
NIST (1997).
Oechsner, H., *Int. J. Mass Spectrom. Ion Proces.*, **143**, 271 (1995).
Physical Electronics, Applications Note **9705** (1997).
Pitton, Y., PhD. thesis, Ecole Polytechnique Fédérale de Lausanne (EPFL), Lausanne (1996).

Powell, C.J. and Jablonski, A., *J. Phys. Chem. Ref. Data*, **28**, 19 (1999).
Ratner, B.D. and Castner, D.G., Electron spectroscopy for chemical analysis. In: *Surface Analysis – The Principal Techniques*, J.C. Vickerman (ed.). John Wiley, Chichester (1997).
Rivière, J., *Surface Analytical Techniques*. Clarendon Press, Oxford (1990).
Schueler, B.W., *Microsc. Microanal. Microstruct.*, **3**, 119 (1992).
Seah, M.P. and Dench, W.A., *Surf. Interface Anal.*, **1**, 2 (1979).
Seah, M.P., *Thin Sol. Films*, **81**, 279 (1981).
Seah, M.P., Mathieu, H.J. and Hunt, C.P., *Surf. Sci.*, **139**, 549 (1984).
Siegbahn, K., Nordling, C., Fahlmann, A., Nordberg, R., Hamrin, K., Hedman, J. and Johansson, G. *et al., ESCA, Atomic, Molecular, and Solid State Structure Studies by Means of Electron Spectroscopy*. Almquist and Wiksells, Uppsala (1967).
Sigmund, P. (ed.), *Sputtering by Particle Bombardment. Physical Sputtering of Single-Element Solids*. Springer, Berlin (1981).
Tanuma, S., Powell, C.J. and Penn, D.R., *Surf. Interface Anal.*, **11**, 577 (1988).
Tanuma, S., Powell, C.J. and Penn, D.R., *Surf. Interface Anal.*, **17**, 911 (1991).
Tanuma, S., Powell, C.J. and Penn, D.R., *Surf. Interface Anal.*, **21**, 165 (1994).
Tougaard, S. and Jansson, C., *Surf. Interface Anal.*, **19**, 171 (1992).
Tyler, B.J., Castner, D.G. and Ratner, B.D., *Surf. Interface Anal.*, **14**, 443 (1989).
Vickerman, J.C., Brown, A. and Reed, N.M. (eds), *Secondary Ion Mass Spectrometry: Principles and Applications*. Oxford Science Publications, Oxford (1989).
Vickerman, J.C., Briggs, D. and Henderson, A. (eds), *The Wiley Static SIMS Library*. John Wiley, Chichester (1996).
Vickerman, J.C. and Swift, A.W., Secondary ion mass spectrometry. In: *Surface Analysis – The Principal Techniques*, J.C. Vickerman (ed.), John Wiley, Chichester, p. 135 (1997).
Wilson, R.G., Stevie, F.A. and Magee, C.W., *Secondary Ion Mass Spectrometry – A Practical Handbook for Depth Profiling and Bulk Impurity Analysis*. John Wiley, New York (1989).

7 Charged particle-induced X-ray emission

R.D. VIS

7.1 Introductory remarks

7.1.1 *History*

The observation of X-rays induced upon bombardment with charged particles is as old as the discovery of X-rays. The discovery of X-rays in 1895 by Röntgen took place while studying cathode rays, a study evidently involving the bombardment of a target with energetic electrons. The X-ray tube was born, leading to a vast amount of fundamental research and applications.

In the first decades of the twentieth century, radioactivity was discovered and was from the beginning linked strongly to X-ray physics. Examples from more recent times are X-ray emission as a result of electron capture decay or the so-called 'shake-off' processes during β-decay. During the early days, X-ray emission by ion bombardment was observed in studies involving α-radioactivity.

The link with nuclear physics, a fast expanding field in the midst of the twentieth century, also caused an impressive progress in instrumentation. Detection of X- and γ-radiation was improved by the advent of proportional and scintillation counters.

This chapter deals with the emission of X-rays during charged particle bombardment for a special purpose, namely elemental analysis. A prerequisite for this application is that the detection system has sufficient energy resolution to discriminate between adjacent elements in the periodic table. This could be done even before the introduction of semiconductor detectors by using X-ray spectrometers comprising crystalline material. Reflection, according to Bragg conditions, disperses the X-rays according to their wavelength and such systems are referred to as wavelength-dispersive spectrometers (WDS). In combination with intense electron bombardment an elemental analytical technique was developed between 1950 and 1960 called EPMA (electron probe micro analysis), essentially a combination of an electron microscope and a WDS (Castaing 1960).

The development of energy-resolving X-ray spectrometers, based initially upon lithium-drifted silicon semiconductors, revolutionised elemental analysis by X-ray spectrometry (Fitzgerald *et al.* 1968). This system enables measurement of all relevant X-rays simultaneously, which implies multielement analysis. These energy-dispersive spectrometers (EDS) have a much higher efficiency than WDS in terms of subtended solid angle leading to an increase of the analytical sensitivity by several orders of magnitude. To compensate partly for this difference, WDS detection is normally performed in combination with a higher current on the sample. It was only after the invention of EDS that ions, in the beginning exclusively protons, from small accelerators became useful for analytical purposes. This application has been called PIXE (particle-induced X-ray emission) and was demonstrated for the first time in Lund,

Sweden by Johansson *et al.* (1970). It is again an example of a development that became available through links with nuclear physics.

7.1.2 *Introduction*

This section covers research in which X-rays are used to extract information from a sample about its elemental composition.

X-rays are induced by charged particles, which, for PIXE, immediately excludes X-ray fluorescence, still frequently used for routine types of elemental analysis. In general, X-ray fluorescence uses an X-ray tube as its primary source. A relatively new source for X-ray fluorescence experiments, however, is the electron storage ring from which the emitted synchrotron radiation is used to obtain a very intense and bright source of X-rays, so as to minimise detectable limits of trace elements and to provide the option of micro-analysis by collimating and/or focusing the X-ray beam. Details about X-ray fluorescence can be found elsewhere in this book.

For the sake of completeness we will briefly mention excitation with electrons, especially in comparison with the main topic of this chapter, excitation with protons and heavier ions. Electrons are available from electron microscopes. These microscopes feature excellent lateral resolution and are used widely at many research establishments. The facilities can be equipped with both WDS and EDS.

Turning now to PIXE, the technique is divided into macro- and micro-PIXE, the former using beams routinely available from small accelerators where the beam dimension is several mm^2, whereas the latter makes uses of so-called nuclear microprobes, beam-guiding systems that are capable of focusing the ion beam down to the micron or sub-micron range.

7.2 Theory

7.2.1 *The cross-section for inner-shell ionisation by ions*

In making available a sensitive method for determining (trace) concentrations of elements in any sample, a high-probability process is needed. For nuclear- and atomic-type reactions, the probability is determined by the ionisation cross-section, expressed in units of area (m^2). However, a more convenient unit in accordance with atomic dimensions called the barn is used to express cross-section (1 barn = 10^{-28} m^2).

7.2.1.1 *The binary encounter approximation (BEA)* The mechanism of inner-shell cross-section, by either electron or ion bombardment, involves a simple Coulomb interaction between the incoming particle and the ejected shell electron. It has been shown previously, mainly by Garcia and coworkers (1970a, b, 1973), that the universal validity of the Rutherford scattering formalism can also be used to describe inner-shell ionisation if the incoming particle and the removed electron are considered to leave the rest of the atom in a 'spectator' role. This model is the binary encounter approximation (BEA), which produces reasonable results for the ionisation cross-sections. The equation describing the process is

$$\sigma_i(v_1) = N_i \int_0^{\infty} \sigma_i(v_1, v_2) f(v_2)\, dv_2 \tag{7.1}$$

in which $f(v_2)$ is the velocity distribution of the bound electron and N_i the number of electrons having binding energy u. The ionisation cross-section, σ_i, is given by

$$\sigma_i = \int_u^{E_p} \left(\frac{d\sigma}{dE'}\right) dE' \tag{7.2}$$

in which E' is the exchange energy between the incoming charged particle and the electron. For reasons of conservation of energy it follows that $u < E' < E_p$, where E_p is the energy of the incoming particle.

The nice feature of BEA is that besides its simplicity, hydrogen-type velocity distributions can be used in Equation (7.1) to scale the low results. It therefore becomes possible to express the cross-section as a universal function of the energy of the incoming particle in units of the binding energy u, i.e.

$$u^2\sigma_i = Z_1^2 f\left[\frac{E_1}{uM_1}, M_1\right] \tag{7.3}$$

where Z_1, E_1 and M_1 are the atomic number, energy and mass of the incoming particle, respectively.

7.2.1.2 *The plane wave Born approximation (PWBA)* The criterion for the validity of the Born approximation is

$$\frac{Z_1 Z_2 e^2}{\hbar v} \ll 1 \tag{7.4}$$

with v being the projectile's velocity relative to the bound electron.

It is assumed in this model that the incident and elastically scattered particles can be described as plane waves. The transition from the initial to the final state is then described as a transition of the initial bound state of the electron to a state given by a continuum wave function, with all other electrons remaining in their initial states. An excellent derivation of the PWBA formalism for inner-shell ionisation is given by Madison and Merzbacher (1970).

7.2.1.3 *The semi-classical approach (SCA)* This model is in fact a mixed form of the others. The motion of the impinging particle in the field of the target nucleus is treated classically, whereas the transition of the electron is described from the point of view of quantum mechanics. The model is based upon first-order time-dependent perturbation theory, in impact parameter form, and the use of unperturbed single-electron wave functions. As the perturbing potential (V), the Coulomb interaction between projectile and electron is obviously used

$$V = \frac{Z_1 e^2}{|\,\mathbf{r} - \mathbf{R}(t)\,|} \tag{7.5}$$

with **R** and **r** the position vectors of the projectile and the electron, respectively. Results of predictions using this approach may be found in Hansteen *et al.* (1974, 1975). In this model, it is assumed that the ions follow straight trajectories. It has been shown that, especially for moderate incoming energies, this model does break down. This was the reason for introducing hyperbolic trajectories for ions.

7.2.1.4 *The ECPSSR theory* Of the above theoretical models the PWBA has formed the basis of modern improvement approaches to the calculation of inner shell cross-sections. Several authors, but notably Brandt and Lapicki (1981), have published a series of improvements and refinements to the original PWBA theory. Factors taken into account were (1) energy loss during the collision (E), (2) the deflection and velocity change of the projectile due to the nuclear Coulomb field (C), (3) perturbation of the atomic stationary states (PSS) by the projectile, and (4) relativistic effects (R).

These corrections are handled as a series of modifications to the effective projectile energy and the effective electron binding energy of the electron to be ionised. The ECPSSR theory incorporating the refinements is now commonly used. The vast amount of experimental data available are usually compared with this theory. Good agreement exists between theory and experiment, especially for the K shell, but also for L ionisation as long as the incoming particle's velocity is not too low. At very low energy (normally expressed as $E/\lambda u$, with λ the mass ratio of the projectile over the electron mass and u the binding energy of the electron), deviations occur owing to multiple ionisation, quasi-molecule formation and/or the proximity of the ion during the X-ray transition. An arbitrary limit to avoid deviations is $E/\lambda u > 0.02$.

Various computer programs used in the evaluation of inner-shell ionisation cross-sections, normally referred to as semi-empirical approaches, have proved to be very useful. Illustrative examples are given in the work of Paul (1984) for K X-rays and of Sow *et al.* (1993) for the L sub-shells.

By far the majority of published data have been obtained for protons. However, scaling laws (see Section 7.2.1.1) enable the conversion of data for any projectile using the following equation:

$$\sigma_{Z_p,A_p}(E_p, Z) = Z_p^2 \sigma_{1,1}\left(\frac{E_p}{A_p}, Z\right) \tag{7.6}$$

The cross-section of any particle is that of a proton with the same energy per nucleon multiplied by the charge of the projectile squared. Although this formula is adequate when applied to, e.g. α particles, the formula breaks down for heavier particles due to multiple ionisation effects and/or quasi-molecule formation.

7.2.2 *The fluorescence yield*

In order to correlate the ionisation cross-section of a given shell with observed X-ray intensities, knowledge of the decay of the ionised state is necessary. Decay occurs either by emission of X-rays or by the emission of an Auger electron. The probability for the emission of X-rays is known as the fluorescence yield (ω) and is a function of Z. Bambynek *et al.* (1972) produced a semi-empirical formula for ω_K:

$$\sqrt[4]{\left(\frac{\omega_K}{1-\omega_K}\right)} = \sum_{i=0}^{3} B_i Z^i \tag{7.7}$$

The same expression was evaluated by Cohen (1987), who found suitable coefficients for the L shell (ω_L).

For evaluation of Equation (7.7), commonly used coefficients are listed in Table 7.1. It should be realised that the L shell consists of three sub-shells, complicating the situation. Each sub-shell has its own ω value. The vacancy created can, moreover, be transferred to a higher sub-shell, the so-called Coster–Kronig effect. Therefore, it is necessary to write

$$\begin{aligned} X(L_1) &= n_1\,\omega_1 \\ X(L_2) &= (n_2 + f_{12}n_1)\omega_2 \\ X(L_3) &= [n_3 + f_{23}n_2 + (f_{13} + f_{12}f_{23})n_1]\omega_3 \end{aligned} \tag{7.8}$$

in which X denotes the X-ray intensities, f the Coster–Kronig coefficients for transitions between the sub-shells indicated and n the initial vacancies in the sub-shell indicated. In PIXE work the effective L fluorescence yield is commonly used, expressed by:

$$\omega_{L_{eff}} = \frac{X(L_1) + X(L_2) + X(L_3)}{n_1 + n_2 + n_3} \tag{7.9}$$

Equation (7.9) is employed in the evaluation of Equation (7.7) using Table 7.1 coefficients.

Table 7.1 Common coefficients for fluorescence yield evaluations

	K shell	L shell
B_0	$(3.70 \pm 0.53 \times 10^{-2}$	0.17765
B_1	$(3.112 \pm 0.044) \times 10^{-2}$	2.98937×10^{-3}
B_2	$(5.44 \pm 0.11) \times 10^{-5}$	8.91297×10^{-5}
B_3	$-(1.25 \pm 0.07) \times 10^{-6}$	-2.67184×10^{-7}

The accuracy of L-shell data is limited by this complication, although the relative emission rates of the sub-shells were calculated theoretically by Scofield using the Dirac–Hartree–Fock predictions (Scofield 1974). The fact that n_i and ω_i depend both on the incoming particle energy and the Z value of the element being irradiated limits very precise calculations. Nevertheless, by having at hand ECPSSR theory for the cross-section when creating a hole in one of the inner shells and values for the fluorescence yield, it is possible to predict the probability for the production of a particular X-ray with a given particle energy. For the K shell, for instance, the K-production cross-section, σ_K is given simply by:

$$\sigma_K = \omega_K \sigma_{ECPSSR} \tag{7.10}$$

Note that for the calculation of the intensity of a particular transition, such as the K_α

line, the branching ratio, or better the relative amount of K_α radiation, should be included in Equation (7.10). For a detailed description of the shell level scheme and the Siegbahn notation for the various X-ray transitions, including the selection rules, see for example, Dyson (1973). In Fig. 7.1, cross-sections for the production of K X-rays are given. For comparison the cross-sections that were obtained for electrons and photons are also depicted.

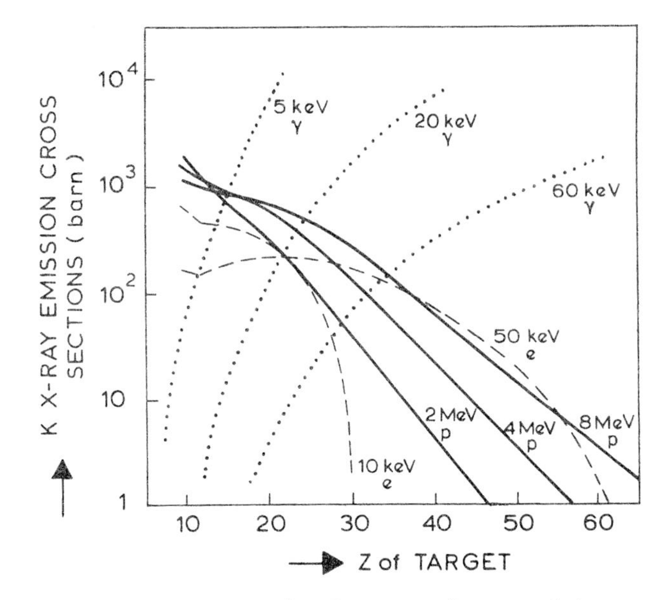

Fig. 7.1 The K X-ray production cross-sections for protons, electrons and photons.

From Fig. 7.1, it is clear that cross-sections for electrons and protons are comparable assuming that their velocities are of the same magnitude. A notable difference of course is that the cross-section for the electron becomes zero if its energy no longer exceeds the atomic binding energy. Photons behave differently – the process differs from Coulomb excitation and is usually the common photoelectric effect.

The popularity of PIXE is not primarily due to its high ionisation cross-sections. Electrons do the job as well as heavier particles and are easily and less expensively available. It is the lower limit of detection in a method that is also important and this is determined by background production. The nature of this background process in PIXE places the technique in a very advantageous position when compared with other modes of excitation.

7.3 The sensitivity and detection limits of PIXE

7.3.1 *The optimal bombarding energy*

From the theory of inner-shell ionisation, it can be concluded that the maximum cross-section occurs if the velocity of the incoming particles equals the orbiting velocity of the electron to be ionised. This is called velocity matching. To achieve the maximum cross-section in ionising an element such as Zn, for instance, 18 MeV protons are needed to profit from this velocity matching. Using this bombardment energy will lead to the highest possible count rate for Zn X-rays, but there is a drawback in that the production of background signals from continuous X-rays is increased enormously when compared with the excitation at lower energies. Moreover, by exceeding the Coulomb barrier for various nuclear reaction channels, the production of γ-rays and neutrons adversely influences the detection of X-rays. It is, therefore, worthwhile compromising in such a way that the peak-to-background ratio reaches an optimal value. To this end, an understanding of the processes leading to background interference would be useful.

7.3.1.1 *The definition of the sensitivity* According to IUPAC conventions, the sensitivity is defined here as the slope of the calibration line describing the relationship between the analytical signal (in the present case the intensity, I, of the relevant X-ray peak) and the concentration, c, of the element to be measured. For thin-layer analysis, this concentration is often expressed in atoms cm^{-2} or g cm^{-2}; for the determination of an element in a given matrix, the units will be $\mu g\ g^{-1}$ or ppm. Hence the sensitivity, S, can be expressed as:

$$S = \frac{\partial I}{\partial c} \tag{7.11}$$

The intensity of the detected X-ray peak emitted by an element Z under bombardment with a number of particles N_p of a given energy, can be described by

$$I_Z = N_P\, \sigma_Z\, \omega_Z f c_Z\, T_{E_X}\, \varepsilon_{E_X}\, \Omega \tag{7.12}$$

where c_Z is the amount of element present (here in atoms cm^{-2}), and σ_Z, ω_Z and f are the cross-section for ionisation, the fluorescence yield and the branching ratio of element Z, respectively. T is the transmission of the emitted X-rays of energy E_X through the absorbing windows of the detector and possibly the sample chamber, whereas ε and Ω are the efficiency and the solid angle of the detector, respectively. Note that it is assumed that the specimen is thin, which means that the energy of the beam will not change resulting in a constant cross-section for ionising element Z. Transmission and efficiency are dependent on the X-ray energy. Equation (7.11) can therefore be written as:

$$S = N_p\, \sigma_Z\, \omega_Z f\, T_{E_X} \varepsilon_{E_X} \Omega \tag{7.13}$$

In Fig. 7.2, the sensitivity for element Z is depicted when 3.0 MeV protons are used with various filters that determine T. It is obvious from this figure that for light elements PIXE has low sensitivity, caused by the absorption of soft X-rays. Although thinner filters are feasible, $Z = 11$ (Na) is often considered as the element of lower limit for PIXE analyses. Windowless or other detector solutions to measure beyond this limit brings the problem that scattered particles from the sample are prevented from entering the detecting system anyway. For heavier elements sensitivity decreases owing to decreasing cross-section. However, what is not indicated in Fig. 7.2 is the high sensitivity for elements of $Z > 60$, a region where it is possible to exploit the

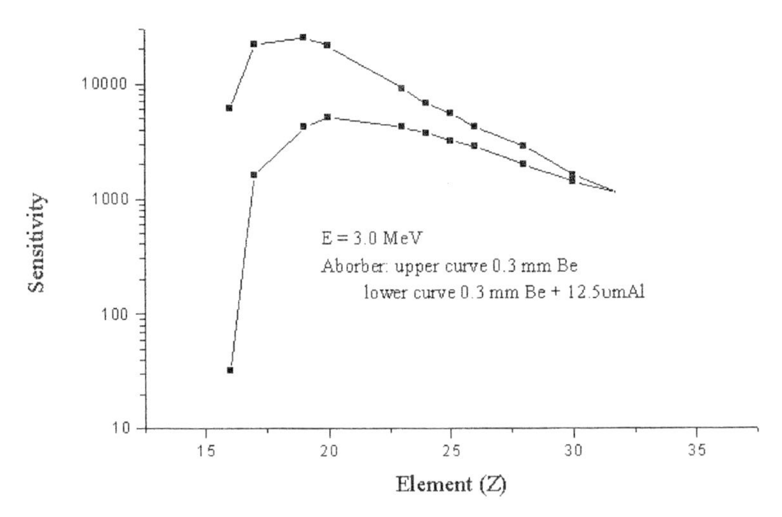

Fig. 7.2 The sensitivity as a function of Z for typical experimental conditions. Units are in counts (10^{16} atoms cm^{-2} μC sr)$^{-1}$.

ionisation of the L shell. Although shape and magnitude of the curves in Fig. 7.2 can be influenced by the choice of beam energy, almost invariably energies between 2 and 3 MeV are used, the motivation for which is described below.

7.3.1.2 *The production of background* The principal contribution to the background is the secondary electron Bremsstrahlung, caused by electrons that were ejected and which decelerate into the Coulomb field of other atoms. As early as 1974, Folkmann *et al.* quantified this process, which involves two steps, namely the production of the secondary electrons and the subsequent emission of a photon. Much later, other contributions to the background were discussed, mainly by Ishii and Morita (1987, 1990), who described apart from secondary electron Bremsstrahlung also quasi-free electron Bremsstrahlung and atomic Bremsstrahlung. The latter occurs when an ejected electron falls back into its initial bound state, whereas quasi-free electron Bremsstrahlung takes place if, in the collision, the electron can be considered as unbound and as such is scattered by the incoming ion's Coulomb field. This contribution to the background is only significant at low X-ray energy. The secondary electron Bremsstrahlung has a cut-off energy at the point of maximum energy transfer, T_M, from the projectile with mass M to an electron, which is given by

$$T_M = 4m_e \frac{E_p}{M_p} \tag{7.14}$$

As the secondary electrons are the dominant source of background the trade-off between high cross-section for inner-shell ionisation of the analyte and low background is determined by Equation (7.14). As mentioned above, calculations show that the optimum value for the incoming energy if protons are used is between 2 and 3 MeV. The theory predicts the background to a high accuracy (Ishii & Morita 1990; Murozono *et al.* 1999). Also the anisotropy of the background has been calculated and measurements show excellent agreement (Gonzalez *et al.* 1988). The anisotropy is caused by the forward preference for the scattered electrons resulting in a preference for photon emission at 90°. Due to relativistic effects this 90° is somewhat shifted. Most workers detect the X-rays at angles around 135° with respect to the beam.

7.3.1.3 *The detection limit* With the known sensitivity and knowledge about the production processes of the background, detection limits can be estimated. These limits are usually defined as the concentration (again in ppm or g cm^{-2}) that will give rise to a significant peak. The level of significance is arbitrarily defined, commonly as three times the standard deviation of the background level. As far as the width of the background is concerned, again arbitrarily, one frequently uses the FWHM value of the peak riding on that background. If then the smallest detectable intensity, I_D, is determined, the limit of detection, c_D, is simply given by (where I_b is the intensity of the local background):

$$c_D = \frac{I_D}{S} = \frac{3\sqrt{I_b}}{S} \tag{7.15}$$

If a cross-section for the production of background is defined as τ_b, substituting Equation (7.13) into (7.15) gives

$$c_D = \frac{3}{\sigma_Z \omega_Z f} \sqrt{\frac{\tau_b c_M}{N_p T_{E_X} \varepsilon_{E_X} \Omega}} \tag{7.16}$$

in which c_M represents the number of atoms on which the background is produced. Note that Equation (7.16) is only valid for thin samples. Note also that σ_Z and τ_b are both a function of the beam energy E. It is in this competition that the optimum beam energy for PIXE is determined. This optimum is influenced by the region of interest in the X-ray spectrum and therefore is a function of Z. Figure 7.3 gives an example of detection limits obtainable in the analysis of dried tissue samples (Maenhaut et al. 1984).

Figure 7.3 serves as a good indication of the analysis possibilities when using PIXE. Detection limits are also influenced by the character of the matrix. In heavy matrices, as for instance those found in geological work, detection limits will be higher. As can be concluded from Equation (7.16), a high ion dose and high detection efficiency, including a large sold angle, will improve the situation.

If samples are thicker, the situation becomes more complex. The slowing down of the ion beam leads to a decreasing cross-section through the sample and X-rays emitted in deeper regions will be partially absorbed on their way to the surface. The usual approach is to sub-divide the sample into layers of small thickness and to add all contributions from all layers towards the total yield. This summation, or integration, is done numerically. An extra complication, however, is that both processes are dependent on the gross composition of the sample, necessitating a measurement of the major elements simultaneously with the trace elements. The experimental procedures necessary to obey these requirements are described below.

For the sake of completeness the equation used for so-called infinitely thick samples, in which the ion beam is stopped completely, is given by:

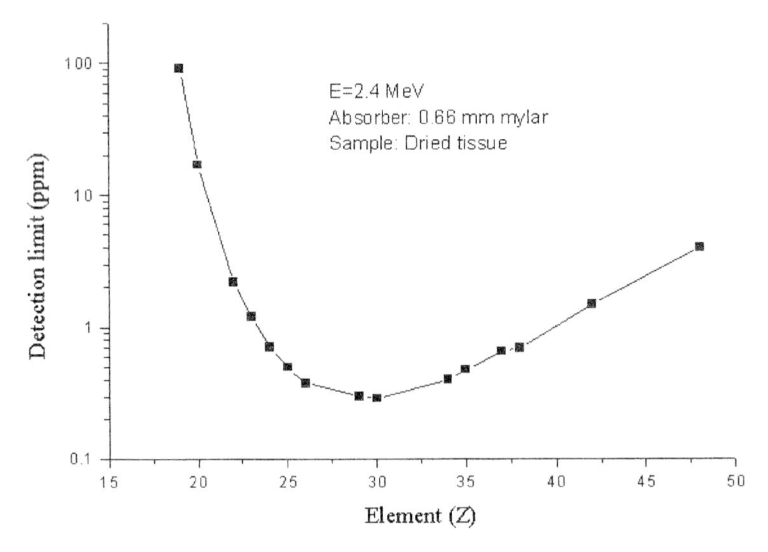

Fig. 7.3 Experimental detection limit as a function of Z. The collected charge was 0.1 mC (N_p); the sample thickness was 4 mg cm^{-2}. Data from Maenhaut *et al.* (1984).

$$I_Z = N_p \omega_Z f c_Z \varepsilon_Z \Omega \int_{E_0}^{0} \frac{\sigma_Z(E) T_{E_X}(E)}{S(E)} \mathrm{d}E \tag{7.17}$$

Note that in contrast to Equation (7.12), integration is necessary over both the cross-section and also the transmission as absorption takes place not only in the instrument window but also in the sample itself. $S(E)$ is the stopping power of the matrix. It is self-evident that in the expression for self-absorption the geometry should be taken into account. Both the incident and exit angles influence this absorption. As mentioned previously, both the stopping power and the mass absorption, normally expressed as μ/ρ (in cm^{-1}), are a function of the matrix composition.

7.4 The accuracy and precision of PIXE

7.4.1 *The sources of error*

7.4.1.1 *Random errors* For the precise determination of an unknown concentration in a given sample, certain quantities found in Equation (7.17) should be known. Required values are:

- full and detailed knowledge of the detector's efficiency and solid angle
- accurate measurement of N_p
- knowledge of the physical quantities describing the production of the relevant X-ray
- knowledge of the energy dependence of these quantities, including the

absorption of the X-ray (for the evaluation of these parameters, the energy loss of the beam should be known)

From the above, statistical errors are caused only by the counting process itself, according to a Poisson distribution, and to the determination of the accumulated charge, which is either a counting process or based upon current integration.

7.4.1.2 *Systematic errors* Other factors, such as the experimental geometry, detector efficiency, the theoretical values of beam cross-sections and stopping power are subject to systematic errors. The statistical errors can be minimised by collecting a sufficient number of counts – even for low concentrations it is generally possible to obtain a precision better than 10%. Only close to the limit of detection will the precision be worse. Computer programs used in the calculation of peak areas in a spectrum also contribute errors through deviations from the fit to the data.

Systematic errors leading to inaccuracies are difficult to avoid. Precise knowledge of detector efficiency, which requires the precise physical dimensions of its crystal and absorbing media, is not easy to obtain. For low-energy analysis (< 5 keV) Orlic *et al.* (1989) determined the efficiency of a Si(Li) detector by using a calibrated ^{55}Fe source for the photoelectric excitation of secondary targets. The accuracy is influenced also by deviations in the theoretical values of cross-section, fluorescence yield and branching ratio of the X-rays. These data are quite good for K X-rays but unsatisfactory for the L shell and also when incident beams other than protons are used. The bottom line is that the use of PIXE in an absolute or standardless way will provide an accuracy no better than about 30%.

The use of standards to obtain sensitivity factors for elements of the periodic table is the way to proceed. As an example, the spectrum given in Fig. 7.4 was obtained from such a standard, in this case SRM 1833, a glass standard comprising a 0.65 μm

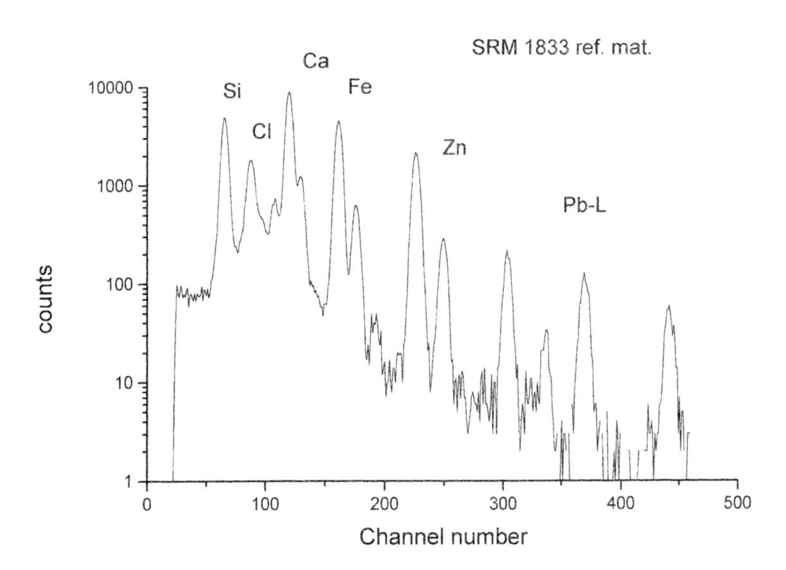

Fig. 7.4 An example of a PIXE spectrum used for calibration purposes.

glass layer deposited on a polymer substrate. Knowing the exact concentrations of the elements present and measuring the relevant peak areas combined with the accumulated charge enables these sensitivity factors to be obtained. If subsequently unknown samples are measured against a standard to determine the same elements, theoretical values and detector characteristics will vanish in the final equations. An extra advantage of PIXE is that it is quite safe to obtain additional sensitivity factors from other elements by interpolation. By doing so, it is possible in favourable cases to obtain an accuracy better than 5% for routine types of analysis.

7.5 The experimental set-up

7.5.1 *The accelerator*

In the early days of PIXE, accelerators originally purchased for fundamental nuclear physics were used. These accelerators became redundant, however, owing to the demand for higher energies in nuclear physics. Most of these machines were van de Graaff electrostatic generators, build in the 1950s and 1960s. A few groups used a cyclotron as the source of ions, normally at the very low energy end. The rapid growth of PIXE and other ion beam analytical techniques gave rise to a new market for small accelerators and an increasing number of research groups acquired such 'second generation' machines. In addition to the technically much improved belt-driven van de Graaff machines, two other types came into use: the Pelletron, built by the National Electrostatics Corporation in Middleton, Wisconsin (USA); and accelerators based upon a diode stack as was used in Cockroft–Walton type machines. In the latter, the diodes used are solid-state devices. These machines are constructed by High Voltage, Amersfoort, The Netherlands. Both machine types have single-end and tandem versions. In the case of the High Voltage machines these versions are referred to as singletrons and tandetrons, respectively.

The requirements for an accelerator to be used in PIXE analyses are obvious: protons of between 1 and 3 MeV should be available having beam currents of several hundred nA, which can be focused to mm dimensions. Normally more ion beam analytical techniques are exploited with one accelerator and the choice will be dependent on its precise use. If it is required to change the ionic species frequently, a tandem with external ion sources is more convenient though this is at the price of available current and a minor deterioration of the phase-space because of the stripping procedure in the terminal. For injection of a nuclear microprobe, high current and maximum brightness are important. Also the choice of ion sources depends on precise analytical requirements and various sources such as the duoplasmatron, r.f., and sputter sources are all in use. Of course, features such as beam stability, energy resolution and running costs are also factors influencing the choice.

7.5.2 *The sample chamber*

As most PIXE research groups have developed their own techniques, a large variety of sample chambers is available. Common features to all systems, however, are detailed below:

(1) A sample changer device, so as to accommodate a relatively large number of samples in a carousel or target ladder. Frames holding the specimen are also non-standard, although some groups use standard slide frames. In a few cases, even the slide-change mechanism of slide projectors is built into a vacuum chamber.

(2) A (set of) diaphragm(s) to shape the beam before it hits the sample. Occasionally, diffuser foils are employed to improve the homogeneity of the beam on the target.

(3) An exit port with a thin window to the X-ray detector. This detector is usually placed backwards to reduce Bremsstrahlung background. In principle the detector window can itself be used as common between the vacuum in the chamber and the detector vacuum so as to reduce the absorption of very soft X-rays. Risks are the changing pressures on the detector window during venting of the sample chamber and penetration of scattered protons into the X-ray detector.

(4) Additional detectors for techniques used simultaneously. Examples are an RBS detector and/or a γ detector for the detection of prompt γ-rays from light elements.

(5) A beam-dump (Faraday-cup) downstream of the target. This is required for beam-tuning, but can also be used for current integration in examples of beam-transparent specimens. For beam integration other solutions are in use. Beam choppers that intercept the beam a known number of times or a monitor foil, transparent for the beam and placed upstream of the sample, are frequently used. Chopper or monitor foil can be observed using a separate detector, where observed counts are directly proportional to incident ion flux. In a few cases, the whole chamber is electrically isolated for current integration.

(6) So-called beam-on-demand systems can be inserted in front of the scattering chamber. This device displaces the beam immediately after registration of an event by the X-ray detector. During a fixed time the beam is then dumped onto a diaphragm. The rationale behind such a device, normally a set of electrostatic deflection plates, is that during amplification, digitalisation and storing of a beam event no other event will be processed by the detector and this beam is just producing radiation damage without generation of a useful signal. The device is triggered by the pre-amplifier of the detector.

For micro-PIXE work, the basics of the scattering chamber are comparable. The microbeam set-up requires a few extras such as an optical microscope to view the specimen (or the beam hitting a scintillating target) and an X,Y,Z translation stage to position the beam on the region of interest. Here also, several technical solutions are in use because most systems are home-built. The production of a microprobe is itself a major undertaking, including the incorporation of high-precision object and collimating slits, probe-forming lenses, for which purpose magnetic quadrupole lenses are the most commonly used, etc. The technique of beam handling, however, is outside the scope of this chapter.

7.5.3 *The detector(s)*

Practically all PIXE work is performed with energy-dispersive detectors. Liquid nitrogen-cooled lithium-compensated Si detectors are the most common featuring adequate resolution (~150 eV at 6.4 keV) and having an efficiency for X-rays in the relevant region of 100%. The transparency for higher energy photons can be advantageous if, apart from the X-rays, γ-radiation is produced on nuclei such as Li or F. Since 1970, germanium crystals have been produced with impurity concentrations of less than 10^{10} atoms cm^{-3}, enabling development of junction detectors with a sufficiently thick depletion layer for X-ray or even γ-ray spectrometry if reverse-biased at hundreds of volts (see Hall *et al.* 1970; Hansen 1971). These so-called high-purity Ge detectors possess totally depleted p–n junctions. The resolution is as good as that for Si(Li) detectors, with larger crystals having a superior energy resolution in comparison with large-area Si(Li) detectors. The higher efficiency for high-energy photons can be a disadvantage, but can be of benefit if K X-rays of heavier elements or the de-excitation of low-lying nuclear levels are to be measured simultaneously. Escape peaks are more pronounced for Ge detectors.

Entrance windows are usually made from beryllium and vary in thickness in the range 10–30 μm. Detectors can be placed outside or constructed as part of the scattering chamber in which case the detector window separates the chamber vacuum from the detector vacuum. Ultra-thin windows enabling measurement of X-rays from elements down to carbon are often used in XRF systems but less so for PIXE work as ions and electrons scattering from the sample must be prevented from entering the detector. To circumvent this problem Kobayashi *et al.* (1999) placed a permanent magnet in front of the detector in order to deflect scattered protons. In their study they used a dual-detector system: a Si(Li) detector with the magnetic protection for light element detection and a cadmium–zinc–telluride detector for heavy elements. Systems with two Si(Li) detectors having differing efficiencies and equipped with different absorbers can also be useful in broadening the elemental range to be measured with higher sensitivity.

Little work has been done with detectors other than those of the energy-dispersive type. A wavelength-dispersive crystal combined with a proportional counter has been used by Folkmann *et al.* (1990, 1993) for high-resolution work. The impressive resolution (6 eV at 3.5 keV) is at the price of efficiency, however, which is only a few counts per μC per μg cm^{-2}. The strength of the system is that it enables the measurement of intensity ratios of satellite peaks so as to extract information about the chemical form in which a particular element is present in the sample. Also the newly developed Peltier-cooled detectors are not used widely yet, the reason being their small dimension which reduces the efficiency.

7.5.4 *Special geometries*

7.5.4.1 *External beams* For a number of applications it is useful to extract the ion beam from the vacuum system of the accelerator. An early survey of these external-beam facilities was presented by Williams (1984). Among the various exit-window materials described, kapton is the most popular and is used with a thickness of between 7 and 25 μm. Be, Al and Ti are also used. So-called differential pumping, in which the beam exits through a pin-hole, has also been used (Mac Arthur *et al.* 1981; Lenglet 1988). The obvious advantage is the decrease of scattering. In most cases the beam enters from the vacuum into some protective atmosphere (N_2, He) to avoid excitation of the impurities in air.

External beams have been utilised in a variety of applications. For instance, fusion reactor limiters have been analysed by external PIXE (Doyle *et al.* 1987) as have various objects of art or archaeology, e.g. ancient coins (Meyer & Demortier 1990). To promote uniform irradiation on aerosol samples, Halder *et al.* (1999) equipped their external-beam facility with a rotating target holder.

7.5.4.2 *Glancing incident beams* In an analogy of total-reflection X-ray analysis, in which the primary beam enters the specimen at angles below the critical angle for total reflection, MeV ion beams are also used in this mode. It was shown by Vis and Van Langevelde (1991) that MeV ions reflect from surfaces at comparable angles to the critical angles for X-rays. They derived that the critical angle for total reflection of a proton beam can be described by

$$\varphi_c = \sqrt{2\delta_p} = \sqrt{V(r)/E} \qquad (7.18)$$

in which δ_p is the analogue of the deviation of the optical index of refraction from unity (Vineyard 1982), $V(r)$ is an average surface potential and E the energy of the incoming beam. If $V = 11.5$ eV, the electron value for Si and a proton energy of 3.0 MeV are assumed, then $\delta_p = 1.92 \times 10^{-6}$, leading to a critical angle of 1.96 mrad, which indeed is very close to the value for X-ray reflection, e.g. 1.78 mrad for Mo K_α radiation.

The obvious advantage here is the huge reduction of Bremsstrahlung background, i.e. the ions are not slowing down in the sample thus avoiding production of secondary electrons. Van Kan and Vis (1995, 1996) showed an increase in surface sensitivity enabling detection at the 10^{10} atoms cm^{-2} level on Si wafers. Additional techniques such as RBS look very promising using this geometry because of an enormous gain in depth resolution approaching one monolayer. In Fig. 7.5, the PIXE spectra obtained with normal incident and glancing incident beams are illustrated. The reduction in background signal is obvious. Although various names are in use for this geometry, GIPIXE seems to be the most appropriate in an analogy with X-ray fluorescence.

7.5.4.3 *Glancing exit geometry* In this geometry, the beam arrives normally, though X-rays are now detected at angles very close to the surface plane. This method can be referred to as GEPIXE and to avoid confusion, both techniques are schematically illustrated in Fig. 7.6.

The glancing exit technique has been used in X-ray spectrometry for some time (see Becker *et al.* 1983), but only recently has the observation of the so-called evanescent wave during proton beam bombardment been reported (Petukhov 1999). The extra surface sensitivity obtained in this geometry arises from the fact that sub-surface X-rays travelling towards the detector are reflected to larger angles with respect to the surface and are unable to pass the slit placed in front of the detector. Bremsstrahlung,

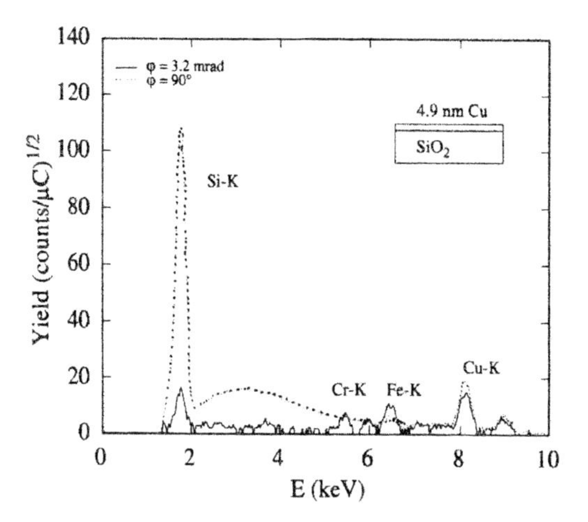

Fig. 7.5 Reduction of background as a result of glancing incident beam. The sample used is shown in the inset.

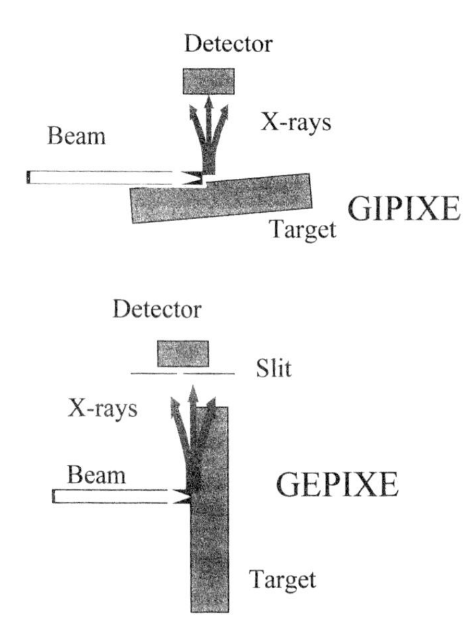

Fig. 7.6 The geometry used in GIPIXE and GEPIXE experiments.

which will be produced along the proton path sub-surface, is suppressed. As the critical angle of the emitted X-rays is a function of their wavelength according to

$$\varphi_c = \sqrt{5.4 \times 10^{10}\, Z\rho\lambda^2/A} \tag{7.19}$$

with Z and A the atomic number and atomic weight, respectively, ρ the density of the scatterer in g cm^{-3} and λ the wavelength of the emitted radiation in cm, it is possible, by rotating the specimen, to detect the X-rays in decreasing order of their energy. This

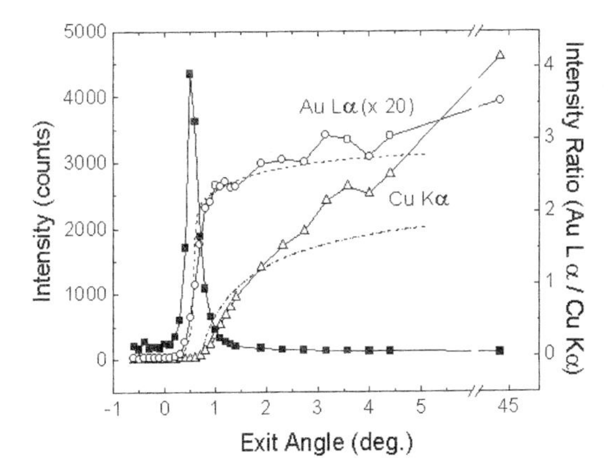

Fig. 7.7 GEPIXE measurements on a target comprising 500 nm Cu and 50 nm Au layers on an Si wafer. Open circles are Au L_α counts. The closed squares represent the ratio of Au/Cu counts showing extra surface sensitivity. For this ratio, the right-hand side ordinate is used. Dotted lines are derived from calculations performed by Tsuji *et al.* (1997).

is nicely demonstrated in Fig. 7.7, in which the yield of Cu and Au radiation is plotted as a function of exit angle. Note also the huge increase in surface sensitivity.

7.5.5 *The data-handling system and software*

7.5.5.1 *Data acquisition* Data acquisition for PIXE measurements is usually simple. Detector signals are pre-amplified. With the pre-amplifier incorporated into the detector system, it is necessary to cool the first amplifying transistor. The signals are subsequently fed into a main amplifier and transmitted to an ADC for digitisation. The spectra are stored for further processing. The hardware may include some extra features, such as systems to prevent pile-up, either electronically or by the use of a beam-on-demand system as previously mentioned. Pre-sets for time, charge (with the aid of a current integrator), and dead-time correction electronics are usually standard. The multi-scaling option to follow the whole spectrum or a particular peak in time is very useful as this enables a check to be done on the integrity of the sample under irradiation.

During micro-PIXE experiments, spectra should be correlated with the area under bombardment, which can be completed in the hardware, but is more conveniently solved by storing data 'event-by-event', which is referred to as the list mode. Every event recorded by the ADCs is labelled with the actual position of the beam and written to disc. The obvious advantage is the possibility to re-play the data. Lateral resolution can be changed subsequently, and the events also contain time information. For monitoring purposes, spectra are sorted on-line to enable the experimenter to follow what is going on. Evidently, in this case, the hardware should provide beam-scanning devices and a trigger system to correlate events to the x,y- coordinates.

7.5.5.2 *The software* From the early days of PIXE, much work has been done on the development of computer programs to fit PIXE spectra in order to obtain reliable peak areas and, with the aid of a sophisticated database, reliable numbers on the amount of (trace) element present in a sample. To account for background peaks, two principal approaches are adopted: use of an analytical model; and removal using digital filters. Although there is good progress towards a full description of the physics of the background (Murozono *et al.* 1999), no computer code is available yet with this physics implemented. An early program was published by Kaufmann *et al.* (1977) and was called REX. The code used a physics-based model for non-linear least-squares analysis of PIXE spectra and was written in Fortran. The code was modified and improved at Lund by Johansson (1982) and is still in use under the name HEX. Some groups use software that was originally developed for X-ray fluorescence. This is called AXIL and was modified to enable fitting of X-ray spectra induced by ions rather than photons. It has been developed and modified at Antwerp and Gent, Belgium (Van Espen *et al.* 1977; Maenhaut & Vandenhaute 1986).

More recently, commercially developed packages have become available. GUPIX was written at Guelph, Canada, and uses top-hat filtering for removal of the background (Maxwell *et al.* 1988) PIXAN uses a polynomial as background or adopts a procedure for iterative removal. It was developed at Lucas Heights, Australia (Clayton 1987). Apart from these two packages, each of which is used at many institutes worldwide, a number of others exist. Without being an exhaustive list, further examples are PIXYKLM, developed at Debrecen, Hungary (Szabó & Borbély-Kiss 1993) and SESAMX, written originally at Marburg, Germany (Bombelka *et al.* 1987). The latter program was also developed in the very early days of PIXE.

Modern programs all run on PCs both for single spectrum fitting and in batch mode. Requirements for good analytical results are the knowledge of the detector response function and the precise experimental conditions in terms of collected charge, solid angle, the presence of escape- and/or sum-peaks, etc. Obviously, the analysis of standards will be mandatory to determine the optimal settings for a given laboratory.

7.6 Sample preparation and sample damage

7.6.1 *General considerations*

Sound analytical work starts with sample taking and sample preparation. Without a good and representative sample, useful data cannot be obtained. Although this is very obvious, sample preparation still seems to be a bit underestimated when compared to the vast amount of effort put into acquiring hardware and software to process data. An immediate two-division separation of techniques is necessary where samples are prepared differently for standard-type PIXE work and micro-PIXE work.

Standard PIXE determines concentrations present in a given sample whereas micro-PIXE aims at unravelling the spatial distribution of elements in a specimen, preferentially in a quantitative manner. This means that in the latter case preservation of the original elemental distribution on a micrometre level is needed. The integrity of the sample under bombardment is also an important issue. For all specimens, the possession of electrical conductivity will be a major advantage, first to prevent charge build-up and second, because a better heat conductivity will reduce the effects of radiation damage. To this end, very often thin layers (20–100 nm) of carbon are evaporated on to the surfaces of samples. At all times, the frequent monitoring of sample integrity during irradiation is recommended. It should be emphasised that PIXE is often referred to as a non-destructive technique in accordance with this book's title. None the less, some sort of sample pre-treatment is invariably necessary to achieve this and shape the material in a form suitable for irradiation. Frequently, sample preparation goes a step further, aiming at homogenisation of materials prior to analysis or by adding an internal standard. Pre-concentration with the aim of improving detection limits obviously changes the character of the sample but is mentioned because of the widespread use of such methods. The irradiation itself, if radiation damage is kept to within acceptable limits, is non-destructive, enabling repeated analysis with various ion-beam techniques.

7.6.1.1 *Standard PIXE* As with all trace analytical techniques, PIXE aims at the accurate and precise measurement of a given material at hand. The first step towards a good result is to take a sample from the bulk that is representative. It should be realised that the ion beam irradiates only at the milligram or microgram level, thus imposing a problem if larger amounts of material are to be sampled. If the average concentration of a larger amount of material is to be determined, it is essential to take a series of samples to check for homogeneity. Before sampling, extra homogenisation is normally completed by grinding, milling, etc. If possible, dissolving in solvent is helpful, both for homogenisation and enabling addition of an internal standard. Drops of the solution can subsequently be freeze-dried on a suitable backing foil. Insoluble solid material can be pressed into pellets and mounted on a substrate. Lytle *et al.* (1993) proposed using fusion with $LiBO_2$ in order to dissolve silicate-rich matrices normally insoluble in acid.

Biological material can be dissolved in a process called wet-ashing involving a strong oxidising acid such as $HClO_4$. Alternatively, material is dry-ashed in a low-temperature, low-oxygen-pressure plasma and subsequently dissolved in dilute acid. Backing foils frequently used are thin (a few μm) polymer foils (mylar, kapton, hostaphane, polyethylene, etc.), which are mounted on home-made or commercially available frames. It is obvious that these foils should be as pure as possible and thin to avoid excessive Bremsstrahlung.

If thick samples are used, the dimensions of the backing are of course less critical. If the material is an insulator, however, problems may arise with charging by the particle beam. This problem can be counteracted either by coating the sample with a conductive layer (normally carbon is used), or spraying electrons from an electron gun onto the specimen to compensate for the beam charge. Passing the beam through a metal foil or grid serves the same goal: electrons ejected from the foil or grid travel to the sample thus avoiding charge build-up. Finally, a poor vacuum can provide a solution where the impacting electrons are provided by collisions with the rest gas. In favoured situations, a good target is produced by the sampling itself, the most obvious case being aerosol filters/impactors. These filters can be irradiated, as such, without any pre-treatment.

Attempts to include the pre-concentration of trace elements in the sample preparation procedure are recorded. The obvious aim of this process is to lower the detection limit. For instance, activated carbon to adsorb trace elements, which were firstly chelated with 8-hydroxy-quinoline from liquid (Johansson & Akselsson 1981) and liquid–liquid extraction of metal carbamates at various pH values (Cecchi *et al.* 1987) have been developed.

Sub-ppb detection limits have been claimed for a wide range of elements in these studies. None the less, the procedures are not used widely, possibly caused by the wide availability of other analytical methods providing ppb-range detection, such as atomisation of the sample in an inductively coupled plasma with subsequent detection of the atoms with atomic absorption spectrometry or mass spectrometry (ICP-AAS and ICP-MS, respectively).

Removal of the organic matrix in analysis of biological material by ashing is also a form of pre-concentration. The gain in terms of detection power is in this case not convincing, as only very light elements are removed, contributing relatively little to the Bremsstrahlung background (Pallon & Malmqvist 1981).

7.6.1.2 *Micro-PIXE* As mentioned above, other requirements prevail in sample preparation for micro-PIXE. The aim here is to measure the distribution of element(s). Preparation problems to this end are much the same as those encountered by electron microscopists. The study of, e.g. biological tissue, requires a preparation procedure that maintains the original features of the sample. It should be realised, however, that structure and trace element distributions are different things. Structure can be maintained by various fixation techniques, whereas the conservation of trace element distributions requires immobilisation of these elements. The following criteria, based on electron microscopy, can be formulated (Coleman & Terepka 1974):

(1) The normal structural relationships of the specimen should be adequately preserved. The morphological spatial resolution should be better than the spatial resolution one expects.

(2) The amount of material lost from or gained by the sample must be known. This is a rather obvious criterion, since this knowledge is necessary for concentration assignment.

(3) The chemical identity of material lost from or gained by the sample must be known. This criterion is set since the loss of weight can be selective for a given element, such as losses of KCl or NaCl reported during freeze-drying. The use of metal-containing fixation agents is an example of selective weight gain.

(4) The amount of elemental redistribution and trans-location within the sample must be known. Although a given preparative procedure may not change the total concentration of an element in the system, it may well have caused gross redistribution of the elements in the sample. Artificial movement of material in the sample is probably the most difficult phenomenon to assess. The observation of unusually high concentrations of an element, either as crystals or precipitates gives some indication that redistribution has occurred. It should be kept in mind that even if the preparation procedure was correct in

this respect, the beam itself can cause migration of elements during the irradiation procedures.

(5) Chemicals used to prepare the sample should not mask the elements being analysed. This criterion is incorporated to avoid spectral interference in the X-ray spectra between the analyte and possible additives.

Additional to these prerequisites, the preparations should be able to withstand the effects of the ion beam and, if applicable, the vacuum.

The above requirements were set with biological sections in mind. For other samples, comparable demands apply. If the sample does not contain appreciable amounts of water, embedding in polymers such as araldite and polishing is often used. Geological samples can be polished as such and mounted on frames. For biological work the technique that is mostly used is cryo-sectioning after shock-freezing.

It is important here that heat is removed rapidly from the specimen to minimise ice-crystal growth. There are various cryo-liquids and one readily available at most laboratories is liquid nitrogen (LN_2). There is a major drawback, however, when using LN_2. If a small tissue sample is plunged into LN_2 at its boiling point, a gaseous layer will form around the specimen. This will severely limit heat transfer from the specimen.

A series of cryo-liquids, usually cooled to their melting point by LN_2, have been used in fixation of biological tissues. Among these are isopentane, monochlorotrifluoromethane, monochlorodifluoromethane, propane, ethane, dichlorofluoromethane and propane/propylene. As mentioned above the cryo-liquids are usually chilled with LN_2. The use of propane chilled with LN_2 is popular and successful. The boiling point of liquid propane is 42°C, which is much higher than its freezing or melting point (−187°C). When cooled by LN_2, the liquid propane is unlikely, therefore, to form a gaseous layer around the specimen. Freezing against a cooled metal block is also worth mentioning because of its success and relative ease of application. The tissue is brought into contact with a cold metal block quite rapidly and owing to the high rate of freezing, no deformation at the cellular level is observed.

Cooling rates in excess of 10^4 K s^{-1} at a specific cell location are required for minimum ice-crystal formation and good quality cryo-fixation. Table 7.2 shows some cooling rates statistics.

Table 7.2 Cooling rates ($\times\ 10^3$ K s^{-1}) of various cryo-liquids

Liquid	Propane	Freon 22	LN_2	Ethane	Propylene
Cooling rate	98	66	16	4.4	5.2

Knives used in the sectioning procedure are often made from glass, diamond, steel or tungsten carbide. Of these, only steel knives may be contributing contamination to the sections. Glass knives produce good sections but they deteriorate and become blunt very quickly. Diamond knives have better durability but are, of course, much more expensive. Most cryo-sectioning is carried out below −60°C in order to avoid, if possible, preliminary freeze-drying, ice re-crystallisation and superficial melting of the sections. Manipulation of the tissue sections after they are cut is difficult because of their small size. Proper specimen support for such small, fragile sections is not a trivial problem, since the rigours of a proton beam irradiation under vacuum must be withstood. Air drying has been shown to result in gross distortion of the tissue, melting artefacts and loss of elements, and critical-point drying while providing good structural preservation is still predominantly a wet chemistry procedure and suffers from similar limitations. Therefore, freeze-drying is recommended for most specimens.

7.6.1.3 *Samples for GIPIXE and GEPIXE* For irradiation in the geometry described earlier, extremely flat surfaces are needed. It is possible to distinguish between two cases, namely the analysis of this surface itself, e.g. the determination of impurities in Si wafers, and the analysis of a deposit, in which case the flat surface is used as a substrate that provides minimum background. In the former case minimal sample pre-treatment is required and the sample can be irradiated as is. The second case aims at a virtual backing-free situation, either by not allowing the protons to enter the backing (GIPIXE) or to prevent X-rays from the backing entering the detector (GEPIXE). It will be clear that only very little material should be deposited on the backing to avoid disturbance of the total reflection conditions from protons or X-rays, respectively. The methods are new and little research exists, but spin-rotating of a solution containing the sample in very fine particle form looks promising.

7.6.2 *Sample damage during irradiation*

7.6.2.1 *Definition* Specimen damage may represent a serious problem during (micro-)PIXE analysis owing to the high linear energy transfer (LET) value of a proton beam combined with a high beam density. In the literature, a division is very often made between specimen damage caused by temperature increase and subsequent loss of the integrity of the preparations, and the radiation damage caused by the effect of the ionising radiation itself. Radiation damage, moreover, is defined in different ways, depending on the field of research. Radiation damage in biology or biophysics, for instance, is often related to the irreparable damage to cells (where there is no occurrence of mitosis). In solid-state physics radiation damage is associated with the characteristics of, for example, a semiconductor in situations where unwanted electron–hole pairs are produced. For trace element analysis the important question is whether or not the material examined is a true representation of the original specimen in its natural state, so far as the trace elements and their dispositions are concerned.

Having overcome the numerous problems in producing an uncontaminated, undistorted, dried more or less thin target, the question arises as to what happens under bombardment by a few MeV protons. Changes in the elemental distributions on the micrometre scale as a result of this bombardment are referred to as specimen damage in this context. As far as these changes are caused by radiation effects, this definition will do in the realm of (micro-)PIXE, since all kinds of molecular rearrangements after breaking of chemical bonds will not necessarily lead to changes in these elemental distributions. Of all specimens, the biological samples are the most vulnerable, and thus most research has been conducted towards detecting radiation damage in these materials.

7.6.2.2 *Heat effects* Usually thin sections will be used as targets. This means that these targets are transparent to the incident beam. If the beam loses an amount of energy ΔE and a beam current I is impinging on the target, the dissipated power in the specimen is simply given by:

$$P = I \times \Delta E \tag{7.20}$$

This energy is transferred to the surrounding environment by conduction and radiation. It has been shown (Vis 1985) that during micro-PIXE, the conduction is the predominant method of heat dissipation from the specimen to the environment. Measures to avoid damage by heating should therefore point towards promoting the heat conductivity of the sample preparations.

7.6.2.3 *Effects of radiation dose* The very high radiation dose delivered by a proton (micro-)beam will cause disruption of chemical bonds along the beam path. Severe structural rearrangements in most organic materials will be the result. This damage can be noticed on irradiated targets in the form of a colour change and loss of mechanical strength. Quite commonly black beam tracks are visible after bombardment. This blackening is explained by the breaking of chemical bonds in organic materials to such an extent that elemental carbon is formed along the beam track. The possible loss of trace elements will strongly depend on the chemical form of the element under investigation. Volatile elements will be more vulnerable to losses; loss of halogens for instance has been reported (see Ishii *et al.* 1975) in a variety of specimens. It was claimed that evaporation of a thin layer of Al on top of the preparation prevents the losses. Whitehead (1979) describes in great detail the changes of human hair under the beam, including shrinking and changes of the total mass leading to problems in the concentration assignment of trace elements. Lenglet *et al.* (1985) used a deuterium beam and the (d,p) nuclear reaction to investigate the loss of C and O and concluded that predominantly O was lost during even mild irradiation.

The major matrix elements of biological systems are found to suffer from severe losses of hydrogen and oxygen, even with very moderate beam intensities. Although these losses do not necessarily influence the distribution of trace elements, concentration assignments should not rely on measuring backscattered protons or the energy loss of the beam in the target. Results actually turn out to be better if the amount of trace element detected is related to the original thickness of the preparation as determined during sectioning with the microtome. It is most likely that these losses are the result of breaking the C–H and C–O chemical bonds; the free H and O atoms diffuse towards the surface and are released into the vacuum system.

7.6.2.4 *Other effects of the beam* It is quite common to see carbon being build-up on top of the preparations. Organic molecules in the rest gas are being adsorbed and 'cracked' in a process where elementary carbon is formed. Counter-measures include a better vacuum in combination with cold traps. The process is normally not very harmful to specimens, but should be avoided during GIPIXE/GEPIXE types of experiment.

7.7 Applications

7.7.1 *General remarks*

Since its discovery in the early 1970s, PIXE has matured into an established analytical technique amidst many other methods available for trace element analysis. Applications can be found in a very wide range of research fields and it is beyond the scope of the present study to provide a full coverage of all applications published over the years. A good impression of the establishment of this process can be obtained by scanning the proceedings of the series of conferences on PIXE and its analytical application (Johansson *et al.* 1977, 1981; Martin 1984; Van Rinsvelt *et al.* 1987; Vis 1990; Uda 1993; Moschini *et al.* 1996; Malmqvist 1999), the first held at Lund, Sweden in 1976 and the most recent one before preparation of this chapter, coincidentally also at Lund, in 1998 enabling an overview of the progress of the last two decades. Noticeable is the increase in applications but also the refinements in theory and instrumentation. Apart from these proceedings two excellent textbooks are available (Johansson & Campbell 1988; Johansson *et al.* 1995), covering not only the fundamentals and instrumentation but a wide variety of illustrative applications too.

As with other techniques PIXE has its advantages and its drawbacks. PIXE's major assets are its speed and possibility for full automation. Disadvantages are (apart from the obvious demand for an accelerator) the detection limits, which although in the ppm range, are now surpassed by novel techniques such as ICP-MS. Also the fact that the analytical requirements in many fields of research are tending towards lower levels of concentration sometimes limits the possibilities for PIXE. This is especially true in biology, a very favourable area in the early days of PIXE but decreasing now, although the minute amounts of sample needed for a full PIXE analysis still represents a major advantage.

In short, PIXE is ideal in situations requiring large amounts of sample where a survey is needed of elements present down to the ppm level. This is quite often the case in geology, but especially so for environmental monitoring in the analysis of aerosols and fly-ash or biological monitoring such as lichens, leaves or algae.

It should be realised that the situation is completely different for micro-PIXE. The nuclear microprobe adds the feature of localising the trace elements, maintaining the sensitivity of PIXE and placing micro-PIXE beyond the competition with respect to other trace methods that do not provide spatial information. Micro-PIXE is applied extensively in the biological field and in geology, but is less suitable for surveying large numbers of samples owing to the relatively long scanning times needed to obtain an elemental map. Consequently, there has been a shift towards micro-applications where PIXE still has unique features. It is for this reason that major effort has been put into improving nuclear microprobes in terms of their spatial resolution and spot brightness.

In the following sections a few applications in various fields of research are discussed for both macro- and micro-PIXE.

7.7.2 *Biomedical applications*

7.7.2.1 *Macro-PIXE* The beginning of PIXE saw biological work to be most accessible. With mg quantities of samples, full analysis was possible with detection limits below 1 ppm; the organic matrix was ideal by not producing interfering X-ray lines. Whole blood and blood serum were frequently sampled, usually to search for correlations between any kind of disorder and the concentration of trace element(s) present. Early examples of the analysis of whole blood and serum can be found in Bearse *et al.* (1974) and Breiter and Roush (1975).

The influence of pregnancy on both healthy and diabetic women with respect to the concentrations of analytes in blood were studied at Debrecen (Hungary) (Borbély-Kiss *et al.* 1984; Gödeny *et al.* 1986a,b). Apart from plasma, erythrocytes also were analysed. A number of studies was published on the influence of a haemodialysis on trace element levels (Vis *et al.* 1979, 1981; Miura *et al.* 1999). Changes in Fe, Zn and Br concentrations were observed in particular. Most papers conclude, however, by saying that in spite of significant changes in concentrations, the biochemical implications are still largely unknown. Moreover, the homeostatic mechanisms that control blood levels may obscure an excess or depletion of an element in other body compartments. The former argument therefore led to the need for speciation, i.e. the determination of the binding sites of elements, whereas the latter argument called for the analysis of tissue rather than whole blood, plasma or serum.

A first attempt to measure protein binding was made by Keszthelyi *et al.* (1984), who used a nuclear reaction on nitrogen to determine the protein content simultaneously with PIXE. In a later stage, the same group employed electrophoresis to separate metal-binding proteins and analysed the electrophoretograms with PIXE. The authors were able to establish the molecular weight of the proteins binding to a given metal (Szökefalvi-Nagy *et al.* 1993).

Appreciating the limitations of blood analysis, the analysis of hair did attract quite some attention. It was a compromise between blood and tissue, but is so accessible that the question was asked whether hair could serve as a bio-monitor for the trace element status of the body. It was also realised that environmental pollutants may influence hair trace element concentrations. Sampling as a function of the distance to the scalp and measuring longitudinal profiles actually revealed this influence. Valkovic *et al.* (1975) measured such longitudinal profiles from schoolchildren in Rijeka, Croatia. In Fig. 7.8 data of Pb are presented, normalised to Zn, which is present in a fairly high concentration in hair. Zn has a rather flat distribution longitudinally, as demonstrated by Campbell *et al.* (1981).

Bascó *et al.* (1982) measured longitudinally the Ca distribution in hair. In combination with washing procedures, it turned out to be possible to show the effect of health conditions on Ca levels in hair. It was clear from the beginning of hair studies that there was a need for a microprobe to be able to measure also radial distributions with the hope of distinguishing between trace elements from the body and those from the environment. In fact one of the first biological applications of the nuclear microprobe using micro-PIXE was on hair (Cookson & Pilling 1975).

Tissue studies have concentrated among other sample types, e.g. brain, skin and kidney. Duflou *et al.* (1989) published a study on the distribution of various elements

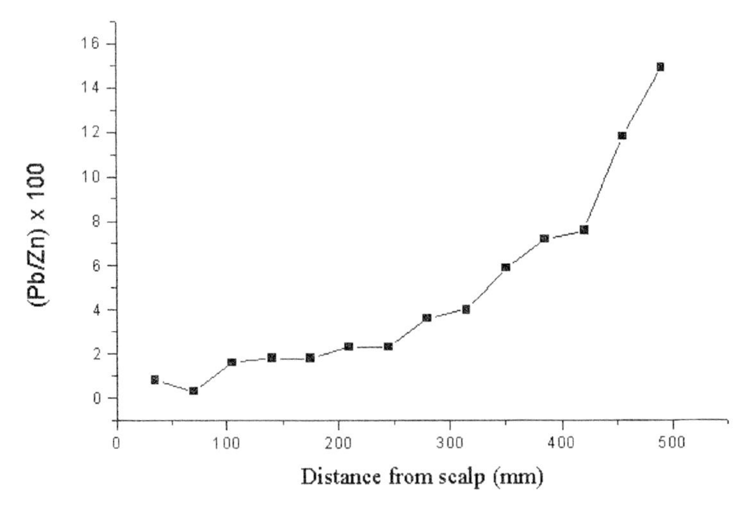

Fig. 7.8 A longitudinal distribution of Pb along hair of schoolchildren living in an urban area. Data from Valkovic *et al.* (1975), reproduced with permission from the author.

in the brain, comparing concentrations in grey and white matter and established variations of these elements during ageing. In Fig. 7.9 this influence is depicted for the elements K, Fe and Se, showing an increase of Fe and Se upon ageing. The same group made very useful efforts in addressing methodological aspects of the analysis of tissue, also in terms of accuracy and precision (Maenhaut *et al.* 1980; Van Lierde *et al.* 1997).

Skin and kidney tissue were analysed with the aim of measuring profiles. Even with a normal beam (macro-PIXE), occasionally referred to as a milli-beam, useful information can be obtained from these sample types. As an example, the work of

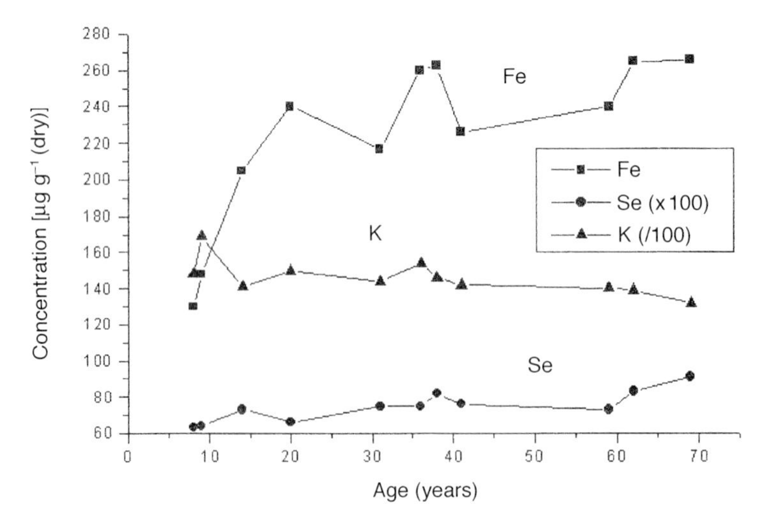

Fig. 7.9 The average concentrations of trace elements in brain tissue as a function of age.

Yukawa and Kitao (1991) is mentioned, in which concentration profiles across the cortex and the medulla of kidney tissue were measured. The well-known accumulation of the toxic element Cd in the cortex was clearly demonstrated (see Fig. 7.10). To obtain average concentrations wet-digestion has been employed. Järnström *et al.* (1983) analysed uterine tissue in this way and showed by using bovine liver reference material that this method leads to quantitative results for Fe, Cu and Zn.

The aforementioned examples are just illustrations and do not claim any completeness. Generally speaking, however, it is fair to say that the number of macro-PIXE applications in biology has not grown as fast as applications in other fields, partially because of the advent of competing methods and partially due to the fact

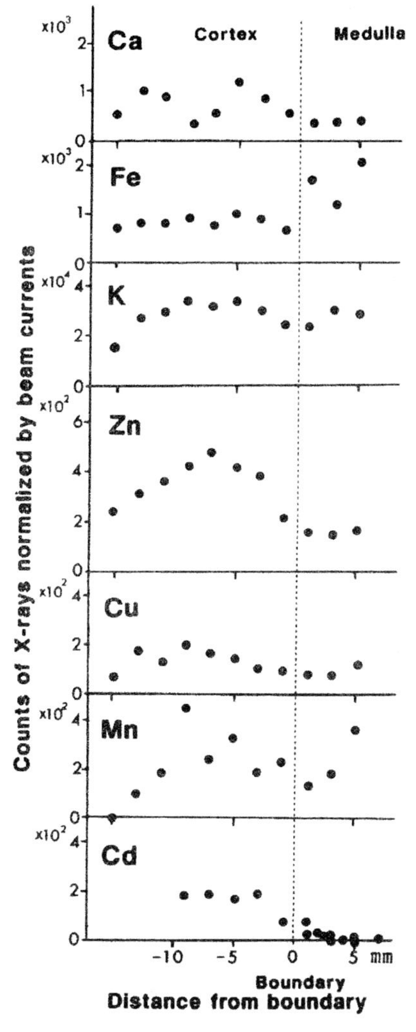

Fig. 7.10 Relative concentrations of some minor and trace elements across kidney tissue samples. Data from Yukawa and Kitao (1991), reproduced with permission of the author.

that average concentrations of trace elements in biological objects are not easily interpretable biochemically.

7.7.2.2 *Micro-PIXE* The introduction of the nuclear microprobe opened up the possibility of analysing biological samples with a lateral resolution at or below cell size. Measuring low concentrations in a single cell was and still is a quite unique feature of this instrument. Together with options to measure precise distributions over various tissue sections, including hard tissue such as hair, teeth and bone, the introduction of the microprobe meant a revival of biological applications. Today the large majority of biomedical PIXE applications include the use of the nuclear microprobe. By following the proceedings of the series of conferences on nuclear microprobe technology and applications (see Demorier *et al.* 1982; Grime *et al.* 1988; Legge *et al.* 1991; Lindh 1993; Yang *et al.* 1995; Doyle *et al.* 1997) substantial biological contributions to various fields, including medical, applications and applications in botany, have been seen. Figure 7.11 illustrates the analytical power of micro-PIXE by displaying the PIXE spectrum of a single red blood cell (O'Brien & Legge 1987). The analytical power is demonstrated by observing Zn at the ppm level in single cells, especially in terms of absolute detection limits, which are, assuming the mass of this cell being 10^{-10} g, of the order of 0.1 fg.

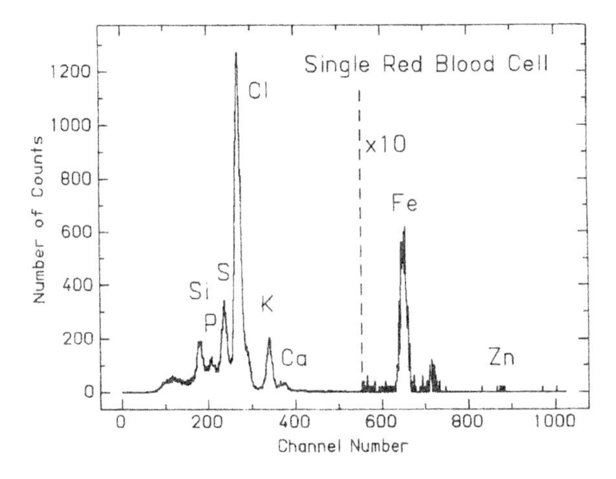

Fig. 7.11 The PIXE spectrum of a single red blood cell. Data taken from O'Brien and Legge (1987).

Since early microprobe work, the technique has improved to such an extent that sub-cellular structures can be analysed at the same level of concentration as shown by Lindh (1995) who detected the enrichment of Zn in mitochondria and the cell nucleus in one neutrophil granulocyte and also in one lymphocyte. The relevant Zn concentrations are of the order of 10 μg g^{-1} dry weight; the nuclear microprobe used had a resolution of 0.7 μm.

Various tissue sections have been studied with micro-PIXE. Malmqvist *et al.* (1987) and Lövestam *et al.* (1988) did experimental studies on the physiology of the human skin epidermis, especially to look into the penetration of Ni and Cr, detected in the stratum corneum. Artery tissue, in most cases in connection with arteriosclerosis patients, were studied (Pallon *et al.* 1985; Watt *et al.* 1995) and revealed a certain role of the elements Zn and Fe in the unwanted calcification process. Brain, both from humans and from test animals, was studied with micro-PIXE (Wensink *et al.* 1987; Van Lierde *et al.* 1997) to search for the influence of Zn uptake on its concentration in the hippocampus.

As mentioned above, human hair was also studied extensively with nuclear microprobes. A combination of longitudinal and radial scans were made by Bos *et al.* (1985) to identify the incorporation routes of various trace elements. Figure 7.12 shows results of such scans for six elements. It can be clearly seen that Cu and Zn, as bivalent ions, can be incorporated into the keratin structure, most likely by way of bridging S atoms, whereas other elements remain in the sheaths of the hair root. From these studies, however, it was concluded that hair also is not, in all cases, a suitable monitor in establishing the trace element status of the body.

Hard tissue analyses have been directed towards the measurement of diffusion

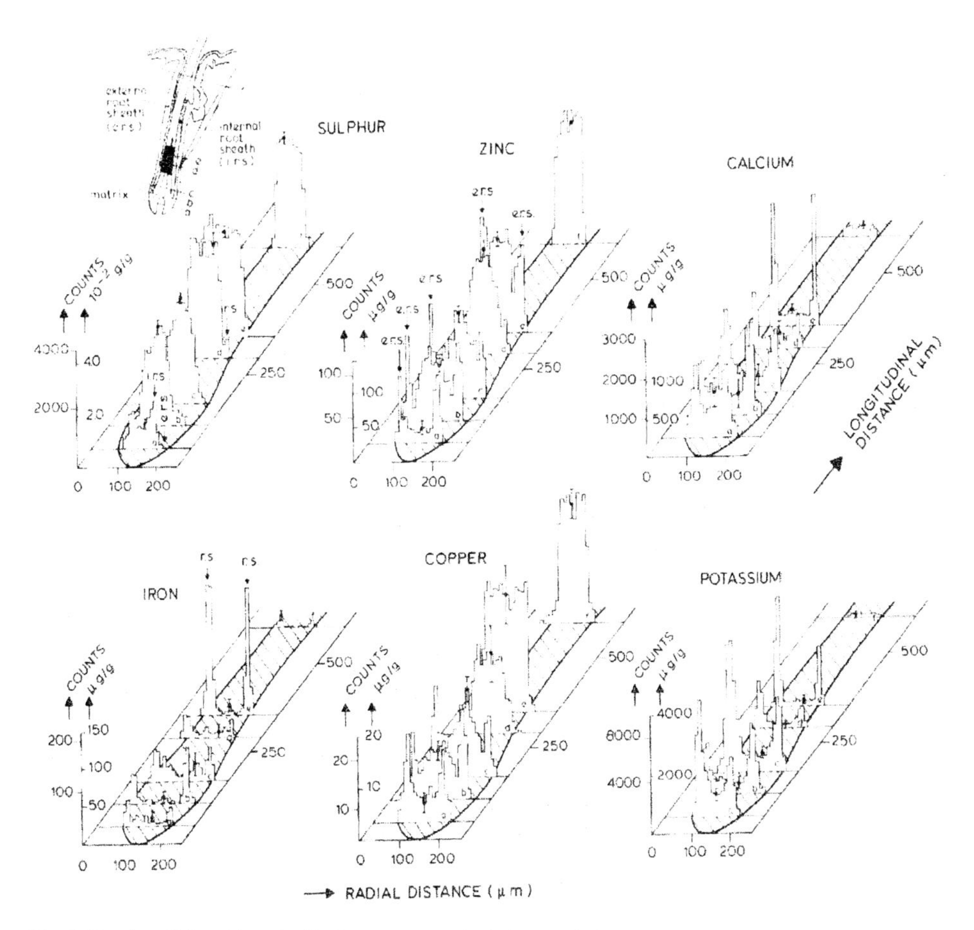

Fig. 7.12 A radial and longitudinal scan over a single human hair (Bos *et al.* 1985).

profiles in calcified tissues, either from implants or other external influences. Lindh (1980) measured concentration profiles of Zn and Pb in osteons of the human femur after uptake of these elements. Coote and Vickridge (1988) studied the influence of fluorine during the calcification process. Tros *et al.* (1993) performed comparable work but on neonatal hamster tooth germs and observed hypercalcification induced by the administration of F^- ions. The leakage of metals from dental amalgams was the subject of an investigation by Lindh *et al.* (1997). These authors were able to detect a different Zn profile and an invasion of Hg in granulocytes from individuals with large quantities of amalgam in their teeth.

In botany, an increasing number of publications on micro-PIXE can be found. A thorough review of the subject was presented by Przybylowicz *et al.* (1997). Quite a number of plant studies serve the purpose of obtaining environmental data.

7.7.3 *Environmental applications*

7.7.3.1 *Macro-PIXE* By far the majority of all environmental work is on aerosols. The reason is that these samples, invariably prepared by drawing a given amount of air through filters, can be analysed with PIXE without any further pre-treatment and with sufficient sensitivity to produce interesting data on the air quality. The high throughput in a fully automated method enables the rapid analysis of a vast number of samples when results are needed for further statistical analysis. Selection of the particles with respect to their size or mass fraction is completed during the sampling itself through use of various types of samplers. Examples are the continuous time sequence filter or the streaker (as developed at Florida State University), impactors using a rotating drum, or stacked filter samples. Also, the pore size of the filters determines the fraction of aerosol to be analysed. Filter materials can be polycarbonate, teflon, cellulose or glass fibre.

Owing to the very advantageous characteristics of this type of application; a large number of PIXE groups embarked on an atmospheric research programme and provided major contributions to the field of environmental studies. In the USA, a fine-aerosol study was started to collect data from a large network of sampling stations with principal contributions coming from the Air Quality Group at Davis, California (see Cahill 1993). The aim of the study was to analyse all elements from Na through to Pb in a quantitative matter, ensuring also that very clean areas where little element mass will be available on the filters were included. To meet these goals, the Davis group used a PIXE chamber with two X-ray detectors, equipped with different filters so as to optimise the detectors for Na–Fe and Cr–U, respectively. Complementary ion-beam techniques were used to detect signals from H and F.

Such aerosol work serves the study of the global exchange of pollutants, for which purpose sampling has even been done in the Antarctic region, and also it correlates sources of pollution with air quality in the near environment. For instance, in the latter category the study of fly-ash fits in as a source of man-made airborne particles, but at the same time it serves as a source for monitoring purposes to follow the air quality over the course of time.

Apart from this atmospheric research not many other sample types have been analysed with macro-PIXE. However tree rings have been proposed as a monitor for the long-term variation of industrial pollution. Harju *et al.* (1996) studied tree rings of spruce and pine to detect seasonal variations in their trace element concentrations. In addition to these variations, the authors observed a decrease in the concentrations of Cl, K and Ca over the five-year period studied (1947–1952).

7.7.3.2 *Micro-PIXE* Nuclear microprobes can be used to analyse individual particles of an aerosol or fly-ash source. For statistical purposes the number of particles analysed in this way is usually prohibitively small, but to gain insight into the character of the particles in relation to their sources, such as sea salt, soil or industrial, these measurements are of great value. Moreover, it is possible to correlate the presence of certain elements with the precise size of the particles, as all microprobes are equipped with an optical microscope. These data can be of importance in evaluating the health risk of a given pollutant as uptake is strongly dependent on particle size. Also the transport behaviour of the pollutant is size dependent.

In a relevant study, Orlic *et al.* (1995) analysed a large number of individual particles with the microprobe at NUS, Singapore and were able to identify nine groups of particles by principal component analysis. The authors claimed that elements could be detected in their method that would have definitely been missed if macro-PIXE was used. A detection limit as low as 10^{-18} g was claimed in favourable cases. Figure 7.13 illustrates the abundance of the identified element groups in their samples taken during a two-week period in Singapore.

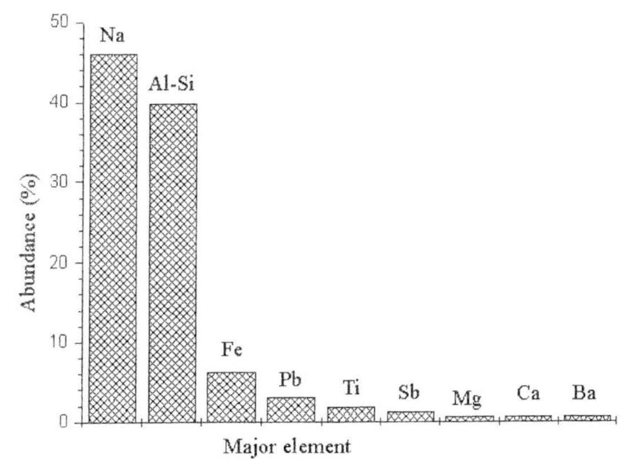

Fig. 7.13 Relative abundance of particle types as identified by principal component analysis. Data taken from Orlic *et al.* (1995).

Fly-ash particles have also been subjected to microprobe analysis (Valkovic *et al.* 1984; Rousseau *et al.* 1997). It was found that trace elements can accumulate in these particles by way of three mechanisms. Firstly, trace elements remain associated with unaltered mineral phases. Secondly, some elements are dissolved in larger particles. Thirdly, volatile trace elements are found to react with molten fly-ash particles, especially when the particles approach eutectic compositions in ternary systems of oxides such as Fe–Al–Si or Ca–Al–Si.

As a final example of single-particle analysis, it is worthwhile mentioning the detection of many elements in cosmic dust (Van der Stap *et al.* 1986). It was concluded that the so-called chondritic aggregates are richer in volatiles than the average composition of a solar nebula, explaining the depletion of these elements in ordinary chondrites in a satisfactory manner.

Other microprobe applications are found on the boundary between biology and environmental science. Bio-monitors such as lichen and tree rings have been mentioned already and can be subject to microanalysis as well. A revealing but rather sad example of a direct source–effect relationship was the study of the cause of death of lactating ewes in New Zealand as a result of the eruption of the volcano Mt. Ruapehu. After measuring the distribution of Ca with PIXE and F with NRA in the developing teeth of sheep it was concluded that F-poisoning was responsible (Coote *et al.* 1997).

Examples of microprobe studies in botany are the uptake of P and S in the roots of *Tagets patula* L. (Makjanic *et al.* 1988a) and the uptake of Si in the sedge *Schoenus nigricans* (Ernst *et al.* 1995). Both examples illustrate nicely the interaction between biological systems and the environment. As final illustration of a study in biology, the work of Carriker *et al.* (1982) is mentioned. The authors measured 16 elements in the shells of oysters. The peculiarity of this work was that, thanks to an external microprobe, the oysters could remain alive, enabling the incorporation of the elements during growth in a more or less controlled environment to be followed and thus assessed in relation to the sea-water composition.

7.7.4 *Applications in geology*

7.7.4.1 *Macro-PIXE* As argued before, macro-PIXE applications are in competition with other analytical methods for the determination of elemental composition in geology. This is even more so for geological work, with a longstanding tradition of elemental analysis using neutron activation and various optical techniques. The quantity of sample available is not normally a limiting factor, but the matrix of mostly silicates is not as ideal for PIXE as the organic matrix of biological materials. None the less, various ion-beam analytical techniques, including PIXE, have been used in geology. The benefit of the MeV ion beams for mineral analysis are reviewed by Sueno (1995), including a number of applications. Optimum experimental conditions will differ from sample to sample, but generally speaking detection limits will be worse compared with those found in biological work. Detection limits of a few ppm are within reach, but filtering of low-energy X-rays is mandatory, for which purpose Al is used of varying thickness from 100 to 300 μm (Campbell *et al.* 1990; Sie *et al.* 1991).

A useful application is in the analysis of soil, either for agricultural purposes to determine nutrients or to measure the degree of contamination from human activities. Cruvinel *et al.* (1999) measured the soil of an experimental field in Pindorama, Brazil and accumulated data on 16 elements. In his paper, a novel technique for soil sample preparation is described in which the sample is aerosolised and collected onto a filter holder sitting in a plastic bag. Rajander *et al.* (1999) used soil PIXE data to determine the degree of contamination from Cr, Cu and As in an area of land near Raisio, Finland. With a special soil sampler, a depth profile was measured and compared with a reference area. Figure 7.14 depicts the results of these measurements. The data show convincingly the degree of contamination, whereas the depth profile is useful in learning about the measures to be taken before agricultural activities can be started.

Another rather novel PIXE application is described by Malmqvist *et al.* (1999). The authors measured the extremely small concentrations of trace elements in an upward gas stream. The method is called geo-gas prospecting. The gas originates from water-filled fractures in bedrock. Collecting the sample gas is done using a plastic funnel placed in a hole dug in the ground and equipped with a polystyrene membrane. The authors published data on 17 elements present in such gas.

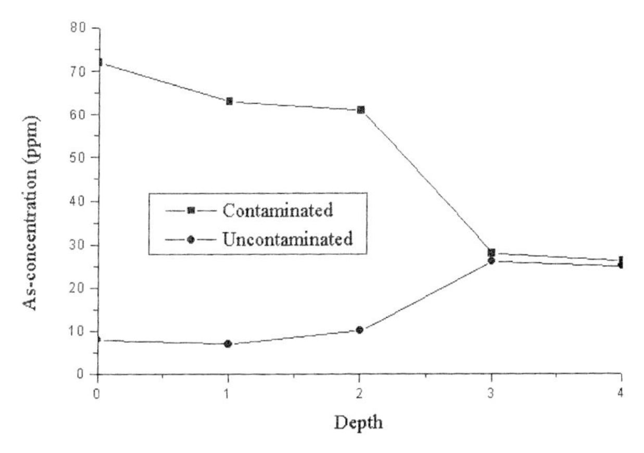

Fig. 7.14 Arsenic contamination of soil as a function of depth. Every step in depth is about 40 m. Results are compared with soil from an uncontaminated area. Data taken from Rajander *et al.* (1999), reproduced with permission of the author.

7.7.4.2 *Micro-PIXE* Most PIXE applications in mineralogy are completed with the nuclear microprobe. The need for data on elemental distributions, to establish diffusion patterns for reconstruction of the temperature/pressure history of rocks, or for exploration purposes, has triggered geologists to exploit various micro-analytical techniques, of which the most heavily used is one referred to as EPMA, electron-probe micro-analysis. Other techniques used include analysis with the Raman microprobe and SIMS.

As soon as nuclear microprobes became available, their value for geological applications was recognised and comparisons with EPMA were made. The most striking differences are the improved detection limits for micro-PIXE and the different penetration depths of the beam, which can be a disadvantage if the grain sizes to be analysed are small, but is normally an advantage because probing beneath the surface is very much desired. The most obvious and widely explored example of the use of this latter advantage is the analysis of unopened fluid or gas inclusions in minerals such as quartz. These studies serve the purpose of tracing back the early equilibrium between magmatic brines and vapour (Ryan *et al.* 1993) or the determination of ore elements in fluids (Damman *et al.* 1996). By micro-PIXE various types of inclusions can be distinguished. The complex geometry and often the presence of precipitates in the fluid complicate quantitative work, although modelling has been attempted to improve the situation. Melt inclusions have also been analysed. Mosbah *et al.* (1995) compared the analyses of natural and artificial volcanic glasses using micro-PIXE and a focused beam of synchrotron radiation.

Another important micro-PIXE application is the exploration for precious metals. The inhomogeneous distribution of these metals in combination with their low abundance promotes micro-PIXE as a unique tool for this purpose. Moreover, one may find positive correlations between a precious metal and other more abundant and therefore easier-to-detect elements. Such a correlation is described by Zhou *et al.* (1995) who found Au to be correlated with As, S, Ag and Fe in the Carlin-type gold deposit in China. Also in China, Li *et al.* (1997) published evidence of such a strong correlation between Pt and As that it was concluded that Pt was present in the Xinjie Cu–Ni deposit as $PtAs_2$.

Micro-PIXE has also been used in planetary science, particularly in the analysis of meteorites. Iron meteorites are studied with the aim of establishing the inter-spacing of the so-called Widmanstätten patterns, originating from a separation of phases of kamacite and taenite. The size of the 'M-profiles', which result from concentrations along a line scan, provide information about the cooling rate and in that way about the size of the original parent planetesimal. Meibom *et al.* (1996) determined these profiles for Ni, Ge and Ga and less convincingly Cu, whereas Makjanic *et al.* (1988b) compared the Ni distribution with carbon, which is more solvable in taenite.

Stony meteorites also, mainly carbonaceous and ordinary chondrites, have been subjected to micro-PIXE. Bajt *et al.* (1991) measured single crystals of olivine from Semarkona and Djahala and developed a simple model to deduce their cooling rate. Glass inclusions from Allende were measured by Makjanic *et al.* (1989) for their chemical composition including C, whereas Kramer (1996) made line-scans over chondrules from Chainpur. Chondrules are the typical spherical inclusions abundantly present in chondrites; their origin is largely unknown. A photograph of a small piece of Tieschitz is shown in Fig. 7.15. Results of a line-scan from Kramer (1996) are depicted in Fig. 7.16, demonstrating the depletion of, in this case, Zn in chondrules as compared to the concentration in the surrounding matrix. All these studies are directed towards collecting information about the temperature history of the early solar system, including the formation of chondrules. As a last example, the measurements on individual particles of cosmic dust are mentioned, principally a planetary application and referred to previously in Section 7.7.3.2.

7.7.5 *Miscellaneous applications*

7.7.5.1 *Applications in arts and archaeology* From the beginning of PIXE, objects of ancient origin and of art have been analysed. These studies were also the main incentive in developing external beams to avoid the risk to these valuable samples of placing them in a vacuum (Horowitz & Grodzins 1975). Swann (1995) reviewed these applications over the period 1990–1994. The variety of objects studied was found to be very wide indeed. Among others, ancient pottery (Ruvalcaba-Sil *et al.* 1999; Uda *et al.* 1999), porcelain (Lin *et al.* 1999), coins (Abraham *et al.* 1999; Hajivaliei *et al.* 1999), ancient

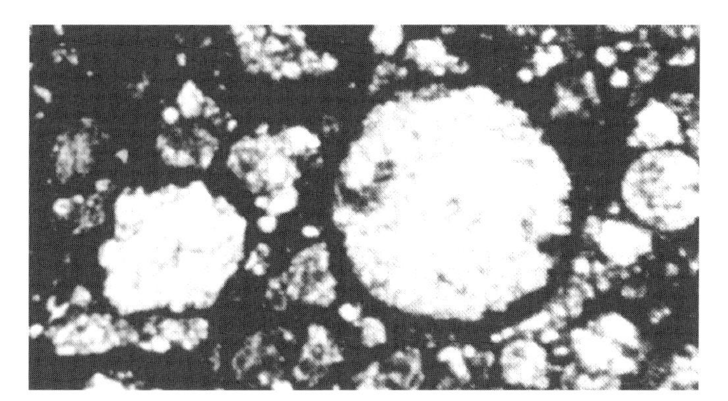

Fig. 7.15 A photomicrograph of a rimmed chondrule in a Tieschitz sample.

gold artefacts (Demortier *et al.* 1999) and jewellery (Calligaro *et al.* 1999) were all studied for their elemental composition. The aim is to reproduce the manufacturing process used in the past. With sufficient knowledge of these processes, dating of unknown objects might be possible. Also paints and ink (Lucarelli & Mandò 1996) have been measured with the same purpose in mind. The value of these analyses are demonstrated by the fact that the Louvre Museum in Paris, France, decided to install a dedicated PIXE set-up (Kusko *et al.* 1990) to examine their objects of art. Pappalardo (1999) has even developed a portable PIXE system to perform field analysis. A ^{210}Po α-source (10 mCi) deposited on a Ag substrate is used to induce the X-rays.

For most of these applications the lateral resolution is not really an important issue; most of the work in this field can be done with a relatively broad beam, and this is sometimes even necessary to avoid possible damage to these irreplaceable objects.

7.7.5.2 *Applications in materials science* In contrast to the previous section, applications here require a good lateral resolution. Materials science is developing towards nano-technology, dimensions at present beyond the detection of nuclear microprobes. None the less, a number of interesting applications is feasible with a micron-sized beam.

Surveying the literature, it is noticed that micro-PIXE is normally part of a complete study. Invariably, other ion-beam techniques are used simultaneously to extract as much information as possible from the sample. An example is the local characterisation of ceramic superconductors, in which the precise local stoichiometry is of importance and thus additional determination of oxygen is required. To this end the (d,p) nuclear reaction on oxygen is used (Berger *et al.* 1995) or Rutherford back scattering is applied, with the extra possibility of channelling studies (Osipowicz *et al.* 1999). Other examples can be found in metallurgy (Songlin *et al.* 1995; Berger *et al.* 1997) and in micro-electronics. In this latter area, the microprobe is often used to induce an electric charge at a given location in the device, a method called IBIC (ion-beam induced charge). This method is not easily combined with micro-PIXE owing to the very low ion dose used in IBIC studies. A better combination is formed between ionoluminescence and micro-PIXE as shown by Yang *et al.* (1997) who studied inorganic materials using this combination.

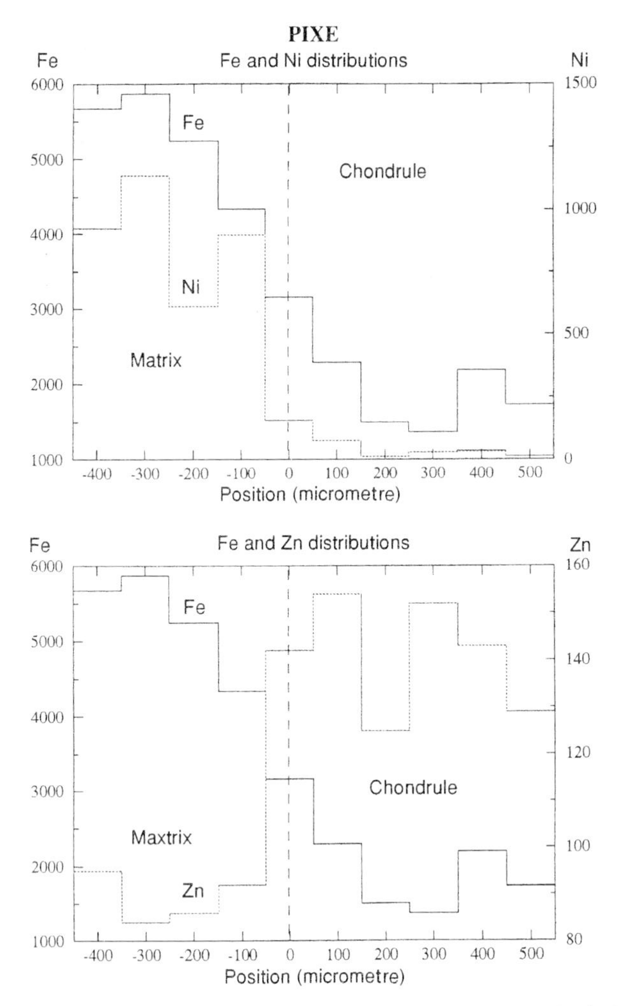

Fig. 7.16 Micro-PIXE results of a line-scan over a chondrule. Clearly seen are the depletion in Fe and Ni (more silicates in chondrules) and also a depletion of the moderately volatile element Zn. Both left-hand and right-hand ordinates are in ppm; $x = 0$ (vertical dashed lines) indicates the chondrule–matrix boundary. Data taken from Kramer (1996).

7.8 Conclusions and future prospects

7.8.1 *PIXE amidst other methods for elemental analysis*

Many techniques for elemental analysis are currently available and established in the laboratory. Methods such as neutron activation analysis and various optical methods, for instance atomic absorption/emission spectrometry in different modes (with or without flame etc.), have a longstanding tradition. All techniques have their advantages in terms of sensitivity, detection limits, accuracy and precision obtainable and whether they are easily automated. Also the sample types that can be analysed (liquid, solid material or both) and whether or not the method is multielemental are of importance. In addition to these so-called objective criteria, there are subjective criteria of which cost and turnaround times are normally mentioned.

Among the techniques, PIXE has its own place. Favourable characteristics are its multielemental ability, its relatively high accuracy and precision and the fact that minute amounts of sample are sufficient. Its detection limits, however, are not particularly good compared with other methods. An important advantage is that PIXE physics is fully understood enabling the simulation of an experiment and assessing by calculation whether a particular analysis is possible. The smooth behaviour of detection sensitivity as a function of atomic number Z facilitates the determination of elements of which a standard is not available. A principal disadvantage of PIXE often mentioned is the need for a particle accelerator. This is of course true, but the argument is not the cost of the accelerator, which is comparable with other analytical equipment, but the need to develop, at least in part, the beam-guiding systems, the scattering chamber, etc. Moreover, a small workshop for maintenance and minor construction is necessary in a PIXE environment.

It is likely that in the future it will be possible to buy a complete PIXE instrument, including a small almost maintenance-free accelerator and all other equipment needed, which is manufactured to a high degree of user-friendliness comparable to conventional analytical equipment. In addition it should be stressed that a major advantage of the PIXE set-up is that it is, apart from suitable detectors, usable immediately in other ion-beam techniques.

The nuclear microprobe adds an important extra feature to PIXE. The combination of a lateral resolution of about 1 μm maintaining ppm detection power sets it aside from quite a number of competing methods. Other probes that aim at elemental analysis on the micron scale are those involving electrons, lasers, low-energy ions and synchrotron radiation. Electrons, although focusable to much smaller spots, lack detection power owing to their primary Bremsstrahlung. Laser beams are used to evaporate and atomise the samples locally with subsequent detection in a mass spectrometer (LIMA). However, the method suffers from strong matrix effects and the physics behind the process is not fully understood. Quantification is, therefore, not straightforward. Much older is the technique in which the sample is bombarded with slow (keV) heavy ions and the fragments from the sample analysed with a mass spectrometer (SIMS). The sensitivity of both LIMA and SIMS is quite good but for SIMS also there are some quantification problems.

The new generation of synchrotron sources has made available high-intensity micro-spots of polarised X-rays of sufficient energy for micro-X-ray fluorescence (SXRF). For a limited number of elements with their absorption edge just below the incident energy, detection limits will definitely be lower than for micro-PIXE as a result of the higher cross-section for the photoelectric effect. The primary beam is usually tuneable, which makes micro-SXRF a good instrument for real trace analysis, but is only available at a limited number of sites around the world. The physics of XRF is of course also known in detail enabling full quantitative work.

Despite this competition, the nuclear microprobe remains a unique instrument, again thanks to the variety of methods to which it can be applied, mostly simultaneously enabling full characterisation of a given sample. The number of these instruments around the world is understandably increasing, helped also by the wide availability of most of the crucial parts of the device (such as slits, lenses and scattering chambers) on the market.

7.8.2 *Future prospects*

It is to be expected that the developments of PIXE will continue along the lines seen at present. The two most important aspects are instrumental improvements and improvements in understanding the finer details of the theory. Both those aspects will lead to a broader applicability of PIXE and to a fuller characterisation of the samples at hand.

Of the instrumental aspects the development of better nuclear microprobes attracts most attention. The main issues are higher intensity at the beam-spot in combination with an improved resolution in order to obtain nanoprobes rather than microprobes. Much progress has been made in the manufacture of more efficient magnetic quadrupole lenses. However, this progress has the effect of inhibiting large-scale investigations into alternative ways of focusing the beam, such as superconducting Einzel lenses, beam cooling by electrons or electrostatic focusing. Aside from lenses, research is underway to improve the brightness of the ion source so as to deliver a beam to as small a phase-space as possible. In addition to improving beam characteristics, improvements can be expected in detection. Optimised energy resolution in combination with a large subtended solid angle is still on the list of developments for many PIXE workers.

Theoretical improvements will involve a full analytical description of the various contributions to the background in PIXE spectra and a more accurate description of cross-sections and fluorescence yields for all relevant atomic shells. Research is also being done on projectiles other than protons. The probability of multiple-vacancy production in the target atom is increasing rapidly with the projectile mass, complicating the underlying theory. Nevertheless, these processes are not only interesting from the point of view of physics. If multiple-vacancy production is combined with high-resolution detection, information can be obtained about the chemical environment of the target atom. This so-called speciation will give added value to PIXE analyses.

The presence of multiple vancancies leads to both changes in the intensity ratio of various satellite lines and to an energy shift of the transition itself. Török *et al.* (1999) describe the energy systematics for the $K_{\alpha}L^{i}$ satellite transitions. The energy shift of the Mg K_{α} lines amounts to between 10 and 50 eV depending on the number of extra vacancies in the L shell during the K transition. The authors propose a semi-empirical description of these shifts. A shift of the K_{β} of Ti has been observed by Wijsman and Vis (1999), who found strong deviations of the $K_{\beta}K_{\alpha}$ intensity ratio in Ti upon impact with heavy ions. It was shown that K_{β} can shift towards an energy exceeding the K absorption edge leading to a strong increase of the absorption of K_{β} in the Ti target and a decrease in the $K_{\beta}K_{\alpha}$ intensity ratio. The authors became aware of this phenomenon during measurements of H in Ti foils with the well-known resonant nuclear reaction $^{15}N(^{1}H, \alpha\gamma)^{12}C$. Ti X-rays were used for monitoring, but demonstrated poor fits with a standard database for the intensity of the diagram lines.

The possibility of speciation was explored by Uda and Yamamoto (1999), using a wavelength-dispersive detection system. Important in this contribution were the reproduction of the fine structures from theory, including shake-off, and resonant orbital-rearrangement processes. In Fig. 7.17 spectra from various fluorides are shown, excited with a 5.5 MeV He^+ beam.

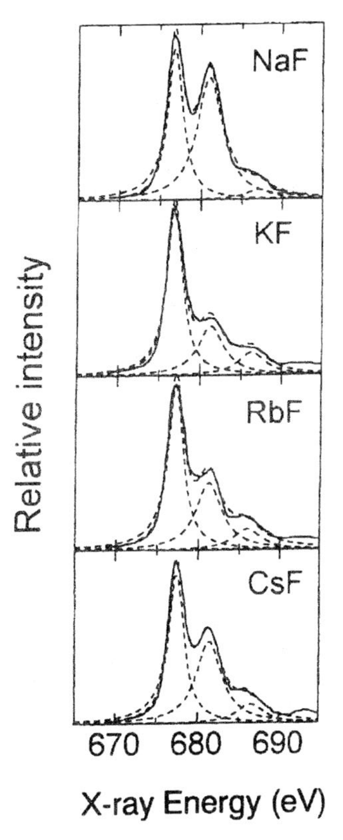

Fig. 7.17 Fluorine K_α spectra for various fluorides. Excitation was achieved with 5.5 MeV He^+ ions. The dotted lines represent calculations of the intensity ratio (see Uda & Yamamoto 1999, reproduced with permission of the author).

From Fig. 7.17 it is seen that the satellite strength is higher for Na and Cs, an irregular behaviour not to be expected from direct Coulomb excitation and necessitating more complicated molecular-orbital calculations. The similarity in the electronic structure of Na^+ and F^- ions leads to extra intensity. After measurements on suitable standards it should be possible to extract information from unknown fluorides showing that these satellites are in future also useful for analytical purposes.

In general and as final conclusion, PIXE at present is alive and developing simultaneously with a wide variety of applications employed in various fields of science.

References

Abraham, M.H., Grime, G.W., Nirthover, J.P., Smith, C.W. (1999) *Nucl. Instrum. Meth. Phys. Res.* **B150**, 651.
Bacsó, J., Sarkadi, L., Koltay, E. (1982) *Int. J. Appl. Radiat. Isot.* **33**, 5.
Bajt, S., Traxel, K. (1991) *Nucl. Instrum. Meth. Phys. Res.* **B54**, 317.
Bambynek, W., Crasemann, B., Fink, R.W., Freund, H.U., Mark, H., Swift, C.D., Price, R.F., Venugopala Rao, P. (1972) *Rev. Mod. Phys.* **44**, 716.
Bearse, R.C., Close, D.A., Malanify, J.J., Umbarger, C.J. (1974) *Anal. Chem.* **46**, 499.
Becker, R.S., Golovchenko, J.A., Patel, J.R. (1983) *Phys. Rev. Letters* **50**, 153.
Berger, P., Berthier, B., Revel, G., Trocellier, P. (1995) *Nucl. Instrum. Meth. Phys. Res.* **B104**, 542.
Berger, P., Moulin, G., Viennot, M. (1997) *Nucl. Instrum. Meth. Phys. Res.* **B130**, 717.
Bombelka, E., König, W., Richter, F.W., Wätjen, U. (1987) *Nucl. Instrum. Meth. Phys. Res.* **B22**, 21.
Borbély-Kiss, I., Koltay, E., László, S., Szabó, Gy., Gödeny, S., Seif El-Nasr, S. (1984) *J. Radioanal. Nucl. Chem.* **83**, 175.
Bos, A.J.J., Van der Stap, C.C.A.H., Valkovic, V., Vis, R.D., Verheul, H. (1985) *Sci. Total Environ.* **42**, 157.
Brandt, W., Lapicki, G. (1981) *Phys. Rev.* **A23**, 1717.
Breiter, D.N., Roush, M.L. (1975) *Am. J. Phys.* **43**, 569.
Cahill, T.A. (1993) *Nucl. Instrum. Meth. Phys. Res.* **B75**, 217.
Calligaro, T., Poirot, J.P., Querré, G. (1999) *Nucl. Instrum. Meth. Phys. Res.* **B150**, 628.
Campbell, J.L., Faiq, S., Gibson, R.S., Russell, S.B., Schulte, W. (1981) *Anal. Chem.* **53**, 1249.
Campbell, J.L., Maxwell, J.A., Teesdale, W.J., Wang, J.X. (1990) *Nucl. Instrum. Meth. Phys. Res.* **B44**, 347.
Carriker, M.R., Swann, C.P., Ewart, J.W. (1982) *Marine Biology* **69**, 235.
Castaing, R. (1960) In: *Advances in Electronics and Electron Physics*, ed. Marton, L. Academic Press, New York, Vol. 13, p. 317.
Cecchi, R., Ghermandi, G., Calvelli, G. (1987) *Nucl. Instrum. Meth. Phys. Res.* **B22**, 460.
Clayton, E.J. (1987) *The Lucas Heights PIXE Analysis Computer Package*. AAIC, Menoi, NSW, Australia.
Cohen, D.D. (1987) *Nucl. Instrum. Meth. Phys. Res.* **B22**, 68.
Coleman, J.R., Terepka, A.R. (1974) In: *Principles and Techniques of Electron Microscopy*, ed. Hayat, M.A. Van Nostrand Reinhold, New York, ch. 8.
Cookson, J.A., Pilling, F.D. (1975) *Phys. Med. Biol.* **20**, 1015.
Coote, G.E., Cutress, T.W., Sucklingm G.W. (1997) *Nucl. Instrum. Meth. Phys. Res.* **130**, 571.
Coote, G.E., Vickridge, I.C. (1988) *Nucl. Instrum. Meth. Phys. Res.* **B30**, 393.
Cruvinel, P.E., Flocchini, R.G., Artaxo, P., Crestana, S., Herrmann Jr., P.S.P. (1999) *Nucl. Instrum. Meth. Phys. Res.* **B150**, 478.
Damman, A.H., Kars, S.M., Touret, J.L.R., Rieffe, E.C., Kramer, J.L.A.M., Vis., R.D., Pintea, I. (1996) *Eur. J. Geology* **8**, 1081.
Demortier, G. (Ed.) (1982) *Nucl. Instrum. Meth.* **197**; Ed. Grime, G.W., Watt, F. (1988) *Nucl. Instrum. Meth. Phys. Res.* **B30**; Ed. Legge, G.J.F., Jamieson, D.N. (1991) *Nucl. Instrum. Meth. Phys. Res.* **B54**; Ed. Lindh, U. (1993) *Nucl. Instrum. Meth. Phys. Res.* **B77**; Ed. Yang, F., Tang, J., Zhu, J. (1995) *Nucl. Instrum. Meth. Phys. Res.* **B104**; Ed. Doyle, B.L., Maggiore, C.J., Bench, G. (1997) *Nucl. Instrum. Meth. Phys. Res.* **B30**.
Demortier, G., Morciaux, Y., Dozot, D. (1999) *Nucl. Instrum. Meth. Phys. Res.* **B150**, 640.
Doyle, B.L., McGrath, R.T., Pontau, A.E. (1987) *Nucl. Instrum. Meth. Phys. Res.* **B22**, 34.
Doyle, B.L., Maggiore, C.J., Bench, G. (1997) *Nuclear Instruments and Methods in Physics Research* **B130**.
Duflou, H., Maenhaut, W., De Reuck, J. (1989) *Neurochem. Res.* **14**, 1099.
Dyson, N.A. (1973) *X-rays in Atomic and Nuclear Physics*. Longman, London, UK.
Ernst, W.H.O., Vis, R.D., Piccoli, F. (1995) *J. Plant Physiol.* **146**, 481.
Fitzgerald, R., Keil, K., Heinrich, K.F.J. (1968) *Science* **159**, 528.
Folkmann, F. (1993) *Nucl. Instrum. Meth. Phys. Res.* **B75**, 9.
Folkmann, F., Frederiksen, F. (1990) *Nucl. Instrum. Meth. Phys. Res.* **B49**, 126.
Folkmann, F., Gaarde, C., Huus, T., Kemp, K. (1974) *Nucl. Instrum. Meth.* **116**, 487.
Gödeny, S., Borbély-Kiss, I., Koltay, E., László, S., Szabó, Gy. (1986a) *Int. J. Gynaecol. Obstet.* **24**, 191.
Gödeny, S., Borbély-Kiss, I., Koltay, E., László, S., Szabó, Gy. (1986b) *Int. J. Gynaecol. Obstet.* **24**, 201.
Garcia, J.D. (1970a) *Phys. Rev.* **A1**, 280.
Garcia, J.D. (1970b) *Phys. Rev.* **A1**, 1970.
Garcia, J.D., Fortner, R.J., Kavanagh, T.M. (1973) *Rev. Mod. Phys.*, **45**, 111.
Gonzalez, A.D., Packer, M.C., Miraglia, J.E. (1988) *Phys. Rev.* **A37**, 4947.
Grime, G.W., Watt, F. (1988) *Nuclear Instruments and Methods in Physics Research* **B30**.
Hajivaliei, M., Garg, M.L., Handa, D.K., Govil, K.L., Kakavand, T., Vijayan, V., Singh, K.P., Govil, I.M. (1999) *Nucl. Instrum. Meth. Phys. Res.* **B150**, 645.
Halder, H., Menzel, N., Hietel, B., Wittmaack, K. (1999) *Nucl. Instrum. Meth. Phys. Res.* **B150**, 90.

Hall, R.N., Baertsch, R.D., Soltys, T.J., Petrucca, J.L. (1970) *US Atomic Energy Commission Report*, NY0-3870–3.
Hansen, W.L. (1971) *Nucl. Instrum. Meth.* **94**, 377.
Hansteen, J.M., Johnsen, G.M., Kocbach, L. (1974) *J. Phys* **B7**, L271.
Hansteen, J.M., Johnsen, G.M., Kocbach, L. (1975) *Atom. Data Nucl. Data Tables* **15**, 307.
Harju, L., Lill, J.O., Saarela, K.E., Heselius, S.J., Hernberg, F.J., Lindroos, A. (1996) *Nucl. Instrum. Meth. Phys. Res.* **B109/110**, 536.
Horowitz, P., Grodzins, L. (1975) *Science* **189**, 795.
Ishii, K., Morita, S. (1987) *Nucl. Instrum. Meth. Phys. Res.* **B22**, 68.
Ishii, K., Morita, S. (1990) *Int. J. PIXE* **1**, 1.
Ishii, K., Morita, S., Tawara, H., Chu, T.C., Kaji, H., Shiokowa, T. (1975) *Nucl. Instrum. Meth.* **126**, 75.
Järnström, S., Dahlbacka, J., Pakarinen, P., Näntö, V., Grönroos, M. (1983) *J. Radioanal. Chem.* **79**, 207.
Johansson, E.M., Akselsson, K.R. (1981) *Nucl. Instrum. Meth.* **181**, 221.
Johansson, G.I. (1982) *X-ray Spectrometry* **11**, 194.
Johansson, S.A.E. (Ed.) (1977) *Nucl. Instrum. Meth.* **142**; Ed. Johansson, S.A.E. (1981) *Nucl. Instrum. Meth.* **181**; Ed. Martin, B. (1984) *Nucl. Instrum. Meth. Phys. Res.* **B3**; Ed. Van Rinsvelt, H., Bauman, S., Nelson, J.W., Winchester, J.W. (1987) *Nucl. Instrum. Meth. Phys. Res.* **B22**; Ed. Vis, R.D. (1990) *Nucl. Instrum. Meth. Phys. Res.* **B49**; Ed. Uda, M. (1993) *Nucl. Instrum. Meth. Phys. Res.* **B75**; Ed. Moschini, G., Valkovic, V. (1996) *Nucl. Instrum. Meth. Phys. Res.* **B109/110**; Ed. Malmqvist, K.G. (1999) *Nucl. Instrum. Meth. Phys. Res.* **B150**.
Johansson, S.A.E., Campbell, J.L. (1988) *PIXE, A Novel Technique for Elemental Analysis*. Wiley, Chichester, UK.
Johansson, S.A.E., Campbell, J.L., Malmqvist, K.G. (1995) *Particle Induced X-ray Emission Spectrometry (PIXE)*. Wiley, New York.
Johansson, T.B., Akselsson, K.R., Johansson, S.A.E. (1970) *Nucl. Instrum. Meth.* **84**, 141.
Kaufmann, H.C., Akselsson, K.R., Courtney, W.J. (1977) *Nucl. Instrum. Meth.* **142**, 251.
Keszthelyi, L., Varga, L., Demeter, I., Hollös-Nagy, K., Szökefalvi-Nagy, Z. (1984) *Anal. Biochem.* **139**, 418.
Kobayashi, K., Koizumi, Y., Nakano, C., Hatori, S., Sunohara, Y. (1999) *Nucl. Instrum. Meth. Phys. Res.* **B150**, 144.
Kramer, J.L.A.M. (1996) *Quantitative micro-distributions of olatile trace elements in pristine meteorites*. PhD thesis, Vrije Universiteit, Amsterdam, The Netherlands.
Kusko, B.H., Menu, M., Calligaro, T., Salomon, J. (1990) *Nucl. Instrum. Meth. Phys. Res.* **B49**, 288.
Lövestam, N.E.G., Pallon, J., Tapper, U.A.S., Themner, K., Hyltén, G., Forslind, K.G., Malmqvist, K.G. (1988) *J. Trace Microprobe Techniques* **6**, 47.
Legge, G.J.F., Jamieson, D.N. (1991) *Nuclear Instruments and Methods in Physics Research* **B54**.
Lenglet, W.J.M. (1988) *On the distribution of trace elements in biological material: studies based on the use of an accelerator*. PhD thesis, Vrije Universiteit, Amsterdam, The Netherlands.
Lenglet, W.J.M., van Langevelde, F., Bos, A.J.J., Vis, R.D., Verheul, H. (1985) *Spectrochimica Acta* **B40**, 763.
Li, X., Zhu, J., Lu, R., Gu, Y., Wu, X., Chen, Y. (1997) *Nucl. Instrum. Meth. Phys. Res.* **B130**, 617.
Lin, E.K., Yu, Y.C., Wang, C.W., Liu, T.Y., Wu, C.M., Chen, K.M., Lin, S.S. (1999) *Nucl. Instrum. Meth. Phys. Res.* **B150**, 581.
Lindh, U. (1980) *Int. J. Appl. Radiat. Isot.* **31**, 737.
Lindh, U. (1993) *Nuclear Instruments and Methods in Physics Research* **B77**.
Lindh, U. (1995) *Nucl. Instrum. Meth. Phys. Res.* **B104**, 285.
Lindh, U., Frisk, P., Nyström, J., Danersund, A., Hudecek, R., Lindvall, A., Thunell, S. (1997) *Nucl. Instrum. Meth. Phys. Res.* **B130**, 406.
Lucarelli, F., Mandò, P.A. (1996) *Nucl. Instrum. Meth. Phys. Res.* **B109/110**, 644.
Lytle, N.W., Silk, J.E., Hill, M.W., Mangelson, N.F., Rees, L.B. (1993) *Nucl. Instrum. Meth. Phys. Res.* **B75**, 144.
Mac Arthur, J.D., Amm, D., Barfoot, K.M., Sayer, M. (1981) *Nucl. Instrum. Meth.* **191**, 204.
Madison, D.H., Merzbacher, E. (1970) In: *Atomic Inner-Shell Processes*, ed. Crasemann, B. Academic Press, New York, ch. 1.
Maenhaut, W., De Reu, L., Van Rinsvelt, H.A., Cafmeyer, J., Van Espen, P. (1980) *Nucl. Instrum. Meth.* **168**, 557.
Maenhaut, W., De Reu, L., Vandenhaute, J. (1984) *Nucl. Instrum. Meth. Phys. Res.* **B3**, 135.
Maenhaut, W., Vandenhaute, J. (1986) *Bull. Soc. Chim. Belg.* **95**, 407.
Makjanic, J., Heymann, D., Van der Stap, C.C.A.H., Vis, R.D., Verheul, H. (1988b) *Nucl. Instrum. Meth. Phys. Res.* **B30**, 466.
Makjanic, J., Touret, J.L.R., Vis, R.D., Verheul, H. (1989) *Meteoritics* **24**, 49.

Makjanic, J., Vis, R.D., Groneman, A.F., Gommers, F.J., Henstra, S. (1988a) *J. Exp. Botany* **39**, 208.
Malmqvist, K.G., Forslind, B., Themner, K., Hyltén, G., Grundin, T., Roomans, G.M. (1987) *Biol. Trace Elem. Res.* **12**, 297.
Malmqvist, L., Kristiansson, K., Kristiansson, P. (1999) *Nucl. Instrum. Meth. Phys. Res.* **B150**, 484.
Martin, B. (1984) *Nuclear Instruments and Methods in Physics Research* **B3**.
Maxwell, J.A., Campbell, J.L., Teesdale, W.J. (1988) *Nucl. Instrum. Meth. Phys. Res.* **B43**, 218.
Meibom, A., Yang, C., Elfman, M., Rasmussen, K.L. (1996) *Proc. XVIIth Lunar and Planterary Science Conf., Houston, USA*, p. 857.
Meyer, M.A., Demortier, G. (1990) *Nucl. Instrum. Meth. Phys. Res.* **B49**, 300.
Miura, Y., Nakai, K., Sera, K., Sato, M. (1999) *Nucl. Instrum. Meth. Phys. Res.* **B150**, 218.
Mosbah, M., Clocchiatti, R., Michaud, V., Piccot, D., Chevallier, P., Legrand, F., Als Nilsen, G., Grübel, G. (1995). *Nucl. Instrum. Meth. Phys. Res.* **B104**, 481.
Moschini, G., Valkovic, V. (1996) *Nuclear Instruments and Methods in Physics Research* **B109/110**.
Murozono, K., Ishii, K., Yamazaki, H., Matsuyama, S., Iwasaki, S. (1999) *Nucl. Instrum. Meth. Phys. Res.* **B150**, 76.
O'Brien, P.M., Legge, G.J.F. (1987) *Biol. Trace Elem. Res.* **13**, 159.
Orlic, I., Lenglet, W.J.M., Vis, R.D. (1989) *Nucl. Instrum. Meth. Phys. Res.* **A276**, 202.
Orlic, I., Osipowicz, T., Watt, F., Tang, S.M. (1995) *Nucl. Instrum. Meth. Phys. Res.* **104**, 630.
Osipowicz, T., Xu, X.Y., Yang, C., Zhou, W.Z., Ong, C.K., Watt, F. (1999) *Nucl. Instrum. Meth. Phys. Res.* **B150**, 543.
Pallon, J., Homman, P., Pinheiro, T., Halpern, M.J., Malmqvist, K.G. (1995) *Nucl. Instrum. Meth. Phys. Res.* **B104**, 344.
Pallon, J., Malmqvist, K.G. (1981) *Nucl. Instrum. Meth.* **181**, 71.
Pappalardo, L. (1999) *Nucl. Instrum. Meth. Phys. Res.* **B150**, 576.
Paul, H. (1984) *Nucl. Instrum. Meth. Phys. Res.* **B3**, 5.
Petukhov, V.P. (1999) *Nucl. Instrum. Meth. Phys. Res.* **B150**, 46.
Przybylowicz, W.J., Mesjasz-Przybylowicz, J., Prozesky, V.M., Pineda, C.A. (1997) *Nucl. Instrum. Meth. Phys. Res.* **B130**, 335.
Rajander, J., Harju, L., Lill, J.O., Saarela, K.E., Lindroos, A., Heselius, S.J. (1999) *Nucl. Instrum. Meth. Phys. Res.* **B150**, 510.
Rousseau, P.D.S., Przybylowicz, W.J., Scheepers, R., Prozesky, V.M., Pineda, C.M., Churms, C.L., Ryan, C.G. (1997) *Nucl. Instrum. Meth. Phys. Res.* **B130**, 582.
Ruvalcaba-Sil, J.L., Ontalba Salamanca, M.A., Manzanilla, L., Miranda, J., Cañetas Ortega, J., López, C. (1999) *Nucl. Instrum. Meth. Phys. Res.* **B150**, 591.
Ryan, C.G., Heinrich, C.A., Mernagh, T.P. (1993) *Nucl. Instrum. Meth. Phys. Res.* **B77**, 463.
Scofield, J.H. (1974) *Phys. Rev.* **A10**, 1507.
Sie, S.H., Griffin, W.L., Ryan, C.G., Suter, G.F., Cousens, D.R. (1991) *Nucl. Instrum. Meth. Phys. Res.* **B54**, 284.
Songlin, F., Minqin, R., Ming, Z., Jiequing, Z., Changyi, Y., Rongron, L., Homman, N.P.O., Malmqvist, K.G., Yu, W., Qiyong, H. (1995) *Nucl. Instrum. Meth. Phys. Res.* **B104**, 557.
Sow, C.H., Orlic, I., Loh, K.K., Tang, S.M. (1993) *Nucl. Instrum. Meth. Phys. Res.* **B75**, 58.
Sueno, S. (1995) *Eur. J. Mineral.* **7**, 1273.
Swann, C.P. (1995) *Nucl. Instrum. Meth. Phys. Res.* **B104**, 576.
Szökefalvi-Nagy, Z., Demeter, I., Hollós-Nagy, K., Räisänen, J. (1993) *Nucl. Instrum. Meth. Phys. Res.* **B75**, 165.
Szabó, Gy., Borbély-Kiss, I. (1993) *Nucl. Instrum. Meth. Phys. Res.* **B75**, 123.
Török, I., Papp, T., Raman, S. (1999) *Nucl. Instrum. Meth. Phys. Res.* **B150**, 8.
Tros, G.H.J., Lyaruu, D.M., Vis, R.D. (1993) *Nucl. Instrum. Meth. Phys. Res.* **B83**, 255.
Tsuji, K., Wagatsuma, K., Hirokawa, K., Yamade, T., Utaka, T. (1997) *Spectrochimica Acta* **B52**, 841.
Uda, M. (1993) *Nuclear Instruments and Methods in Physics Research* **B75**.
Uda, M., Akiyoshi, K., Nakamura, M. (1999) *Nucl. Instrum. Meth. Phys. Res.* **B150**, 601.
Uda, M., Yamamoto, T. (1999) *Nucl. Instrum. Meth. Phys. Res.* **B150**, 1.
Valkovic, V., Makjanic, J., Jaksic, M., Popovic, S., Bos, A.J.J., Vis, R.D., Wiederspahn, K.J., Verheul, H. (1984) *Fuel* **63**, 1357.
Valkovic, V., Rendic, D., Phillips, G.C. (1975) *Environ. Sci. Technol.* **9**, 1150.
Van der Stap, C.C.A.H., Vis, R.D., Verheul, H. (1986) *Proc. XVIIth Lunar and Planterary Science Conf., Houston, USA*, p. 1013.
Van Espen, P., Nullens, H., Adams, F. (1977) *Nucl. Instrum. Meth.* **142**, 243.
Van Kan, J.A., Vis, R.D. (1995) *X-ray Spectrometry* **24**, 58.
Van Kan, J.A., Vis, R.D. (1996) *Int. J. PIXE* **6**, 89.
Van Lierde, S., Maenhaut, W., De Reuck, J., Vis, R.D. (1997) *Nucl. Instrum. Meth. Phys. Res.* **B130**, 444.

Van Rinsvelt, H., Bauman, S., Nelson, J.W., Winchester, J.W. (1987) *Nuclear Instruments and Methods in Physics Research* **B22**.

Vineyard, G.H. (1982) *Phys. Rev.* **B26**, 4146.

Vis, R.D. (1985) *The Proton Microprobe: Applications in the Biomedical Field.* CRC Press, Boca Raton, FL.

Vis, R.D. (1990) *Nuclear Instruments and Methods in Physics Research* **B49**.

Vis, R.D., Oe, O.L., Bos, A.J.J., Verheul, H. (1979) *Radiochem. Radioanal. Letters* **41**, 245.

Vis, R.D., Oe, P.L., Bos, A.J.J., Verheul, H. (1981) *Nucl. Instrum. Meth.* **181**, 293.

Vis, R.D., van Langevelde, F. (1991) *Nucl. Instrum. Meth. Phys. Res.* **B61**, 515.

Watt, F., Selley, M., Thong, P.S.P., Tang, S.M. (1995) *Nucl. Instrum. Meth. Phys. Res.* **B104**, 356.

Wensink, J., Lenglet, W.J.M., Vis, R.D., Van den Hamer, C.J.A. (1987) *Histochemistry* **87**, 65.

Whitehead, N.E. (1979) *Nucl. Instrum. Meth.* **164**, 381.

Wijsman, W. Vis, R.D. (1999) *Nucl. Instrum. Meth. Phys. Res.* **B150**, 14.

Williams, E.T. (1984) *Nucl. Instrum. Meth. Phys. Res.* **B3**, 211.

Yang, C., Malmqvist, K.G., Elfman, M., Kristiansson, P., Pallon, J., Sjöland, A., Utui, R.J. (1997) *Nucl. Instrum. Meth. Phys. Res.* **B130**, 746.

Yang, F., Tang, J., Zhu, J. (1995) *Nuclear Instruments and Methods in Physics Research* **B104**.

Yukawa, M., Kitao, K. (1991) *Int. J. PIXE* **1**, 339.

Zhou, S., Ren, C., Tang, J., Yang, F. (1995) *Nucl. Instrum. Meth. Phys. Res.* **B104**, 437.

8 Charged particle-induced γ-ray emission

J. RÄISÄNEN

8.1 Introduction

Charged particle-induced reactions leading to the emission of γ-rays as a tool for sensitive microanalytical determination of elemental concentrations has been well established. Particle-induced γ-ray emission (PIGE or PIGME) is normally considered a method complementary to particle-induced X-ray emission (PIXE, see Chapter 7), capable of detecting the low-atomic-number elements not seen in PIXE measurements. Other acronyms besides PIGE are used, namely prompt radiation analysis (PRA) and proton-induced prompt photon spectrometry (PIPPS). In PIGE the prompt γ-rays from short-lived reaction products are detected during particle bombardment (in contrast to activation analyses). This method can often be used in distinguishing between different isotopes of a given element.

The principles of the technique have been known since the 1930s. The first actual elemental determinations, however, were only carried out in the 1960s (Sippel & Glover 1960; Pierce *et al.* 1964, 1965, 1967). At this time protons or deuterons were used as bombarding particles and very often a single element was only determined. The obvious reason for lack of multielemental analyses was the poor energy resolution of the NaI(Tl) scintillators that were used. The invention of the Ge(Li) detector has made it possible to perform multielemental analyses since the 1970s. Today PIGE has gained wide interest as a complementary method to other ion beam techniques. An especially useful feature of the PIGE method is its capability for light element depth profiling by resonance reactions. This subject is discussed in detail in Chapter 10 in connection with nuclear reaction analysis.

PIGE is a sub-surface method with the analysable depth being few tens of micrometres from the surface. When PIGE is used together with PIXE, elements from Li can be detected. An advantage of the PIGE method is the high penetrability of the γ-rays, which reduces any matrix effects. Thus, absorption problems are not so important as they are in PIXE. Another difference when compared with PIXE is the nature of the cross-sections (excitation curves), which do not vary smoothly as a function of particle energy nor with the atomic number of the target material, but show in most cases distinct resonances. A fingerprint (the γ spectrum) is obtained from every element which enables them to be distinguished. The obtained spectra are not so straightforwardly readable as PIXE and RBS (Rutherford backscattering) spectra since no systematic patterns exist. The number of γ-rays emitted during particle bombardment is proportional to the element concentration in the sample. The sample preparation procedures for PIGE are very similar to those of PIXE including the need for various standards.

8.2 Nuclear reactions

In PIGE the sample is bombarded with charged particles obtained from a particle accelerator. Nuclear reactions or Coulomb excitation occurs and the target nuclei are left in excited states followed by γ-ray emission. The cross-sections for lightweight elements are typically of the order of several millibarns to hundreds of millibarns. The most important nuclear processes in PIGE are discussed below. Since mostly protons are employed as bombarding particles, the examples given relate to proton-induced reactions.

8.2.1 *Resonance capture*

In resonance capture reactions the projectile is captured by the target nucleus to form a nucleus B in an excited state. The reaction can be written as

$$\mathrm{a} + \mathrm{A} \rightarrow \mathrm{B}^{*} \rightarrow \mathrm{B} + \gamma$$

where 'a' represents the bombarding particle and A the target nucleus. B* denotes the compound nucleus in an excited state, which then decays to B via γ emission. Typical examples are (p,γ) proton capture reactions. The reaction cross-sections may show sharp resonances that are associated with the formation of compound nuclei. The excitation curves are of the Breit–Wigner type in which the excited states of the compound nucleus correspond to reaction resonances. Many capture reactions are useful for light element depth profiling. A typical reaction, which is often applied to aluminium detection and depth profiling, is $^{27}\mathrm{Al(p,\gamma)}^{28}\mathrm{Si}$.

8.2.2 *Inelastic scattering*

When no change of particles occurs, but the target remains in an excited state, the process is called inelastic scattering. The reaction can be formally written as:

$$\begin{aligned} &\mathrm{a} + \mathrm{A} \rightarrow \mathrm{A}^{*} + \mathrm{a}' \\ &\rightarrow \mathrm{A} + \gamma \end{aligned}$$

Typical inelastic scattering reactions involved with PIGE are (p,p′γ) reactions. With light ($Z \leqslant 20$) target nuclei the excitation curves show resonance features on top of a continuum. The resonances are not in this instance of the pure Breit–Wigner type. With increasing energy the resonances overlap rapidly and the cross-section curve becomes a continuum. An example of such a reaction is $^{23}\mathrm{Na(p,p'\gamma)}^{23}\mathrm{Na}$. With heavier target nuclei the inelastic scattering occurs well below the Coulomb barrier and the mechanism is Coulomb excitation, i.e. electromagnetic interaction. In such reactions the cross-sections increase smoothly as a function of bombarding energy. As no exact distinction between the two situations in practical elemental analyses can be given, both reaction types are marked similarly.

8.2.3 *Rearrangement collision*

In rearrangement reactions particles other than the bombarding particles are emitted. The reaction can be written as:

$$\begin{aligned} a + A &\rightarrow B^* + b \\ &\rightarrow B + \gamma \end{aligned}$$

Typical sequences of this type involved in PIGE are (p,nγ), (p,$\alpha\gamma$) and (p,n) reactions. With increasing bombardment energy the most intense outgoing channel observed becomes eventually neutron emission. As an example the very useful reaction for fluorine analysis, $^{19}F(p,\alpha\gamma)^{16}O$, can be given:

$$\begin{aligned} ^{19}F + p &\rightarrow {}^{16}O^* + \alpha \\ &\rightarrow {}^{16}O + \gamma \end{aligned}$$

Rearrangement reactions also show resonance features which can be taken advantage of in light element depth profiling.

8.3 Thick target γ-ray yields

8.3.1 *γ-ray yield from a thick target*

A sample can be considered as a thick target when its thickness exceeds the bombarding particle range. In the following treatment absorption effects of the γ-rays are not taken into account. This is reasonable, since the effective range of the charged particles is only few tens of micrometres and the energy of the observed γ-rays is usually high. Straggling effects are also ignored and smooth cross-section curves are considered without resonance structures. An approximate formula for the γ-ray yield (μC·sr) from a thick target is

$$Y_i = n\varepsilon f_{wi} \int_0^{E_p} \sigma_i(E)/S_m(E)\,dE \tag{8.1}$$

where: i refers to the measured nuclide and m to the matrix; n is the number of bombarding ions per μC (i.e. 0.625×10^{13} per charge of the ion); ε is the detection efficiency including the solid angle; E_p is the incident bombarding energy; f_{wi} is the weight fraction of nuclide i; S_m is the stopping power for the matrix in units of eV $(g\,cm^{-2})^{-1}$; and $\sigma_i(E)$ is the cross-section for the specific reaction.

The formula is valid when nuclide i is evenly spread through the matrix. At low ion energies the resonances cause abrupt increases in the thick target γ-ray yields, so the excitation curves are composed of successive plateaux. In Fig. 8.1 the proton-induced excitation curve obtained from a thick aluminium target is shown. The sharpness of the edges depends on the resonance width and energy spread of the initial beam. At proton energies above about 2 MeV the yield increases fast and the resonances begin to overlap. A continuum is reached and the occurrence of isolated resonances vanishes. Such behaviour is typical for protons and light elements ($Z \leqslant 20$). For heavy elements the yield curves are continuous. It should be noted that when performing measurements with ion energies close to resonance energies, caution should be adopted since a small change in the energy leads to significant changes in the yields. Equation (8.1) is not, however, useful for practical analytical determinations. Simpler approximate formulae are thus needed.

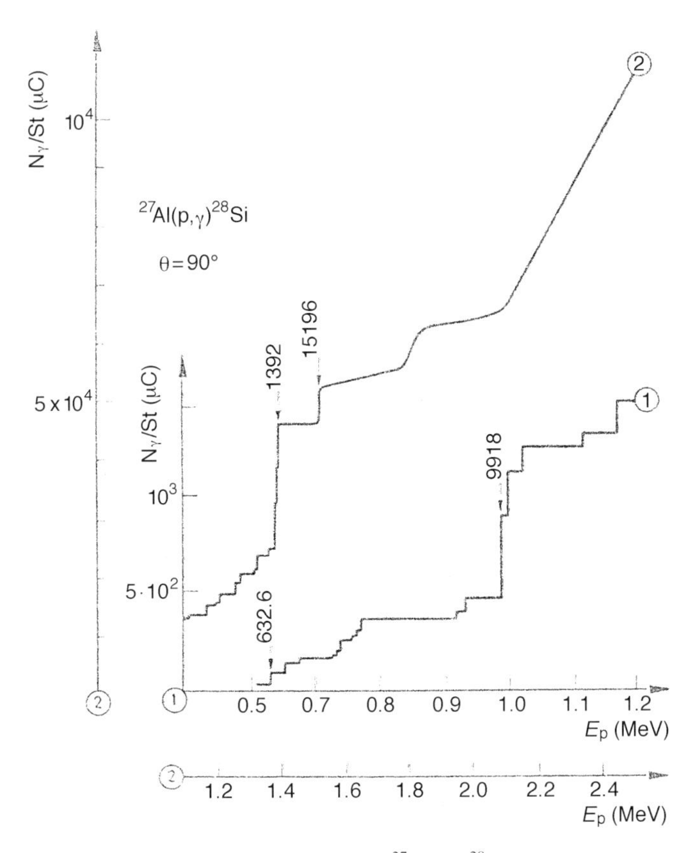

Fig. 8.1 Aluminium excitation function for the reaction $^{27}Al(p,\gamma)^{28}Si$ (1779 keV γ-ray line) from a thick target. The figure has been adopted from Deconninck and Demortier (1972). Reproduced with permission.

8.3.2 *Stopping power correction*

The aim of the formulation is to find an equation by which concentrations can be calculated using thick target γ-ray yields of pure elements. The main concern is finding a procedure that accounts for the relationship between stopping power and γ-ray yield as a function of bombarding energy. The procedure described by Kenny et al. (1980) is adopted and for other, similar procedures the reader is referred to Ishii *et al.* (1978a, b, c). According to Equation (8.1), the pure element yield under the same experimental conditions is

$$Y_{\text{pure}} = n\varepsilon f_{\text{wi}}^{\text{pure}} \int_{0}^{E_{\text{p}}} \sigma_{\text{i}}(E)/S_{\text{i}}(E)\,\text{d}E \tag{8.2}$$

where $S_{\text{i}}(E)$ is the stopping power for the pure element. The weight fraction, $f_{\text{wi}}^{\text{pure}}$, is now equal to unity. By differentiating Equation (8.2) it follows that:

$$\partial Y_{\text{pure}}(E)/\partial E = \varepsilon\, \sigma_{\text{i}}(E)/S_{\text{i}}(E) \tag{8.3}$$

Inserting this into Equation (8.1) gives:

$$Y_i = f_{wi} \int \partial Y_{pure}(E)/\partial E \cdot S_i(E)/S_m(E)\, dE \tag{8.4}$$

By integrating Equation (8.4) the following is obtained:

$$Y_i = f_{wi}\{Y_{pure}(E_{HI}) \cdot S_i(E_{HI})/S_m(E_{HI}) - Y_{pure}(E_{LO}) \cdot S_i(E_{LO})/S_m(E_{LO}) - \int Y_{pure}(E) \cdot d/dE[S_i(E)/S_m(E)]\, dE\} \tag{8.5}$$

For a thick target, E_{LO} is 0. The exact solution of Equation (8.5) is obtained numerically. Unfortunately, the pure element yields, $Y_{pure}(E)$, are not known accurately as a function of bombarding energy. For practical use a simplified rule of thumb has been suggested (Kenny *et al.* 1980):

$$Y_i = f_{wi}\{Y_{pure}(E_{HI}) \cdot S_i(E_{HI})/S_m(E_{HI})\} \tag{8.6}$$

Another rule of thumb, introduced in Deconninck and Demortier (1972), is

$$Y_i = f_{wi}\, Y_{pure}(E_{HI}) \cdot S_i(E_{1/2})/S_m(E_{1/2}) \tag{8.7}$$

where $E_{1/2}$ is the energy for which the yield $Y_{pure}(E_{1/2}) = \frac{1}{2} Y_{pure}(E_{HI})$. The difference in the yield values calculated from Equation (8.5) compared to those obtained by the rules of thumb of Equations (8.6) and (8.7) for various concentrations of F, Na and Al in matrices of different Z are shown in Fig. 8.2.

As can be observed, both rules of thumb provide yield values within 1.5% of the calculated yields; the $E_{1/2}$ rule gives slightly better agreement. In the following the practical use of the rules of thumb will be examined.

The procedure is based on the knowledge of one element concentration in the sample. This can be obtained using a standard – a known amount of the element is mixed into a similar matrix as the one studied. Extra care should then be paid to the homogeneity of the standard. The concentration, f_{wA}, for element A in the sample is thus obtained. The unknown concentration of element B in the sample can be calculated using the rule of thumb given by Equation (8.6) [or alternatively (8.7)]. For element A:

$$Y_A = f_{wA}\, Y^A_{pure}(E_{HI}) \cdot S_A(E)/S_m(E) \tag{8.8}$$

By solving for $S_m(E)$ and substituting it into the corresponding equation for element B then the following is obtained:

$$f_{wB} = Y_B/Y_A \cdot Y^A_{pure}/Y^B_{pure} \cdot S_A(E)/S_B(E) \cdot f_{wA} \tag{8.9}$$

The yields, Y, are obtained from the γ-ray peak areas. The pure element yields, Y_{pure}, should be measured or taken from literature (see Section 8.5). In applying the literature values one should be cautious since they should correspond to the same measuring angle. This point will be treated in the next section. Furthermore, as the values may include significant errors, for accurate elemental analyses relevant standard spectra should always be taken using an appropriate geometrical set-up. If an order of magnitude result is sufficient, the literature data procedure can well be

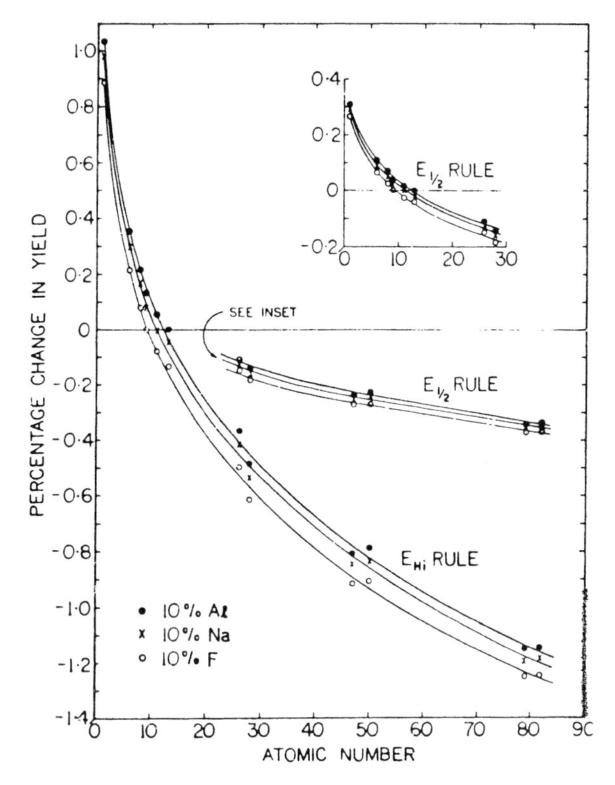

Fig. 8.2 Percentage yield change between the rule of thumb and the calculated yield. The figure has been taken from Kenny *et al.* (1980). Reproduced with permission.

applied. The stopping power values needed can be deduced, for example, by the SRIM 2000 computer program (Ziegler, private communication).

8.4 Experimental considerations

The sensitivity of the PIGE method depends on several factors: reaction cross-sections; beam current; irradiation time; detection system; and the sample matrix. Specific points affecting the sensitivity and which should therefore be kept in mind are discussed next.

8.4.1 *Points to be taken into account*

8.4.1.1 *Geometry* Generally it can be expected that thick target γ-ray yields are isotropic. However, in cases when the yield is mainly due to strong resonances, noticeable deviations from isotropic distributions are observed. In such cases the angular distributions should be taken into account if Equation (8.9) is employed and the measurements are carried out at a different angle. In Table 8.1 the thick target γ-

Table 8.1 Angular distribution effects in magnesium. The values are from Kenny *et al.* (1980). Reproduced with permission

		γ-ray yield per (μC · sr)			
		$E_p = 2.00$ MeV		$E_p = 2.514$ MeV	
E_γ (keV)	Reaction	90°	135°	90°	135°
390	^{25}Mg(p,p′γ)^{25}Mg	4.0×10^3	4.0×10^3	3.5×10^4	3.7×10^4
585	^{25}Mg(p,p′γ)^{25}Mg	3.0×10^4	3.1×10^4	1.1×10^5	1.1×10^5
844	^{26}Mg(p,γ)^{27}Al	3.7×10^2	5.0×10^2	1.0×10^3	1.0×10^3
975	^{25}Mg(p,p′γ)^{25}Mg	4.7×10^3	5.0×10^3	4.4×10^4	4.8×10^4
1014	^{26}Mg(p,γ)^{26}Al	5.6×10^2	6.3×10^2	1.2×10^3	1.1×10^3
1369	^{24}Mg(p,p′γ)^{24}Mg	9.8×10^3	1.4×10^4	1.1×10^3	2.4×10^5

ray yields of magnesium are presented for angles of 90° and 135° at two proton energies. Note the significant difference in the yield of the 1369 keV γ-ray line obtained by 2.514 MeV protons at 135°.

The angular distribution of the emitted γ-rays is defined as

$$W(\theta) = 1 + a_2P_2(\cos\theta) + a_4P_4(\cos\theta) + \text{higher order terms}$$

where θ is the angle between the incident beam and the sample–detector line. P_2 and P_4 are the Legendre polynomials. In nuclear spectroscopy the angle of 55° is commonly applied. Since $P_2 = \frac{1}{2}(3\cos^2\theta - 1) \sim 0$, when $\theta = 55°$ and as the higher order a_i factors are small, $W(55°) \sim 1$. When performing measurements at the angle of 55° the angular distribution effects can be ignored. Another factor affecting the measurement geometry choice is peak broadening (see Section 8.4.1.3).

8.4.1.2 *Background spectra*

8.4.1.2.1 *Laboratory background* Interfering γ-rays originate from natural radioactivity in the surrounding materials and they can be taken into account by using spectra obtained without the beam. But very often the intensity ratios of γ-ray peaks have to be compared in order to accurately subtract the background contribution. It should be remembered that the background spectra could vary with laboratory. In Table 8.2 typical laboratory background γ-ray lines are listed. An important issue is to determine the laboratory background with the same detector as used for the analyses. The most intense laboratory background peaks in spectra generally are the 511 keV annihilation peak, the 1461 keV peak from ^{41}K, and the γ-rays originating from the ^{232}Th and ^{238}U series, for example, the peak at 2615 keV. Also the bismuth K X-ray peaks are often noticeable. Some radionuclides are produced by cosmic radiation but have too short half-lives to register as naturally occurring radioactivity. Such isotopes in the atmosphere are ^{3}H and ^{14}C which are not γ-active. But in the decay of ^{7}Be, 478 keV γ-rays are emitted. This isotope is produced via the cosmic ray-induced spallation reaction of oxygen and nitrogen. Significant

Table 8.2 The γ-ray lines of a typical laboratory background. The relative intensity values, corrected by detector efficiency, have been taken from Anttila *et al.* (1981)

E_γ (keV)	Relative intensity	Source
239	5	^{212}Pb, ^{214}Pb
352	3.6	^{214}Pb
478	—	^{7}Be
511	—	annihilation
585	5.9	^{208}Tl
609	7.6	^{214}Bi
662	—	^{137}Cs
911	8.4	^{228}Ac
969	5.3	^{228}Ac
1120	4.4	^{214}Bi
1461	100	^{40}K
1765	7.6	^{214}Bi
2204	3.1	^{214}Bi
2448	1.1	^{214}Bi
2615	31.4	^{208}Tl, ^{208}Bi

background also originates from other sources, e.g. activated collimators and other beam-line equipment.

8.4.1.2.2 *Prompt background* Prompt background has two sources: the detector and the experimental set-up. The most important origin of the detector-induced background peaks is the neutrons produced in the sample. The main contribution comes from (p,n) reactions and from (α,n) reactions if α particles are employed. These fast neutrons induce reactions with the Ge isotopes of the detector crystal. For example, significant peaks at 600 and 693 keV arise from the reactions ^{74}Ge(n,n′) and ^{73}Ge(n,γ), and the reaction ^{72}Ge(n,n′), respectively. This background cannot be avoided. The neutron-induced peaks in the γ spectra are significantly broader than normal γ-ray peaks. Also reactions with Fe, Al, Cu and B present in the detector end cap occur and the resulting γ-ray peaks are visible in the measured spectra. The resulting interferences in the spectra depend upon the target material and the ion energy. Even at proton energies below 2 MeV significant neutron yields may result. With deuterons the situation is even more difficult (see Section 8.5.2). A list of neutron-induced γ-ray lines is provided in Bird (1989).

8.4.1.3 *Peak broadening* When γ-rays are emitted while the residual nucleus is moving, a shift in the γ-ray energy is observed. This phenomenon is the Doppler effect and it is significant especially in the case of light target nuclei. The Doppler-shifted energy of γ-rays is determined by the velocity of the recoiling nucleus relative to the observer according to the following equation (up to second order)

$$E_\gamma = E_\gamma^0\,[1 + v/c \cdot \cos\theta + (v/c)^2(\cos^2\theta - 1/2)]$$

where θ is the angle between the beam direction and the sample–detector line, v is the velocity of the excited nucleus and c is the velocity of light in vacuum. E_γ^0 is the energy of the γ-ray if the nucleus decays at rest. The recoil nucleus must decay during flight before it is stopped. In the situation of solid materials the lifetime of the excited state should be between 5×10^{-15} and 2×10^{-12} s in order to cause Doppler broadening. The effect is generally taken advantage of in determining the lifetimes of excited states (the stopping power for the medium is assumed to be known). In the inverse use, assuming the lifetime known, the phenomenon can be applied to stopping power determinations (Arstila, 2000). Typical examples of Doppler-broadened peaks are the 7.115 and 6.916 MeV γ-ray lines of the $^{19}F(p,\alpha\gamma)^{16}O$ reaction, as shown in Fig. 8.3.

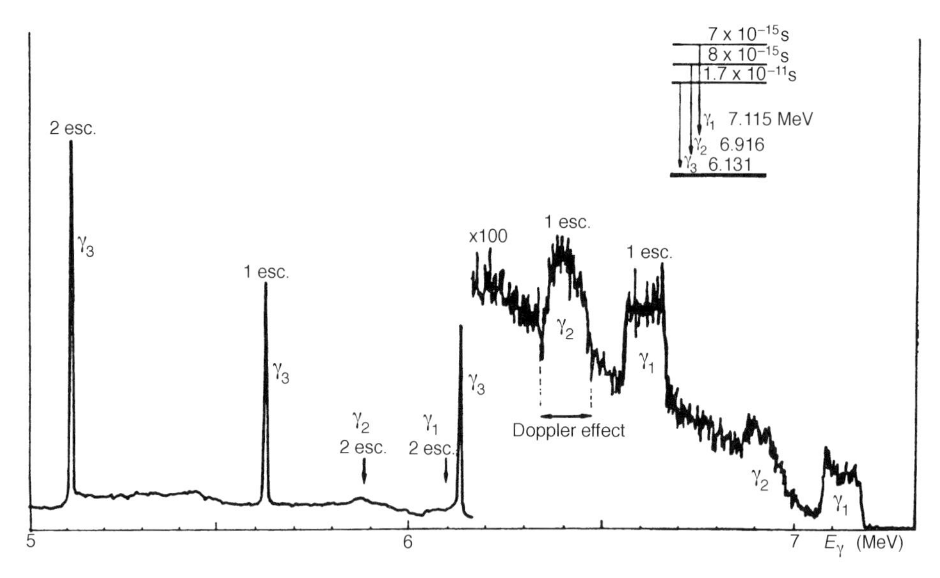

Fig. 8.3 The γ-ray spectrum of fluorine obtained using 2.5 MeV protons. A significant Doppler effect is observed in the 7.115 MeV and 6.916 MeV lines from the reaction $^{19}F(p,\alpha\gamma)^{16}O$. Note the excited state lifetimes (fs). The γ-ray line at 6.129 MeV is not Doppler broadened due to the long lifetime of the excited state (ps). The figure has been reproduced from Stroobants *et al.* (1976), with permission.

The shape of the Doppler-shifted γ-ray peak depends on the lifetime of the excited state and the stopping power of the slowing-down medium. The constant product of the average stopping power and the mean lifetime give similar line-shapes. The effect is most prominent at 0° and not observable at 90°.

In (p,p′γ) and (p,αγ) reactions the recoil nuclei are ejected in all directions with a fixed angular distribution. The resulting γ-ray peaks are broadened and this is the case especially with (p,αγ) reactions.

8.4.1.4 *Energy and absolute efficiency calibration* The energy and absolute efficiency calibration of the detection system can be determined using, for example, ^{60}Co, ^{56}Co and ^{152}Eu sources. To take into account absorption and geometry effects the sources should be placed at the target position. At high energies (> 3.5 MeV), γ-rays from known γ decays of radiative capture reactions (Anttila *et al.* 1977) or an Am–Be source can be used.

8.4.2 *Experimental arrangements*

Slightly varying experimental arrangements for PIGE measurements have been reported in the literature. The trend has been to construct multipurpose chambers for simultaneous PIGE, PIXE and RBS measurements. A typical chamber for such experiments is shown in Fig. 8.4. A lead absorber is often used in front of the detector to reduce the count rate and improve the sensitivity. HPGe (high purity) detectors with good energy resolution are usually applied in the detection of γ-rays. In special cases, when the energy resolution is not of prime importance, the high-efficiency BGO (bismuth–germanate) or NaI(Tl) detectors can be used. When very low concentrations are to be determined, minimal system background is essential. In choosing the construction and lining materials for the experimental set-up, thick target γ-ray yield tables from the literature can be taken advantage of (see Section 8.5).

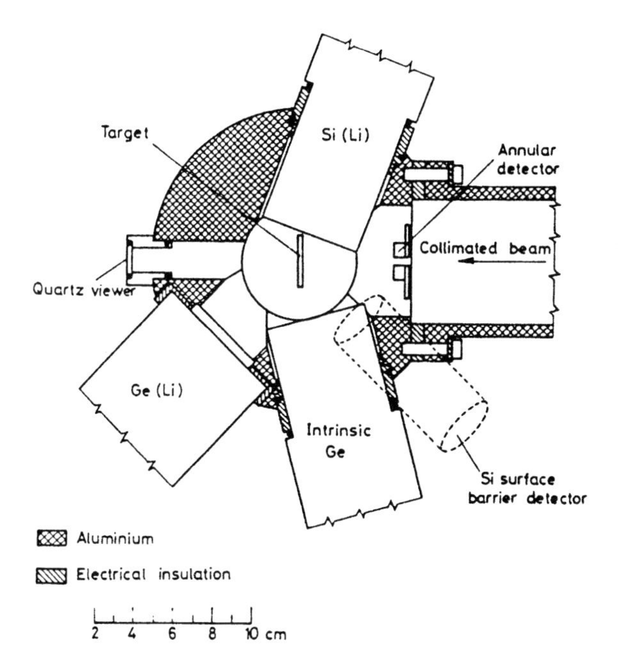

Fig. 8.4 An experimental set-up for simultaneous ion beam analysis by various techniques. The figure has been reproduced from Giles and Peisach (1976), with permission.

In principle there are two experimental arrangement possibilities; in vacuum measurements and external beam measurements.

8.4.2.1 *In vacuum arrangements* In all experimental considerations, careful shielding and lining are needed owing to particle scattering behaviour. Copper is often used as a construction material since it is a good conductor of heat. A drawback of copper though is the contribution of spurious peaks to the measured γ spectra. Tantalum is commonly used for shielding from the scattered particles. With tantalum, low-energy γ-ray lines occur at energies of 136, 165 and 302 keV. Better materials are tin and lead. Tin is ideal when the proton energy is below ~2.5 MeV and lead can be used at higher proton energies. It should be noted that a background spectrum with the beam alone should always be taken. This then takes into account the background induced by scattered ions and any impurities in the shielding and lining materials.

8.4.2.2 *External beams* External-beam PIGE can be applied to the analysis of sensitive materials. In such arrangements care should again be taken with the selection of the construction materials. Low-yield materials should be used in an attempt to obtain increased sensitivity. Unfortunately, the best materials for PIXE are not the best for PIGE. Lead, however, remains a good and easily obtainable material for lining the PIGE set-up. Special attention should be paid also to the exit foil material. Polyimide ($H_{10}C_{22}N_2O_5$) can be used as is commonly done in external-beam PIXE. It yields a low background although some interference may result from the C, N and O. Also exit-foil cooling may be required in some cases. With organic exit foils and ions heavier than protons a cautious approach should be adopted due to the large energy losses involved which may lead to rapid foil rupture. In PIGE, high beam currents are often used and therefore metallic exit foils, such as havar (a high-strength alloy) or nickel, are more convenient. Furthermore, the use of metallic exit foils allows analysis of carbon, nitrogen and oxygen (see Section 8.6.3 for examples).

The γ-ray peaks originating from the exit foil can be used to normalise the beam current (Räisänen 1986). The γ-ray yields are generally sufficiently high for such purposes but not so high that they provide a disruptive background signal. In some laboratories the external-beam PIXE and PIGE methods are used simultaneously. This is practical in some cases, although the price to be paid for the convenience of detecting the elements in a single run is a slight loss in the sensitivity of the methods. This is because of the need for different optimal materials in the PIXE and PIGE set-ups. An external-beam PIGE set-up is described in Fig. 8.5. Facilities that use external-beam PIGE have been previously reviewed in Räisänen (1989).

8.4.2.3 *PIGE with micro-beams and other specialised equipment* The particle beam can be focused down to micrometre sizes by magnetic quadrupoles. Use of such a

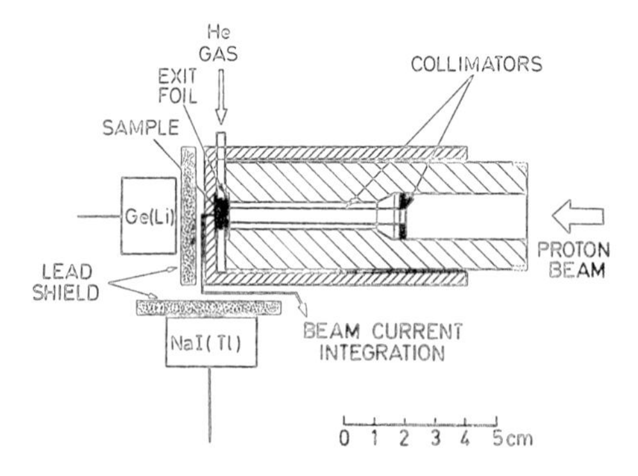

Fig. 8.5 An external beam set-up for PIGE measurements. The figure has been reproduced from Hänninen and Räisänen (1984), with permission.

micro-beam in conjunction with PIGE has been applied in some laboratories to the determination of light elements (Hinrichsen *et al.* 1988; Mosbah & Duraud 1997). The high efficiency of the HPGe and Compton-suppression detector systems was used to advantage in the study of Elekes *et al.* (1999). The better peak-to-background ratio obtained allows sensitive elemental analysis and is an important factor for non-destructive microprobe measurements. In their study, 1.7 and 3.2 MeV protons, with a beam spot of $10 \times 10\ \mu m^2$, were used for the scanning measurements. The detection limits for F, Na, Al and Si were improved by a factor of 2–3 compared with traditional Ge-detector measurements. The technique shows potential for use in geological studies by micro-beam PIGE.

Another improvement in the PIGE technique was introduced by Badica *et al.* (1996) where the measurement of γ–n coincidences was applied. In the determination of the minor elements in steel by 5.5 MeV protons a significant improvement in the sensitivity for certain elements was obtained.

8.5 Particle-induced thick target γ-ray yields and bombarding particle choice

Systematic thick target γ-ray yields have been tabulated in the literature for protons, deuterons and tritons, and for ^{4}He, ^{7}Li, ^{12}C, ^{14}N, ^{16}O and ^{55}Cl ions. The available references are provided in Table 8.3. The yields for low-energy (< 300 keV) γ-rays tend to be more inaccurate due to self-absorption in the target material. The very low energy lines are usually excluded because of absorption effects. When targets are not available in elemental form compounds have been applied. For such elements the results have higher error limits owing to the stopping power correction needed to correspond with a pure elemental target. Generally, accurate γ-ray yields are difficult to obtain because of inaccurate charge collection (secondary electrons, insulators), errors in detection efficiency and, in the case of compound targets, the stopping power correction. When the literature yield values are used therefore, only informative results can be secured. For accurate analyses, standards are needed.

In the sections that follow, various particle-induced γ-ray yields are reviewed. As it is often equally important to have a knowledge of the elements possessing low γ-ray yields, the end of each section provides a listing of suggested low-yield elements for use in backings and construction. Original studies are cited and considerations of the detection sensitivities for elemental analyses are discussed. In situations when different isotopes can be detected, the relevant reactions are provided even if the yields for them are not high.

8.5.1 *Proton-induced γ-ray yields*

In most PIGE studies protons have been used as bombarding particles. In the following the lightweight elements are discussed keeping in mind their analyses by the PIGE method. The analysis of hydrogen has been presented elsewhere in this book in connection with nuclear reaction analysis (Chapter 9). The γ-ray lines relevant to elemental analysis are given in Table 8.4 where the corresponding nuclear reactions

Table 8.3 Systematic particle-induced γ-ray yields available in the literature. Only the references where more than one element has been studied are included

Bombarding particle	Energy (MeV)	Measurement angle	Elements or range of elements	Reference
Protons	2 and 2.5	135°	Li–Cl	Bird *et al.* (1978)
	1.2–2.9	90°	Rh, Pd, Ag, Pt, Au	Deconninck & Demortier (1975)
	0.6–3.2	90°	Ti–Zn	Demortier (1978)
	2 and 2.5	135°	F–Si, Ti–Zn, Mo, Ag, Au	Kenny *et al.* (1980)
	1, 1.7, 2.4	55°	Li–Sc	Anttila *et al.* (1981)
	1.7 and 2.4	55°	Ga–Pb	Räisänen & Hänninen (1983)
	2.4–4.2	55°	Li–Sc	Kiss *et al.* (1985)
	7 and 9	55°	Li–Pb	Räisänen *et al.* (1987)
	0.4–1.5	90°	Li–F	Golicheff *et al.* (1972)
	4.5	45°	F–Co	Gihwala & Peisach (1982)
	1.0–4.1	90°	Li, B, F–P	Savidou *et al.* (1999)
Deuterons	0.7–3.4	135°	Li–Ca	Kiss *et al.* (1994)
Tritons	2, 3 and 3.5	90°	Li, O–Mg, Si–Au	Borderie & Barrandon (1978)
^{4}He ions	2.4	55°	Li–Si	Lappalainen *et al.* (1983)
	3.5	90°	Li–Na, Ti–Fe	Borderie & Barrandon (1978)
	5	90°	Li–Bi	Giles & Peisach (1979)
^{7}Li ions	12 and 18	55°	Li–Pb	Räisänen (1990)
^{12}C ions	22 and 28	55°	Li–Ca, Ti–Zn	Seppälä *et al.* (1998)
^{14}N ions	28	55°	Li–F	Seppälä *et al.* (1998)
^{16}O ions	28 and 33	55°	Li–F	Seppälä *et al.* (1998)
^{35}Cl ions	55	90°	Li, B, F–Si, Ti–Th	Borderie *et al.* (1979)

are also indicated. In many situations, peak overlapping occurs and this should also be taken into account.

Lithium: has two stable isotopes, ^{6}Li (7.5%) and ^{7}Li (92.5%).

^{7}Li can be easily detected using the 478 keV γ-ray line, which is slightly Doppler broadened. For ^{6}Li analysis no usable γ-ray emitting reactions exist. When analysing samples with high lithium concentrations it should be kept in mind that neutrons are easily emitted.

Beryllium: has only one stable isotope, ^{9}Be.

The analysis of beryllium with protons is not as sensitive as that of lithium. With α particles a better sensitivity for beryllium is obtained. The most intense γ-ray line is

Table 8.4 Most useful proton-induced reactions for the elemental analysis of light elements ($Z \leq 30$)

Element	E_γ (keV)	Reaction	Remarks
Li	429	$^{7}Li(p,n\gamma)^{7}Be$	E_p should be > 3 MeV
	478	$^{7}Li(p,p'\gamma)^{7}Li$	Provides best sensitivity
Be	3562	$^{9}Be(p,\alpha\gamma)^{6}Li$	
B	429	$^{10}B(p,\alpha\gamma)^{7}Be$	
	718	$^{10}B(p,p'\gamma)^{10}B$	
	2125	$^{11}B(p,p'\gamma)^{11}B$	For ^{11}B detection at E_p > 3 MeV
C	3089	$^{13}C(p,p'\gamma)^{13}C$	For ^{13}C detection at E_p > 3.5 MeV
	4439	$^{12}C(p,p'\gamma)^{12}C$	For ^{12}C detection at E_p > 5.4 MeV
N	2313	$^{14}N(p,p'\gamma)^{14}N$	E_p should be > 4 MeV
	4439	$^{15}N(p,\alpha\gamma)^{12}C$	For ^{15}N detection
O	495	$^{16}O(p,\gamma)^{17}F$	For ^{16}O detection at low energies
	871	$^{17}O(p,p'\gamma)^{17}O$	For ^{17}O detection
	1982	$^{18}O(p,p'\gamma)^{18}O$	For ^{18}O detection
	6129	$^{16}O(p,p'\gamma)^{16}O$	E_p should be > 7.5 MeV
F	110	$^{19}F(p,p'\gamma)^{19}F$	Absorption problems possible
	197	$^{19}F(p,p'\gamma)^{19}F$	Absorption problems possible
	6129	$^{19}F(p,\alpha\gamma)^{16}O$	
Na	440	$^{23}Na(p,p'\gamma)^{23}Na$	
	1634	$^{23}Na(p,\alpha\gamma)^{20}Ne$	Overlapping with 1636 keV γ-line
	1636	$^{23}(Na(p,p'\gamma)^{23}Na$	
Mg	390	$^{25}Mg(p,p'\gamma)^{25}Mg$	
	585	$^{25}Mg(p,p'\gamma)^{25}Mg$	
	975	$^{25}Mg(p,p'\gamma)^{25}Mg$	
	1369	$^{24}Mg(p,p'\gamma)^{24}Mg$	For ^{24}Mg detection
	1612	$^{25}Mg(p,p'\gamma)^{25}Mg$	E_p should be > 3 MeV
	1809	$^{26}Mg(p,p'\gamma)^{26}Mg$	For ^{26}Mg detection at E_p > 3 MeV
Al	844	$^{27}Al(p,p'\gamma)^{27}Al$	
	1014	$^{27}Al(p,p'\gamma)^{27}Al$	
	1369	$^{27}Al(p,\alpha\gamma)^{24}Mg$	
	1779	$^{27}Al(p,\gamma)^{28}Si$	Only at E_p < 3 MeV
Si	755	$^{29}Si(p,p'\gamma)^{29}Si$	E_p should be > 3 MeV
	1273	$^{29}Si(p,p'\gamma)^{29}Si$	
	1779	$^{28}Si(p,p'\gamma)^{28}Si$	For ^{28}Si detection
	2233	$^{30}Si(p,\gamma)^{31}P$	For ^{30}Si detection
	2235	$^{30}Si(p,p'\gamma)^{30}Si$	Overlapping with 2233 keV γ-line
P	1266	$^{31}P(p,p'\gamma)^{31}P$	
	1779	$^{31}P(p,\alpha\gamma)^{28}Si$	
	2230	$^{31}P(p,\gamma)^{32}S$	Overlapping with 2233 keV γ-line
	2233	$^{31}P(p,p'\gamma)^{31}P$	
S	841	$^{33}S(p,p'\gamma)^{33}S$	For ^{33}S detection
	1219	$^{34}S(p,\gamma)^{35}Cl$	For ^{34}S detection
	2230	$^{32}S(p,p'\gamma)^{32}S$	E_p should be > 4.9 MeV
Cl	1219	$^{35}Cl(p,p'\gamma)^{35}Cl$	
	1410	$^{37}Cl(p,n\gamma)^{37}Ar$	E_p should be > 3.5 MeV
	1611	$^{37}Cl(p,n\gamma)^{37}Ar$	E_p should be > 3.5 MeV
	1763	$^{35}Cl(p,p'\gamma)^{35}Cl$	
	2126	$^{37}Cl(p,\alpha\gamma)^{38}Ar$	

Contd

Table 8.4 *Contd*

Element	E_γ (keV)	Reaction	Remarks
K	1294	$^{41}K(p,p'\gamma)^{41}K$	E_p should be > 3.5 MeV
	1943	$^{41}K(p,n\gamma)^{41}Ca$	E_p should be > 3.5 MeV
	2010	$^{41}K(p,n\gamma)^{41}Ca$	E_p should be > 3.5 MeV
	2168	$^{41}K(p,\alpha\gamma)^{38}Ar$	
	2522	$^{39}K(p,n\gamma)^{41}Ca$	For ^{39}K detection at E_p > 4 MeV
Ca	371	$^{48}Ca(p,n\gamma)^{48}Sc$	For ^{48}Ca detection
	373	$^{43}Ca(p,p'\gamma)^{43}Ca$	For ^{43}Ca detection
	1157	$^{44}Ca(p,p'\gamma)^{44}Ca$	For ^{44}Ca detection
	1525	$^{42}Ca(p,p'\gamma)^{42}Ca$	For ^{42}Ca detection
Sc	364	$^{45}Sc(p,p'\gamma)^{45}Sc$	
	719	$^{45}Sc(p,p'\gamma)^{45}Sc$	
Ti	308	$^{47}Ti(p,\gamma)^{48}V$	For ^{47}Ti detection
		$^{48}Ti(p,n\gamma)^{48}V$	
	889	$^{46}Ti(p,p'\gamma)^{46}Ti$	For ^{46}Ti detection
	983	$^{48}Ti(p,p'\gamma)^{48}Ti$	For ^{48}Ti detection
V	319	$^{51}V(p,p'\gamma)^{51}V$	
	749	$^{51}V(p,n\gamma)^{51}Cr$	
	808	$^{51}V(p,n\gamma)^{51}Cr$	Useful at E_p > 3.5 MeV
Cr	379	$^{52}Cr(p,\gamma)^{53}Mn$	Strongest line from Cr
	565	$^{53}Cr(p,p'\gamma)^{53}Cr$	For ^{53}Cr detection
	783	$^{50}Cr(p,p'\gamma)^{50}Cr$	For ^{50}Cr detection
Mn	411	$^{55}Mn(p,n\gamma)^{55}Fe$	
	931	$^{55}Mn(p,n\gamma)^{55}Fe$	Strongest line from Mn
Fe	847	$^{56}Fe(p,p'\gamma)^{56}Fe$	Strongest line from Fe
	1408	$^{54}Fe(p,p'\gamma)^{54}Fe$	For ^{54}Fe detection at E_p > 7 MeV
Co	339	$^{59}Co(p,n\gamma)^{59}Ni$	
	1332	$^{59}Co(p,\gamma)^{60}Ni$	
Ni	1172	$^{62}Ni(p,p'\gamma)^{62}Ni$	For ^{62}Ni detection at E_p > 3.5 MeV
	1332	$^{60}Ni(p,p'\gamma)^{60}Ni$	Strongest line from Ni
	1454	$^{58}Ni(p,p'\gamma)^{58}Ni$	For ^{58}Ni detection at E_p > 3.5 MeV
Cu	962	$^{63}Cu(p,p'\gamma)^{63}Cu$	For E_p > 3 MeV
	992	$^{63}Cu(p,\gamma)^{64}Zn$	Best at E_p < 3 MeV
	865	$^{65}Cu(p,n\gamma)^{65}Zn$	For ^{65}Cu detection at E_p > 3.5 MeV
Zn	360	$^{67}Zn(p,n\gamma)^{67}Ga$	For ^{67}Zn detection
	508	$^{70}Zn(p,n\gamma)^{70}Ga$	For ^{70}Zn detection at E_p > 3.5 MeV
	874	$^{68}Zn(p,p'\gamma)^{68}Zn$	For ^{68}Zn detection
	992	$^{64}Zn(p,p'\gamma)^{64}Zn$	For ^{64}Zn detection
	1039	$^{66}Zn(p,p'\gamma)^{66}Zn$	For ^{66}Zn detection

at 3562 keV, which is also Doppler broadened. High neutron yields are obtained for samples with elevated Be concentrations.

Boron: has two stable isotopes, ^{10}B (19.9%) and ^{11}B (80.1%).

Boron can be analysed easily with good sensitivity. The most intense γ-ray line for ^{10}B is at 429 keV, which is Doppler broadened and is thus easy to distinguish. A significant sodium concentration in the sample will cause interference with boron analysis. At energies above 3 MeV the γ-ray line at 718 keV of the reaction

$^{10}B(p,p'\gamma)^{10}B$ is recommended. Boron isotope ^{11}B can be detected at proton energies above 3 MeV by the 2125 keV γ-ray line of the reaction $^{11}B(p,p'\gamma)^{11}B$.

Carbon: has two stable isotopes, ^{12}C (98.9%) and ^{13}C(1.1%).
Detection of carbon at low proton energies is not sensitive. Good sensitivity for ^{12}C is obtained at higher bombarding energies employing the reaction $^{12}C(p,p'\gamma)$ ^{12}C and the 4439 keV line. At energies below the strong 5.370 MeV resonance the sensitivity is poor and above about 8 MeV the background usually increases significantly. The optimum energy region is thus between 7.5 and 8.0 MeV. In some cases nitrogen can cause interference (see below). ^{13}C can be determined with moderate sensitivity at low bombarding energies.

Nitrogen: has two stable isotopes, ^{14}N (99.63%) and ^{15}N (0.37%).
The most significant γ-ray line originating from ^{15}N at low proton energies is the 4439 keV line due to the reaction $^{15}N(p,\alpha\gamma)^{12}C$. The Doppler-broadened peak in the spectrum can be easily distinguished, but the sensitivity obtained is moderate. The optimum proton energy is close to 2 MeV (Hänninen & Räisänen 1984) and the obtained detection limit is about 0.2 wt%. At higher proton energies the sensitivity for nitrogen analysis can be increased significantly using the reaction $^{14}N(p,p'\gamma)^{14})N$ and the 2313 keV line. Owing to the strong resonance at 3.903 MeV the optimum energy for this reaction is at 4.0–4.5 MeV (Hänninen & Räisänen 1984). At higher energies the neutron yield increases which results in a higher background and poorer sensitivity.

Oxygen: has three stable isotopes, ^{16}O (99.76%), ^{17}O (0.038%) and ^{18}O (0.20%).
Unfortunately the analysis of oxygen in general by PIGE is not sensitive. In principle the various isotopes can be determined, but not with good sensitivity. At low bombarding energies ^{16}O can be detected by the 495 keV γ-rays resulting from the reaction $^{16}O(p,\gamma)^{17}F$. A significantly better sensitivity for ^{16}O detection is obtained at energies above 7 MeV using the reaction $^{16}O(p,p'\gamma)^{16}O$ and the 6129 keV line. The optimum energy region for this reaction is at about 7.5–8.5 MeV. Possible interference due to sample fluorine and the resulting γ-rays from the reaction $^{19}F(p,\alpha\gamma)^{16}O$ should be properly taken into account.

Fluorine: has only one stable isotope, ^{19}F.
Fluorine detection can be carried out by using either the 6129 keV γ-rays from the reaction $^{19}F(p,\alpha\gamma)^{16}O$ or the 110 and 197 keV γ-rays from the reaction $^{19}F(p,p'\gamma)^{19}F$. The 110 and 197 keV γ-rays offer a better sensitivity (by a factor of about 5; see Shroy *et al.* 1978), but the detector should face the front side of the sample to avoid absorption effects. This is a disadvantage, since for multielemental analysis, an absorber in front of the detector is often needed to reduce the proportion of low-energy γ-rays. The detector solid angle is also smaller compared with the situation when the detector is positioned immediately behind the sample.

Sodium: has one stable isotope, ^{23}Na.
A good sensitivity for sodium analysis is obtained by detecting the 440 keV γ-rays. The γ-ray line at 1636 keV is somewhat Doppler broadened.

Magnesium: has three stable isotopes, ^{24}Mg (78.99%), ^{25}Mg (10.00%) and ^{26}Mg (11.01%).

Magnesium can be analysed with a relatively good sensitivity. At low proton energies the most prominent lines are at 390, 585, 975 (from ^{25}Mg) and 1369 keV (from ^{24}Mg). Unfortunately the 585 keV line has interference owing to background from ^{208}Tl, which disturbs the signal during long measurements and when the magnesium concentration is low. It should be noted that the 1369 keV line also originates from aluminium. In principle with proton energies above 3 MeV all magnesium isotopes can be detected.

Aluminium: has one stable isotope, ^{27}Al.

Aluminium has several strong γ-ray lines at 844, 1014, 1369 and 1779 keV. The 1369 keV line may incorporate interference from sample magnesium. The 1779 keV line may contain interference from sample silicon and phosphorus. Also, the 844 and 1014 keV lines, which are most suitable for aluminium analysis, have interference from magnesium, ^{26}Mg(p,γ)^{27}Al, but the magnesium yields are fortunately rather low.

Silicon: has three stable isotopes, ^{28}Si (92.23%), ^{29}Si (4.67%) and ^{30}Si (3.10%).

Silicon can be analysed by PIGE with moderate sensitivity. The γ-ray line at 1779 keV, ^{28}Si(p,p′γ)^{28}Si, has interference resulting from reactions ^{27}Al(p,γ)^{28}Si and ^{31}P(p,$\alpha\gamma$)^{28}Si. The 1273 keV line is the most useful for silicon analysis. All silicon isotopes can be determined employing protons of over 3 MeV energy.

Phosphorus: has one stable isotope, ^{31}P.

In the analysis of phosphorus several interfering and overlapping γ-ray lines and reaction channels occur. The 1266 keV line, ^{31}P(p,p′γ)^{31}P, overlaps with the lines originating from silicon and sulphur, namely ^{30}Si(p,γ)^{31}P and ^{34}S(p,$\alpha\gamma$)^{31}P, respectively. The 2233 keV γ-ray line does not encounter any overlapping and is the most safe to use. Unfortunately it has a rather low yield.

Sulphur: has four stable isotopes, ^{32}S (95.02%), ^{33}S (0.75%), ^{34}S (4.21%) and ^{36}S (0.002%).

Sulphur can be detected by employing the reaction ^{32}S(p,p′γ)^{32}S (E_γ = 2230 keV) at the level of 100 ppm (by weight) from thick organic samples and at the 100 ng cm^{-2} level from aerosol samples (Räisänen & Lapatto 1988). The optimum bombarding energy is about 4.9 MeV. The reaction shows two strong resonances at energies 4.77 and 5.10 MeV. At low bombarding energies the analysis of sulphur by PIGE is unpractical. When analysing sulphur care should be taken to compensate for the possible overlapping γ-ray peaks in the spectrum. Such peaks are possible from the elements phosphorus [2230 keV ^{31}P(p,γ)^{32}S; 2233 keV ^{31}P(p,p′γ)^{31}P], silicon [2233 keV ^{30}Si(p,γ)^{31}P; 2235 keV ^{30}Si(p,p′γ)^{30}Si] and chlorine [2230 keV ^{35}Cl(p,$\alpha\gamma$)^{32}S]. Fortunately, these elements have more dominating lines at other energies, which can be used for checking their presence in the samples analysed. For sulphur analysis PIXE is more sensitive, but due to overlapping X-ray peaks and absorption effects, PIGE is often a useful complementary method to PIXE.

Chlorine: has two stable isotopes, ^{35}Cl (75.77%) and ^{37}Cl (24.23%).

The strongest line from chlorine (at proton energies above 3 MeV) is the 1219 keV line which is due to the reaction $^{35}Cl(p,p'\gamma)^{35}Cl$. The same γ-rays originate also from the reaction $^{34}S(p,\gamma)^{35}Cl$. The 1763 keV line does not have any overlapping.

Potassium: has three stable isotopes, ^{39}K (93.2581%), ^{40}K (0.0117%) and ^{41}K (6.7302%).

The analysis of potassium is feasible using the 2168 keV γ-rays. In general the PIGE method is not sensitive for potassium analyses. It should be noted that neutrons are emitted from potassium samples even at low proton energies.

Calcium: has six stable isotopes, ^{40}Ca (96.941%), ^{42}Ca (0.647%), ^{43}Ca (0.135%), ^{44}Ca (2.086%), ^{46}Ca (0.004%) and ^{48}Ca (0.187%).

The most intense γ-ray line at low bombarding energies is at 373 keV. Otherwise the same remarks are valid here as for potassium, including neutron emission at low proton energies.

Scandium: has one stable isotope, ^{45}Sc.

Several γ-ray lines originate from scandium. As the presence of scandium in ordinary samples is generally very rare its analyses are not significant.

Titanium: has five stable isotopes, ^{46}Ti (8.0%), ^{47}Ti (7.3%), ^{48}Ti (73.8%), ^{49}Ti (5.5%) and ^{50}Ti (5.4%).

In some applications (for example in the analysis of painting pigments) it is useful to determine the transition elements by PIGE. Titanium can be analysed by taking advantage of the 983 keV γ-ray line.

Vanadium: has two stable isotopes, ^{50}V (0.25%) and ^{51}V (99.75%).

Vanadium can be analysed by employing the γ-ray lines at 319 and 749 keV. The 319 keV γ-ray line overlaps with a line originating from zinc.

Chromium: has four stable isotopes, ^{50}Cr (4.345%), ^{52}Cr (83.79%), ^{53}Cr (9.50%) and ^{54}Cr (2.365%).

Chromium has a prominent line at 379 keV. For the analysis of the different isotopes, refer to Table 8.4.

Manganese: has one stable isotope, ^{55}Mn.

Manganese has a significant line at 931 keV. At the bombarding energy of 5.5 MeV the line at 1408 keV is also useful for analytical purposes (Badica *et al.* 1996).

Iron: has four stable isotopes, ^{54}Fe (5.9%), ^{56}Fe (91.72%), ^{57}Fe (2.1%) and ^{58}Fe (0.28%).

Iron has a significant line at 847 keV that can be used for practical purposes. It is often found as a background peak in PIGE spectra because of scattering from stainless steel materials. Unfortunately, the 847 keV line often overlaps with the strong 844 keV line derived from aluminium.

Cobalt: has one stable isotope, ^{59}Co.

Cobalt has two distinct lines at 339 and 1332 keV. The overlapping of the latter with a nickel signal should be noted.

Nickel: has five stable isotopes, ^{58}Ni (68.077%), ^{60}Ni (26.233%), ^{61}Ni (1.140%), ^{62}Ni (3.634%) and ^{64}Ni (0.926%).
The strongest γ-ray line that is useful for nickel analysis is at 1332 keV. Possible overlapping with cobalt should be noted.

Copper: has two stable isotopes, ^{63}Cu (69.17%) and ^{65}Cu (30.83%).
At low proton energies copper can be best detected using the 992 keV γ-ray line and at high energies by the 962 keV line.

Zinc: has five stable isotopes, ^{64}Zn (48.6%), ^{66}Zn (27.9%), ^{67}Zn (4.1%), ^{68}Zn (18.8%) and ^{70}Zn (0.6%).
Zinc has a prominent γ-ray line at 360 keV. The various isotopes can be determined with moderate sensitivity (see Table 8.4).

When analysing thin targets the resonance structures of the excitation curves should be taken into account. In Boni *et al.* (1988) useful thin target excitation curves for PIGE analysis of Li, B, F, Mg, Al, Si and P are given. The data are provided for proton energies from 2.2 to 3.8 MeV at a measurement angle of 90°.

To acquire quantitative information on the method sensitivity of PIGE, typical detection limits obtained under practical conditions (~30 min measurement) for biomedical and organic samples are provided for selected elements in Table 8.5. Sensitivity is defined, in this context, as the minimum detectable concentration. The minimum detectable peak is assumed to be three times the square root of the background at FWHM (full width at half maximum) of the peak.

For Li, B, C, N, O and S, the E_p values referred to are the optimum proton energies for the analyses, and the detection limits stated are the best values obtainable by the

Table 8.5 Detection limits for some elements detectable in a typical biomedical and organic sample by external-beam PIGE. The values have been taken from Räisänen (1987, 1989) and Räisänen and Lapatto (1988)

Element	E_p (MeV)[a]	Reaction	E_γ (keV)	Detection limit (ppm by weight)
Li	1.8	^{7}Li(p,p′γ)^{7}Li	478	0.15
B	1.8	^{10}B(p,αγ)^{7}Be	429	0.18
C	7.6	^{12}C(p,p′γ)^{12}C	4439	100
N	4.1	^{14}N(p,p′γ)^{14}N	2313	10
O	7.6	^{16}O(p,p′γ)^{16}O	6129	300
F	2.4	^{19}F(p,αγ)^{16}O	6129	0.5
Na	2.4	^{23}Na(p,p′γ)^{23}Na	440	0.3
Mg	3.0	^{25}Mg(p,p′γ)^{25}Mg	585	15
Al[b]	3.1	^{27}Al(p,p′γ)^{27}Al	1014	40
P	3.0	^{31}P(p,p′γ)^{31}P	1266	15
Ca	2.4	^{40}Ca(p,p′γ)^{40}Ca	3736	4000
S	5.0	^{32}S(p,p′γ)^{32}S	2230	100
Cl	3.0	^{35}Cl(p,p′γ)^{35}Cl	1219	100

[a] Energy on the exit foil.
[b] Sample is ash-diluted in graphite and the values are from an in vacuum measurement (Boni *et al.* 1990).

technique. As can be noted, the optimum proton energies for the analysis of all elements do not coincide. The detection limit values depend strongly on the sample composition, especially on the concentrations of the elements for which the method detection sensitivity is good (e.g. Li, B, Na and F).

8.5.1.1 *Low-yield elements* The γ-ray yields of the heavy elements are generally rather low. Several elements find use as low-background materials for apparatus construction purposes. With proton energies below 3 MeV the best low-yield elements are tin (no γ-ray lines) and lead (though the disadvantage of lead is the intense Pb K X-ray yield, otherwise no γ-ray lines). Tantalum is also a good candidate, although the rather strong γ-ray lines at 136, 165 and 302 keV increase slightly the background signal during analyses.

8.5.2 *Deuteron- and triton-induced γ-ray yields*

8.5.2.1 *Deuterons* One systematic study of deuteron-induced thick target γ-ray yields has been presented (Kiss *et al.* 1994). The studied elements included those of $Z = 3–19$ in the energy region of 0.7–3.4 MeV. In this work comparison with the corresponding proton bombardment is also provided. In employing deuteron-induced γ-ray emission (DIGE), the possibility of an associated neutron hazard should be kept in mind. This fact demands, for instance, careful beam alignment.

The most intense γ-ray lines of the light elements following 1.8 MeV deuteron bombardment, and the nuclear reactions involved, are listed in Table 8.6. The selected energy of 1.8 MeV is ideal for analysing samples containing high oxygen concentrations as the energy is just below the 1.828 MeV neutron threshold of the reaction $^{16}O(d,n)^{17}F$. At this energy the best peak-to-background ratios are obtained (Kiss *et al.* 1994).

The detection limits for the transition elements in steel when bombarding with 5 MeV deuterons have been determined by Peisach *et al.* (1994). The reactions yielding the best detection limits for the transition elements are provided in Table 8.7. The values correspond to a collected charge of 1 mC.

The total cross-section of the $^{14}N(d,p\gamma)^{15}N$ nuclear reaction for analytical applications has been determined recently in the energy region of 500–1500 keV (van Bebber *et al.* 1998). The presented data are useful for nitrogen analysis in thin films. An example of its use for nitrogen content determination in an ultra-thin silicon oxynitride film is provided. The high-precision detection of ^{16}O in high T_C superconductors by 1.8 MeV deuterons has been presented by Vickridge *et al.* (1994). Oxygen was determined by employing the reaction $^{16}O(d,p\gamma)^{17}O$ and detecting the prompt 871 keV γ-rays emitted from the first excited state of ^{17}O. The uncertainty obtained for the analyses was less than 1%.

The DIGE method is sensitive for C, N and O determinations by 1.8 MeV deuterons. It is also rather sensitive in the detection of Na and Si. If proton energies above 4 MeV are not available, DIGE can also be used for S, Cl and K analyses. The detection limits obtainable depend on the sample and the background caused by neutron-induced γ-rays in the detector.

Table 8.6 The most intense γ-ray lines of the light elements induced by 1.8 MeV deuterons. The values have been compiled from Kiss *et al.* (1994)

Element	E_γ (keV)	Reaction	Remarks
Li	478	^{6}Li(d,pγ)^{7}Li	
Be	718	^{9}Be(d,n)^{10}B	
B	953 1674	^{11}B(d,pγ)^{12}B ^{11}B(d,pγ)^{12}B	
C	3089	^{12}C(d,pγ)^{13}C	The line is Doppler broadened
N	7301 8313	^{14}N(d,pγ)^{15}N ^{14}N(d,pγ)^{15}N	The γ-ray energies are above most lines induced from other elements
O	871	^{16}O(d,pγ)^{17}O	Presence of fluorine must be taken into account. The yield ratios should be applied in such cases
F	871 1057	^{19}F(d,α)^{17}O ^{19}F(d,pγ)^{20}F	Yield from F is lower than from O
Na	472	^{23}Na(d,pγ)^{24}Na	
Mg	585 1809	^{24}Mg(d,pγ)^{25}Mg ^{25}Mg(d,pγ)^{26}Mg	Overlapping from aluminium
Al	585 975 983	^{27}Al(d,αγ)^{25}Mg ^{27}Al(d,αγ)^{25}Mg ^{27}Al(d,pγ)^{28}Al	Al yield is lower than Mg yield The sum of the 975 keV and 983 keV γ yields is given in Kiss *et al.* (1994)
Si	4935	^{28}Si(d,pγ)^{29}Si	
S	841	^{32}S(d,pγ)^{33}S	
Cl	1970 2168	^{35}Cl(d,nγ)^{36}Ar ^{37}Cl(d,nγ)^{38}Ar	
K	770	^{39}K(d,nγ)^{40}Ca	

Table 8.7 Detection limits for the transition elements in steel samples using 5 MeV deuterons. The values are from Peisach *et al.* (1994). Reproduced with permission

Element	E_γ (keV)	Reaction	Detection limit (mg g^{-1})
Ti	153	^{48}Ti(d,nγ)^{49}V	0.51
V	1434	^{51}V(d,nγ)^{52}Cr	0.04
Cr	379	^{52}Cr(d,nγ)^{53}Mn	0.21
Mn	847	^{55}Mn(d,nγ)^{56}Fe	0.11
Fe	352	^{56}Fe(d,pγ)^{57}Fe	0.35
Co	1332	^{59}Co(d,nγ)^{60}Ni	0.12
Ni	339	^{58}Ni(d,pγ)^{59}Ni	0.26
Cu	992	^{63}Cu(d,nγ)^{64}Zn	0.20
Zn	360	^{66}Zn(d,nγ)^{67}Ga	1.30

8.5.2.1.1 *Low-yield elements* The elements Fe, Ni, Cu, Zn, Sn, Ta, W and Pb can be recommended as low-yield backing or construction materials when deuteron energies are below 1.8 MeV (Kiss *et al.* 1994). At energies above 3 MeV only Ta, W and Pb should be applied. As pointed out by Kiss *et al.* (1994), the use of aluminium in measuring chambers (Al is frequently used in set-ups constructed for simultaneous analyses by PIXE) should be avoided if deuterons are employed.

8.5.2.2 *Tritons* To the knowledge of the author, only one systematic study concerned with triton-induced thick target γ-ray yields can be found in the literature (Borderie & Barrandon 1978). In that study 2.0, 3.0 and 3.5 MeV tritons were employed for selected elements. The best sensitivity values for relevant elements are collected in Table 8.8.

Table 8.8 Sensitivity values obtained by triton analysis of a niobium sample. The data have been collected from Borderie and Barrandon (1978) and Borderie (1980)

Element	Reaction	Energy (MeV)	E_γ (keV)	Detection limit (ppm)
Li	$^{7}Li(t,t'\gamma)^{7}Li$	3	478	2
C	$^{12}C(t,n\gamma)^{14}N$	3	1632	5
O	$^{16}O(t,p\gamma)^{18}O$	3	1982	1
F	$^{19}F(t,p\gamma)^{21}F$	3	280	5
Na	$^{23}Na(t,\alpha\gamma)^{22}Ne$	2	1275	30
Mg	$^{24}Mg(t,n\gamma)^{26}Al$	3	417	5
Si	$^{28}Si(t,n\gamma)^{30}P$	3	709	20
S	$^{31}S(t,n\gamma)^{34}Cl$	3	461	15
Ti	$^{48}Ti(t,n\gamma)^{50}V$	3.5	226	2
V	$^{51}V(t,d\gamma)^{52}V$	3	124	50
Cr	$^{52}Cr(t,n\gamma)^{54}Mn$	3.5	157	15
Mn	$^{55}Mn(t,\alpha\gamma)^{54}Cr$	3.5	835	15
Fe	$^{56}Fe(t,n\gamma)^{58}Co$	3.5	116	55
Co	$^{59}Co(t,d\gamma)^{60}Co$	3.5	945	220
Ni	$^{58}Ni(t,d\gamma)^{59}Ni$	3.5	460	290
Cu	$^{63}Cu(t,n\gamma)^{65}Zn$	3.5	115	260
Zn	$^{66}Zn(t,n\gamma)^{68}Ga$	3.5	175	350

8.5.2.2.1 *Low-yield elements* The best low-yield materials are the heavy elements, e.g. Mo, Pt and Au can be recommended.

8.5.3 *α particle-induced γ-ray yields*

As was noted in Table 8.3, three systematic γ-ray yield studies exist for α particles (Borderie & Barrandon 1978; Giles & Peisach 1979; Lappalainen *et al.* 1983). The light-element γ-ray yields induced by 2.4 MeV α particles and the sensitivities obtained for the analyses of a volcanic stone are provided in Lappalainen *et al.* (1983). A selection of the highest γ-ray yields induced by 3.5 MeV α particles is given in Borderie & Barrandon (1978). Also, the detection limits obtained for a pure niobium sample with an irradiation time of one hour are stated. In Constantinescu *et al.* (1986), 4.7 MeV α particles were employed in the analysis of zeolite and geochemical reference samples. The sensitivity values obtained in these studies are collected in Table 8.9. As can be observed, for several elements the γ-ray energies involved are rather low. This can cause problems with absorption.

Table 8.9 Detection limits obtained by 2.4 MeV (volcanic stone sample), 3.5 MeV (Nb sample) and 4.7 MeV (zeolite and geochemical samples) α-particle bombardment. The values have been collated from Borderie and Barrandon (1978), Lappalainen *et al.* (1983) and Constantinescu *et al.* (1986)

			Detection limit (ppm)		
Element	Reaction	E_γ (keV)	Volcanic stone	Nb sample	Zeolite
Li	$^{7}Li(\alpha,\alpha'\gamma)^{7}Li$	478	0.45	0.165	
Be	$^{9}Be(\alpha,n\gamma)^{12}C$	4439	0.42	0.5	
B	$^{10}B(\alpha,p\gamma)^{13}C$	169		15	
		3854	10		
N	$^{14}N(\alpha,\gamma)^{18}F$	4525	1%		
O	$^{18}O(\alpha,n\gamma)^{21}Ne$	350	2%	875	0.5%
F	$^{19}F(\alpha,\alpha'\gamma)^{19}F$	197	22	10	250
Na	$^{23}Na(\alpha,\alpha'\gamma)^{23}Na$	440	260	30	
	$^{23}Na(\alpha,p\gamma)^{26}Mg$	1809			100
Mg	$^{25}Mg(\alpha,n\gamma)^{28}Si$	1779	1.3%		
	$^{26}Mg(\alpha,n\gamma)^{29}Si$	1273			0.5%
Al	$^{27}Al(\alpha,p\gamma)^{30}Si$	2235	3.2%		200
Ti	$^{47}Ti(\alpha,\alpha'\gamma)^{47}Ti$	159		755	0.35%
V	$^{51}V(\alpha,\alpha'\gamma)^{51}V$	319		570	
Mn	$^{55}Mn(\alpha,\alpha'\gamma)^{55}Mn$	126		70	
Fe	$^{57}Fe(\alpha,\alpha'\gamma)^{57}Fe$	122		3650	

In a study by Giles and Peisach (1979), the spectra obtained by 5 MeV α-particle bombardment of light weight elements were presented. Unfortunately, absolute yield values were not given (only relative intensities and net counts per collected charge), but a useful atlas of spectra was provided. The results of this study are summarised in Table 8.10.

Table 8.10 Sensitivities obtained with 5 MeV α-particle irradiation (Giles & Peisach 1979)

Over 1%	0.1–1%	Below 0.1%
Sc, Cr, Cu, Zn, Rb, Zr Er, Hf, Hg	O, Mg, Si, Cl, K, Ti Fe, Br, Mo, Ru, Pd, Ag Cd, Ta, W, Re, Ir, Pt Au	Li, B, N, F, Na, Al P, V, Mn, Rh, Be

It should be noted that with α-particle energies below 5 MeV no γ-ray lines originate from carbon. The elements that are easily detectable by α-particle bombardment are Li, Be, B and F. When comparing the yields obtained by 2.4 MeV α particles with those obtained by 2.4 MeV protons it can be concluded that the α particle-induced yields are higher for beryllium only. For beryllium the sensitivity obtained by α particle-induced γ-ray emission (sometimes referred to as AIGE) is about 1000 times better than that obtained by protons.

AIGE could be of specific interest in the analysis of Li, Be, B and F from samples which are rich in Na, Mg or Al (for which AIGE is not sensitive). Such analyses are not sensitive with proton bombardment. Additional advantages of the use of α particles are the low neutron yields and the simplicity of the γ spectra. Due to the short lifetime of the de-exciting state and the high recoil of the nucleus, a clear Doppler broadening occurs in several γ-ray lines. The broadening is useful in identifying the origin of peaks in γ spectra.

8.5.3.1 *Low-yield elements* Of the light elements, carbon can be used as a low-yield material. Of the heavier elements from Co and Ni no γ-rays originate. Also the yields for Y, Nb, In and Sn are sufficiently low to be employed as construction materials.

8.5.4 *Lithium-induced γ-ray yields*

Energetic ^{7}Li ions are occasionally used in RBS measurements, substituting for ^{4}He ions. Lithium ion-induced γ-ray emission can be used in connection with other ion beam analytical techniques for simultaneous detection of the light elements. The γ-ray yields induced by 12 and 18 MeV ^{7}Li ions are provided in Räisänen (1990). The yields for the light elements are listed in Table 8.11.

Table 8.11 The reactions leading to the highest light element γ-ray yields following 18 MeV ^{7}Li-ion bombardment. The data have been compiled from Räisänen (1990)

Element	E_γ (keV)	Reaction
C	871	^{12}C(^{7}Li,pnγ)^{17}O; ^{13}C(^{7}Li,p2nγ)^{17}O
	937	^{12}C(^{7}Li,nγ)^{18}F; ^{13}C(^{7}Li,2nγ)^{18}F
N	1233	^{14}N(^{7}Li,2nγ)^{19}Ne
	1633	^{14}N(^{7}Li,nγ)^{20}Ne; ^{15}N(^{7}Li,2nγ)^{20}Ne
O	350	^{16}O(^{7}Li,pnγ)^{21}Ne
	937	^{16}O(^{7}Li,αnγ)^{18}F
F	350	^{19}F(^{7}Li,αnγ)^{21}Ne
	1636	^{19}F(^{7}Li,p2nγ)^{23}Na
Na	585	^{23}Na(^{7}Li,αnγ)^{25}Mg
	1369	^{23}Na(^{7}Li,pnγ)^{28}Al
Mg	1779	^{24}Mg(^{7}Li,p2nγ)^{28}Si
	1809	^{24}Mg(^{7}Li,αpγ)^{26}Mg; ^{25}Mg(^{7}Li,αpnγ)^{26}Mg
Al	1323	^{27}Al(^{7}Li,pnγ)^{32}P
	2028	^{27}Al(^{7}Li,αnγ)^{29}Si
Si	1966	^{28}Si(^{7}Li,pnγ)^{33}S; ^{29}Si(^{7}Li,p2nγ)^{33}S
	2233	^{28}Si(^{7}Li,αγ)^{31}P; ^{29}Si(^{7}Li,αnγ)^{31}P
	2235	^{28}Si(^{7}Li,αpγ)^{30}Si; ^{29}Si(^{7}Li,αpnγ)^{30}Si
P	788	^{31}P(^{7}Li,pnγ)^{36}Cl
	1966; 1970	^{31}P(^{7}Li,αnγ)^{36}Ar; ^{31}P(^{7}Li,2nγ)^{36}Ar

Although the γ-ray yields increase with increasing particle energy (from 12 to 18 MeV), the background in the spectra will be also elevated due to higher neutron yields and Bremmstrahlung. As a result the minimum detectable amount of an element is not necessarily lower for increased particle energy. The exact optimum bombarding energy depends on the matrix and the element to be determined. A limitation in the use of ^{7}Li ions is that Li, Be and B cannot be detected because no significant γ-ray lines originate from these elements.

8.5.4.1 *Low-yield elements* From the values provided by Räisänen (1990) it can be concluded that lead is a good low-yield material when used with ^{7}Li ions. Because of its rather high neutron yield, carbon is not usable as a construction material in experimental set-ups when ^{7}Li ions are employed. Tungsten, hafnium and tantalum are good backing materials, for example, in nuclear physics experiments with ^{7}Li beams. No significant γ-ray lines originate either from zirconium.

8.5.5 *Heavy ion-induced γ-ray yields*

8.5.5.1 *^{12}C, ^{14}N and ^{16}O ion-induced γ-ray yields* Backscattering spectrometry (BS) and elastic recoil detection analyses (ERDA) can be carried out using ^{12}C, ^{14}N and ^{16}O ions. In such determinations the simultaneous use of the ion-induced γ-ray emission technique can be taken advantage of. Absolute γ-ray yields induced by 22 and 28 MeV ^{12}C ions, 28 MeV ^{14}N ions and 28 and 33 MeV ^{16}O ions for the light elements are available in the literature (see Seppälä *et al.* 1998). The bombarding energies are well above the Coulomb barriers of the light elements ($Z < 9$) for which several strong γ-ray lines are reported. The highest yields obtained by all ions were for beryllium. The yields induced by ^{12}C, ^{14}N and ^{16}O ions can be compared with the values obtained by other ions. With ^{12}C, ^{14}N and ^{16}O ions the values are systematically higher for beryllium, carbon, nitrogen and oxygen than those obtained with 2.4 MeV protons (equivalent energy per nucleon). The γ-ray yields induced by 2.4 MeV α particles are systematically lower for all elements except for beryllium. It can be concluded that the ^{12}C, ^{14}N and ^{16}O ion-induced yields are sufficiently high for practical detection of carbon, nitrogen and oxygen. The relevant reactions for such analyses are provided in Table 8.12.

8.5.5.1.1 *Low-yield elements* The yields are low for elements heavier than titanium and they can all be used as construction materials. Also, the neutron yields induced by the heavy ions are low.

8.5.5.2 *^{55}Cl ion-induced γ-ray yields* The usability of 55 MeV $^{35}Cl^{7+}$ ions for elemental analyses has been demonstrated by Borderie *et al.* (1979). For all other elements, except helium, lithium and beryllium, 55 MeV is below the Coulomb barrier. The method was found to be suitable also for analyses of heavier elements in special cases. The detection limits obtainable for the light elements from a niobium sample are listed in Table 8.13. The values refer to irradiation times of one hour with a beam current of 0.5 μA.

Table 8.12 Analyses of carbon, nitrogen and oxygen by ^{12}C, ^{14}N and ^{16}O ion bombardment. The data have been collated from Seppälä *et al.* (1998)

Element	E_γ (keV)	E_{ion} (MeV)	Reaction
C	440	28	^{12}C(^{12}C,pγ)^{23}Na
	1636	28	^{12}C(^{12}C,pγ)^{23}Na
	3333	28	^{12}C(^{12}C,αγ)^{20}Ne
	350	28	^{12}C(^{14}N,αpγ)^{21}Ne
	583; 585	28	^{12}C(^{14}N,αγ)^{22}Na; ^{12}C(^{14}N,pγ)^{25}Mg
	1369	28	^{12}C(^{14}N,αγ)^{22}Na; ^{12}C(^{14}N,pnγ)^{24}Mg
	2788; 2794; 2798	28	^{12}C(^{14}N,αpγ)^{21}Ne; ^{12}C(^{14}N,αnγ)^{21}Na
	417	33	^{12}C(^{16}O,pnγ)^{26}Al
	440	33	^{12}C(^{16}O,αpγ)^{23}Na
	1369; 1371	33	^{12}C(^{16}O,αpnγ)^{22}Na; ^{12}C(^{16}O,pnγ)^{26}Al
N	350	28	^{14}N(^{12}C,αpγ)^{21}Ne
	1369	28	^{14}N(^{12}C,αγ)^{22}Na; ^{16}O(^{12}C,pnγ)^{24}Mg
	2794	28	^{14}N(^{12}C,αpγ)^{21}Ne
	417	28	^{14}N(^{14}N,pnγ)^{26}Al
	440	28	^{14}N(^{14}N,αpγ)^{23}Na
	1636	28	^{14}N(^{14}N,αpγ)^{23}Na
	1809	28	^{14}N(^{14}N,2pγ)^{26}Mg
	585	33	^{14}N(^{16}O,αpγ)^{25}Mg
	975	33	^{14}N(^{16}O,αpγ)^{25}Mg
	1779; 1780	33	^{14}N(^{16}O,pnγ)^{28}Si; ^{16}O(^{12}C,αγ)^{26}Al
O	440	28	^{16}O(^{12}C,αpγ)^{23}Na
	1369; 1371	28	^{16}O(^{12}C,αpnγ)^{22}Na; ^{16}O(^{12}C,αγ)^{24}Mg ^{16}O(^{12}C,pγ)^{27}Al
	1821	28	^{16}O(^{12}C,pnγ)^{26}Al
	583; 585	28	^{16}O(^{14}N,2αγ)^{22}Na; ^{16}O(^{14}N,αpγ)^{25}Mg
	1779; 1780	28	^{16}O(^{14}N,pnγ)^{28}Si; ^{16}O(^{14}N,αγ)^{26} Al
	2887; 2904	28	^{16}O(^{14}N,2pγ)^{28}Al; ^{16}O(^{14}N,α2pγ)^{24}Na
	709	33	^{16}O(^{16}O,pnγ)^{30}P
	1369	33	^{16}O(^{16}O,αpγ)^{22}Na
	2233; 2235	33	^{16}O(^{16}O,pγ)^{31}P; ^{16}O(^{16}O,2pγ)^{30}Si

8.5.5.2.1 *Low-yield elements* As expected, heavy elements are the best low-yield materials. The reactions, in all cases, involve Coulomb excitation and the resulting γ-ray energies are low.

8.6 Selected applications of PIGE

In the following sub-sections some selected examples of the many and varied applications of PIGE are given.

Table 8.13 Detection limits for light elements with 55 MeV ^{35}Cl ions. The data are collected from Borderie *et al.* (1979)

Element	Reaction	E_γ (keV)	Detection limit (ppm)
Li	^{7}Li(^{35}Cl,nγ)^{41}Ca	2606	8
B	^{10}B(^{35}Cl,npγ)^{43}Sc	350	175
F	^{19}F Coul. ex.[a]	197	7
Na	^{23}Na Coul. ex.	440	17
Mg	^{24}Mg Coul. ex.	1369	350
Al	^{27}Al Coul. ex.	844	310
Ti	^{47}Ti Coul. ex.	159	105
V	^{51}V Coul. ex.	319	35
Mn	^{55}Mn Coul. ex.	126	10
Fe	^{56}Fe Coul. ex.	847	60

[a] Coulomb excitation.

8.6.1 *Fluorine determination*

The analysis of fluorine by PIGE has been very popular since the analysis of F from small sample amounts is not feasible by other methods. Fluorine determination from teeth has been studied in several odontological publications. Another advantage of PIGE in such studies is that Ca, P, Mg, Na and O can be determined simultaneously. A new chemically enriched hydroxylapatite standard for fluorine determinations has also been developed and tested (Zschau *et al.* 1992). It has been found suitable for fluorine determinations in tooth enamel at the ppm level.

8.6.2 *Channelling measurements with PIGE*

The reaction ^{7}Li(p,p$'\gamma$)^{7}Li has been frequently applied, as a complementary method, to particle–particle reactions in lithium lattice site location determinations by channelling (William *et al.* 1994). The reaction ^{15}N(p,$\alpha\gamma$)^{12}C has been applied successfully in channelling studies of nitrogen (Brenn 1998).

8.6.3 *Analysis of the major structural elements in biomedical samples*

A special and useful advantage of the PIGE method with proton bombardment is the ability to determine accurately at the same time carbon, nitrogen and oxygen. These are the major structural elements of biomedical and organic samples. When analysing such samples the use of an external beam is essential. The method is most sensitive and practical for the simultaneous analysis of C, N and O with 7.6 MeV protons (see also Table 8.5).

A general problem in determining the absolute concentrations of the major structural elements from biomedical and organic samples is the lack of suitable standard samples. With PIGE the concentration values for carbon, nitrogen and oxygen can be achieved using organic compounds as standards (when proper standards are not available). The procedure is based on the measurement of pure chemical compounds containing C, N and O. The known stoichiometry of the compounds can be used directly. As a restriction the selected organic compounds should not have very high concentrations of any single element (>40 wt%). The suitability of such a procedure is based on the insignificant difference in the stopping power values for typical organic samples and the compounds used as standards, although the relative concentrations of C, N and O may differ significantly.

The PIGE method has been used successfully in the analysis of human placental and serum samples (Nikkinen *et al.* 1985; Lapatto & Räisänen 1988). A significant advantage of the method is that very little sample preparation is needed, freeze-drying is generally sufficient. In addition to C, N and O, the elements of Na, Mg and P can be detected simultaneously from biomedical samples. An example of a typical spectrum from a bone sample obtained by an external beam system is given in Fig. 8.6.

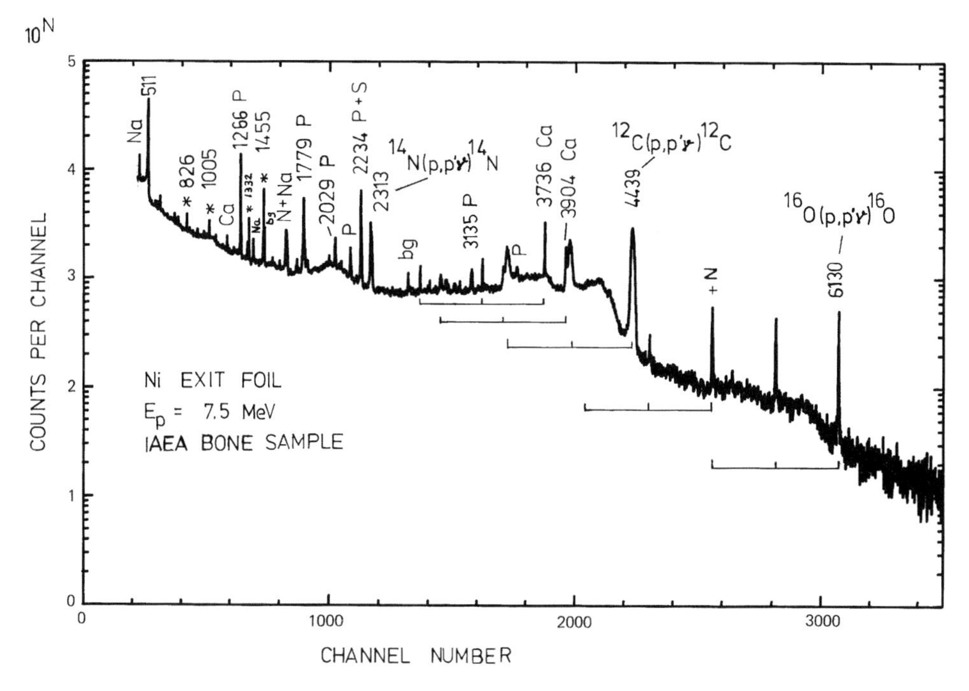

Fig. 8.6 A typical spectrum from a standard bone sample obtained by 7.5 MeV protons. The measuring time was about 30 min. The γ energies are given in keV. The γ-ray peaks due to the nickel exit foil are marked with an asterisk. The figure has been adopted from Räisänen (1986), with permission.

8.6.4 *Miscellaneous applications*

Some recent applications of PIGE are cited below. For detailed information the reader is referred to the References. The rule of thumb introduced in Section 3.2 has been applied successfully in the determination of Na, Mg, Al, P and Cl from a geological sample (Olabanji *et al.* 1996). PIGE has been also popular in characterising archaeological samples, see, e.g. Climent-Font *et al.* (1998). As an example of medical studies, the external beam work performed by Robertson *et al.* (1992) can be cited where bone tissue of Alzheimer's disease patients were analysed simultaneously with PIXE.

8.7 Advantages and limitations of PIGE

An advantage of PIGE is the high penetrability of the γ-rays, thus diminishing matrix effects. The detector can be positioned immediately behind the sample, providing a maximum solid angle for improved sensitivity. The experimental set-ups are usually simple. Bringing the charged particle beam (mainly protons) into the atmosphere through a thin exit foil offers increased applications, e.g. in biomedical work. Volatile materials can be analysed and the number of sample preparation steps is reduced. As a rule, very small amounts of sample material only are needed. Also, surface topography does not significantly influence the determinations. The method is considered non-destructive and the measurements are relatively fast. Furthermore, it can be used as a complementary method to PIXE allowing for the simultaneous detection of nearly all elements. The use of resonance reactions allows the taking of depth distribution measurements for several light elements.

Unfortunately, PIGE can only be used for sensitive determination of selected isotopes. The detection sensitivities for many elements are only moderate. The choice of the optimum physical parameters, e.g. bombarding energy, depends on the matrix and the elements to be determined. Therefore, no universal 'best' conditions and physical parameter choices can be provided.

References

Anttila, A., Keinonen, J., Hautala, M. & Forsblom, I., Use of the $^{27}Al(p,\gamma)^{28}Si$, E_p = 992 keV resonance as a gamma-ray intensity standard, *Nucl. Instrum. Meth.* **147** (1977), 501–505.

Anttila, A., Hänninen, R. & Räisänen, J., Proton-induced thick-target gamma-ray yields for the elemental analysis of the Z = 3–9, 11–21 elements, *J. Radioanal. Chem.* **62** (1981), 293–306.

Arstila, K., An experimental method for precise determination of electronic stopping powers for heavy ions, *Nucl. Instrum. Meth. B* **168** (2000), 473–483.

Badica, T., Besliu, C., Ene, A., Olariu, A. & Popescu, I., Coincidence method for the determination of minor elements in steel by proton-induced prompt gamma-ray spectrometry (PIGE), *Nucl. Instrum. Meth.* **B111** (1996), 321–324.

Bebber, H. van, Borucki, L., Farzin, K., Kiss, Á.Z. & Schulte, W.H., Total cross section of the $^{14}N(d,p\gamma)^{15}N$ nuclear reaction for analytical applications, *Nucl. Instrum. Meth.* **B136–138** (1998), 72–76.

Bird, J.R., Scott, M.D. Russel, L.H. & Kenny, M.J., Analysis using ion induced gamma-rays, *Aust. J. Phys.* **31** (1978), 209–213.

Bird, J.R., in: Ion beams for materials analysis, eds. J.R. Bird & J.S. Williams (Academic Press, Sydney, 1989), p. 180.

Boni, C., Cereda, E., Braga Marcazzan, G.M. & DeTomasi, V., Prompt gamma emission excitation functions for PIGE analysis of Li, B, F, Mg, Al, Si and P in thin samples, *Nucl. Instrum. Meth.* **B35** (1988), 80–86.

Boni, C., Caridi, A., Cereda, E., Braga Marcazzan, G.M. & Redaelli, P., Simultaneous PIXE–PIGE analysis of thin and thick samples, *Nucl. Instrum. Meth.* **B49** (1990), 106–110.

Borderie, B. & Barrandon, J.N., New analytical developments in prompt gamma-ray spectrometry with low-energy tritons and alpha particles, *Nucl. Instrum. Meth.* **156** (1978), 483–492.

Borderie, B., Barrandon, J.N., Delaunay, B. & Basutcu, M., Use of Coulomb excitation by heavy ions (^{35}Cl, 55 MeV) for analytical purposes: possibilities and quantitative analysis, *Nucl. Instrum. Meth.* **163** (1979) 441–451.

Borderie, B., Present possibilities for bulk analysis in prompt gamma-ray spectrometry with charged projectiles, *Nucl. Instrum. Meth.* **175** (1980), 465–482.

Brenn, R., in: *Abstracts for the Fifteenth International Conference on the Application of Accelerators in Research and Industry*, eds. J.L. Duggan & I.L. Morgan (University of North Texas, Denton, USA, 1998), p. 158.

Climent-Font, A., Demortier, G., Palacio, C., Montero, I., Ruvalcaba-Sil, J.L. & Diaz, D., Characterisation of archaeological bronzes using PIXE, PIGE, RBS and AES spectrometries, *Nucl. Instrum. Meth.* **B134** (1998), 229–236.

Constantinescu, B., Dima, S., Florescu, V., Ivanov, E.A., Plostinaru, D. & Sârbu, C., Use of 4.7 MeV alpha particles in elemental analysis and fusion reactor materials studies, *Nucl. Instrum. Meth.* **B16** (1986), 488–492.

Deconninck, G. & Demortier, G., Quantitative analysis of aluminium by prompt nuclear reactions, *J. Radioanal. Chem.* **12** (1972), 189–208.

Deconninck, G. & Demortier, G., Thick-target excitation yields of prompt gamma-radiation from proton-bombardment of Rh, Pd, Ag, Pt and Au, *J. Radioanal. Chem.* **24** (1975), 437–445.

Demortier, G., Prompt gamma-ray yields from proton bombardment of transition elements (Ti to Zn), *J. Radioanal. Chem.* **45** (1978), 459–496.

Elekes, Z., Kiss, Á.Z., Gyürky, Gy., Somorjai, E. & Uzonyi, I., Application of a clover–Ge–BGO detector system for PIGE measurements at a nuclear microprobe, *Nucl. Instrum. Meth.* **B158** (1999), 209–213.

Gihwala, D. & Peisach, M., Proton-induced prompt gamma-ray spectrometry: a survey of its analytical significance and some applications, *J. Radioanal. Chem.* **70** (1982), 287–309.

Giles, I.S. & Peisach, M., Determination of fluorine by the spectrometry of prompt gamma-rays, *J. Radioanal. Chem.* **32** (1976), 105–116.

Giles, I.S. & Peisach, M., A survey of the analytical significance of prompt gamma-rays induced by 5 MeV alpha-particles, *J. Radioanal. Chem.* **50** (1979), 307–360.

Golicheff, L., Loeuillet, M. & Engelmann, C., Analytical application of the direct observation of nuclear reactions induced by low-energy protons and leading to the emission of gamma-photons, which are measured, *J. Radioanal. Chem.* **12** (1972), 233–250.

Hänninen, R. & Räisänen, J., Nitrogen determination with external beam proton induced gamma-ray emission analysis, *Nucl. Instrum. Meth.* **B5** (1984), 43–48.

Hinrichsen, P.F., Houdayer, A., Kajrys, G. & Belhadfa, A., The microbeam facility at the university of Montreal, *Nucl. Instrum. Meth.* **B30** (1988), 276–279.

Ishii, K., Sastri, C.S., Valladon, M., Borderie, B. & Debrun, J.L., Two reaction methods for accurate analysis by irradiation with charged particles, *Nucl. Instrum. Meth.* **153** (1978a), 507–509.

Ishii, K., Valladon, M., Sastri, C.S. & Debrun, J.L., Accurate charged particle activation analysis: calculation of the average energy in the average stopping power method, *Nucl. Instrum. Meth.* **153** (1978b), 503–505.

Ishii, K., Valladon, M. & Debrun, J.L., The average stopping power method for accurate charged particle activation analysis, *Nucl. Instrum Meth.* **150** (1978c), 213–219.

Kenny, M.J., Bird, J.R. & Clayton, E, Proton induced γ-ray yields, *Nucl. Instrum. Meth.* **168** (1980), 115–120.

Kiss, Á.Z., Koltay, E., Nyakó, B., Somorjai, E., Anttila, A. & Räisänen, J., Measurements of relative thick-target yields for PIGE analysis on light elements in the proton energy interval 2.4–4.2 MeV, *J. Radioanal. Nucl. Chem.* **89** (1985), 123–141.

Kiss, Á.Z., Biron, I., Calligaro, T. & Salomon, J., Thick target yields of deuteron induced gamma-ray emission from light elements, *Nucl. Instrum. Meth.* **B85** (1994), 118–122.

Lapatto, R. & Räisänen, J., A rapid method for the quantification of C, N and O in biomedical samples by proton induced gamma-ray emission analysis applied to human placental samples, *Phys. Med. Biol.* **33** (1988), 75–81.

Lappalainen, R., Anttila, A. & Räisänen, J., Absolute α-induced thick-target gamma-ray yields for the elemental analysis of light elements, *Nucl. Instrum. Meth.* **212** (1983), 441–444.

Mosbah, M. & Duraud, J.-P., Proton microbeam induced damages in glasses: Ca and Na distribution modification, *Nucl. Instrum. Meth.* **B130** (1997), 182–187.

Nikkinen, P., Räisänen, J., Hänninen, R., Hyvönen-Dabek, M. & Dabek, J.T., A new rapid method for quantification of nitrogen in human serum employing the $^{14}N(p,p'\gamma)^{14}N$ reaction: application to human pregnancy, *Phys. Med. Biol.* **30** (1985), 921–928.

Olabanji, S.O., Haque, A.M.I., Zandolin, S., Ajayi, R.T., Buoso, M.C., Ceccato, D., Cherubini, R., Zafiropoulos, D. & Moschini, G., Applications of PIXE, PIGE, SEM and EDAX techniques to the study of geological samples, *Nucl. Instrum. Meth.* **B109/110** (1996), 262–265.

Peisach, M., Pineda, C.A. & Pillay, A.E., The application of high resolution deuteron-induced prompt gamma-ray spectrometry to the analysis of some transition elements, *Nucl. Instrum. Meth.* **B85** (1994), 142–144.

Pierce, T.B., Peck, P.F. & Henry, W.M., Determination of carbon in steels by measurement of the prompt γ-radiation emitted during proton bombardment, *Nature* **204** (1964), 571–572.

Pierce, T.B., Peck, P.F. & Henry, W.M., The rapid determination of carbon in steels by measurement of the prompt radiation emitted during deuteron bombardment, *Analyst* **90** (1965), 339–345.

Pierce, T.B., Peck, P.F. & Cuff, D.R.A., Application of inelastic proton scattering to the rapid determination of silicon in steels, *Anal. Chim. Acta* **39** (1967), 433–436.
Räisänen, J. & Hänninen, R., Heavy element ($Z > 30$) thick-target gamma-ray yields induced by 1.7 and 2.4 MeV protons, *Nucl. Instrum. Meth.* **205** (1983), 259–268.
Räisänen, J., A rapid method for carbon and oxygen determination with external beam proton induced gamma-ray emission analysis, *Nucl. Instrum. Meth.* **B17** (1986), 344–348.
Räisänen, J., External-beam methods in biomedical work, *Biol. Trace Elem. Res.* **12** (1987), 55–64.
Räisänen, J., Witting, T. & Keinonen, J., Absolute thick-target gamma-ray yields for elemental analysis by 7 and 9 MeV protons, *Nucl. Instrum. Meth.* **B28** (1987), 199–204.
Räisänen, J. & Lapatto, R., Analysis of sulphur with external beam proton induced gamma-ray emission analysis, *Nucl. Instrum. Meth.* **B30** (1988), 90–93.
Räisänen, J., Elemental analysis with external beam methods, *Nucl. Instrum. Meth.* **B40/41** (1989), 638–642.
Räisänen, J., Thick-target gamma-ray yields induced by 12 and 18 MeV ^{7}Li ions, *Nucl. Instrum. Meth.* **A299** (1990), 387–393.
Robertson, J.D., Samudrawar, D.L. & Markesbery, W.R., Ion beam analysis of the bone tissue of Alzheimer's disease patients, *Nucl. Instrum. Meth.* **B64** (1992), 553–557.
Savidou, A., Aslanoglou, X, Paradellis, T. & Pilakouta, M., Proton induced thick target gamma-ray yields of light nuclei at the energy region E_p = 1.0–4.1 MeV, *Nucl. Instrum. Meth.* **B152** (1999), 12–18.
Seppälä, A., Räisänen, J., Rauhala, E. & Szökefalvi-Nagy, Z., Absolute thick-target gamma-ray yields for the light elements induced by ^{12}C, ^{14}N and ^{16}O ions, *Nucl. Instrum. Meth.* **B143** (1998), 233–243.
Shroy, R.E., Kraner, H.W., Jones, K.W., Jacobson, J.S. & Heller, L.I., Determination of fluorine in food samples by the ^{19}F(p,p′γ)^{19}F reaction, *Nucl. Instrum. Meth.* **149** (1978), 313–316.
Sippel, R.F. & Glover, E.D., Sedimentary rock analysis by charged particle bombardment, *Nucl. Instrum. Meth.* **9** (1960), 37–48).
Stroobants, J., Bodart, F., Deconninck, G., Demortier, G. & Nicolas, G., Analysis of fluorine by nuclear reactions and application to human dental enamel, in: *Ion Beam Surface Layer Analysis*, Vol. 2, eds. O. Meyer, G. Linker & F. Käppeler (Plenum Press, New York, 1976), pp. 933–943.
Vickridge, I., Tallon, J. & Presland, M., High precision determination of ^{16}O in high T_C superconductors by DIGME, *Nucl. Instrum. Meth.* **B85** (1994), 95–99.
William, E.K., Ila, D. & Zimmerman, R.L., Simultaneous application of (p,γ) and (p,α) channeling to the detection of light elements in crystals, *Nucl. Instrum. Meth.* **B85** (1994), 537–540.
Zschau, H.-E., Plier, F., Vogt, J., Otto, G., Duschner, H., Arends, J., Grambole, D., Herrmann, F., Klabes, R., Salomonovic, R., Hulek, Z. & Setvak, M., Test of a new standard for fluorine determination with PIGE, *Nucl. Instrum. Meth.* **B68** (1992), 158–160.

Exercises and questions

(1) Which are the interaction processes of electromagnetic radiation with matter relevant to the PIGE method?

(2) Explain qualitatively the excitation curves for thin and thick targets.

(3) Explain qualitatively the main background formation processes in PIGE.

(4) Explain qualitatively the importance of escape peaks in PIGE.

(5) Identify the origin of the γ-ray peaks in the spectrum given below. The energies are in keV, and the bombarding particles are 2.4 MeV protons. Be aware of possible overlapping peaks.

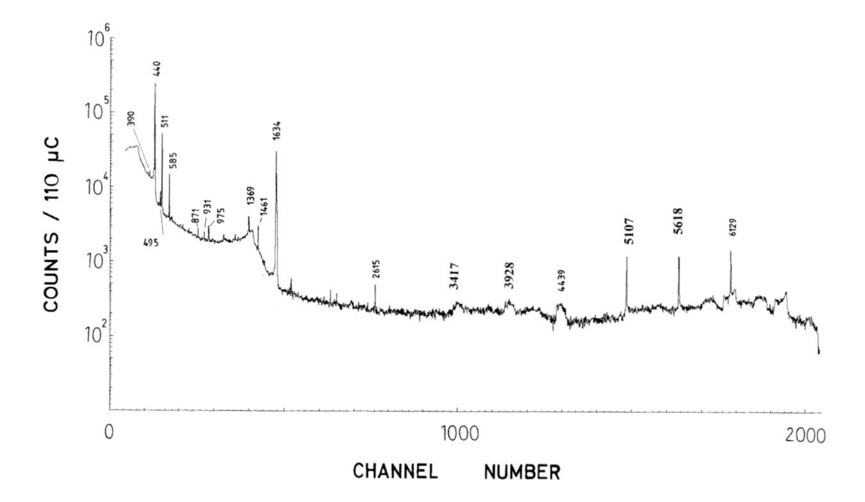

(6) Consulting the data presented in the yield references would you use protons or α particles for boron determination from (a) tantalum, or (b) silicon substrates? Explain your reasoning.

9 Nuclear reaction analysis

D.J. CHERNIAK and W.A. LANFORD

9.1 Introduction

In nuclear reaction analysis (NRA), samples are bombarded by a beam of charged particles, with energies commonly in the range of a few hundred keV to several MeV. Nuclear reactions with low-Z nuclei in the sample are induced by the incident ion beam. Products of these reactions (typically protons, deuterons, ^3He, α particles, or γ-rays) are detected, producing a spectrum of particle yield *versus* energy. Depth information is obtained from the spectrum using energy loss rates for incident and product ions travelling through the sample. Particle yields are converted to concentrations through the use of experimental parameters and nuclear reaction cross-sections.

Nuclear reaction analysis (NRA) is used to determine the concentration and depth distribution of light elements in the near surface (the first few micrometres) of solids. Because this method relies on nuclear reactions, it is largely insensitive to solid-state matrix effects. Hence, it can be made quantitative without reference to standard samples. This method is isotope specific, making it ideal for isotopic tracer experiments. This characteristic also makes NRA less vulnerable than some other methods to interference effects that may overwhelm signals from elements present in low abundance. In addition, measurements are rapid and 'non-destructive'.

NRA can be highly sensitive, with typical detection limits of ~10–100 ppm, depending on the reaction used. Depth resolutions typically range from a few nanometres to tens of nanometres; lateral resolutions down to a micrometre can be achieved with microbeams.

An especially significant application of NRA is the measurement of quantified hydrogen depth profiles, which is difficult with most other analytical techniques. Hydrogen concentrations can be measured to levels down to a few tens or hundreds of ppm (atomic) and with near-surface depth resolutions on the order of 10 nm.

As NRA is sensitive only to the nuclei present in the sample, it does not provide information on chemical bonding or microscopic structure. Hence, it is often used in conjunction with other techniques that do, such as ESCA, optical absorption, Auger, or electron microscopy. NRA is generally limited to light nuclei, because at energies around a few MeV the Coulomb barrier between the projectile and heavy target nuclei preclude nuclear reactions with these heavy nuclei. Heavy element profiling is often done by Rutherford backscattering (RBS). Hence, NRA and RBS are complementary methods, with the former well suited to profiling light nuclei and the latter in profiling distributions of heavy nuclei.

NRA has a wide range of applications, including use in investigations of metals, insulators, and semiconductor materials, and in such diverse fields as physics, archaeology, biology and geology.

9.2 Q values and cross-sections

NRA exploits the body of data accumulated through research in low-energy nuclear physics to determine concentrations and distributions of specific elements or isotopes in a material. Two parameters key to interpreting NRA spectra are reaction Q values and cross-sections.

Q values are the energies released in specific nuclear reactions and are used to calculate the energies of particles produced in the reaction. Reactions with large positive Q values are generally most suitable for NRA, because this maximizes the energy of reaction products and thus separates this signal from other possible 'background' signals. However, nuclear reactions with small or negative Q values have been successfully employed, for example (α,p) reactions in profiling ^{14}N, ^{32}S, ^{11}B and ^{31}P (e.g. Doyle *et al.* 1985; McIntyre *et al.* 1988, 1992; Soltani-Farsi *et al.* 1996). Table 9.1 presents Q values for a number of nuclear reactions. More comprehensive compilations of Q values exist elsewhere (e.g. Everling *et al.* 1961).

Table 9.1 Selected nuclear reactions and Q values. The references give information on reaction cross-sections and examples of applications (in italics)

Reaction	Q value (MeV)	References
Deuterium		
D(d,p)T	4.033	*Toulhoat et al. 1991*; Jarmie & Seagrove 1957
D(^{3}He,p)^{4}He	18.352	*Toulhoat et al. 1991*; Möller & Besenbacher 1980
Lithium		
^{6}Li(p,α)^{3}He	4.02	*Pretorius & Peisach* 1978; Marion *et al.* 1956
^{6}Li(d,α)^{4}He	22.374	*Courel et al. 1991*; *Räisänen 1992*
^{6}Li(^{3}He,p)^{8}Be	16.786	*Räisänen 1992*; Schiffer *et al.* 1956
^{7}Li(p,α)^{4}He	17.347	*Sagara et al. 1988*; *Courel et al. 1991*
^{7}Li(^{3}He,p)^{9}Be	11.2	*Schulte et al. 1986*; *Räisänen 1992*
Beryllium		
^{9}Be(p,α)^{6}Li	2.126	*Reichle et al. 1990*; Sierk & Tombrello 1973
^{9}Be(d,p)^{10}Be	4.857	*Reichle et al. 1990*; *McMillan 1992*
^{9}Be(d,t)^{8}Be	4.592	*Reichle et al. 1990*; Biggerstaff *et al.* 1962
^{9}Be(d,$α_0$)^{7}Li	7.152	*Reichle et al. 1990*; Biggerstaff *et al.* 1962
^{9}Be(^{3}He,p)^{11}B	10.322	*Reichle et al. 1990*; Wolicki *et al.* 1959
^{9}Be(^{3}He,α)^{8}Be	18.912	*Peacock et al. 1990*; Bilwes *et al.* 1978
Boron		
^{10}B(p,α)^{7}Be	1.147	*Deconninck 1973*; Jenkin *et al.* 1964
^{10}B(d,p_0)^{11}B	9.231	*Toulhoat et al. 1993*; *Moncoffre 1992*
^{10}B(d,$α_0$)^{8}Be	17.818	*Coetzee et al. 1975*; Purser & Wildenthal 1963
^{10}B(d,$α_1$)^{8}Be	14.878	*Coetzee et al. 1975*; Purser & Wildenthal 1963
^{10}B(^{3}He,p)^{12}C	19.693	*Pillay & Peisach 1992*; McIntyre *et al.* 1996
^{10}B(α,p)^{13}C	4.0626	*McIntyre et al. 1992*; Giorginis *et al.* 1996a
^{11}B(p,α)^{8}Be	8.582	*Moncoffre et al. 1990*; *Lu et al. 1989*; Mayer *et al.* 1998
^{11}B(d,p_0)^{12}B	1.138	*Coetzee et al. 1975*; *Moncoffre 1992*
^{11}B(d,$α_0$)^{9}Be	8.022	*Coetzee et al. 1975*; *Moncoffre 1992*
^{11}B(^{3}He,p)^{13}C	13.185	McIntyre *et al.* 1996; Holmgren *et al.* 1956
^{11}B(^{3}He,d)^{12}C	10.464	*Pillay & Peisach 1992*; Holmgren *et al.* 1956
^{11}B(α,p)^{14}C	0.7839	*McIntyre et al. 1992*; Giorginis *et al.* 1996a

Cont

Table 9.1 *Contd*

Reaction	Q value (MeV)	References
Carbon		
$^{12}C(d,p)^{13}C$	2.722	*Chou et al. 1990*; Lennard *et al.* 1991
$^{12}C(^{3}He,p)^{14}N$	4.7789	*Terwagne et al. 1995*; Kuan *et al.* 1964
$^{12}C(^{3}He,\alpha)^{11}C$	1.857	*Gosset 1981*; Kuan *et al.* 1964
$^{13}C(d,p)^{14}C$	5.951	Marion & Weber 1956
Nitrogen		
$^{14}N(d,p_0)^{15}N$	8.610	*Vickridge 1988*; *Bethge 1992*
$^{14}N(d,\alpha_0)^{12}C$	13.574	*Bethge 1985*; *Bethge 1992*
$^{14}N(\alpha,p)^{17}O$	−1.1914	*Doyle et al. 1985*; Giorginis *et al.* 1996b
$^{14}N(^{3}He,p)^{16}O$	15.242	*Bethge 1992*; McIntyre *et al.* 1996
$^{14}N(^{3}He,\alpha)^{13}N$	10.024	*Wielunski 1996*; Terwagne *et al.* 1994
$^{15}N(p,\alpha)^{12}C$	4.966	*Massiot et al. 1992*; Sawicki *et al.* 1986
$^{15}N(d,p)^{16}N$	0.266	*Bethge 1992*; Bohne *et al.* 1972
$^{15}N(d,\alpha)^{13}C$	7.687	*Sawicki et al.* 1986; *Bethge 1992*
Oxygen		
$^{16}O(d,p_0)^{17}O$	1.917	*Seah et al. 1988*; Amsel & Semuel 1967
$^{16}O(d,\alpha_0)^{14}N$	3.116	*Hirvonen & Lucke 1978*; Amsel & Samuel 1967
$^{16}O(^{3}He,p)^{18}F$	2.0321	*Abel et al. 1990*; *Cohen & Rose 1992*
$^{16}O(^{3}He,\alpha)^{15}O$	4.9139	*Abel et al. 1990*; *Cohen & Rose 1992*
$^{18}O(p,\alpha)^{15}N$	3.9804	*Reddy & Cooper 1982*; Amsel & Samuel 1967; Alkemade *et al.* 1988
$^{18}O(d,\alpha_0)^{16}N$	4.237	*Cohen & Rose 1992*; Amsel & Samuel 1967
Fluorine		
$^{19}F(p,\alpha)^{16}O$	8.115	*Grambole et al. 1992*; Dieumagard *et al.* 1980
$^{19}F(d,p_0)^{20}F$	4.379	*Coote 1992*; Maurel *et al.* 1981
$^{19}F(d,\alpha_0)^{17}O$	10.038	*Coote 1992*; Maurel *et al.* 1981
$^{19}F(\alpha,p)^{22}Ne$	1.6746	*McIntyre et al. 1990*; Borgardt *et al.* 1998
Sodium		
$^{23}Na(p,\alpha)^{20}Ne$	2.379	*Carnera et al 1977*; Endt & van der Leun 1973
Silicon		
$^{28}Si(d,p)^{29}Si$	6.253	*Mathot & Demortier 1990*
Phosphorus		
$^{31}P(p,\alpha)^{28}Si$	1.917	*Della Mea 1986*; Riley *et al.* 1967
$^{31}P(\alpha,p)^{34}S$	0.6316	*McIntyre et al. 1988*
Sulfur		
$^{32}S(d,p)^{33}S$	6.418	*Lane et al. 1998*; Healey & Lane 1998
$^{32}S(\alpha,p)^{35}Cl$	−1.87	Soltani-Farsi *et al.* 1996

Reaction cross-sections as functions of incident ion energy and beam–detector angle have also been measured and tabulated. Cross-sections can vary widely with even small changes in these parameters. When reaction cross-sections over a narrow range of incident ion energies are significantly larger than those at energies just above and just below this energy range, it is referred to as a resonance.

Because reaction particle yields are directly proportional to reaction cross-sections, knowledge of the variation of reaction cross-sections with angle and energy permits the experimenter to select an incident beam energy and detector angle that will maximize sensitivity. In addition, concentrations can be calculated directly from particle yields without reference to standards if cross-sections are reliably known. Similarly, yields and energies of γ-rays resulting from nuclear reactions have been determined.

Table 9.1 includes references for cross-sections and applications of selected nuclear reactions, along with Q values. Table 9.2 contains references for applications of selected γ-emitting reactions, and a listing of some resonance energies used in NRA. A number of printed compilations of reaction cross-sections and γ-ray yields exist, which provide more comprehensive information (e.g. Mayer & Rimini 1972; Bird & Williams 1989; Tesmer & Nastasi 1995), and regularly updated cross-section databases can also be accessed electronically (e.g. Sigma-Base at Idaho State University [http://www.physics.isu.edu/sigmabase/], maintained by George Vizkelethy).

Table 9.2 (p,γ) and other γ-emitting reactions, with energies of some resonances used in NRA. References include examples of applications

Reaction	Resonance energies (MeV)	References
Hydrogen		
$^{1}H(^{7}Li,\gamma)^{8}Be$	3.07	Adler & Shulte 1978
$^{1}H(^{15}N,\alpha\gamma)^{12}C$	6.385, 13.35	Lanford *et al.* 1979
$^{1}H(^{19}F,\alpha\gamma)^{16}O$	6.418, 16.44	Xiong *et al.* 1987
Helium		
$^{4}He(^{7}Li,\gamma)^{11}B$	1.677, 1.433, 0.701	Schulte 1978
Lithium		
$^{7}Li(p,\gamma)^{8}Be$	0.441	Toivanen *et al.* 1985
$^{7}Li(\alpha,\gamma)^{7}Li$	0.950, 0.819	Zinke-Allmang *et al.* 1986
Beryllium		
$^{9}Be(p,\gamma)^{10}B$	0.319, 0.992, 1.083	Anttila *et al.* 1981
Boron		
$^{10}B(p,\alpha\gamma)^{7}Be$	3.5	Toulhoat *et al.* 1993
$^{11}B(p,\gamma)^{12}C$	0.163	Toivanen *et al.* 1985
Carbon		
$^{13}C(p,\gamma)^{14}N$	0.551, 1.150, 1.748	Hirvonen *et al.* 1990
Nitrogen		
$^{14}N(p,\gamma)^{15}O$	0.278, 1.74	Bethge 1992
$^{14}N(\alpha,\gamma)^{18}F$	1.531	Gosset 1985
$^{15}N(p,\alpha\gamma)^{12}C$	0.429, 0.897, 1.210	Maurel & Amsel 1983
Oxygen		
$^{18}O(p,\gamma)^{19}F$	1.167	Maurel *et al.* 1982
Fluorine		
$^{19}F(p,\alpha\gamma)^{16}O$	0.340, 0.484, 0.872, 1.37	Kumar *et al.* 1999
Neon		
$^{20}Ne(p,\gamma)^{21}Na$	1.169, 1.311	Deconninck & van Oystaeyen 1983
$^{22}Ne(p,\gamma)^{23}Na$	0.640, 0.851	Lappalainen 1986
Sodium		
$^{23}Na(p,\gamma)^{24}Mg$	0.512, 0.677	Anttila *et al.* 1977
$^{23}Na(p,\alpha\gamma)^{20}Ne$	1.011, 1.164	Della Mea 1986
Magnesium		
$^{26}Mg(p,\gamma)^{27}Al$	1.548	Anttila *et al.* 1977
Aluminium		
$^{27}Al(p,\gamma)^{28}Si$	0.992	Amsel *et al.* 1998

Contd

Table 9.2 *Contd*

Reaction	Resonance energies (MeV)	References
Silicon		
$^{29}Si(p,\gamma)^{30}P$	0.416	Anttila *et al.* 1977
$^{30}Si(p,\gamma)^{31}P$	0.620	Väkeväinen *et al.* 1998
Phosphorus		
$^{31}P(p,\gamma)^{32}S$	0.811, 1.251	Kido *et al.* 1982
Sulfur		
$^{34}S(p,\gamma)^{35}Cl$	1.019	Anttila *et al.* 1977
Titanium		
$^{48}Ti(p,\gamma)^{49}V$	1.007, 1.362	El Khakani *et al.* 1992
Chromium		
$^{52}Cr(p,\gamma)^{53}Mn$	1.005	Ager *et al.* 1998
Manganese		
$^{55}Mn(p,\gamma)^{56}Fe$	1.537, 1.531	Ager *et al.* 1998
Nickel		
$^{58}Ni(p,\gamma)^{59}Cu$	1.424	Gosset 1980

Nuclear reaction analysis can be done in either resonant or non-resonant modes, depending on the type of reaction involved and the variation of reaction cross-section with incident ion energy. Both methods are discussed in the sections that follow.

9.3 Resonant nuclear reaction analysis

Resonant nuclear reaction analysis makes use of resonant cross-sections in nuclear reactions at the resonance energy, E_r. The ideal reaction has a large cross-section at E_r and negligible cross-section at other energies. Perhaps the best example is the $^1H + ^{15}N \rightarrow ^{12}C + ^4He + \gamma$-ray resonant reaction, which is used to profile ^1H (with a ^{15}N beam) and to profile ^{15}N (with a proton beam). The cross-section *versus* energy for this reaction is shown in Fig. 9.1 (Horn & Lanford 1988). The use of this reaction to profile hydrogen in materials will be discussed as an illustration of this general method.

9.3.1 *^{15}N resonant nuclear reaction analysis measurement of hydrogen profiles*

The use of the $^1H + ^{15}N \rightarrow ^{12}C + ^4He + \gamma$-ray nuclear reaction to measure H profiles is illustrated in Fig. 9.2 (Lanford *et al.* 1976). The sample being analysed is bombarded with ^{15}N ions of energy $E(^{15}N) \geq E_r = 6.385$ MeV and the yield of characteristic γ-rays from this nuclear reaction is measured. If the beam energy is equal to E_r, the ^{15}N ions are at the resonant energy at the surface of the sample, and the yield of characteristic γ-rays is proportional to the hydrogen concentration at the surface of the sample. If the ^{15}N beam energy is increased, the ions are no longer at the resonant energy for hydrogen at the surface, but as the ions penetrate the sample, they lose energy and reach the resonant energy at some depth in the sample. Now the yield

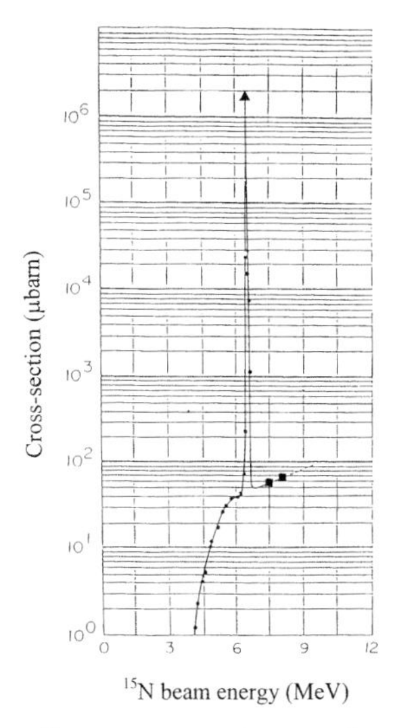

Fig. 9.1 Cross section of the $^{15}N(p,\alpha\gamma)^{12}C$ resonant nuclear reaction used for profiling H (with ^{15}N beam) and ^{15}N (with a proton beam). Note that in this energy range, there is a single isolated resonance with a maximum cross-section several orders of magnitude larger than the off-resonance cross-section. From Horn and Lanford (1988).

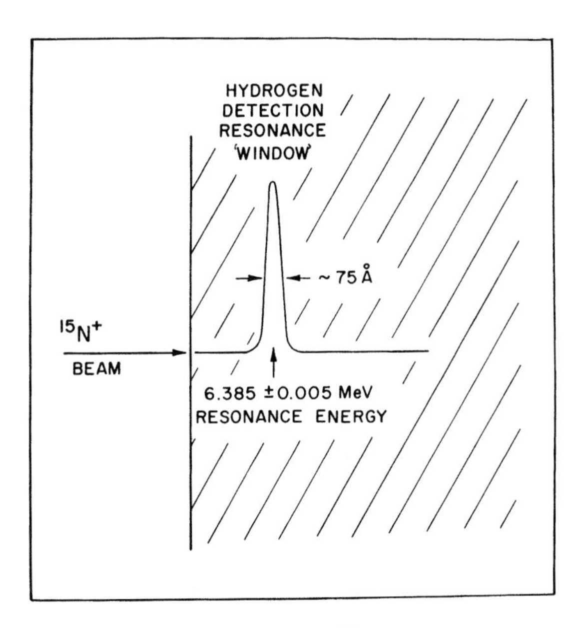

Fig. 9.2 A schematic representation of the use of the ^{15}N resonance profiling method. Samples to be analyzed are bombarded with ^{15}N. This beam creates a 'Hydrogen Detection Resonance Window' when the nuclear reaction is on resonance. If the ^{15}N beam energy is above resonance, as it loses energy penetrating the sample it will be on resonance at some depth in the sample. Hence, by measuring nuclear reaction yield *versus* beam energy, a hydrogen concentration profile is determined. From Lanford *et al.* (1976).

of characteristic γ-rays is proportional to the hydrogen concentration at this depth. Hence, by measuring the yield of γ-rays *versus* ^{15}N beam energy, the hydrogen concentration *versus* depth is determined.

Figure 9.3 shows a typical experimental chamber used for making such resonant NRA measurements (Lanford 1992). Samples to be analysed are mounted around the circumference of the sample holder, facilitating rapid changing of samples. The bombarding beam enters the chamber from the right and is focused to hit the sample (and only the sample). The number of incident ions is measured by integrating the electrical current carried to the target by the ion beam. The number of characteristic γ-rays coming from the target is measured with a large BGO (bismuth germanate) scintillation detector mounted directly behind the target.

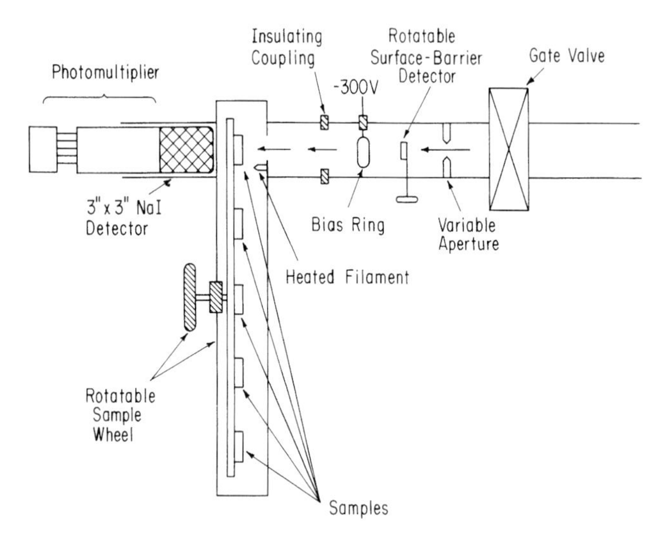

Fig. 9.3 A schematic representation of a typical NRA chamber. Samples to be analyzed are mounted on a rotatable sample wheel and are bombarded with ions. Nuclear reaction products are detected with a large NaI or BGO scintillation detector behind the samples. The number of incident ions is recorded by measuring the electrical current these ions bring to the target. From Lanford (1992).

Figure 9.4 illustrates the use of this method. This figure is from a study by Bedell and collaborators (Bedell & Lanford 1998; Bedell 1999) who studied the use of H ion implantation in the controlled splitting of single-crystal semiconductors. It has been observed that if silicon (and some other semiconductors) is ion implanted with H above a critical dose and subsequently annealed, the silicon splits along a plane at about the depth of the mean ion implant range. In order to study this phenomenon it is necessary to be able to measure the hydrogen distribution immediately after implantation and determine how this distribution changes during the annealing stage. More details are available in the works cited above.

Both the raw data (counts of characteristic γ-rays *versus* beam energy) and the final hydrogen profiles (H concentration *versus* depth) derived from these raw data are shown in Fig. 9.4. This profile shows a very thin surface layer of H (from surface contaminations of water or hydrocarbons and from H termination of surface bonds)

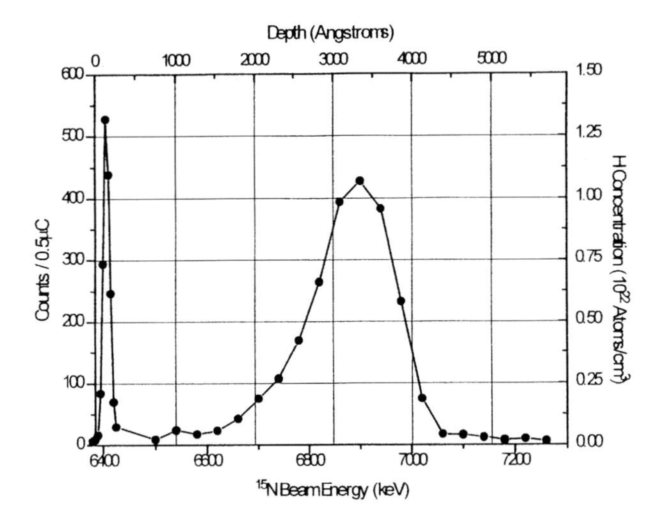

Fig. 9.4 Typical H profile data recorded by the ^{15}N method. This is for a sample of Si, ion implanted with 10^{17} H/cm^2. This figure shows both the raw data (counts/0.5 microcoulomb) *versus* beam energy and final data (H content *versus* depth). From Bedell (1999).

universally observed on samples that have not been cleaned *in situ*. Within the sample, there is a broad H distribution peaking at about 3400 Å, which is due to the H-ion implantation.

The H profile shown in Fig. 9.4 has the typical shape for an ion implant. Figure 9.5 shows the H profile measured in a homogeneous thin film of hydrogenated amorphous Si (Lanford 1980). This is the material widely used to make solar cells to power hand-held calculators and other low-power devices. Again, both raw data (counts *versus* beam energy) and final data (H concentration *versus* depth) are shown. This figure shows a uniform H content through the film until the beam energy is raised so

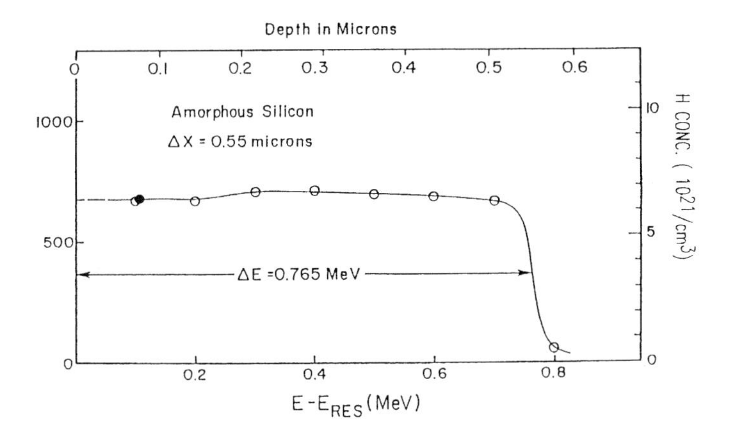

Fig. 9.5 Typical H profile data recorded by the ^{15}N method. This is for a sample of thin film hydrogenated amorphous silicon solar cell material. This figure shows both the raw data (counts/0.5 microcoulomb) *versus* beam energy and final data (H content *versus* depth).

high that H in the substrate is being measured. Also shown is a repeat measurement (solid dot) taken after the entire profile had been measured. This repeat was to check that H was not being lost during the measurement. The ability to make such repeat measurements to check the stability of the sample during analysis is another important feature of NRA. Such repeat measurements are not possible for profiling methods that use sputtering to expose the interior of a sample.

The conversion of the raw data into H profiles simply requires the multiplication of axes of the raw data by known constants. The energy axis becomes a depth scale and the counts axis becomes H concentration. The conversions are discussed below.

9.3.1.1 *Depth scale* Since for most materials the rate of energy loss ($\mathrm{d}E/\mathrm{d}x$) for ^{15}N ions with energy $6.4 < E(^{15}\mathrm{N}) < 10$ MeV remains essentially constant over this energy range, the depth scale (x) is derived from $E(^{15}\mathrm{N})$ simply by the following equation:

$$x = (E_{^{15}\mathrm{N}} - E_{\mathrm{r}})/(\mathrm{d}E/\mathrm{d}x) \tag{9.1}$$

9.3.1.2 *Concentration scale* The concentration scale is slightly more complicated. For bombarding energies greater than E_{r}, the yield of characteristic γ-rays (Y) is given by

$$Y = P\varepsilon \int \sigma(x) c_{\mathrm{H}}(x)\,\mathrm{d}x \tag{9.2}$$

where P is the number of incident ions, ε is the detection efficiency, $\sigma(x)$ is the resonant cross-section at depth x in the target, and $c_{\mathrm{H}}(x)\mathrm{d}x$ is the number of H cm^{-2} between x and $x + \mathrm{d}x$. Changing variables to ion energy E, using $\mathrm{d}x = \mathrm{d}E/(\mathrm{d}E/\mathrm{d}x)$ gives the expression:

$$Y = \mathrm{P}\varepsilon \int \{\sigma(E) c_{\mathrm{H}}(E)/(\mathrm{d}E/\mathrm{d}x)\}\,\mathrm{d}E \tag{9.3}$$

Since the cross-section is large only for a very narrow range of energies around E_{r}, c_{H} and $\mathrm{d}E/\mathrm{d}x$ are approximately constant over this energy range and can therefore be taken outside the integral sign, giving:

$$Y = \{\varepsilon P c_{\mathrm{H}}/(\mathrm{d}E/\mathrm{d}x)\} \int \{\sigma(E)\,\mathrm{d}E\} \tag{9.4}$$

For isolated resonant reactions with small off-resonant cross-sections, such as we have in this instance, the resonance cross-section is given by the well-known Breit–Wigner formula. Integrating this cross-section over energy results in $(\pi/2)\sigma_0\Gamma$, where σ_0 is the peak resonant cross-section and Γ is the energy width of the cross-section.

Solving for the concentration, c_{H}, Equation (9.5) is obtained

$$c_{\mathrm{H}} = K(\mathrm{d}E/\mathrm{d}x)Y \tag{9.5}$$

where $K = 2/(P\varepsilon\pi\sigma_0\Gamma)$ is a constant independent of the material being analysed. In practice, K is determined by analysing materials containing known H concentrations (e.g. ion-implanted samples, plastics, metal hydrides).

9.3.1.3 *Depth resolution* Two important characteristics of any profiling method are depth resolution and sensitivity. For resonant nuclear reaction profiling, assuming a monoenergetic beam, the depth resolution (Δx) is simply determined by the energy width of the resonance (Γ) divided by the energy loss rate (dE/dx), i.e.

$$\Delta x = \Gamma/(dE/dx) \tag{9.6}$$

This expression is appropriate for calculating the near-surface depth resolution in most cases. Deeper into the sample, energy loss straggle causes a spread in the energy (ΔE_s) of the bombarding ions, which causes an increase in depth resolution. ΔE_s can be estimated using the well-known Bohr formula (Bohr 1948). In the limit where energy straggle dominates (i.e. deep in the sample), the depth resolution becomes:

$$\Delta x = \Delta E_s/(dE/dx) \tag{9.7}$$

The transition from the near-surface region (dominated by resonance width) to deep in the sample (dominated by straggle) requires the convolution of a Breit–Wigner and Gaussian distribution and is described in the literature (Amsel & Maurel 1983).

9.3.1.4 *Sensitivity* The sensitivity of resonant profiling methods depends on the reaction yield (which depends on beam current, cross-section and detection efficiency) compared with background events. Most resonant profiling methods detect γ-rays as a measure of the number of reactions occurring in the target, typically measured with a large bismuth germanate or sodium iodide scintillation detector. Such detectors have backgrounds from either long-lived radionuclides in the environment (potassium, uranium/thorium decay series) or cosmic rays. Typical sensitivity for this method is ~ 100 ppm (atomic). If special background suppression methods are used, sensitivities of 1 ppm (atomic) are possible (Endisch *et al.* 1994).

9.3.2 *Other light elements profiled by resonant nuclear reaction analysis*

The results illustrated above with the ^{15}N nuclear reaction analysis method for measuring hydrogen profiles can be generalized and applied in the use of any resonant nuclear reaction for profiling. [For example, see Lanford *et al.* (1995) for a discussion of profiling H, Li, F and Na.] For any resonant nuclear reaction, the conversion of reaction yield *versus* beam energy to concentration *versus* depth is accomplished using the expression

$$c_z = K_z(dE/dx)Y_z \tag{9.8}$$

where c_z is the concentration for isotope z, Y_z is the measured nuclear reaction yield and dE/dx is the energy loss of the bombarding ion in the target. And,

$$x = (E_{beam} - E_r)/(dE/dx) \tag{9.9}$$

where x is the depth in the target, E_{beam} is the bombarding beam energy and E_r is the energy of the resonance. In this expression for x, the assumption is made that dE/dx approximates to a constant. In most applications, experience indicates that this is generally a good approximation. In some cases, however, energy loss rates vary significantly over the range of ion energies between E_{beam} and E_r so it is necessary to integrate dE/dx to relate the energy difference $E_{beam} - E_r$ to the depth in the sample. The expression for c_z is correct even when dE/dx is not constant with energy; the dE/dx that enters this equation is the value at the resonance energy.

Table 9.2 lists some commonly used resonance reactions for profiling H, Li, ^{15}N, F, Na, Al and other elements.

9.3.3 *Applications – resonant profiling*

^{15}N nuclear reaction analysis has been used to measure H concentration profiles in a wide variety of materials, including thin films of amorphous Si, plasma silicon nitride, diamond-like carbon, diamond, silicon dioxide, various low-κ dielectrics, and various metallic single-crystal superlattices. It has been used to study the transport of hydrogen or water into a variety of materials. The applications presented here (and cited in the tables) represent only a small sampling of the many uses of NRA in investigations encompassing a broad range of research fields.

9.3.3.1 *Kinetics of H-ion implantation-induced blistering of semiconductors (H profiles)* Figure 9.4, discussed above, is part of a study by Bedell (Bedell 1999) and collaborators on the use of H-ion implantation followed by annealing of slices of thin single-crystal films of semiconductors. This method is presently under intensive research as a possible means of making inexpensive semiconductor on insulator (SOI) wafers as the initial starting material for modern microelectronic circuits. NRA profiling of H plays a key role, as does channelling, in investigating the defects in the implanted single crystals.

9.3.3.2 *H in amorphous Si solar cell materials* In the 1970s it was discovered that dopable semiconducting amorphous Si could be made by plasma-assisted chemical vapour deposition from SiH_4 gas. This was a surprise, since all previous work on amorphous Si indicated that the electrical properties of this material were dominated by the Si 'dangling bonds' inevitable in non-crystalline material. While not at first recognized, it was soon understood that hydrogen present in large concentration from the precursor gas could electrically passivate the Si dangling bonds. Clearly, it was important to be able to make quantitative measurements of the H content of this material, such as that illustrated in Fig. 9.5. The first quantitative measurements of the H content of this thin film material were made by ^{15}N NRA (Brodsky *et al.* 1977).

9.3.3.3 *Study of the reaction between water and glass (Na and H profiles)* Figure 9.6 shows H and Na profiles, both measured by resonant NRA, of a sample of soda-lime glass exposed to water at 90°C (Lanford *et al.* 1979). As seen in this figure, water exposure results in the transport of H into the glass and the removal of Na from the region with high H content. NRA measurements, such as those illustrated in Fig. 9.6, have been used to investigate the time and temperature dependence of this reaction between water and glass. More recently, exposure of glass to moist air has been used to study the dependence of the water–glass reaction on relative humidity (Cummings *et al.* 1998).

Data such as those shown in Fig. 9.6 have led to development of a detailed model of

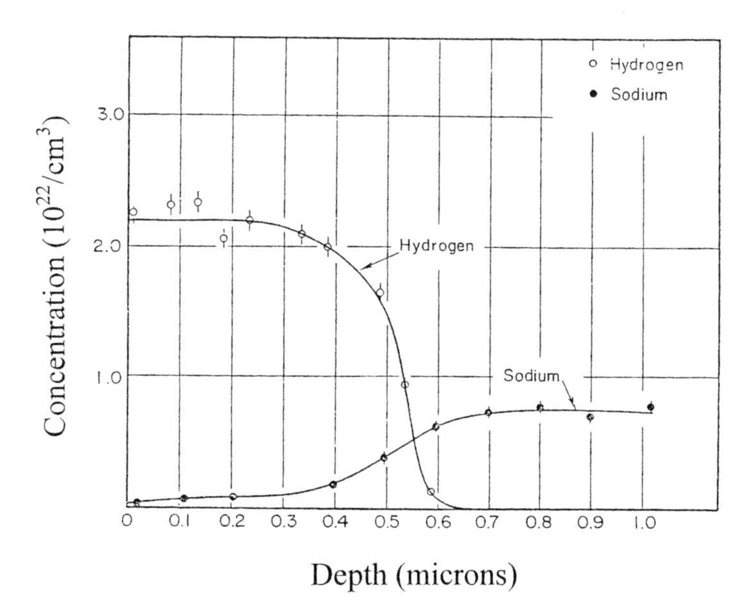

Fig. 9.6 Hydrogen and sodium concentration profiles measured by nuclear reaction analysis. The sample was soda-lime glass exposed to water at 90°C. This water exposure leads to an ion exchange and interdiffusion reaction which replaces the Na^+ ions in the glass with hydrogen ions (perhaps, hydronium, H_3O^+). From Lanford *et al.* (1979).

the microscopic mechanism responsible for this transport of H from water into glass. The model is based on the fact that soda-lime glass is a good Na^+ ionic conductor. Because of the high Na^+ concentration gradient at the glass surface, Na^+ would like to diffuse out of the glass but cannot because of its charge. However, if there is another positive ion available at the surface of the glass, it can ion-exchange and interdiffuse with that ion. Water at the surface provides just such an ion through the following reaction

$$Na^+ \text{ (glass)} + 2H_2O \text{ (surface)} \rightarrow H_3O^+ \text{ (glass)} + NaOH \text{ (surface)}$$

where Na^+ (glass) represents sodium ions in the glass and H_2O (surface) represents water at the surface of the glass, with similar notation on the right-hand side of the expression. Hence, the initial mechanism resulting in the reaction between water and glass is ionic exchange, an interdiffusion process that can be numerically modelled.

This type of study has wide application in areas including basic glass science, art history and archaeology (glass conservation and obsidian dating), and energy and environmental studies (consolidation of radioactive reaction wastes in glass).

9.3.3.4 *Fluorine in low-κ dielectrics (F profiles)* In an effort to reduce the RC delays intrinsic to signal propagation in microelectronic wiring interconnects, there is presently a large effort to develop electrical insulating materials with a low dielectric constant (low-κ dielectrics). Historically, by far the most commonly used dielectric has been SiO_2, with dielectric constant $\kappa = 4.0$. Promising low-κ materials include both 'fluorinated SiO_2' and fluorinated polymers. Kumar and collaborators (Kumar *et al.* 1997, 1999) at the University at Albany have been using resonant nuclear reaction analysis to study both the fluorine content of these materials as they are deposited and the transport of F that occurs when these materials are annealed. Figure 9.7 shows a typical F profile for a fluorinated SiO_2 film, showing that these materials can contain very large amounts of F. Figure 9.8 shows F profiles in an as-deposited and an annealed fluorinated polymer, both with a 1000 Å aluminium cap. As can be seen in Fig. 9.8, the F in the polymer can move significant distances (in this case into the Al cap) under thermal anneals.

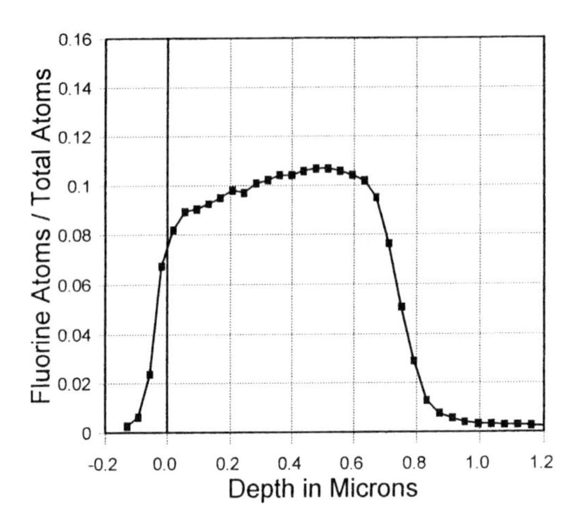

Fig. 9.7 A fluorine profile measured by nuclear reaction analysis of a fluorinated low-κ dielectric film. Reproduced from Kumar *et al.* (1997), with permission from the author.

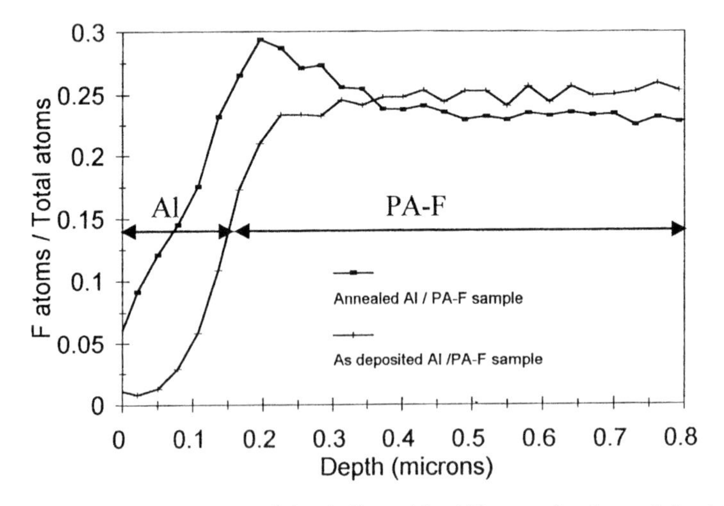

Fig. 9.8 Fluorine profiles for two low-κ dielectric films with a 100 nm surface layer of aluminium. One of the films had been annealed at 450°C, resulting in the transport of fluorine into the aluminium capping layer. Reproduced from Kumar *et al.* (1999), with permission from the author.

9.3.3.5 *Study of N-ion implantation into metal (^{15}N profiles)* The presence of nitrogen near the surface of many metals can have significant effects on both the chemical and mechanical properties of these metal surfaces. Clearly, it is useful to have a method to measure quantitative profiles of N. While the $^{14}N(p,\gamma)^{15}O$ reaction can be used to measure the naturally more abundant isotope, the $^{15}N(p,\alpha\gamma)^{12}C$ has better depth resolution and sensitivity. Figure 9.9 shows ^{15}N profiles for a sample of Al ion implanted with 6×10^{16} N ions cm^{-2} at 50 keV (Miyagawa *et al.* 1996). The top diagram (a) shows the raw data (counts *versus* beam energy) and the bottom (b) shows a concentration profile. Overlaid on the experimental N profile is a numerical calculation of the expected profile.

9.3.3.6 *Migration of Li in electrochromic windows (Li profiles)* Electrochromic windows are clear glass windows with a multilayer thin-film coating that can be

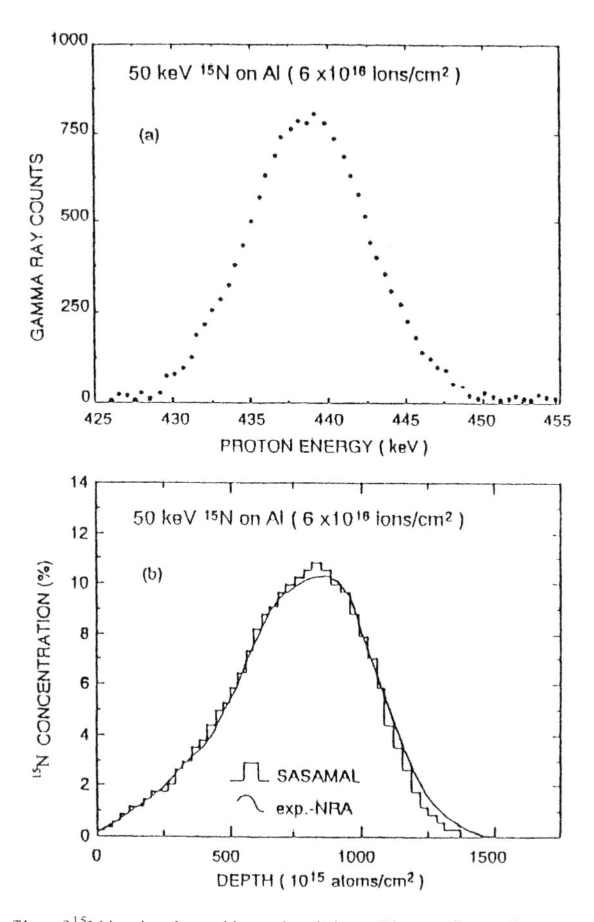

Fig. 9.9 Depth profiles of ^{15}N ion implanted into aluminium. The top figure shows raw data (counts *versus* beam proton energy) and the bottom shows the final profiles (^{15}N concentration *versus* depth). The bottom figure also shows a calculated profile. Reprinted from Miyagawa *et al.* (1996), with permission from Elsevier Science.

reversibly changed from clear to dark by applying an electrical bias. One such device consists of two electrochromic layers (WO_3 and $LiCoO_2$) separated by ion-conducting solid electrolyte. The expectation is that by applying an electric field, Li ions are transported from one electrochromic layer to the other through the solid electrolyte, changing the layer from transparent to opaque (a reversible process). In order to test this model, Li profiles were measured in an electrochromic window and then the window switched from clear to dark. These profiles were measured using the $^7Li(p,\gamma)^8Be$ resonant nuclear reaction which occurs at a proton energy of 440 keV. Such profiles are shown in Fig. 9.10. The electrochromic layers are layers II and IV; layer III is the solid electrolyte ($LiNbO_3$). Layers I and VI are conducting electrode layers, and layer V is a Li blocking layer. For our purposes, the important observation is the large change in the Li profile as measured in the clear state (open squares) and the dark state (closed squares). [Taken from Goldner *et al.* (1993).]

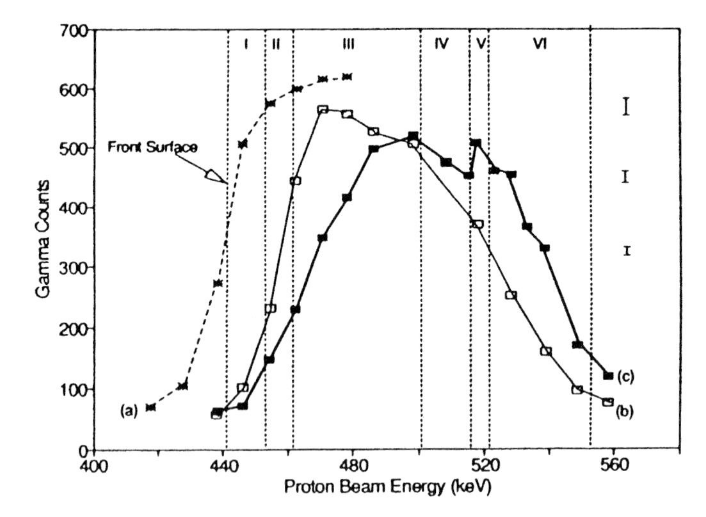

Fig. 9.10 Li NRA yield curves measured on 'smart glass', a glass that can be turned from clear to dark by applying an electric field. What these data show is that Li ions move as the glass switches between the clear state (open squares) and the dark state (filled squares). Those data labelled (a) are for a sample with Li present up to the surface of the sample. From Goldner *et al.* (1993).

9.4 Non-resonant profiling

When reaction cross-sections are sufficiently large and vary smoothly over an extended energy range, the entire depth profile may be obtained using a single incident beam energy. This is referred to as non-resonant profiling. Non-resonant profiling can be used with many reactions in which particles (e.g. protons, deuterons, ^3He, α particles), rather than γ-rays, are the emitted product. An example of this technique is the profiling of ^{18}O using the nuclear reaction $^{18}O(p,\alpha)^{15}N$. Figure 9.11 shows the cross-section of this reaction as a function of incident proton energy, illustrating the large and smoothly varying cross-section in the vicinity of 800 keV. It should be noted that ^{18}O can also be profiled with this reaction using the resonant technique, employing, for example, the sharp resonance at 629 keV.

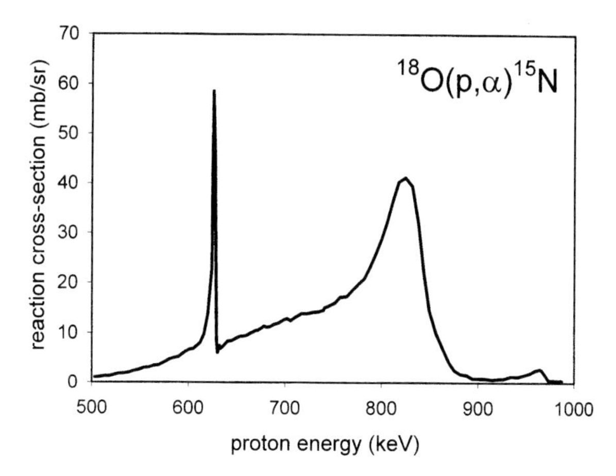

Fig. 9.11 Reaction cross-section of the $^{18}O(p,\alpha)^{15}N$ reaction as a function of proton energy for a laboratory angle of 165°. In this energy region, there is both a narrow resonance around 629 keV suitable for resonant NRA and a broader region of enhanced cross-section around 800 keV suitable for non-resonant NRA. Data from Amsel and Samuel (1967).

For non-resonant profiling, a sample is bombarded with protons at a suitable energy and the α particles resulting from the reaction of the protons with ^{18}O are detected. A spectrum of α particles over a range of energies is collected, representing contributions from ^{18}O at various depths in the material.

Since non-resonant profiling uses the energy of outcoming ions to deduce the depth at which a nuclear reaction occurred, a brief description of the detector used to measure ion energy is perhaps useful. While magnetic spectrographs with much better energy resolution have recently begun to be used (Lanford *et al.* 1999), almost all experimenters use solid-state silicon detectors. These detectors, which are cheap and reliable (see, for example, the catalogue of EG&G Ortec, Oak Ridge, Tennessee), consist of a biased semiconductor (Si) connected to linear electronics. An ion stopping in the semiconductor loses energy by scattering electrons from below the gap to above the gap, creating electron–hole pairs. Hence, an ion stopped in a detector results in a charge pulse the amplitude of which is proportional to the energy of the stopped ion. The typical energy resolution of such a detector is 12–14 keV (measured full width at half maximum).

9.4.1 *Fundamental aspects of non-resonant profiling*

9.4.1.1 *Depth scale* The α spectra are converted to depth profiles in a manner analogous to that outlined above for resonant profiling. However, it must be noted that the ingoing protons as well as the outgoing α particles lose energy when travelling through the sample (unlike the γ-rays produced in the reaction in the examples of resonant profiling cited above). The detected α energy for ^{18}O at the sample surface can be calculated kinematically given the incident proton energy, the detection angle, and the Q value for the reaction (3.97 MeV in this case). As the protons travel deeper into the sample, they lose energy. When these protons of lowered energy interact with the ^{18}O at depth in the material and the nuclear reaction occurs, resultant α particles will have a lower energy than those produced closer to the sample surface. The α particles lose additional energy as they travel out of the material, so α particles originating from a certain depth will have a characteristic energy. This is schematically illustrated in Fig. 9.12. In order to construct the depth scale from this information, the rate of energy loss for both protons and α particles in the material must be determined. Energy loss information is tabulated for most elements and a number of ions (e.g. Ziegler *et al.* 1985), and values for compound targets can be calculated by weighting the elemental contributions according to their abundance in the material. This is referred to as the Bragg Rule. Energy loss in compounds can be readily calculated using the Bragg Rule, for which it is assumed that the interaction processes between the elements of the target compound sample and ions (either incident or produced in the reaction) are independent of the surrounding atoms in the sample. The stopping cross-section (ε) for a compound of form A_mB_n (where $m + n$ is normalized to unity) can be written:

$$\varepsilon^{AB} = m\varepsilon^{A} + n\varepsilon^{B} \tag{9.10}$$

Stopping cross-section is defined as $\varepsilon = (1/N)\,dE/dx$, with N equal to the number or volume density (i.e. number of atoms cm^{-3}), and dE/dx the energy loss per distance x traversed (stopping power, S). As an example, we consider the silicate mineral zircon ($ZrSiO_4$). The stopping cross-section can be written as:

$$\varepsilon^{zirc} = (\varepsilon^{Zr} + \varepsilon^{Si} + 4\varepsilon^{O})/6 \tag{9.11}$$

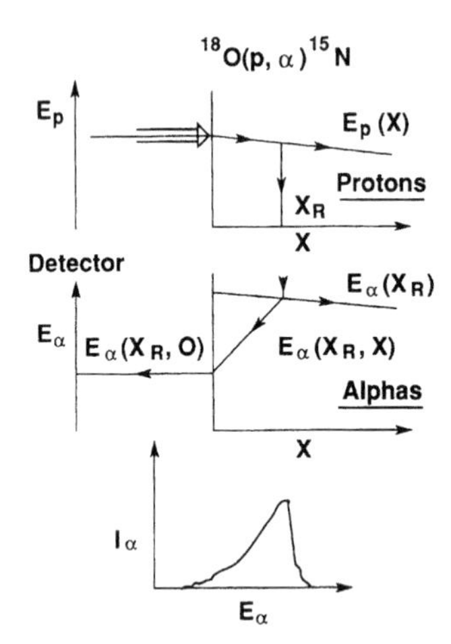

Fig. 9.12 Schematic illustrating energy-depth relations for non-resonant nuclear reaction analysis, using the $^{18}O(p,\alpha)^{15}N$ reaction as an example. Ingoing protons lose energy as they travel through the material. When the reaction occurs at depth x_R, the alpha product will have energy $E_\alpha(x_R,x)$. The α particle loses additional energy travelling out of the sample to the detector, where an α particle energy $E_\alpha(x_R, 0)$ is recorded. The contributions from α particles produced in the $^{18}O(p,\alpha)^{15}N$ reaction from various depths in the material result in a spectrum of detected particle intensity as a function of α energy, with lower α energies corresponding to ^{18}O at greater depths in the sample.

For 3 MeV ^4He, stopping cross-sections are 75.53, 39.17 and 27.54 for Zr, Si, and O, respectively, in units of $eVcm^2/10^{15}$ atoms. From the expression above a stopping cross-section of 37.48 $eVcm^2/10^{15}$ atoms is obtained. The number density of zircon is 9.24×10^{22} atoms cm^{-3}, which yields a stopping power of 346 keV μm^{-1}.

9.4.1.2 *Concentration scales* The number of detected α particles corresponding to a particular depth is a function of the detector solid angle, the total proton flux delivered to the sample, the reaction cross-section at the appropriate energy and the concentration of ^{18}O at that depth, such that

$$C(x_i) = \frac{Y_i}{N_{ox} P\sigma(x_i)\Delta x_i \, \Delta\Omega} \tag{9.12}$$

where $C(x_i)$ is the ^{18}O concentration (as a fraction of total oxygen) at depth x_i, Y_i is the expected count yield at that depth, $\sigma(x_i)$ is the reaction cross-section at depth x_i (which is determined by the energy incident protons have at this depth after travelling the distance x_i through the material), Δx_i is the channel width over which the counts Y_i are collected (determined from the stopping powers in the material for the ingoing and outgoing particles), $\Delta\Omega$ is the solid angle subtended by the particle detector, P is the integrated proton flux to the sample and N_{ox} is the number density of oxygen atoms in the sample (this example considers an oxide sample with no change in total oxygen concentration, only an alteration of the oxygen isotope ratio). In our example of ^{18}O analysis of diffusional uptake of a tracer from an ^{18}O-enriched source, near-surface concentrations of ^{18}O typically range from 10 to 80% of total oxygen. For a standard experimental configuration (detector solid angle 2.5×10^{-3} sr; channel width 5 nm), count yield from the $^{18}O(p,\alpha)^{15}N$ reaction in zircon for a depth at which the reaction cross-section is 30 mbarn sr^{-1} and ^{18}O comprises 20% of the total oxygen, would be 290 counts in that channel (of course, signal from other depths is collected simultaneously in the other energy channels in the analyser) for a total proton dose of 100 μC delivered to the sample (6.25×10^{14} incident protons, requiring about an hour of acquisition time with a beam current of 25 nA).

9.4.1.3 *Depth resolution* The observed profile, however, is a convolution of the actual concentration profile with a 'spreading' or energy resolution function which takes into consideration such factors as the energy spread of the incident beam, proton and α straggling in the sample (i.e. the spread in beam energy due to statistical variation in the number of interactions each individual ion experiences as it travels through the sample), and detector resolution. In the near-surface region, the dominant contributor to depth resolution is the energy resolution of the detector. For a typical surface barrier detector (15 keV FWHM resolution for ^4He ions), this translates to a few tens of nanometres for protons and 10–15 nm for ^4He in most targets. Deeper into the sample, the effects of energy straggle become much more significant as ions travel a greater distance through the sample. They are most pronounced for proton beams, because the ratio of energy straggle to energy loss decreases with increasing ion mass. These effects may be quite substantial; for example, depth resolutions worse than 1000 Å are typical for 1 MeV protons a few microns into a material.

9.4.1.4 *Sensitivity* The sensitivity of non-resonant NRA is affected by the cross-sections of individual reactions, interfering reactions and other background effects. Incident beam energies should be selected to maximize yields from reactions with the species being profiled, with due consideration of the possibility that interfering nuclear reactions might be induced with other species in the material. However, Q values of respective reactions generally differ enough so that energy separation of signals from reaction products is adequate, and reaction cross-sections of the species of interest and potential interfering reactions will rarely be simultaneously large at a given energy.

The surface barrier detectors, which are commonly used to detect the products of the nuclear reaction in non-resonant profiling, may also detect the signal from incident ions that have been backscattered from the sample. Figure 9.13 shows an α particle spectrum due to the reaction $^{18}O(p,\alpha)^{15}N$ along with the signal produced by backscattered protons. The yield of backscattered particles (which is proportional to Z^2) may overwhelm the electronics in the detecting system, resulting in pile-up, which greatly reduces sensitivity. This difficulty is compounded by the fact that cross-sections for nuclear reactions are generally much smaller than backscattering cross-sections and either (or both) long acquisition times or relatively large incident ion beam currents are necessary to collect reasonable count yields of the reaction products of interest. A solution to this problem is to shield the detector with a thin-film absorber (such as Mylar). The absorber (of appropriate thickness) stops the backscattered incident ions while permitting the higher-energy ions from the nuclear reaction to pass through to the detector. However, the presence of an absorbing material degrades depth resolution, since additional energy spread occurs as the ions travel through the absorber to the detector.

Because of this trade-off between depth resolution and sensitivity, the experimenter

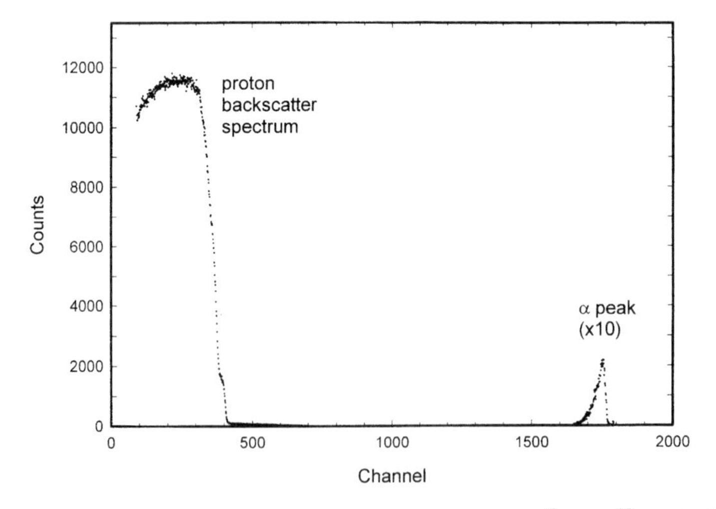

Fig. 9.13 Typical spectrum for non-resonant NRA using the reaction $^{18}O(p,\alpha)^{15}N$. The distribution of counts below ~ channel 500 is due to elastic scattering of incident protons from target atoms. The peak above channel 1500 is α particles produced in the nuclear reaction between incident protons and target ^{18}O.

should consider whether the use of absorbers is necessary in each case, or whether another type of detection method might be more effective. In ^{18}O profiling, for example, absorbers would be needed when profiling tantalum oxide but may not be required during analysis of most glasses, minerals and oxides with low-Z matrices, where proton ion backscattering is much reduced.

However, other sorts of detectors may mitigate this difficulty. For example, electrostatic detectors or magnetic spectrometers, which rely on the differences in the trajectories of energetic charged particles travelling in magnetic or electrostatic fields due to differences in particle mass, charge and energy, or time-of-flight detectors, may warrant use for some applications despite their greater complexity compared with surface barrier detectors. In cases where surface barrier detectors are used, sometimes a judicious choice of detector thickness can resolve analytical problems. For example, particle energies in a particular reaction may be such that α and signals from lighter ions overlap. If the α signal is of greater interest, a thin detector, which will capture and fully stop the α particles but permit most of the lighter ions to pass through with only a fraction of their energy lost, may be successfully employed.

9.4.2 *Standards*

Concentrations may be determined without reference to standards if all of the above experimental parameters are well known or readily measured, but this is not always practical. Standards are often useful, especially if cross-sections are poorly known for the beam–detector geometry used. Good standards should possess lateral homogeneity over the area of the beam, have well-characterized concentration of the species of interest over the relevant depth range and remain stable under vacuum and when bombarded by the ion beam. Measurements on standards and samples of unknown composition should be performed under conditions as near to identical as possible (same incident beam energy, same detector geometry and absorber foil, etc.). For a bulk standard and bulk specimen (i.e. neither a thin film), the surface concentration of the unknown can be determined from the standard by using the equation

$$C_{\mathrm{u}} = \frac{H_{\mathrm{A}}^{\mathrm{u}}}{H_{\mathrm{A}}^{\mathrm{std}}} \frac{[\varepsilon]_{\mathrm{A}}^{\mathrm{u}}}{[\varepsilon]_{\mathrm{A}}^{\mathrm{std}}} \frac{P_{\mathrm{std}}}{P_{\mathrm{u}}} C_{\mathrm{std}} \tag{9.13}$$

where the subscripts and superscripts 'u' and 'std' denote the unknown and standard, respectively. H_{A} is surface height of the signal due to species A, C is the concentration of species A, $[\varepsilon]_{\mathrm{A}}$ is the stopping cross-section factor, and P is the number of incident ions. It is assumed that the experimental system set-up remains the same for analyses of the standard and the unknown.

9.4.3 *Applications – non-resonant NRA*

Since NRA induces specific nuclear reactions, it permits selective observation of certain isotopes. This makes it ideal for tracer experiments using stable isotopes. For example, the $^{18}O(p,\alpha)^{15}N$ reaction has been used for oxygen diffusion studies in a variety of oxides, including silicon dioxide, zinc oxide, rutile, alumina, and the minerals olivine and zircon. The following section presents a few examples of applications of non-resonant NRA in investigating a range of materials. Additional examples are cited in Table 9.1.

9.4.3.1 *Oxygen diffusion in rutile* Rutile (TiO_2) shows semiconducting properties when doped with various elements, including H and Nb, and also finds technological application as a refractory, photocatalyst and substrate for thin-film growth. Rutile is present in nature in minor amounts in many geologic environments, and it is often employed in determining thermal histories through the measurement of its oxygen isotope ratios. Moore *et al.* (1998) measured oxygen diffusion in synthetic and natural rutile under both hydrothermal and dry conditions, using an ^{18}O tracer and the nuclear reaction $^{18}O(p,\alpha)^{15}N$. Two mechanisms for oxygen diffusion were observed, with diffusion slower when rutile was annealed under reducing conditions in the presence of hydrous species. The authors suggest that in the faster process diffusion is dominated by the migration of oxygen vacancies, while the slower process primarily involves the migration of Ti^{3+} interstitials, with a subordinate contribution from O vacancies. The difference in diffusion rates with annealing environments can be seen in Fig. 9.14, where a much shorter ^{18}O profile was found for the sample annealed under hydrothermal conditions.

9.4.3.2 *Oxygen diffusion in sapphire* Alumina is a versatile material, whose properties can be significantly affected by doping. Reddy and Cooper (1982) have investigated oxygen diffusion in sapphire using an ^{18}O tracer and the $^{18}O(p,\alpha)^{15}N$ reaction. Diffusion of oxygen was found to proceed somewhat more slowly in Ti-doped alumina when compared with undoped or Mg-doped specimens (Fig. 9.15).

9.4.3.3 *Residual carbon in ceramic substrates* Multilayer ceramic substrates are used widely as multiple chip carriers in high-performance microelectronic packaging technologies. These substrates, however, may contain residual carbon after processing, which can have adverse effects on their mechanical and electrical properties, even if C is present at ppm levels. Chou *et al.* (1990) have investigated the carbon content of these ceramics, and the effects of surface cleaning, with the reaction $^{12}C(d,p)^{13}C$. Figure 9.16 shows the spectra of ceramic substrates, illustrating the significant reduction in carbon following surface cleaning.

9.4.3.4 *Sulfur in bulk CdTe and CdTe/CdS thin-film heterojunctions* CdTe/CdS heterojunctions have potential as materials for solar cells of low cost and high efficiency. Cell efficiencies can be reduced by the formation of intermediate CdS_xTe_{1-x}layers during the annealing step in fabrication. Lane *et al.* (1998) have measured sulfur diffusion in bulk CdTe and CdTe/CdS thin-film heterojunctions using the nuclear reaction $^{32}S(d,p_0)^{33}S$. (A subscript for p is used in the expression for the nuclear reaction because more than one group of protons is produced in the reaction. These proton groups, of differing energy, are named in increasing numerical sequence with decreasing particle energy (i.e. p_0, p_1, ..., etc.). Selection of a particular proton group for detection and measurement will depend on interferences and particle yields.) Figure 9.17 shows NRA spectra for a single-crystal CdTe sample annealed with CdS

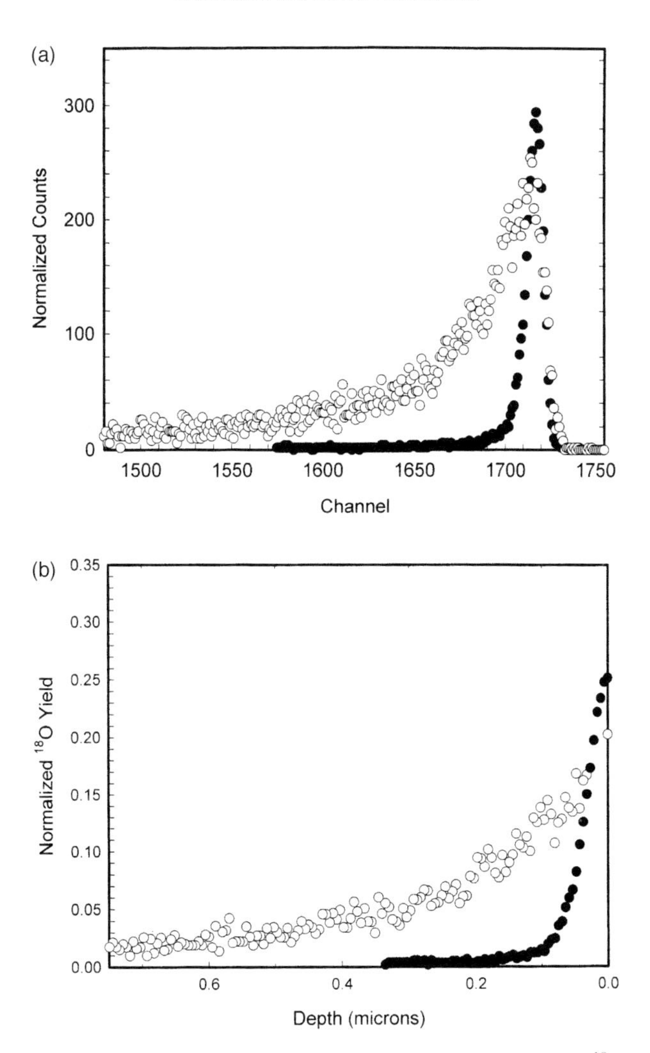

Fig. 9.14 Measurements of oxygen diffusion in rutile (TiO_2) using the reaction $^{18}O(p,\alpha)^{15}N$. The open circles are for an experiment run under dry conditions, the solid symbols for one done under hydrothermal conditions. Penetration depth (and therefore diffusion coefficient) is considerably smaller for the hydrothermal experiment, despite similar temperatures and duration for diffusion anneals. Data from Moore *et al.* (1998).

vapour at two different temperatures. The penetration depth of S (as evidenced by higher count yield at lower energy) is greater for the higher temperature anneal, and S diffusion rates measured with NRA were comparable to those determined by SIMS.

9.4.3.5 *Silicon diffusion in gold* Mathot and Demortier (1990) have investigated Si diffusion from a thin Si film into polycrystalline gold foils with microbeam NRA. The $^{28}Si(d,p)^{29}Si$ reaction was employed, with the deuteron beam penetrating directly through the foil (typically about 16 μm in thickness) to the detector, to take advantage

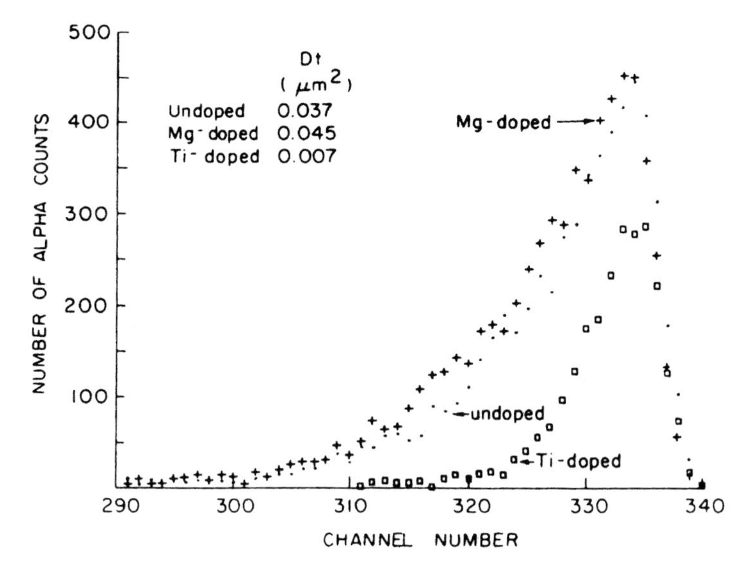

Fig. 9.15 Oxygen diffusion in alumina both undoped, and doped with Mg or Ti (Reddy and Cooper, 1982). Alpha count yields from the reaction $^{18}O(p,\alpha)^{15}N$ show a much narrower distribution of ^{18}O in the Ti-doped alumina, indicating a slower oxygen diffusion rate than in the Mg-doped or undoped specimens. The Dt values in the figure inset are the oxygen diffusion coefficients multiplied by experimental anneal times, in units of microns squared.

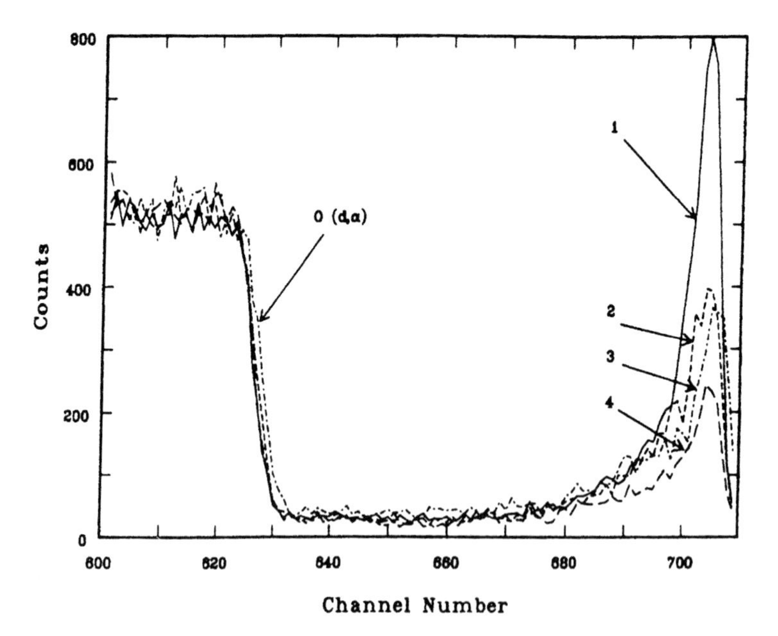

Fig. 9.16 NRA spectra of ceramic substrate materials, investigating the effect of surface cleaning on carbon concentration with the reaction $^{12}C(d,p)^{13}C$. (1) Spectrum of a specimen before cleaning; (2) spectrum of the same specimen after cleaning; (3,4) spectra of other surface-cleaned specimens. Reprinted from Chou *et al.* (1990), with permission from Elsevier Science.

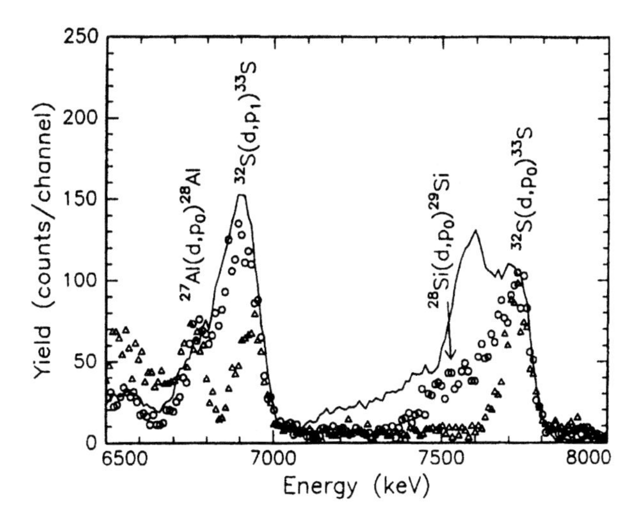

Fig. 9.17 $^{32}S(d,p)^{33}S$ NRA spectra for single crystal CdTe exposed to CdS vapour for one week at 450°C (Δ) or 550°C (O). The solid line is the spectrum of a ZnS standard (yield ÷ 10). Also indicated are contributions to the spectrum from (d,p) reactions with ^{27}Al and ^{28}Si. Reprinted from Lane *et al.* (1998), with permission from Elsevier Science.

of increased reaction cross-section at 0° relative to the incident beam. The geometric arrangement also permitted analyses from both sides of the foil, to better characterize grain-boundary diffusion of Si. Figure 9.18(a) shows proton yields from the $^{28}Si(d,p)^{29}Si$ reaction for samples annealed for various times, showing the increase in Si content in grain boundaries with diffusion for longer times. With the microbeam, three-dimensional mapping of Si concentrations is also possible. Figure 9.18(b) shows Si distributions around and in Au grain boundaries.

9.4.3.6 *Uranium dioxide corrosion* An important means of assessing the safety of disposal of spent nuclear fuel is to investigate the alteration of UO_2 under conditions similar to those it would be exposed to in repositories. Trocellier *et al.* (1998) have investigated the corrosion of UO_2 exposed to a leachant comparable to thermal granitic groundwaters by using NRA. Oxygen profiles were measured with microbeams, using the reactions $^{16}O(^3He,p_0)^{18}F$ and $^{16}O(^3He,\alpha_0)^{15}O$. Figure 9.19 compares the spectra of leached and unleached samples, indicating the degree of alteration of the near-surface region. Because of the greater depth range of the protons compared with the α particles, the α peak from the $^{16}O(^3He,\alpha_0)^{15}O$ reaction contains a greater contribution from near-surface ^{16}O than does the proton peak from the $^{16}O(^3He,p_0)^{18}F$ reaction, providing added information to better characterize the processes involved in UO_2 alteration.

9.5 Multiple probes for light element analysis

Nuclear reaction analysis is isotope specific. While this is often an advantage (e.g. in avoiding background interferences), it generally implies the need to design an

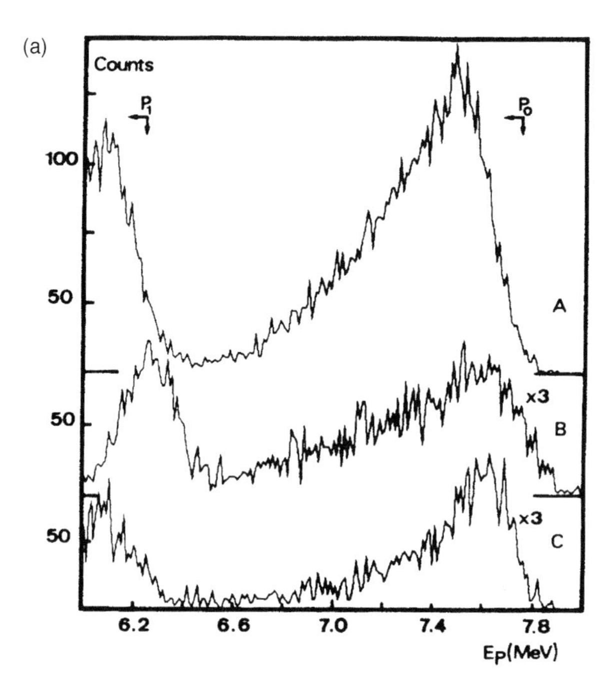

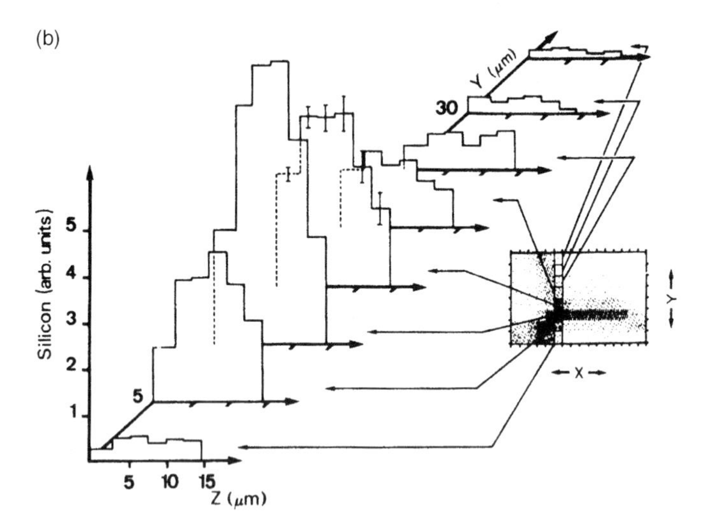

Fig. 9.18 Investigation of silicon diffusion in polycrystalline gold foil using the nuclear reaction $^{28}Si(d,p)^{29}Si$ (reprinted from Mathot and Demortier, 1990, with permission from Elsevier Science) and a microbeam. (a) Spectra of protons from a grain boundary after annealing for 1 hour (A), 30 min (B), and 15 min (C), showing the increase in Si concentration with increasing anneal time. (b) 3D distribution of Si in and around grain boundaries by microbeam mapping. Both lateral and depth information can be obtained with the microbeam.

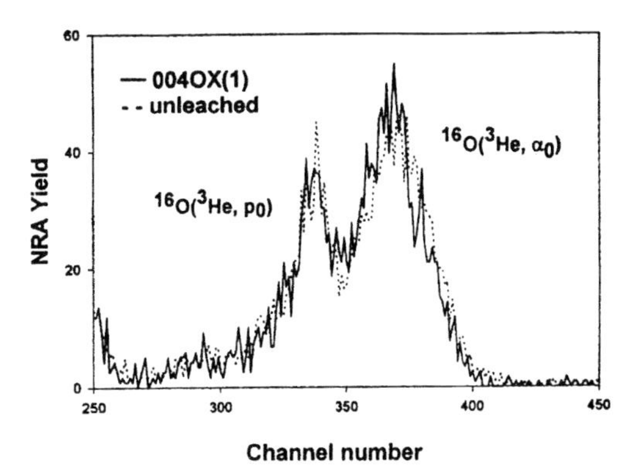

Fig. 9.19 ^{3}He microbeam analysis of a UO_2 pellet leached in a granite groundwater at 96°C, compared with the spectrum of an unleached specimen. Both proton and α yields from the $^{16}O(^3He,p_0)^{18}F$ and $^{16}O(^3He,\alpha_0)^{15}O$ reactions appear in the spectrum. Because α particles have higher energy loss rates per unit depth in the material than protons, contributions from near-surface ^{16}O comprise a greater percentage of the yield in the α peak; the proton yield samples ^{16}O to greater depth in the material. Reprinted from Trocellier *et al.* (1998), with permission from Elsevier Science.

experiment to probe for a specific isotope. This is unlike methods (such as RBS, PIXE or ERDA) which automatically probe for a broad range of elements. For the analysis of light elements, it clearly would be an advantage if there were a probe that would provide analysis for multiple light elements in a single measurement. Further, it would be preferable if such a measurement could be made using commonly available experimental chambers. In consideration of these factors, the use of He elastic scattering at energies above the Coulomb barrier (where the scattering cross-sections can be orders of magnitude larger than the Rutherford cross-section) is increasingly being used to probe for light elements (Davies *et al.* 1994). Typical experimental chambers are exactly the same as those used in conventional ~2 MeV He RBS measurements, but conducted with higher He beam energies.

The ideal situation would be to locate an energy at which the elastic cross-section was large for all light elements simultaneously. Unfortunately, there is no such energy. However, for He ions at 5.75 MeV, the backscattering cross-section from both C and O are about 100 times larger than Rutherford; hence beams of this energy can be used quite effectively to probe for these important light elements (Davies *et al.* 1994).

As an example of this approach, consider the analysis of thin $Si_{1.0}O_xC_yH_zAr_{zz}$ films. The H content of such material can be measured using NRA, as described above. However, conventional 2 MeV He RBS provides only weak C and O signals on a large background making accurate compositional analysis impossible. Figure 9.20(a) shows a 2 MeV spectrum from such a film. However, making a measurement at 5.75 MeV, where the cross-sections for C and O are very large, gives good signals for these two elements. Figure 9.20(b) shows such a 5.75 MeV spectrum. The analysis of such non-Rutherford data follows exactly the same procedures as for conventional RBS (except for the cross-sections) and will not be repeated here (see Chapter 10). In practice, because the cross-sections may not be reliably known from the literature, it is

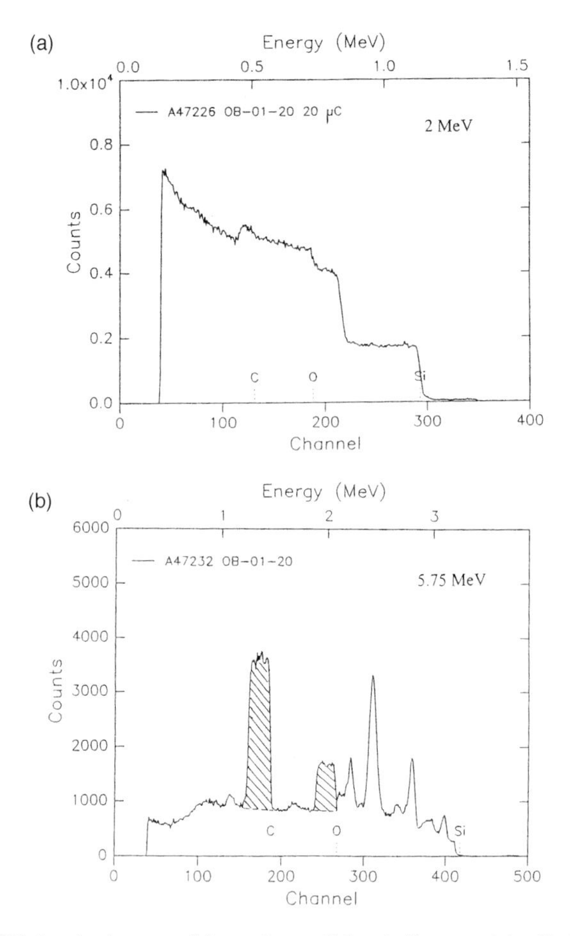

Fig. 9.20 Spectra of He ions backscattered from a low-κ dielectric film containing Si, O, C, H, and Ar. The top spectrum was recorded at 2 MeV, where the scattering cross-sections are given by the Rutherford formula. The spectrum at the bottom was recorded at 5.75 MeV where the cross-sections for both C and O are orders of magnitude larger than that given by the Rutherford formula. These large cross-sections greatly enhance the sensitivity for these elements. As seen in the 5.75 MeV spectrum, there are large peaks from the C and O in the film.

convenient to measure the C and O yields from materials of known composition (e.g. SiO_2, various plastics) to establish the absolute elastic cross-sections.

It is also clear from the spectrum that the 5.75 MeV cross-section on Si is not Rutherford either. The cross-section on Si can also be determined from measurement on material of known composition. An alternative is to determine the Si content of the material by measurement at a lower beam energy where the cross-section is known to be Rutherford. Using NRA (to determine the H content), 2 MeV He backscattering (to determine Si and Ar contents) and 5.75 MeV He (to determine the C and O contents) the composition of the film used in Fig. 9.20 was determined to be $Si_{1.00}O_{1.00}C_{0.55}H_{0.90}Ar_{0.02}$. As a check of this composition, the RBS simulation shown in Fig. 9.20(a) is done assuming this composition. Note that while the 2 MeV RBS data do not determine the film composition very precisely (when used alone), the RBS simulation using the correct film composition must be consistent with the 2 MeV measurement.

This last example also illustrates another feature becoming more common in MeV ion beam analysis of thin films. Namely, more researchers are using multiple measurements on a single sample to determine composition. As the analytical problems become more complex, such as that illustrated in Figure 9.20, these multiple probe approaches are necessary to obtain high-quality analysis for all the elements present. In addition, multiple measurements on a sample often provide redundancy that can be useful in detecting any errors in an analysis.

9.6 Summary

NRA is an effective technique for measuring near-surface (the first few micrometres) depth profiles of light elements in a variety of solid materials. Depth resolutions typically range from a few nanometres to tens of nanometres, and lateral resolutions down to a micrometre can be achieved with microbeams. The sensitivity (typical detection limits of $\sim$10–100 ppm, depending on the reaction used) and the isotope-selective character of NRA make it deal for isotopic tracer experiments such as diffusion studies. An especially important application of NRA is the measurement of quantified hydrogen depth profiles, which is difficult with most other analytical techniques.

NRA is largely insensitive to solid-state matrix effects since it relies on nuclear reactions, so it can be made quantitative without reference to standard samples. In addition, measurements are rapid and 'non-destructive'. However, since NRA is sensitive only to the nuclei present in the sample, it cannot provide information on chemical bonding or microscopic structure. It is therefore frequently employed along with other techniques such as ESCA, optical absorption, Auger, or electron microscopy, which can provide this information. NRA is often used in conjunction with RBS, another ion beam technique more sensitive when profiling heavier elements, in materials characterization.

Prospects for these techniques include further application of microbeams for three-dimensional elemental and isotope mapping, the use of NRA in a broader range of fields, the increased use of multiple measurements (a range of beams and/or beam energies) on a single sample to better characterize composition, and continued development of easily accessible electronic and print databases of reaction cross-sections.

References

Abel, F., Amsel, G., d'Artemare, E., Ortega C., Siejka, S. & Vizkelethy, G. (1990) *Nuclear Instruments and Methods*, **B45**, 100–104.
Adler, P.N. & Schulte, R.L. (1978) *Scripta Metallurgica*, **12**, 669–672.
Ager, F.J., Respaldiza, M.A., Paúl, A., Odriozola, J.A., da Silva, M.F. & Soares, J.C. (1998) *Nuclear Instruments and Methods*, **B136–138**, 1045–1051.

Alkemade, P.F., Stap, C.A.M., Habraken, F.H.P.M. & van der Weg, W.F. (1988) *Nuclear Instruments and Methods*, **B35**, 135–139.

Amsel, G. & Maurel, B. (1983) *Nuclear Instruments and Methods*, **218**, 183–189.

Amsel, G. & Samuel, D. (1967) *Analytical Chemistry*, **39**, 1689–1698.

Amsel, G., d'Artemare, E., Battistig, G., Girard, E., Gosset, L.G., & Révész, P. (1998) *Nuclear Instruments and Methods*, **B136–138**, 545–550.

Anttila, A., Bister, M., Fontell, A. & Winterbon, K.B. (1977) *Radiation Effects*, **33**, 13–19.

Anttila, A., Hänninen, R. & Räisänen, J. (1981) *Journal of Radioanalytical Chemistry*, **62**, 293–306.

Bedell, S. & Lanford, W.A. (1998) *MRS Conference Proceedings*, **513**, 369.

Bedell, S. (1999) Ph.D. thesis, University at Albany, SUNY.

Bethge, K. (1992) *Nuclear Instruments and Methods*, **B66**, 146–157.

Bethge, K. (1985) *Nuclear Instruments and Methods*, **B10–11**, 633–638.

Biggerstaff, J.A., Hood, R.F., Scott, H. & McEllistrom, M.T. (1962) *Nuclear Physics*, **32**, 631–641.

Bilwes, B., Bilwes, R., Ferrero, J.L. & Garcia, A. (1978) *J. Physique*, **39**, 805–814.

Bird, J.R. & Williams, J.S. (1989) *Ion Beams for Materials Analysis*. Academic Press, Sydney.

Bohne, W., Bommer, J., Fuchs, H., Grabisch, K., Kluge, H. & Roschert, G. (1972) *Nuclear Physics*, **A196**, 41–57.

Bohr, N. (1948) *Mat. Fys. Medd. Dan. Vid. Selsk*, **18**, 8.

Borgardt, J.D., Ashbaugh, M.D., McIntyre, jr., L.C., Stoner, jr., J.O., Gregory, R.B., Azrak, M. & Wetzel, J. (1998) *Nuclear Instruments and Methods*, **B136–138**, 528–532.

Brodsky, M., Frisch, M.A., Ziegler, J.F. & Lanford, W.A. (1977) *Applied Physics Letters*, **30**, 561.

Carnera, A., Della Mea, G., Drigo, A.V., Lo Russo, S. & Mazzoldi, P. (1977) *Journal of Non-Crystalline Solids*, **23**, 123–128.

Chou, N.J., Zabel, T.H., Kim, J. & Ritsko, J.J. (1990) *Nuclear Instruments and Methods*, **B45**, 86–90.

Coetzee, P.P., Pretorius, R. & Peisach, M. (1975) *Journal of Radioanalytical Chemistry*, **25**, 283–292.

Cohen, D.D. & Rose, E.K. (1992) *Nuclear Instruments and Methods*, **B66**, 158–190.

Coote, G.E. (1992) *Nuclear Instruments and Methods*, **B66**, 191–204.

Courel, P., Toulhoat, N. & Cuney, M. (1991) *Vacuum*, **42**, 821–822.

Cummings, K., Lanford, W.A. & Feldman, M. (1998) *Nuclear Instruments and Methods*, **B136–138**, 858–863.

Davies, J.A., Almeida, F.J.D., Haugen, H.K., Siegele, R., Forster, J.S. & Jackman, T.E. (1994) *Nuclear Instruments and Methods*, **B85**, 28–33.

Deconninck, G. (1973) *Journal of Radioanalytical Chemistry*, **17**, 29–43.

Deconninck, G. & van Oystaeyen, B. (1983) *Nuclear Instruments and Methods*, **218**, 165–170.

Della Mea, G. (1986) *Nuclear Instruments and Methods*, **B15**, 495–501.

Dieumagard, D., Maurel, B. & Amsel, G. (1980) *Nuclear Instruments and Methods*, **168**, 93–103.

Doyle, B.L., Follstaedt, D.M., Picraux, S.T., Yost, F.G., Pope, L.E. & Knapp, J.A. (1985) *Nuclear Instruments and Methods*, **B7/8**, 166–170.

El Khakani, M.A., Moncoffre, N., Marest, G. & Tousset, J. (1992) *Nuclear Instruments and Methods*, **B64**, 443–447.

Endisch, D., Sturm, H. & Rauch, F. (1994) *Nuclear Instruments and Methods*, **B84**, 380–392.

Endt, P.M. & van der Leun, C. (1973) *Nuclear Physics*, **A214**, 1–625.

Everling, E., Koenig, L.A., Mattauch, J.H.E. & Wapstra, A.H. (1961) 1960 *Nuclear Data Tables. Part I.* National Academy of Sciences, Washington, DC.

Giorginis, G., Misaelides, P., Crametz, A. & Conti, M. (1996a) *Nuclear Instruments and Methods*, **B118**, 224–227.

Giorginis, G., Misaelides, P., Crametz, A. & Conti, M. (1996b) *Nuclear Instruments and Methods*, **B113**, 396–398.

Goldner, R.B., Haas, T.E., Arntz, F.O., Slaven, S., Wong, K.K., Wilkens, B., Shepard, C. & Lanford, W.A. (1993) *Applied Physics Letters*, **62**, 1699–1701.

Gossett, C.R. (1980) *Nuclear Instruments and Methods*, **168**, 217–221.

Gossett, C.R. (1981) *Nuclear Instruments and Methods*, **191**, 335–340.

Gossett, C.R. (1985) *Nuclear Instruments and Methods*, **B10/11**, 722–726.

Grambole, D., Herrman, F. & Klabes, R. (1992) *Nuclear Instruments and Methods*, **B64**, 399–402.

Healy, M.J.F. & Lane, D.W. (1998) *Nuclear Instruments and Methods*, **B136–138**, 66–71.

Hirvonen, J.K. & Lucke, W.H. (1978) *Nuclear Instruments and Methods*, **149**, 295–299.

Hirvonen, J.-P., Nastasi, M., Lappalainen, R. & Sickafus, K. (1990) *Mat. Res. Soc. Symp. Proc.*, **157**, 203.

Holmgren, H.D., Wolicki, E.A. & Johnston, R.L. (1956) *Physical Review*, **114**, 1281–1285.

Horn, K. & Lanford, W.A. (1988) *Nuclear Instruments and Methods*, **29**, 609–614.

Jarmie, N. & Seagrove, J. (1957) Los Alamos Scientific Laboratory Report, LA-2014.

Jenkin, J.G., Earwaker, L.G. & Titterton, E.W. (1964) *Nuclear Physics*, **50**, 516–524.

Kido, Y., Kakeno, M., Yamada, K., Hioki, T. & Kawamoto, J. (1982) *Journal of Applied Physics*, **53**, 4812–4816.
Kuan, H.-M., Bonner, T.W. & Risser, J.R. (1964) *Nuclear Physics*, **51**, 481–517.
Kumar, A., Bakhru, H., Wang, B., Yang, G.R., Fortin, J., McDonald, J. & Lu, T.M. (1999) *Materials Chemistry and Physics*, **59**, 135–139.
Kumar, A., Bakhru, H., Haberl, A.W., Carpion, R.A. & Ricci, A. (1997) Characterization of fluorinated silicon dioxide films by nuclear reaction analysis and optical techniques. In: *Applications of Accelerators in Research and Industry* (eds. J.L. Duggan & I.L. Morgan), pp. 697–705. AIP Press, New York.
Lane, D.W., Conibeer, G.J., Romani, S., Healy, M.J.F. & Rogers, K.D. (1998) *Nuclear Instruments and Methods*, **B136–138**, 225–230.
Lanford, W.A., Trautvetter, H.P., Ziegler, J.F. & Keller, J. (1976) *Applied Physics Letters*, **28**, 566–576.
Lanford, W.A., Davis, K., LaMarche, P., Lauresen, T., Groleau, R. & Doremus, R.H. (1979) *Journal of Non-Crystalline Solids*, **33**, 249–256.
Lanford, W.A. (1980) *Solar Cells*, **2**, 351–355.
Lanford, W. A. (1992) *Nuclear Instruments and Methods*, **B66**, 65–75.
Lanford, W.A., Cummings, K., Haberl, A. & Shepard, C. (1995) Prompt gamma-ray resonant nuclear reaction analysis for light elements H, Li, F and Na. In: *Application of Particle and Laser Beams in Materials Technology* (ed. P. Misaelides), pp. 375–395. Kluwer, Dordrecht.
Lanford, W.A., Bedell, S., Amadon, S., Haberl, A., Skala, W. & Hjorvarsson, B. (2000) *Nuclear Instruments and Methods*, **B161–163**, 202–206.
Lappalainen, R. (1986) *Physical Review*, **B34**, 3076–3085.
Lennard, W.N., Alkemade, P.F.A., Mitchell, I.V. & Tong, S.Y. (1991) *Nuclear Instruments and Methods*, **B61**, 1–7.
Lu, X., Yuan, Y., Zheng, Z., Jaing, W. & Liu, J. (1989) *Nuclear Instruments and Methods*, **B43**, 565–569.
Marion, J.B., Weber, G. & Mozer, F.S. (1956) *Physical Review*, **104**, 1402–1407.
Marion, J.B. & Weber, G. (1956) *Physical Review*, **103**, 167–171.
Massiot, P., Sommer, F., Thellier, M. & Ripoli, C. (1992) *Nuclear Instruments and Methods*, **B66**, 250–257.
Mathot, S. & Demortier, G. (1990) *Nuclear Instruments and Methods*, **B50**, 52–56.
Maurel, B. & Amsel, G. (1983) *Nuclear Instruments and Methods*, **218**, 159–164.
Maurel, B., Amsel, G. & Dieumegard, D. (1981) *Nuclear Instruments and Methods*, **191**, 349–356.
Maurel, B., Amsel, G. & Nadai, J.P. (1982) *Nuclear Instruments and Methods*, **197**, 1–13.
Mayer, M., Annen, A., Jacob, W. & Grigull, S. (1998) *Nuclear Instruments and Methods*, **B143**, 244–252.
Mayer, J.M. & Rimini, E. (eds.) (1972) *Ion Beam Handbook for Materials Analysis*. Academic Press, New York.
McIntyre, jr., L.C., Leavitt, J.A., Ashbaugh, M.D., Borgardt, J., Andrade, E., Rickards, J. & Oliver, A. (1996) *Nuclear Instruments and Methods*, **B118**, 219–223.
McIntyre, jr., L.C., Leavitt, J.A., Ashbaugh, M.D., Lin, Z. & Stoner, jr., J.O. (1992) *Nuclear Instruments and Methods*, **B66**, 221–225.
McIntyre, jr., L.C., Leavitt, J.A., Ashbaugh, M.D., Dezfouly-Arjomady, B., Lin, Z., Oder, J., Farrow, R.F.C. & Parkin, S.S.P. (1990) *Nuclear Science and Techniques (China)*, **1**, 56–61.
McIntyre, jr., L.C., Leavitt, J.A., Dezfouly-Arjomady, B. & Oder, J. (1988) *Nuclear Instruments and Methods*, **B35**, 446–450.
McMillan, J.W. (1992) *Nuclear Instruments and Methods*, **B66**, 118–125.
Miyagawa, Y., Saitoh, K., Ikeyama, M., Nakao, S. & Miyagawa, S. (1996) Convention depth profiling using narrow nuclear resonances, *Nuclear Instruments and Methods*, **B118**, 209–213.
Möller, W. & Besenbacher, F. (1980) *Nuclear Instruments and Methods*, **168**, 111–114.
Moncoffre, N. (1992) *Nuclear Instruments and Methods*, **B66**, 126–138.
Moncoffre, N., Millard, N., Jaffrazic, H. & Tousset, J. (1990) *Nuclear Instruments and Methods*, **B45**, 81–85.
Moore, D.K., Cherniak, D.J. & Watson, E.B. (1998) *American Mineralogist*, **83**, 700–711.
Pillay, A.E. & Peisach, M. (1992) *Nuclear Instruments and Methods*, **B66**, 226–229.
Pretorius, R. & Peisach, M. (1978) *Nuclear Instruments and Methods*, **149**, 69–72.
Purser, K.H. & Wildenthal, B.H. (1963) *Nuclear Physics*, **44**, 22–33.
Peacock, A.T., Coad, J.P., Lama, F., Behrisch, R., Martinelli, A.P., Mills, B.E., Pick, M., Partridge, J., Simpson, J.C.B. & Zhu, Y.K. (1990) *Journal of Nuclear Materials*, **176/177**, 326–331.
Räisänen, J. (1992) *Nuclear Instruments and Methods*, **B66**, 107–117.
Reddy, K.P.R. & Cooper, A.R. (1982) *Journal of the American Ceramic Society*, **65**, 634–638.
Reichle, R., Behrisch, R. & Roth, J. (1990) *Nuclear Instruments and Methods*, **B50**, 68–73.
Riley, P.J., Lock, G.A., Rawlins, J.A. & Shin, Y.M. (1967) *Nuclear Physics*, **A96**, 641–657.
Sagara, A., Kamada, K. & Yamaguchi, S. (1988) *Nuclear Instruments and Methods*, **B34**, 465–469.
Sawicki, J.A., Davies, J.A. & Jackman, T.E. (1986) *Nuclear Instruments and Methods*, **B15**, 530–534.

Schiffer, J.P., Bonner, T.V., Davis, R.H. & Prosser, F.W. (1956) *Physical Review*, **104**, 1064–1068.
Schulte, R.L. (1978) *Nuclear Instruments and Methods*, **149**, 65–68.
Schulte, R.L., Papazian, J.M. & Adler, P.N. (1986) *Nuclear Instruments and Methods*, **B15**, 550–554.
Seah, M.P., David, D., Davies, J.A., Jackman, T.E., Jeynes, C., Ortega, C., Read, P.M., Sofield, C.J. & Weber, G. (1988) *Nuclear Instruments and Methods*, **B30**, 140–151.
Sierk, A.J. & Tombrello, T.A. (1973) *Nuclear Physics*, **A210**, 341–354.
Soltani-Farsi, M., Mayer, J.D., Misaelides, P. & Bethge, K. (1996) *Nuclear Instruments and Methods*, **B113**, 399–402.
Terwagne, G., Bodart, F., Blanpain, B., Mohrbacher, H. & Celis, J.-P. (1995) *Nuclear Instruments and Methods*, **B104**, 266–270.
Terwagne, G., Cohen, D.D. & Collins, M.G.A. (1994) *Nuclear Instruments and Methods*, **B84**, 415–420.
Tesmer, J.R. & Nastasi, M. (eds.) (1995) *Handbook of Modern Ion Beam Materials Analysis*. Materials Research Society, Pittsburgh, PA.
Toivanen, R.O., Hirvonen, J.-P. & Lindroos, V.K. (1985) *Nuclear Instruments and Methods*, **B7/8**, 200–204.
Toulhoat, N., Trocellier, P., Massiot, P., Gosset, J., Trabelsi, K. & Rouaud, T. (1991) *Nuclear Instruments and Methods*, **B54**, 312–316.
Toulhoat, N., Courel, P., Trocellier, P. & Gosset, J. (1993) *Nuclear Instruments and Methods*, **B77**, 436–443.
Trocellier, P., Cachoir, C. & Guilbert, S. (1998) *Nuclear Instruments and Methods*, **B136–138**, 357–361.
Väkeväinen, K., Ahlgren, T., Rauhala, E. & Räisänen, J. (1998) *Nuclear Instruments and Methods*, **B136–138**, 563–567.
Vickridge, I.C. (1988) *Nuclear Instruments and Methods*, **B34**, 470–475.
Wielunski, L.S. (1996) *Nuclear Instruments and Methods*, **B118**, 167–171.
Wolicki, E.A., Holmgren, H.D., Johnston, R.L. & Illsley, E.G. (1959) *Physical Review*, **116**, 1585–1591.
Xiong, F., Rauch, F., Sho, S., Zhou, Z., Livi, R.P. & Tombrello, T.A. (1987) *Nuclear Instruments and Methods*, **B27**, 432–441.
Ziegler, J.F., Biersack, J.P., Littmark, U. (1985) *The Stopping and Range of Ions in Solids. Vol. 1.* Pergamon Press, New York.
Zinke-Allmang, M., Kessler, V. & Kalbitzer, S. (1986) *Nuclear Instruments and Methods*, **B15**, 563–568.

10 Backscattering and forward scattering for depth profiling of solids

G. DEMORTIER and G. TERWAGNE

10.1 Introduction

Rutherford backscattering spectroscopy (RBS) is probably the oldest elemental analysis technique involving ion beam bombardment. The method has been continuously improved for about 50 years owing to the development of particle accelerators, particle detectors and their associated electronics and computers. RBS has now become a routine method mostly suitable to make accurate compositional determinations of heavy elements in thin layers at the top of any substrate containing lighter elements. In these favourable cases, the spectroscopy of backscattered particles allows the determination of the depth profiles of any composition (except for hydrogen) in the top layer from 5 nm (the minimum achievable resolution) up to a few microns. The technique is non-destructive, rapid (about 10 min irradiation), sensitive (down to 100 ppm for the heaviest nuclei) and, above all, it does not require any reference material to give quantitative results because the technique is often self-quantitative.

RBS is based on the fundamental explanation given by Rutherford [*Nobel Prize in chemistry in 1908 for discovering that atoms can be broken apart by α-rays and for radioactivity*] of the unexpected large deviation of α particles emitted by a radium source passing through a thin metal foil. The explanation takes into account the classical elastic scattering of rigid balls and the assumption that the incident α particles do not hit the atomic nucleus. Fortunately, the necessary quantum mechanical interpretation of the phenomenon gives the same result as far as only the Coulomb field around the nucleus is taken into account. The model is to be modified at low incident energy due to screening effects by shell electrons and at high incident energy when the distance of closest approach between the projectile and the nucleus is near to or less than the sum of their nuclear radii. For most practical situations, the model based on pure Coulomb deviation is valid for incident protons and α particles in the region of 1–4 MeV and for target nuclei heavier than calcium. For higher incident energies and for lighter nuclei, a correction for competing nuclear processes is to be applied, as will be discussed in this chapter.

As the physical principle of RBS is reversible by exchange of the projectile with the target nucleus, the results could also be applied when a heavy projectile hits a light nucleus which is ejected only in the forward direction. This technique named elastic recoil detection (ERD) will also be discussed here.

In the following section, the fundamentals of RBS and ERD, the selectivity in mass and in depth in order to perform profile measurements, the typical experimental set-up for RBS and ERD, and the capability of the method in practical cases will all be presented.

RBS and ERD applications are reported annually in hundreds of papers published in the most important journal for the subject, *Nuclear Instruments and Methods in Physics Research: Part B*. A selection of works published since 1994 in the *Proceedings of Ion Beam Analysis* (IBA) and *Proceedings of the International Conference on Nuclear Microprobe Technology and Applications* (ICNMTA) meetings is described in Section 10.10. In well-equipped laboratories, RBS and/or ERD are combined with PIXE, PIGE, NRA, STIM, etc., all the technologies involving the detection of signals induced by charged-particle irradiation. Several of the examples given below will refer to these complementary methodologies the fundamentals of which are presented in other chapters of this volume.

10.2 Position of RBS and ERD in comparison with other IBA techniques

When a material is irradiated with charged particles like protons, deuterons, α particles or other heavier ions, of energy ranging from a few hundred keV to a few MeV, the emission of photons, charged particles and neutrons gives rise to various spectroscopic methods of elemental analysis. The great majority of the interactions of the incident charged particles with the atoms of the irradiated material take place with atomic electrons: the incident particle progressively loses its energy in the material. The energy loss $\mathrm{d}E$ is expressed by Bethe's law of stopping power $S(E)$

$$S(E) = -\frac{\mathrm{d}E}{\rho\,\mathrm{d}x} = k\frac{Z_1^2 M_1}{E_\mathrm{i}}\frac{Z_2}{A_2}\ln\frac{aE}{\bar{I}} \qquad (10.1)$$

with Z_1, M_1 and E_i being, respectively, the charge, the mass and the energy of the incident particle; ρ, $\bar{I}$, Z_2 and A_2, the density, the mean ionisation potential, the atomic number and the atomic mass of the material, and finally, a, an adjustable parameter. The term k contains only physical constants. When the incident particle has sufficient energy to ionise atoms from any shell, $S(E)$ decreases with the increase of the incident energy (see Table 10.1). Exceptions to this general rule occur at low incident particle energy (as seen for ^{4}He and ^{3}He ions in Table 10.1) when the projectile cannot eject inner-shell electrons, with as the main consequence that $Z_\mathrm{eff} < Z$. Therefore $S(E)$, in those cases, is smaller than the expected value. The incident particle assumes rest after crossing a distance R, known as the range. By interaction with atomic electrons, this incident particle does not experience an appreciable deviation. A deviation of the incident particle from the straight line by interaction with one atomic nucleus can only occur rarely. The range of an incident particle is then nearly the total distance that the particle travels following a straight line in the material. The range increases with the incident projectile energy ($R \sim E^{1.8}$), and decreases with an increase of its charge and its mass. Table 10.2 gives some data on ranges (R) of protons, deuterons, ^{3}He and α particles for a selection of light, medium and high-Z elements. It is to be noted that the particle range increases with the atomic weight of the material owing to the dependence on Z_2/A_2.

The energy loss in the material can also be calculated in terms of stopping cross-section defined by the relationship

$$\varepsilon(E) = -\frac{\mathrm{d}E}{N\,\mathrm{d}x} \qquad (10.2)$$

where N is the atomic density (atoms cm^{-3}). The two quantities of Equations (10.1) and (10.2) can be converted into each other by the relationship:

$$\varepsilon(E) = \frac{AS(E)}{N_0} 10^{21} \left[\text{eV}/10^{15} \text{ atoms cm}^{-2}\right] \quad (10.3)$$

The stopping power of a compound A_xB_y can be calculated with Bragg's Rule, which is based on a simple assumption that the constituent elements act independently in the energy loss process

$$-\left(\frac{dE}{\rho\, dx}\right)_{AB} = S_{AB} = xS_A + yS_B \quad (10.4)$$

where S_A and S_B are the stopping cross-sections of the constituents and x or y are the mass proportions of each element with $x + y = 1$. This procedure neglects the effect of chemical bonding between A and B, which would affect the $\bar{I}$ value in Equation (10.1), but gives a reasonably accurate value for MeV light ions. For an accurate calculation of the stopping power of oxides and nitrides, some measurement with known reference samples would be recommended.

The stopping cross-section of a compound A_mB_n can be directly calculated by using the atomic fractions m and n of elements A and B, respectively:

$$-\left(\frac{dE}{N\, dx}\right)_{AB} = \varepsilon_{AB} = m\varepsilon_A + n\varepsilon_B \quad (10.5)$$

During the slowing down of the projectile, a transfer of energy to the atomic electrons of the material statistically takes place. This statistical behaviour of the energy loss of the projectile gives rise to an energy straggling: an unequal energy loss of identical particles crossing the same target material under identical conditions. The relation between the energy loss and the achieved depth is therefore affected by some statistical inaccuracy. For a very small energy loss, the straggling is to be calculated by using the Landau and Vavilov theory (Vavilov 1957). This energy straggling is not symmetrical around the average value. For monoenergetic protons of 1 MeV crossing a thickness of 2.4×10^{-5} g cm^{-2} (≈ 90 nm) of pure Al, the mean energy loss is 4.2 keV and the straggling on this energy loss has a width at half maximum of 2.7 keV [see Fig. 10.1(a) and Deconninck (1978)]. For 10^{-3} g cm^{-2}, the energy loss of these 1 MeV protons is 400 keV and the straggling gives rise to a Gaussian distribution with a width at half maximum of 29 keV [Fig. 10.1(b)]. The relation between energy and depth achieved in a material is then defined, in this instance, with an accuracy of 7%, which is relatively much lower than the inaccuracy at lower incident energy – the greater the energy loss, the smaller the statistical uncertainty in the relation of depth versus loss. In most practical cases, the depth profiling analysis is limited to regions of a few hundred nanometres (where the straggling is governed by the Landau and Vavilov calculation theory) to a few microns (where the straggling exhibits a Gaussian shape). In this last situation, the straggling, δ, is simply given by

$$\delta = 0.93Z_1\sqrt{\frac{Z_2}{A_2}} \cdot \sqrt{x} \quad (10.6)$$

where x is the achieved depth in μm, Z_1 is the atomic number of the incident particle, and Z_2 and A_2 are the atomic number and the atomic weight of the target, respec-

Table 10.1 Stopping power of light ions in materials (keV mg^{-1} cm^2)

				Protons			
Energy (MeV)	**C**	**O**	**Al**	**Si**	**Cu**	**Ag**	**Au**
0.05	8.12 (E+2)	6.59 (E+2)	4.49 (E+2)	4.42 (E+2)	2.53 (E+2)	1.85 (E+2)	1.34 (E+2)
0.1	7.45 (E+2)	6.14 (E+2)	4.24 (E+2)	4.17 (E+2)	2.37 (E+2)	1.70 (E+2)	1.20 (E+2)
0.2	5.94 (E+2)	5.04 (E+2)	3.60 (E+2)	3.55 (E+2)	2.05 (E+2)	1.56 (E+2)	1.09 (E+2)
0.5	3.68 (E+2)	3.29 (E+2)	2.54 (E+2)	2.53 (E+2)	1.55 (E+2)	1.11 (E+2)	7.64 (E+1)
1	2.34 (E+2)	2.14 (E+2)	1.75 (E+2)	1.75 (E+2)	1.16 (E+2)	8.65 (E+1)	6.00 (E+1)
2	1.42 (E+2)	1.32 (E+2)	1.11 (E+2)	1.13 (E+2)	8.06 (E+1)	6.38 (E+1)	4.50 (E+1)
3	1.05 (E+2)	9.81 (E+1)	8.38 (E+1)	8.50 (E+1)	6.37 (E+1)	5.03 (E+1)	3.70 (E+1)
5	7.04 (E+1)	6.64 (E+1)	5.75 (E+1)	5.85 (E+1)	4.45 (E+1)	3.64 (E+1)	2.80 (E+1)
7	5.39 (E+1)	5.11 (E+1)	4.45 (E+1)	4.54 (E+1)	3.50 (E+1)	2.93 (E+1)	2.78 (E+1)
				Deuterons			
Energy (MeV)	**C**	**O**	**Al**	**Si**	**Cu**	**Ag**	**Au**
0.05	7.51 (E+2)	6.06 (E+2)	4.14 (E+2)	4.07 (E+2)	2.38 (E+2)	1.76 (E+2)	1.31 (E+2)
0.1	8.13 (E+2)	6.60 (E+2)	4.50 (E+2)	4.42 (E+2)	2.53 (E+2)	1.85 (E+2)	1.34 (E+2)
0.2	7.45 (E+2)	6.14 (E+2)	4.24 (E+2)	4.17 (E+2)	2.37 (E+2)	1.70 (E+2)	1.20 (E+2)
0.5	5.37 (E+2)	4.62 (E+2)	3.36 (E+2)	3.32 (E+2)	1.93 (E+2)	1.37 (E+2)	9.47 (E+1)
1	3.67 (E+2)	3.38 (E+2)	2.54 (E+2)	2.53 (E+2)	1.55 (E+2)	1.11 (E+2)	7.64 (E+1)
2	2.34 (E+2)	2.14 (E+2)	1.74 (E+2)	1.75 (E+2)	1.16 (E+2)	8.65 (E+1)	6.00 (E+1)
3	1.76 (E+2)	1.63 (E+2)	1.35 (E+2)	1.36 (E+2)	9.49 (E+1)	7.24 (E+1)	5.10 (E+1)
5	1.24 (E+2)	1.12 (E+2)	9.54 (E+1)	9.66 (E+1)	7.04 (E+1)	5.57 (E+1)	4.05 (E+1)
7	9.30 (E+1)	8.73 (E+1)	7.50 (E+1)	7.60 (E+1)	5.66 (E+1)	4.58 (E+1)	3.41 (E+1)

				^{3}He			
Energy (MeV)	**C**	**O**	**Al**	**Si**	**Cu**	**Ag**	**Au**
0.05	1.10 (E+3)	8.97 (E+2)	6.25 (E+2)	6.17 (E+2)	3.72 (E+2)	2.82 (E+2)	2.13 (E+2)
0.1	1.64 (E+3)	1.34 (E+3)	9.29 (E+2)	9.26 (E+2)	5.42 (E+2)	4.03 (E+2)	3.00 (E+2)
0.2	2.13 (E+3)	1.75 (E+3)	1.22 (E+3)	1.20 (E+3)	7.03 (E+2)	5.16 (E+2)	3.73 (E+2)
0.5	2.22 (E+3)	1.88 (E+3)	1.35 (E+3)	1.33 (E+3)	7.81 (E+2)	5.64 (E+2)	3.98 (E+2)
1	1.78 (E+3)	1.67 (E+3)	1.27 (E+3)	1.16 (E+3)	7.02 (E+2)	5.07 (E+2)	3.54 (E+2)
2	1.22 (E+3)	1.10 (E+3)	8.73 (E+2)	8.74 (E+2)	5.58 (E+2)	4.10 (E+2)	2.86 (E+2)
3	9.34 (E+2)	8.55 (E+2)	6.94 (E+2)	7.00 (E+2)	4.67 (E+2)	3.50 (E+2)	2.45 (E+2)
5	6.50 (E+2)	6.02 (E+2)	5.04 (E+2)	5.08 (E+2)	3.58 (E+2)	2.76 (E+2)	1.97 (E+2)
7	5.06 (E+2)	4.72 (E+2)	4.00 (E+2)	4.05 (E+2)	2.94 (E+2)	2.32 (E+2)	1.69 (E+2)
				^{4}He			
Energy (MeV)	**C**	**O**	**Al**	**Si**	**Cu**	**Ag**	**Au**
0.05	9.08 (E+2)	7.43 (E+2)	520 (E+2)	5.13 (E+2)	3.11 (E+2)	2.38 (E+2)	1.82 (E+2)
0.1	1.41 (E+3)	1.15 (E+3)	8.00 (E+2)	7.90 (E+2)	4.70 (E+2)	3.52 (E+2)	2.63 (E+2)
0.2	1.95 (E+3)	1.60 (E+3)	1.11 (E+3)	1.10 (E+3)	6.43 (E+2)	4.74 (E+2)	3.46 (E+2)
0.5	2.29 (E+3)	1.92 (E+3)	1.34 (E+3)	1.34 (E+3)	7.80 (E+2)	5.66 (E+2)	4.02 (E+2)
1	2.00 (E+3)	1.73 (E+3)	1.27 (E+3)	1.26 (E+3)	7.46 (E+2)	5.38 (E+2)	3.77 (E+2)
2	1.45 (E+3)	1.29 (E+3)	1.00 (E+3)	1.00 (E+3)	6.21 (E+2)	4.52 (E+2)	3.14 (E+2)
3	1.13 (E+3)	1.02 (E+3)	8.28 (E+2)	8.20 (E+2)	5.31 (E+2)	3.92 (E+2)	2.73 (E+2)
5	7.96 (E+2)	7.34 (E+2)	6.05 (E+2)	6.10 (E+2)	4.17 (E+2)	3.16 (E+2)	2.23 (E+2)
7	6.25 (E+2)	5.80 (E+2)	4.96 (E+2)	4.91 (E+2)	3.47 (E+2)	2.69 (E+2)	1.93 (E+2)

Table 10.2 Range of light ions in materials (mg cm^{-2})

Energy (MeV)	C	O	Al	Si	Cu	Ag	Au
			Protons				
0.05	1.04 (E − 1)	1.28 (E − 1)	1.83 (E − 1)	1.85 (E − 1)	3.05 (E − 1)	3.98 (E − 1)	5.14 (E − 1)
0.1	1.62 (E − 1)	1.99 (E − 1)	2.88 (E − 1)	2.92 (E − 1)	4.96 (E − 1)	6.64 (E − 1)	8.92 (E − 1)
0.2	3.07 (E − 1)	3.72 (E − 1)	5.35 (E − 1)	5.43 (E − 1)	9.37 (E − 1)	1.29 (E 0)	1.79 (E 0)
0.5	9.61 (E − 1)	1.12 (E 0)	1.53 (E 0)	1.55 (E 0)	2.63 (E 0)	3.66 (E 0)	5.24 (E 0)
1	2.71 (E 0)	3.04 (E 0)	3.94 (E 0)	3.95 (E 0)	6.37 (E 0)	8.79 (E 0)	1.27 (E + 1)
2	8.38 (E 0)	9.17 (E 0)	1.13 (E + 1)	1.12 (E 1)	1.69 (E + 1)	2.26 (E + 1)	3.22 (E + 1)
3	1.67 (E + 1)	1.80 (E + 1)	2.18 (E + 1)	2.16 (E 1)	3.10 (E + 1)	4.05 (E + 1)	5.69 (E + 1)
5	4.05 (E 1)	4.33 (E + 1)	5.11 (E + 1)	5.05 (E 1)	6.96 (E + 1)	8.77 (E + 2)	1.20 (E + 2)
7	7.33 (E + 1)	7.80 (E + 1)	9.10 (E + 1)	8.97 (E 1)	1.21 (E + 2)	1.49 (E + 2)	2.00 (E + 2)
			Deuterons				
Energy (MeV)	**C**	**O**	**Al**	**Si**	**Cu**	**Ag**	**Au**
0.05	1.56 (E − 1)	1.91 (E − 1)	2.70 (E − 1)	2.73 (E − 1)	4.33 (E − 1)	5.48 (E − 1)	6.83 (E − 1)
0.1	3.08 (E − 1)	2.56 (E − 1)	3.67 (E − 1)	3.72 (E − 1)	6.11 (E − 1)	7.95 (E − 1)	1.03 (E 0)
0.2	3.24 (E − 1)	3.98 (E − 1)	5.77 (E − 1)	5.85 (E − 1)	9.91 (E − 1)	1.32 (E 0)	1.78 (E 0)
0.5	7.87 (E − 1)	9.46 (E − 1)	1.35 (E 0)	1.37 (E 0)	2.36 (E 0)	3.27 (E 0)	4.58 (E 0)
1	1.92 (E 0)	2.23 (E 0)	3.06 (E 0)	3.10 (E 0)	5.25 (E 0)	7.31 (E 0)	1.05 (E + 1)
2	5.24 (E 0)	6.08 (E 0)	7.88 (E 0)	7.91 (E 0)	1.28 (E + 1)	1.76 (E + 1)	2.54 (E + 1)
3	1.04 (E + 1)	1.15 (E + 1)	1.45 (E + 1)	1.44 (E 1)	2.22 (E + 1)	3.03 (E + 1)	4.36 (E + 1)
5	2.44 (E + 1)	2.66 (E + 1)	3.23 (E + 1)	3.21 (E 1)	4.71 (E + 1)	6.20 (E + 1)	8.79 (E + 1)
7	4.35 (E + 1)	4.69 (E + 1)	5.62 (E + 1)	5.56 (E 1)	7.89 (E + 1)	1.02 (E + 2)	1.42 (E + 2)

^{3}He

Energy (MeV)	C	O	Al	Si	Cu	Ag	Au
0.05	1.51 (E − 1)	1.82 (E − 1)	2.52 (E − 1)	2.53 (E − 1)	3.90 (E − 1)	4.83 (E − 1)	5.89 (E − 1)
0.1	1.77 (E − 1)	2.14 (E − 1)	3.99 (E − 1)	3.02 (E − 1)	4.77 (E − 1)	6.03 (E − 1)	7.58 (E − 1)
0.2	2.17 (E − 1)	2.63 (E − 1)	3.72 (E − 1)	3.76 (E − 1)	6.09 (E − 1)	7.89 (E − 1)	1.02 (E 0)
0.5	3.32 (E − 1)	4.02 (E − 1)	5.70 (E − 1)	5.77 (E − 1)	9.61 (E − 1)	1.28 (E 0)	1.73 (E 0)
1	5.67 (E − 1)	6.74 (E − 1)	9.40 (E − 1)	9.51 (E − 1)	1.60 (E 0)	2.17 (E 0)	3.00 (E 0)
2	1.24 (E 0)	1.43 (E 0)	1.91 (E 0)	1.92 (E 0)	3.16 (E 0)	4.32 (E 0)	6.12 (E 0)
3	2.18 (E 0)	2.45 (E 0)	3.18 (E 0)	3.19 (E 0)	5.10 (E 0)	6.94 (E 0)	9.87 (E 0)
5	4.78 (E 0)	5.27 (E 0)	6.59 (E 0)	6.57 (E 0)	1.00 (E + 1)	1.34 (E + 1)	1.90 (E + 1)
7	8.29 (E 0)	9.04 (E 0)	1.11 (E + 1)	1.10 (E 1)	1.62 (E + 1)	2.13 (E + 1)	3.00 (E + 1)

^{4}He

Energy (MeV)	C	O	Al	Si	Cu	Ag	Au
0.05	1.89 (E − 1)	2.27 (E − 1)	3.13 (E − 1)	3.15 (E − 1)	4.88 (E − 1)	5.89 (E − 1)	7.11 (E − 1)
0.1	2.19 (E − 1)	2.64 (E − 1)	3.69 (E − 1)	3.71 (E − 1)	5.80 (E − 1)	7.28 (E − 1)	9.04 (E − 1)
0.2	2.62 (E − 1)	3.19 (E − 1)	4.49 (E − 1)	4.53 (E − 1)	7.126 (E − 1)	9.32 (E − 1)	1.19 (E 0)
0.5	3.76 (E − 1)	4.56 (E − 1)	6.48 (E − 1)	6.56 (E − 1)	1.09 (E 0)	1.43 (E 0)	1.91 (E 0)
1	5.88 (E − 1)	7.05 (E − 1)	9.94 (E − 1)	1.00 (E 0)	1.69 (E 0)	2.27 (E 0)	3.12 (E 0)
2	1.16 (E 0)	1.35 (E − 1)	1.85 (E 0)	1.87 (E 0)	3.11 (E 0)	4.25 (E 0)	5.97 (E 0)
3	1.94 (E 0)	2.21 (E 0)	2.94 (E 0)	2.95 (E 0)	4.82 (E 0)	6.59 (E 0)	9.34 (E 0)
5	4.06 (E 0)	4.53 (E 0)	5.79 (E 0)	5.78 (E 0)	9.06 (E 0)	1.23 (E + 1)	1.74 (E + 1)
7	6.91 (E 0)	7.61 (E 0)	9.48 (E 0)	9.44 (E 0)	1.43 (E + 1)	1.92 (E + 1)	2.71 (E + 1)

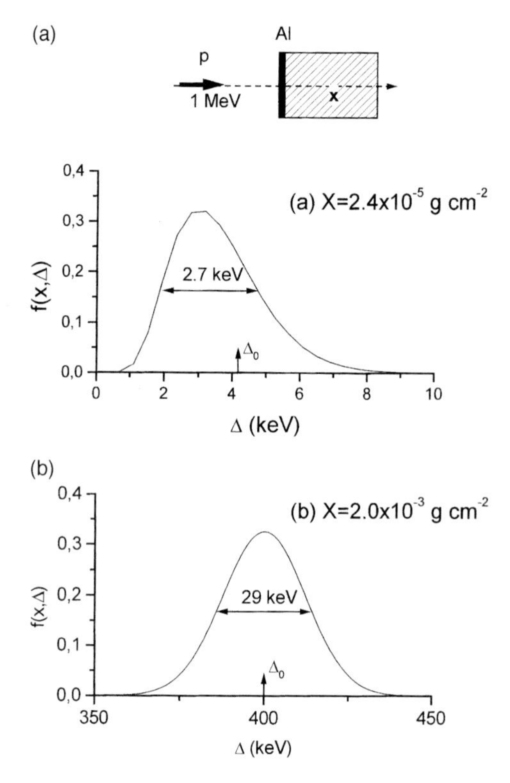

Fig. 10.1 Energy straggling distribution of a proton beam in a pure aluminium sample ($Z = 13$, $A = 27$): (a) small depth, asymmetrical Vavilov distribution; (b) large depth, Gaussian distribution (Deconninck 1978).

tively. The straggling is therefore the same for all particles of the same Z_1 (deuteron or proton, α or ^{3}He). The dependence of the straggling on the energy of the particle is totally included in the value of x. For compound materials, one simply replaces Z_2 and A_2 by the mean weighed values $\overline{Z}_2$ and $\overline{A}_2$.

The interaction of the incident particle with the atomic electrons of the target induces ionisation. When ionisation takes place in an inner electronic shell, a subsequent rearrangement to return to an atomic equilibrium leads to the emission of characteristic X-rays: the analytical technique based on this atomic process is called PIXE (particle-induced X-ray emission). PIXE is a very sensitive method for bulk and trace analysis of elements with an atomic number Z greater than 20. PIXE may also be used for the study of elements from Na to K, but the absorption of the low-energy characteristic X-rays (1–3 keV) of those light elements into thick samples generally restricts the potential of the method. Analysis of low-Z elements could then be satisfactory for thin samples only or to probe these elements in a very narrow layer at the surface of a bulk material.

When the incident particle passes very closely to the atomic nucleus, it may be elastically scattered. At very low impact parameter, a particle of mass M_1 lower than the mass M_2 of the collided nucleus, is scattered at an angle close to 180°. The spectroscopy of this backscattered particle (RBS) may easily be used for depth profiling of elements up to thicknesses close to 30% of their own range:

$$E_S = \left(\frac{M_2 - M_1}{M_2 + M_1}\right)^2 E_i \tag{10.7}$$

In contrast, if M_1 is greater than M_2, only a forward emission of both M_1 and M_2 is allowed and the detection of the light target component M_2 gives rise to another elastic scattering spectroscopy known as ERD. Both RBS and ERD characteristics will be discussed in subsequent sections.

The probability for the Coulomb excitation of a target nucleus leading to the emission of a characteristic γ-ray is much lower than the backscattering and inner-shell ionisation phenomena. Consequently, PIGE (particle-induced γ-ray emission) is of particular interest for the study of light elements where PIXE and RBS suffer from a lack of sensitivity and accuracy. Coulomb repulsion of the incident projectile does not induce γ-ray emission of sufficient intensity for most of the heavy elements.

The transmutation of a light element to another one may also induce charged-particle emission. This analytical process, known as NRA (nuclear reaction analysis) is described in Chapter 9 of this book.

If the energy of the incident particle is maintained below 5–10 MeV, no delayed radioactivity may be induced like in CPAA (charged-particle activation analysis) typically performed at incident energies greater than 10 MeV.

By using simultaneously (or sequentially) all these analytical techniques, it is possible to achieve very powerful characterisations of many materials. In the following sections of this presentation, complementary applications of RBS, ERD, PIXE, PIGE and nuclear reactions for the study of solids will be discussed.

10.3 Kinematics of RBS

When a charged projectile of mass M_1 is scattered from a target nucleus of mass M_2 greater than M_1 at an angle θ, the ratio of the scattered ion energy E_S to the incident energy E_0 is calculated by using the laws of conservation of energy (assuming that M_1 and M_2 form an isolated system interacting by the Coulomb force only) and momentum (neglecting all external forces):

$$K = \frac{E_S}{E_0} = \left[\frac{(M_2^2 - M_1^2 \sin^2 \theta)^{1/2} + M_1 \cos\theta}{M_1 + M_2}\right]^2 \tag{10.8}$$

The binding energy of the target nucleus with the surroundings and the thermal vibration are neglected. In practice, these latter effects are negligibly small in collisions between MeV ions and target atoms. The ratio K, called the kinematic factor, is then only determined by the mass ratio M_2/M_1 and the scattering angle θ. Figure 10.2 shows a three-dimensional diagram of the kinematic factor as a function of the mass ratio and of the angle of detection. The kinematic factor increases monotonically with the mass ratio indicating that the target atom mass can be determined from the observed energy of the scattered ion. For a fixed angle, θ, the energy separation for particles scattered on nuclei with a mass difference ΔM is:

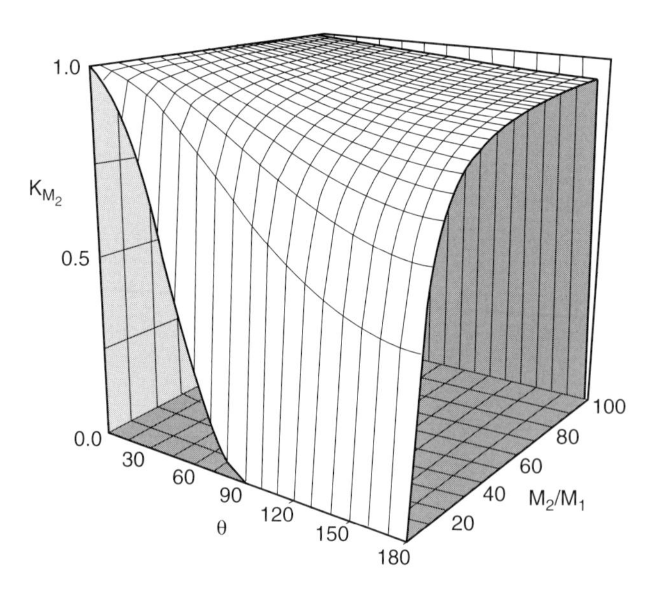

Fig. 10.2 Three-dimensional plot of kinematic factors K for various scattering angles and for various mass ratios.

$$\Delta E_S = E_0 \left(\frac{dK}{dM}\right) \Delta M \tag{10.9}$$

ΔE_S must be greater than the energy resolution of the experimental set-up, δE_{exp}, to distinguish the different components of mass M.

For $\theta = 90°$, the simplified expression of Equation (10.8) is

$$K_{90} = \frac{M_2 - M_1}{M_2 + M_1}\Gamma \tag{10.10}$$

and for 180°

$$K_{180} = \left(\frac{M_2 - M_1}{M_2 + M_1}\right)^2 \tag{10.11}$$

The effective mass resolution is then:

$$\delta M = \frac{\delta E_{exp}}{E_0} \left(\frac{dK}{dM}\right)^{-1} \tag{10.12}$$

The dominant contribution to δE_{exp} comes from the energy resolution of the detector: a surface barrier made from a crystal of Si. To date, detector resolution is around 10 keV for protons and α particles, but around 50 keV for heavier incident particles like C and O. Figures 10.3(a) and 10.3(b) display, respectively, the calculated mass resolution for protons and α particles (2 MeV and 4 MeV) at three different scattering angles (90°, 135° and 175°), for a total energy resolution of 15 keV for the whole set-up. The effect of scattering angles is also shown for 2 MeV He ions. A mass difference of one atomic mass unit (amu) or less can be expected for

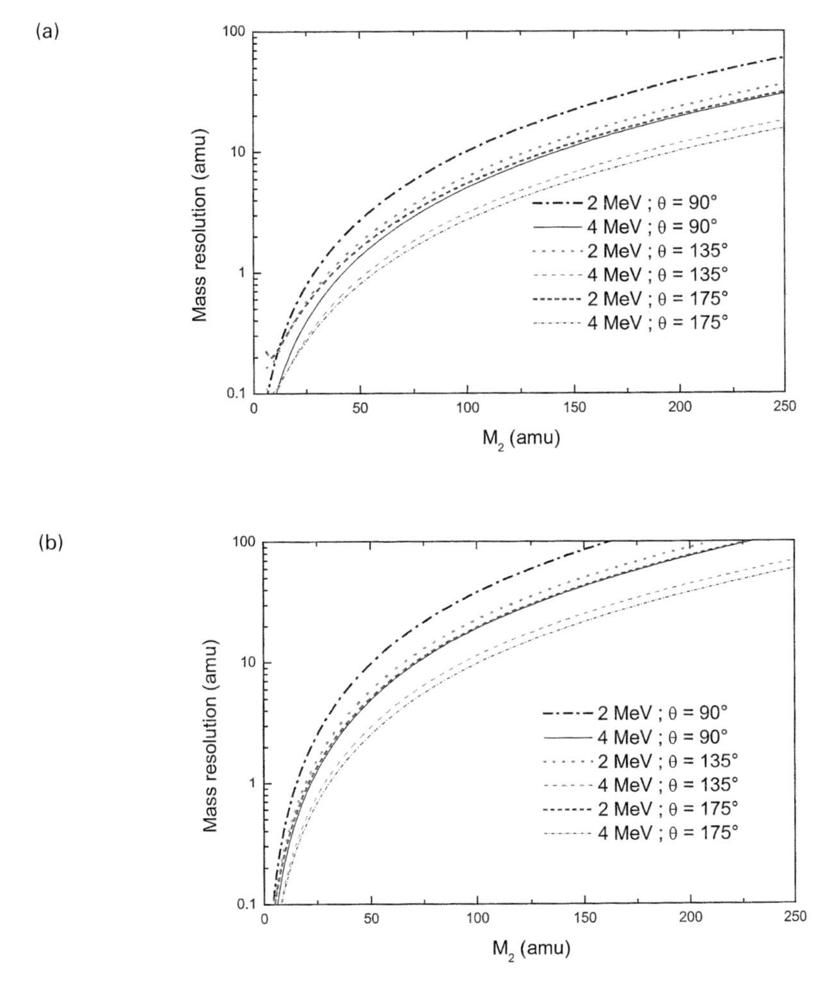

Fig. 10.3 Mass resolution calculated with Equation (10.12) at two energies for three different angles: (a) for α particles; (b) for protons.

light elements ($M < 40$ amu) with 2 MeV He ions. The mass resolution is worse for a higher target mass M_2, and reaches 20 atomic masses for the heaviest elements ($M_2 \approx 200$). For a better mass resolution, heavy incident ions should be used. The energy resolution of the particle detectors for those heavy ions is worse and the radiation damage in those detectors is much more important. Nevertheless, a mass resolution of two atomic masses may be achieved with 8 MeV Si ions. The mass resolution can also be achieved by using the time-of-flight (TOF) technique for energy analysis instead of a solid-state barrier detector in heavy-ion RBS. Other particular problems related to stopping power of heavy ions, their higher damage in the target material and the expected deviation from the Rutherford cross-section could give rise to less accuracy on the depth profile measurement by using only a standardless technique.

10.4 The scattering cross-sections

10.4.1 *Rutherford cross-sections*

If the force between the incident ion (M_1, Z_1e, E_0) and the target nucleus (M_2, Z_2e, initially at rest) is assumed to be the Coulomb force only, the Rutherford cross-section in the laboratory system is:

$$\left(\frac{d\sigma}{d\Omega}\right)_R = \left(\frac{Z_1 Z_2 e^2}{4\pi\varepsilon_0 4E}\right)^2 \frac{4\left\{\sqrt{(M_2^2 - M_1^2 \sin^2\theta)} + M_2\cos\theta\right\}^2}{M_2 \sin^4\theta\sqrt{M_2^2 - M_1^2\sin^2\theta}} \tag{10.13}$$

If $M_1 \ll M_2$, and using the values of physical constants e and ε_0, the Rutherford cross-section becomes:

$$\frac{d\sigma(\theta)}{d\Omega} = 1.296\left(\frac{Z_1 Z_2^2}{E_0}\right)^2\left[\frac{1}{\sin^4\theta/2} - 2\left(\frac{M_1}{M_2}\right)^2\right] \tag{10.14}$$

The unit in this last expression is the mbarn sr^{-1} or is 10^{-27} cm^2 sr^{-1}, when E_0 is given in MeV.

Figure 10.4 shows a three-dimensional diagram of the Rutherford cross-section versus the scattering angle and the ratio between the mass of the target nucleus and the mass of the incident particle. The cross-section is proportional to $(Z_1Z_2)^2$ decreases with the incident energy of the projectile as E_0^{-2}, and is monotically decreasing with the fourth power of sin $\theta/2$. Nevertheless, to achieve the best mass separation, the detector is generally situated close to $\theta = 180°$ where the cross-section reaches its minimum value.

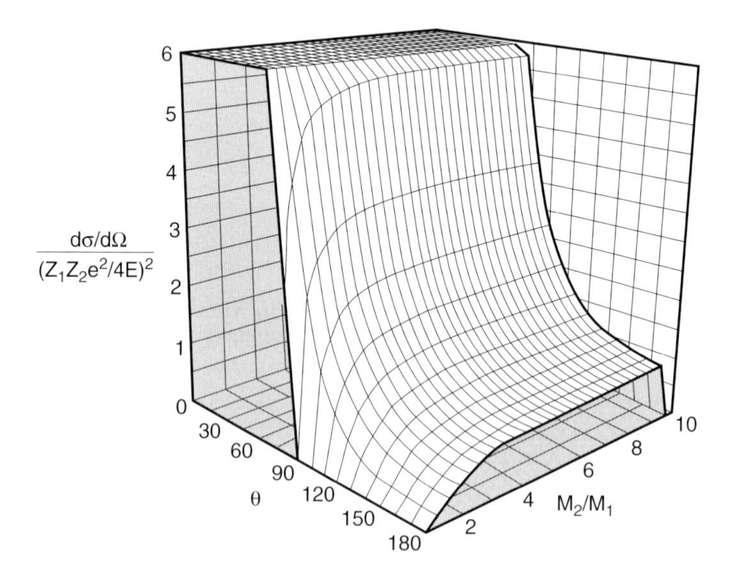

Fig. 10.4 Three-dimensional plot of the Rutherford cross-section for various angles (greater than 80°) and for various mass ratios.

10.4.2 *Non-Rutherford cross-sections*

Experimental measurements indicate that the actual cross-sections depart from Rutherford at both high and low energies for all projectile–target pairs. The low-energy departures are caused by partial screening of the nucleus charge by the electron shells. The results of several investigations indicate that these low-energy corrections have been given with adequate accuracy by L'Ecuyer *et al.* (1979)

$$\sigma/\sigma_R = 1 - \frac{0.049 Z_1 Z_2^{4/3}}{E_{CM}} \tag{10.15}$$

or by Wenzel and Whaling (1952)

$$\sigma/\sigma_R = 1 - \frac{0.0326 Z_1 Z_2^{7/2}}{E_{CM}} \tag{10.16}$$

for light-ion analysis beams with MeV energies. In practice, replacing E_{CM} by E_{Lab} produces a negligible error.

The high-energy departures of the cross-sections from Rutherford behaviour are caused by the presence of short-range nuclear forces. Recent measurements and calculations by Bozoian *et al.* (1990, 1991b) gave the limits for pure Rutherford cross-sections as

for protons:

$$E_{H.E.}^{NR} \approx (0.12 \pm 0.01)Z - (0.5 \pm 0.1) \tag{10.17}$$

for α particles:

$$E_{H.E.}^{NR} \approx (0.25 \pm 0.01)Z - (0.4 \pm 0.2) \tag{10.18}$$

A comparison of these experimental data (Table 10.3) with the computation of geometrical parameters involved in (a pure) Coulomb interaction of the incident projectile and the nucleus clearly indicates that the quantum behaviour is to be taken into account. Classically, the distance of closest approach, R_0, may be postulated when the full energy E_0 of the incident projectile is converted into potential energy in the Coulomb field of the nucleus:

$$E_0 = \frac{Z_1 Z_2 e^2}{4\pi\varepsilon_0 R_0} \tag{10.19}$$

Then, $R_0 = 1.44\, zZ/E_0$ (in fm when E is expressed in MeV).

By using the liquid drop model of the atomic nucleus, the nuclear radius of a nucleus of atomic mass A may also be calculated:

$$R_N = 1.5\, A^{1/3} (\text{fm}) \tag{10.20}$$

Let us suppose that a nuclear contribution in the collision takes place when the incident particle classically hits the nucleus, then:

$$R_0 \leq R_N + R_i \tag{10.21}$$

The comparison of both approaches produces the data of Table 10.3.

One directly observes that a non-Rutherford behaviour takes place at lower energy

Table 10.3 Limits for pure Rutherford cross-section. Data from Bozoian *et al.* (1990, 1991b)

	Protons				α Particles			
	$E^{NR}_{H.E.}$ (MeV) Equation (10.17)	R_0 (fm)	$R_N + R_i$ (fm)	E^{NR}_{clas} (MeV) Equation (10.19)	$E^{NR}_{H.E.}$ (MeV) Equation (10.18)	R_0 (fm)	$R_N + R_i$ (fm)	E^{NR}_{clas} (MeV) Equation (10.19)
C	0.22	39.3	4.9	1.8	1.9	9.1	5.8	3.0
Al	1.1	17.0	6.0	3.1	3.6	10.4	6.9	5.4
Cu	3.0	13.9	7.5	5.6	7.6	11.0	8.4	10.0
Ag	5.1	13.3	8.7	7.8	12.1	11.2	9.6	14.1
Au	9.0	12.6	10.2	11.2	20.1	11.3	11.1	20.5

than the expected classical one. The discrepancy of the classical model is much more evident for protons than for α particles, especially for light nuclei where the influence of nuclear forces extends very far from the nuclear surface.

10.5 Typical RBS spectra for thin layers

The best way to approach RBS is to go from typical academic situations to real problems. Firstly, simulations using protons and α particles of energy ranging from 1 to 4 MeV will be discussed for the study of thin layers of Au, Ag, Cu, and Al on thick substrates of carbon and silicon. The contributions of the layers are the only concern here with respect to pure Rutherford cross-sections. However, note that the carbon substrate will eventually be reached by very energetic protons or α particles and the Al layer could also be affected by a non-Rutherford contribution as seen in Table 10.3. These contributions could give erroneous concentration values in a quantitative aspect for Si, Al and C, but the kinematic factor used to identify the energy does not suffer from possible non-Rutherford contributions.

Figures 10.5–10.12 typify several RBS spectra calculated using pure Rutherford cross-sections of superposed, but not intermixed, Au, Ag, Cu and Al layers for incident 2 and 4 MeV α particles, for 1, 2 and 4 MeV protons and for a scattering angle of 175° (detector position with respect to the incident beam). All calculated spectra were simulated for the same number of incident particles (1 μC) and the same solid angle of detection (1 msr). The energy scale has been normalised with the incident energy. Then, the horizontal scale is the kinematic factor. Data are given for an equal number of atoms derived from each component. Thicknesses in units of length (nm or μm) are not utilised because RBS results are only sensitive to areal densities. A contribution of 5×10^{17} atoms cm^{-2} roughly represents a thickness of 75 nm for each layer.

In simulations of Fig. 10.5–10.8, gold is supposed to provide the top yield of all the layered structures followed sequentially by Ag, Cu and Al. Each layer contains 5×10^{17} atoms cm^{-2} and the substrate is carbon or silicon. Each element gives rise to well-separated information. At the top of each figure, the expected position of the right-step of the RBS signal for the element at the surface is given. The gold surface signal appears at the right position but silver, copper and aluminium appear at lower

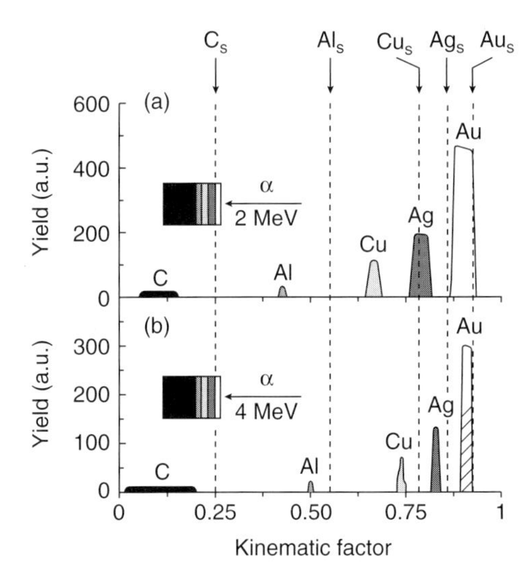

Fig. 10.5 α-RBS (2 MeV and 4 MeV) spectra for a multilayered structure consisting of a succession of Au, Ag, Cu, Al (5×10^{17} atoms cm^{-2} of each species) deposited on a carbon substrate. Note the greater separation of peaks at higher energy (4 MeV) but the wider contribution of each peak at lower energy (2 MeV).

energies: the incident α particles indeed suffer progressive slowing down before reaching the successively buried layers. For 4 MeV α particles (Fig. 10.5), spectrum (b) shows the signals are more separated (because E_S is proportional to E_i) and narrower (because stopping powers are lower at higher energies, as seen in Table 10.1).

The contribution of the thick carbon substrate gives rise to a large continuum, which should extend to zero energy, but the simulated curves are cut off at low energies. Furthermore, copper signals indicate the presence of isotopes (63 and 65). The contribution of RBS on the carbon substrate is completely outside the region of the layers.

For 1, 2 and 4 MeV protons (Fig. 10.6) the peaks are much more narrower and only well separated at the highest energy. The reason is the limited energy resolution of the experimental set-up, namely 15 KeV. The relative intensity of each contributing element (same number of atoms of Au, Ag, Cu, Al) is due to the Z^2 dependence of the Rutherford cross-section. The general trend is similar for a silicon substrate except for the separation of Al signals from those of the substrate. As Al and Si have very similar masses, the right edge of the Si contribution cannot be separated from the peak of Al, which is then superposed to the Si continuum.

If the layered structure is reversed with, successively, Al, Cu, Ag and Au on a carbon substrate, one often obtains much more complicated situations. The separation of elemental contributions is clear for 4 MeV α particles (see Fig. 10.9, spectrum (b)) but is nearly inextricable for 2 MeV α particles (see Fig. 10.9, spectrum (a)) where Au and Ag signals are completely mixed. The situation for incident protons is only clear when the energy also reaches 4 MeV. For the silicon substrate, it is now possible to separate Al from the substrate in each spectrum. In comparison with Figs 10.7 and

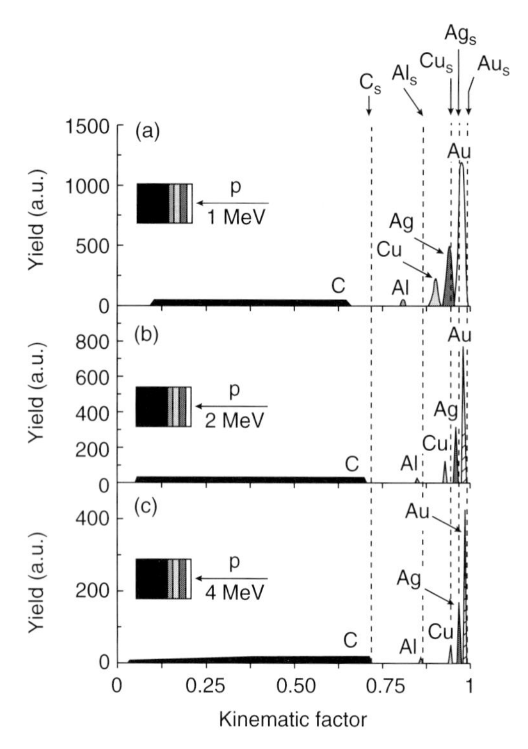

Fig. 10.6 p-RBS (1, 2 and 4 MeV) spectra for a multilayered structure consisting of a succession of Au, Ag, Cu, Al (5×10^{17} atoms cm^{-2} of each species) deposited on a carbon substrate. Note the incomplete separation of peaks at 1 MeV.

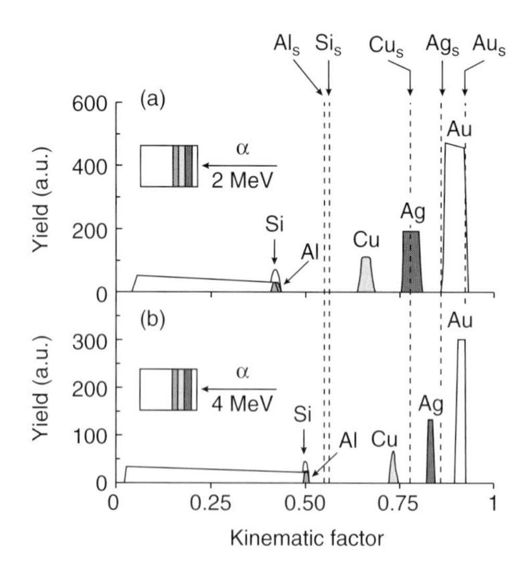

Fig. 10.7 Same as Fig. 10.5 but for a silicon substrate. Note the superposition of Al and Si signals.

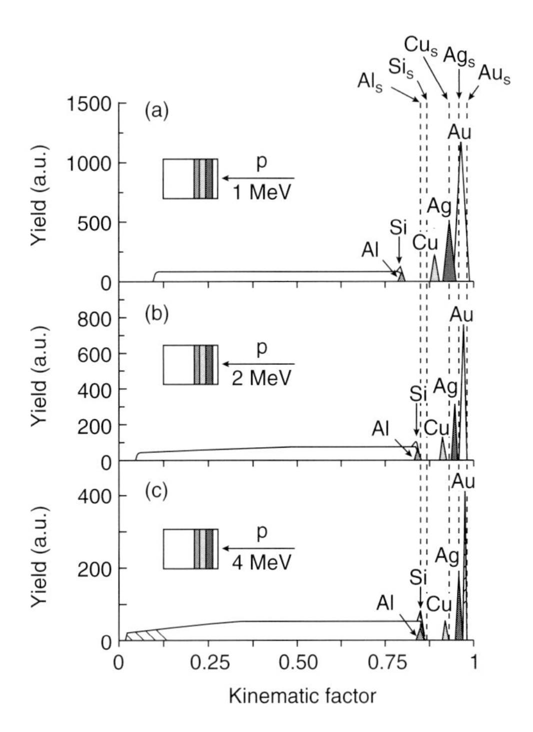

Fig. 10.8 Same as Fig. 10.6 but for a silicon substrate. The separation of Si and Al is still more difficult with protons than with α particles at all energies.

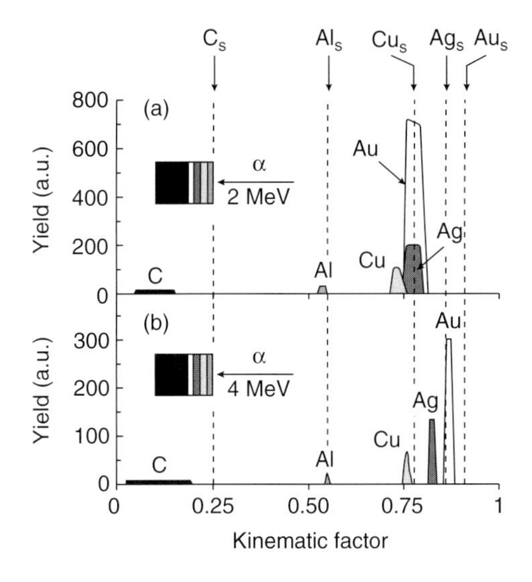

Fig. 10.9 α-RBS (2 and 4 MeV) spectra for a multilayered structure consisting of a succession of Al, Cu, Ag and Au (5×10^{17} atoms cm^{-2} for each species) deposited on a carbon substrate (reverse situation by comparison with Fig. 10.5). As the light element is at the surface, the heavy ones are intermixed at 2 MeV but separated at higher energy.

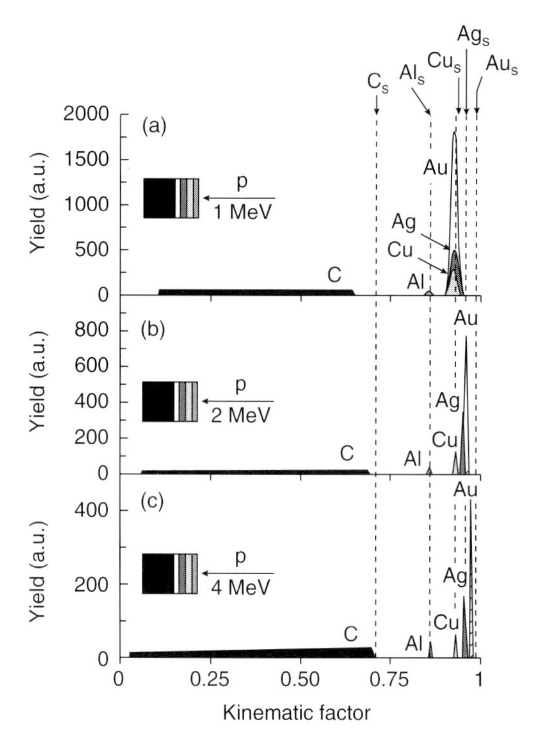

Fig. 10.10 p-RBS (1, 2 and 4 MeV) spectra for a multilayered structure consisting of a succession of Al, Cu, Ag and Au (5×10^{17} atoms cm^{-2} for each species) deposited on a carbon substrate (reverse situation by comparison with Fig. 10.6). The separation of heavy elements is only achieved at high energy.

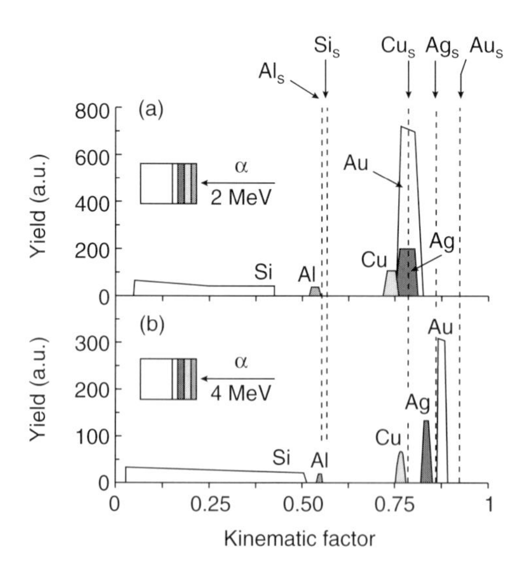

Fig. 10.11 Same situation as in Fig. 10.9 but for a silicon substrate. Note that the Al contribution is now separated from the silicon backing.

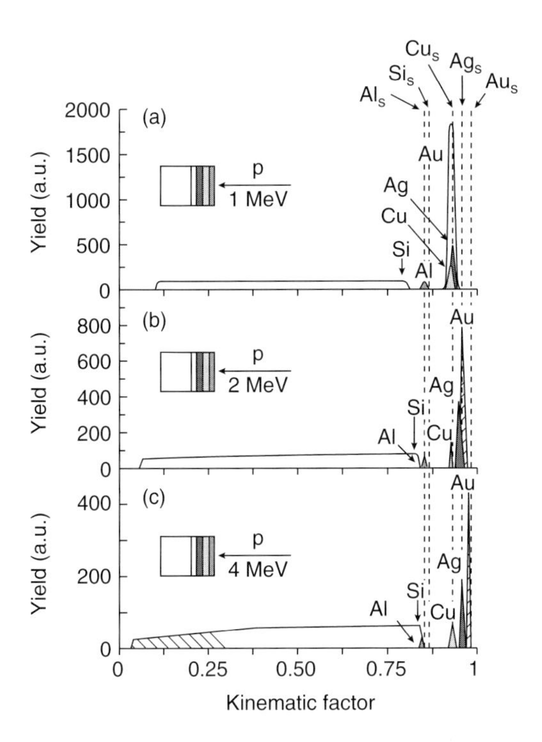

Fig. 10.12 Same situation as in Fig. 10.10 but Al is now only separated from the Si at low energy.

10.8, Al signals appear at higher energy but the continuum from the Si substrate is nearly at the same place for both sequences of layers: Al–Cu–Ag–Au or Au–Ag–Cu–Al.

If the layered structure is now replaced by an alloy containing the same atomic concentration of each species (1.25×10^{18} of each), the spectra given in Figs 10.13 and 10.14 are obtained for a carbon substrate. Note that the carbon substrate is not visible on the spectrum derived from 2 MeV α particles: the scattered energy is not sufficient to permit backscattered particles to cross the layers again. Finally, in Figs 10.15 and 10.16 the corresponding results for a silicon substrate are given.

10.6 Experimental arrangement

For standard RBS, beams of light ions (generally hydrogen and helium isotopes) of energy in the MeV range and a solid-state silicon detector are commonly used. The mass and the depth resolutions are better for He ions than H (see Fig. 10.2), but H ions are occasionally used in RBS in order to analyse deeper layers (see Tables 10.1 and 10.2). Single-ended or tandem accelerators of the van de Graaff or Cockcroft and Walton types produce these ions. After being accelerated, the ions are deflected by a magnet to eliminate contaminant species and to select the desired ion energy. The calibration of the whole system is usually performed by using resonant nuclear

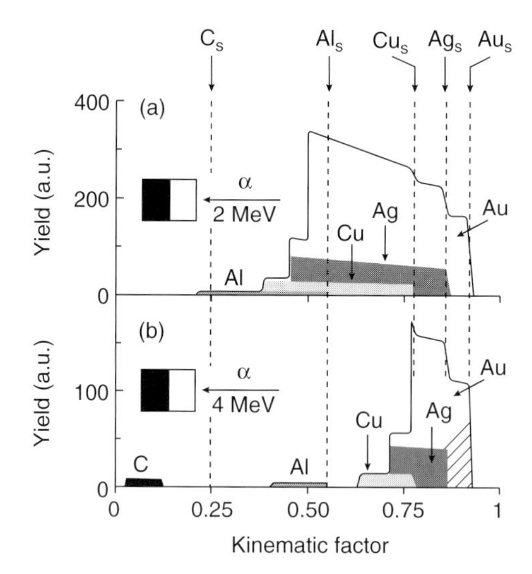

Fig. 10.13 α-RBS spectra for an alloy of 25% Au – 25% Ag – 25% Cu – 25% Al (5×10^{18} atoms cm^{-2} for the whole thickness) on a carbon substrate. The decomposition of each contributing element is only obtained after careful investigation. At 4 MeV, Au, Ag and Cu are superposed in the region of $K = 0.8$.

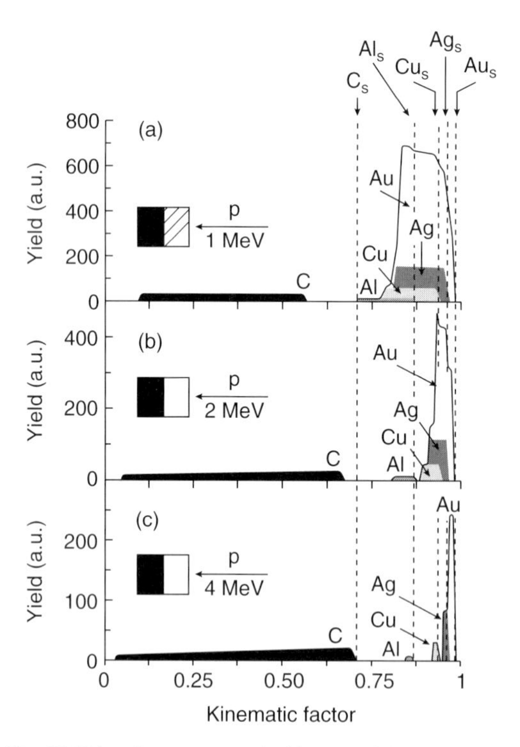

Fig. 10.14 Same as in Fig. 10.13 but for protons as incident particles. The contribution of each element is difficult to observe at 1 MeV. When the incident energy is increased, the peaks due to each element are narrower and more separated.

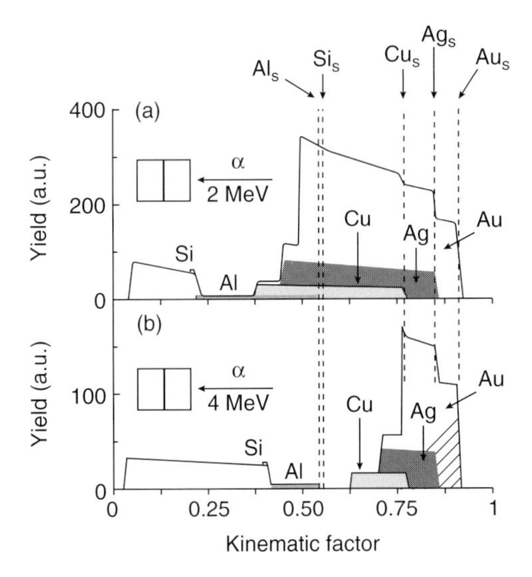

Fig. 10.15 Same as Fig. 10.13 but for a silicon substrate.

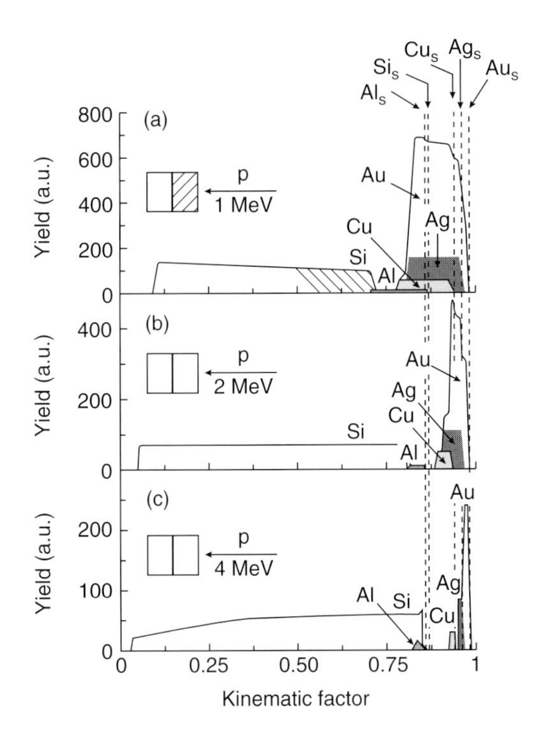

Fig. 10.16 Same as Fig. 10.14 but for a silicon substrate.

reactions with an accuracy of 1 keV or less on the incident energy in the MeV range. The ion beam is finally collimated by apertures to a size of 0.5–1 mm, but broader beams may also be used if the target material is fragile. Beam currents ranging from a few nA to a fraction of a μA are measured directly on the target (in the presence of an electron suppressor) or by some monitoring system installed between the final aperture and the target.

The solid-state detector is installed inside the evacuated chamber. The signal from the detector is amplified and sent to a pulse height analyser (HPA). This analyser attributes a given channel to a signal of a given height. The relation between the channel number and the ion energy is calibrated using α particles from a radioactive source such as ^{241}Am or by using a thick reference material containing at least two elements of very different atomic weight like Au–Al, or W–C. From the kinematic factors it is possible to calibrate rapidly the detector system. A very convenient method is to use simultaneously the peak at channel C of the ^{241}Am α particles ($E_\alpha = 5.486$ MeV) and channels A and B of the high-energy edges in the RBS spectrum on a known thick binary target.

The system of equations

$$C = E_\alpha f + b \tag{10.22}$$

$$B = k_B E_0 f + b \tag{10.23}$$

$$A = k_A E_0 f + b \tag{10.24}$$

immediately gives the precise energy E_0 of the incident particle, the amplifying factor f of the electronic system and the channel number corresponding to zero energy. This last parameter may be of importance if a dead layer is present at the surface of the solid-state detector, especially when the RBS experiment is performed with ions heavier than He. A typical arrangement is given in Fig. 10.17.

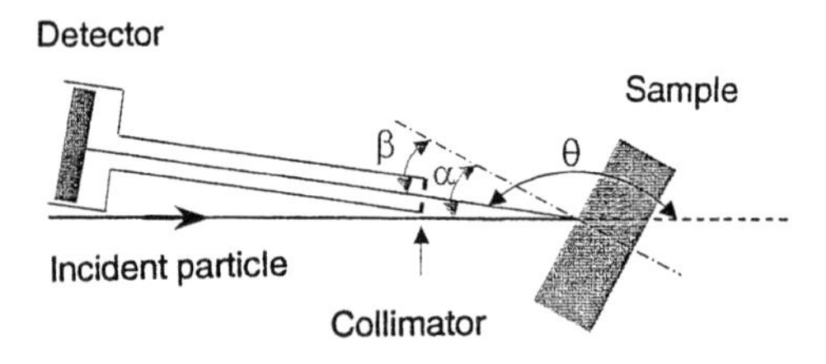

Fig. 10.17 Classical experimental arrangement for RBS analysis. The collimator avoids collecting multiscattered particles in the surrounding of the target. θ is the scattering angle while α and β are, respectively, the incident and the detection angles relative to the target normal.

10.7 Calculation of RBS spectra

The relationship between the energy of the scattered ion and the depth where the ion is scattered can be calculated with the stopping power, the kinematic factor and the geometry of the scattering. The energy of a scattered particle at a thickness below the flat surface is (see Fig. 10.17):

$$E_t = K(E_0 - \Delta E_1) - \Delta E_2 \tag{10.25}$$

ΔE_1 is the energy loss of the incident particle before the collision including the corresponding kinematic factor in the collision region:

$$\Delta E_1 = \int_0^t \left(\frac{dE}{dx}\right)_{in} dx/\cos\alpha \tag{10.26}$$

The energy loss of the scattered particle crossing the backward path to the detector is given by ΔE_2:

$$\Delta E_2 = \int_t^0 \left(\frac{dE}{dx}\right)_{out} dx/\cos\beta \tag{10.27}$$

For very thin samples, corresponding to an energy loss equal or less than the resolution of the system ($\approx$ 15 keV), the integrals ΔE_1 and ΔE_2 are simply $(dE/dx)_{in}$ $t/\cos\alpha$ and $(dE/dx)_{out}$ $t/\cos\beta$, where t represents the thickness of the sample. Therefore, the areas S_A, S_B of the peaks at A, B, etc., are simply related to the relative atomic concentration C_A, C_B of the elements by

$$S_A/Z_A^2 = C_A$$

$$S_B/Z_B^2 = C_B \tag{10.28}$$

$$\vdots$$

with $\Sigma C_i = 1$.

The cross-section is the only parameter differing in the terms S_A and S_B, and the unique parameter which varies in this cross-section is the charge Z of the collided nucleus.

For thick samples, Equations (10.26 and 10.27) may be calculated by a heuristic method. The spectrum is divided firstly into slices dE of arbitrary value (say of the order of the energy resolution). The various spectrum areas are then compared to the calculation and iterations are performed by gradually changing the relative concentrations C_i in each set of slices to achieve 'one' satisfactory fit. But this fit is not necessarily the single solution if overlapping of peaks is observed as shown in the simulations given in Section 10.5.

The computation of an RBS spectrum using various algorithms [e.g. RUMP (Doolittle 1986), GISA (Saarilahti & Rauhala 1992), SCATT/HYPRA (Strathman & Bauman 1992)] requires associating the energy of the backscattered particles to the depth at which the scattering took place, and also to the mass of the interacting nucleus. A signal at a low or medium energy in the RBS spectrum may be attributed to either the scattering on a light nucleus close to the surface or the scattering on a heavy nucleus deeper in the material. An easy way to obtain qualitative information on which situation should be considered, is to collect, simultaneously, the X-rays emitted by the sample. Simultaneous PIXE-RBS is indeed a very convenient method to avoid misinterpretation of an RBS spectrum and to give a rapid starting situation in order to compute the RBS spectrum. Furthermore, characteristic X-rays of the elements of neighbouring masses are easily differentiated by PIXE, even when the

RBS technique is unable to perform the mass separation. Relative intensities of X-ray lines may also be advantageously used for depth profiling measurements of binary and ternary targets containing large concentrations of elements having very close atomic numbers (Demortier 1999).

In addition to the relationship of energy loss with the achieved thickness and/or the mass of the interacting nucleus in the samples, some consideration on the energy spread is to be made. Sources of energy spread include detector resolution, energy spread in the incident beam, straggling and kinematic effects. Since energy straggling increases as the ion beam traverses the sample, the depth resolution degrades with depth in the sample. A common practice is to quote depth resolution at the surface, which need not include a straggling contribution. A convenient approximation is to assume all sources of energy spread are Gaussian and to add them in quadrature.

The depth resolution can be improved by tilting the sample normal relative to the incoming beam (i.e. increasing α and/or β). The effect is to increase the path length required to reach a given depth (measured perpendicular to the surface) in the sample. This increases the scattered particle energy difference for a given depth difference. It should be noted that the use of large tilt angles introduces additional sources of energy broadening and requires that the sample surface be reasonably flat.

10.8 Contribution of non-Rutherford scattering

Advantages may be associated with the use of non-Rutherford cross-sections, such as an improved accuracy in the determination of stoichiometric ratios and an increased sensitivity for detection of light elements in heavy-element matrices. The improvements justify the additional work required to make the measurements of cross-sections in the special geometry used in the laboratory. According to Equation (10.28), the stoichiometric ratio for two elements in a thin film only depends on the ratio of integrated peak counts A_A/A_B, and the cross-section ratio, σ_B/σ_A. These stoichiometric ratios may be determined as accurately as a fraction of one per cent by acquisition of sufficient data if the backscattering peaks are well separated and if the cross-section ratios are accurately known. However, the use of analysis-ion energies so that the cross-sections are purely Rutherford (and therefore accurately known) frequently produces backscattering peaks that overlap. The uncertainties in the peak count ratios (resulting from deconvolution/simulation techniques) often severely limit the accuracies of the stoichiometric ratios, particularly in cases of non-uniform composition in the studies layers or films.

Irradiation with higher incident ion energies would result in a desired reduction of peak overlap since the energy loss of the analysis ions in matter decreases with increasing energy in the energy range normally used. However, the accuracy of the cross-section ratio may be adversely affected if the cross-sections of interest are non-Rutherford at the higher ion energy. The analysis ion energy should usually be chosen to be as high as possible to take advantage of the reduction in peak overlap, but with such a value that the relevant cross-sections are accurately known and that they do not vary wildly in the region just below the incident ion energy. If the ratio of measured-to-Rutherford cross-section slowly varies with energy, the non-Rutherford effect may be easily included in calculations. The non-Rutherford enhancement factor $(\sigma/\sigma_R)_i$ at the mean energy of the projectile in the film would then be introduced in the computation. Since non-Rutherford cross-sections for light target elements are frequently many times the Rutherford value, while corresponding cross-sections for the heavier target elements may remain Rutherford, the relative sensitivity for light element detection results will be increased. An enhancement of several orders of magnitude may be obtained for C and O when the incident energy of the projectile is carefully chosen.

Strong, narrow, isolated resonances in the non-Rutherford cross-sections may be used for depth profiling of light elements in or on heavy matrices. Examples are the 3.045 MeV resonance in the ^{4}He–^{16}O cross-section (Cameron 1953), the 2.525 MeV resonance in the ^{1}H–^{9}Be cross-section (Mozer 1956; Leavitt *et al.* 1994), and the 4.26 MeV resonance in the ^{4}He–^{12}C cross-section (Bittner & Moffat 1954).

There is a lot of information in the literature regarding measured non-Rutherford cross-sections for ^{1}H and ^{4}He projectiles. Most of these data were acquired during the 1950s and 1960s in connection with nuclear-level structure studies. Data have usually been presented in graphical form only, as differential cross-sections in the centre-of-mass system versus projectile energy in the laboratory system (Jarjis 1979). A compilation (see Jarjis n.d.) gives some tables of numerical values produced by the use of a computer digitiser (Knox 1992).

10.9 Fundamentals of ERD

The study of the depth distribution of light elements in a matrix containing heavy atoms is very often not possible by RBS, mainly for two fundamental reasons:

(1) The signals of backscattered particles on light nuclei are always superposed to those of higher atomic mass owing to a lower kinematic factor [Equation (10.8)].
(2) The Rutherford cross-section is much lower for light nuclei than for heavier ones due to its dependence on Z^2 [Equation (10.13)].

Less ambiguous determinations of the content of light elements are then only possible by RBS if we use resonance scattering, but this procedure does not apply to all light nuclei. An alternative would be to use nuclear reactions (Demortier 2000), which often offer good selective parameters to achieve good mass selectivity and depth resolution and sometimes an ultimate limit of detection for trace analysis.

Although for many circumstances ERD has poorer sensitivity and depth resolution than NRA, experiments not involving trace element analysis can be performed with ERD more simply than with NRA. Compared to depth profiling of heavier atoms using RBS with protons or α particles, ERD uses high energy ($\approx$1 MeV amu^{-1}) heavy-ion beams to kinematically recoil and depth profile low atomic number target atoms, mainly for $Z \leq 10$. However, as the heavy-ion projectile need only have a mass greater than the target atom, α particles are commonly used to obtain recoil spectra for hydrogen and its isotopes. ERD has become the technique of choice for many light element profiling applications because of the simplicity, the possibility of simultaneous multielement profiling, acceptable sensitivity and sufficient depth resolution.

Although ERD is becoming a relatively common tool in materials analysis, most researchers use ERD only qualitatively (i.e. for elemental identification or profile shapes), in contrast to the quantitative analyses performed with RBS. The physical concepts governing ERD and RBS are identical, and through the use of a computer, ERD spectra are now easily manipulated to obtain full quantitative analyses. The standard geometry to profile light elements at the surface of thick samples is illustrated in Fig. 10.18.

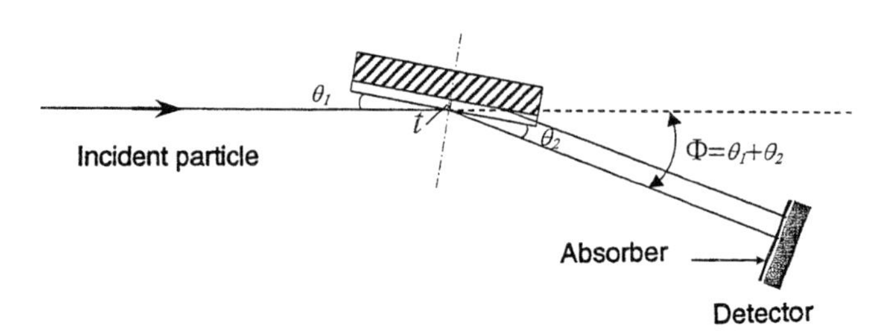

Fig. 10.18 Classical experimental arrangement for ERD. A mylar foil is placed in front of the detector to stop the scattered particles.

Similarly to RBS, ERD is governed by fundamental physical concepts:

(1) The kinematic factor describes the energy transfer from a projectile to a target nucleus in an elastic two-body collision. Choosing M_1 for the mass of the incident particle and M_2 for the collided nucleus ($M_2 < M_1$):

$$K = \frac{4M_1M_2\cos^2\phi}{(M_1+M_2)^2} \tag{10.29}$$

(2) The differential scattering cross-section gives the probability for the scattering event to occur. The cross-section is governed by Coulomb scattering as also modelled by Rutherford and expressed in the laboratory reference frame

$$\sigma_r(E'_0,\phi) = \frac{[Z_1Z_2e^2(M_1+M_2)]^2}{[4\pi\varepsilon_0 E_0]^2\cos^3\phi} \tag{10.30}$$

where $(e^2/4\pi\varepsilon_0 E_0)^2 = 0.020731$ barns for $E_0 = 1$ MeV. For cases where the cross-section is non-Rutherford (see below), the simplest solution for the thin-film approximation is to multiply σ_r by a scaling factor, f, in order to model the true cross-section, i.e. $\sigma = f\sigma_r$.

(3) The stopping powers give the average energy loss of the projectile and the recoiled target atom as they traverse the sample (thereby giving a method for establishing a depth scale). The energy E_1 of the recoiled light nucleus of mass M_2 at depth x[o < x < t] into the material is then:

$$E_1 = KE'_0 \tag{10.31}$$

Here E'_0 is the actual energy of the incident (heavy) ion after crossing a distance $x/\cos\theta_1$. They finally have an energy E_2 (lower than E_1) after crossing the outgoing $x/\cos\theta_2$ path in the layer. In order to stop the elastically scattered ions of mass M_1 (with a maximum energy when they are scattered by the heavy components of the studied material), a 'range foil' (absorber of thickness d) is put on the recoiling path. These ions of mass $M_1 > M_2$ are more easily stopped in any material owing to the higher stopping power [Equation (10.1)]. Finally, the energy, E_d, of the collected ions is used for the depth profiling of light atoms. As the forward scattering angle ϕ is in the range of 20–30°, α and β are very small and the effective distance crossed by the ions is large compared to the position of the depth layer t under analysis.

(4) The energy straggling gives the statistical fluctuation in the energy loss. The energy straggling of both incident and emitted ions is very important and this parameter is the most limiting factor of ERD as far as depth profiling is concerned. For fine-depth profiling, NRA would then be preferred to ERD. The yield (Y_r) for the recoil atoms detected in energy channel E_d with a channel width of δE_d is given by

$$Y_r(E_d) = \frac{QN_r(x)\sigma_r(E'_0, \phi)\Omega\delta E_d}{\cos\Theta_1 \, dE_d/dx} \tag{10.32}$$

where Q is the incident projectile fluence, $N_r(x)$ is the atomic number density of the recoiled atom at depth x measured normal to the surface, Ω is the detector solid angle, and dx is the increment of depth at x corresponding to an increment of energy dE_d. These quantities Q, Ω and Θ_1 are fixed by the experimental set-up.

Deviations from Rutherford scattering cross-sections resulting from Coulomb barrier penetration occur most often for low-Z projectiles at high energies, while screening effects, which can cause deviations from the Rutherford formula, occur most often for high-Z projectiles at low energies. Bozoian (1991a, b) modelled the situation of the Coulomb barrier penetration and determined the energies where the cross-sections may become non-Rutherford. In the laboratory reference frame and for a scattering angle of 30°, the formulation of this energy boundary is given by:

$$E(\text{MeV amu}^{-1}) = \frac{-1.1934(Z_1Z_2)(M_1+M_2)}{M_1M_2A_1^{1/3}\ln/(0.001846\,Z_1A_2/Z_2)} \tag{10.33}$$

The best way to ascertain, in each practical case, as to whether or not a non-Rutherford cross-section is to be used, is to compute the ERD spectrum with that of a reference material.

The study of hydrogen isotopes in a material is usually performed by using 2–4 MeV α particles as incident projectiles. Figures 10.19 and 10.20 give some comparison of measured and fitted cross-sections (Baglin *et al.* 1992) for light hydrogen measurements. For a full treatment of ERD applications the reader is recommended to consult two pertinent studies (Barbour & Doyle 1995; Tirira *et al.* 1996). For other light nuclei, incident beams of Cl and Si ions are the most widely

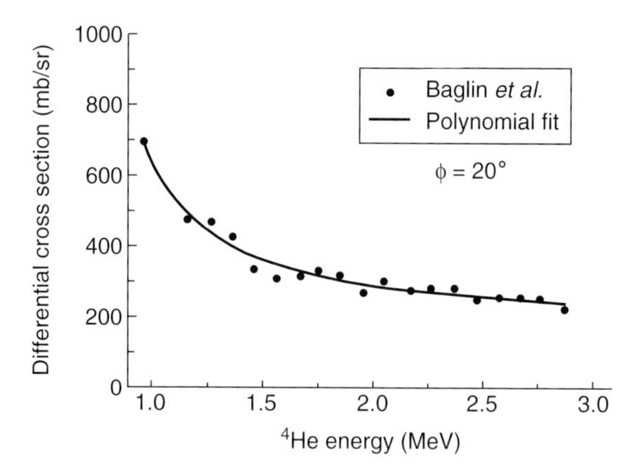

Fig. 10.19 Comparison of experimental data due to elastic recoil of hydrogen at 20° with α particles (Baglin *et al.* 1992) and a polynomial fit (Barbour & Doyle 1995).

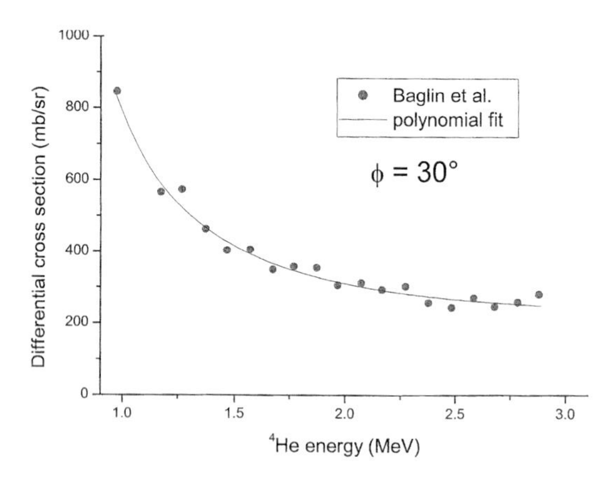

Fig. 10.20 Comparison of experimental data due to elastic recoil of hydrogen at 30° with α particles (Baglin *et al.* 1992).

used particles; the reason lies mainly in their availability in classical tandem accelerator technology.

10.10 Selected applications

10.10.1 *Characterisation of the interdiffusion in Au–Al layers*

The non-destructive RBS standardless analysis technique was used to study interdiffusion phenomena in aluminium and gold layers with a high depth resolution (10–20 nm) at the interfaces. A multilayered system consisting of alternate aluminium and gold layers was deposited under high vacuum conditions on polished glassy carbon and single crystalline silicon substrates to investigate the interdiffusion of gold and aluminium in as-deposited layers (Markwitz & Demortier 1997; Markwitz *et al.* 1997). The energy of the incident beam was adjusted at 2.0 MeV. The aluminium signal was measured in a nearly background-free region.

RBS spectra are fully calculated by using the known physical parameters (e.g. ion type, ion energy, backscattering angle, charge, beam current, etc.), while the depth distributions of gold and aluminium can be extracted from the experimental spectra by using the RUMP simulation program (Doolittle 1986).

Figure 10.21 shows the RBS spectrum of a multilayered Au–Al system consisting of three layers: gold top-layer (channel (ch): 1600–1350; *E*: 1.85–1.55 MeV; layer thickness 250 nm), intermediate aluminium layer (ch: 730–620; *E*: 0.85–0.73 MeV; layer thickness 220 nm) and gold layer (ch: 1200–900; *E*: 1.4–1.05 MeV; layer thickness 270 nm) on silicon substrate. Part of the RBS spectrum shown in Fig. 10.21 is plotted in Fig. 10.22. The interdiffusion zones are emphasised by an additional simulation of the RBS spectrum that does not include an Au–Al interdiffusion (arrows point to the interdiffusion zones). The data for the simulation, including interdiffusion, are listed in Table 10.4. An astonishing result is that both interdiffusion zones (Au–Al and Al–Au) are identical in the composition and thickness of the sub-layers used for simulating the RBS spectrum. This indicates that the interdiffusion observed in as-deposited specimens is reproducible and is strongly dependent on the conditions in the evaporation chamber (e.g. substrate temperature) and independent of whether gold is deposited on aluminium or aluminium on gold.

The accuracy of the gold and aluminium concentrations listed in Table 10.4 is about ±2 at.%. Figure 10.23 shows the Au–Al interdiffusion behaviour in a multilayered system consisting of five layers (170 nm Au, 120 nm Al, 150 nm Au, 120 nm

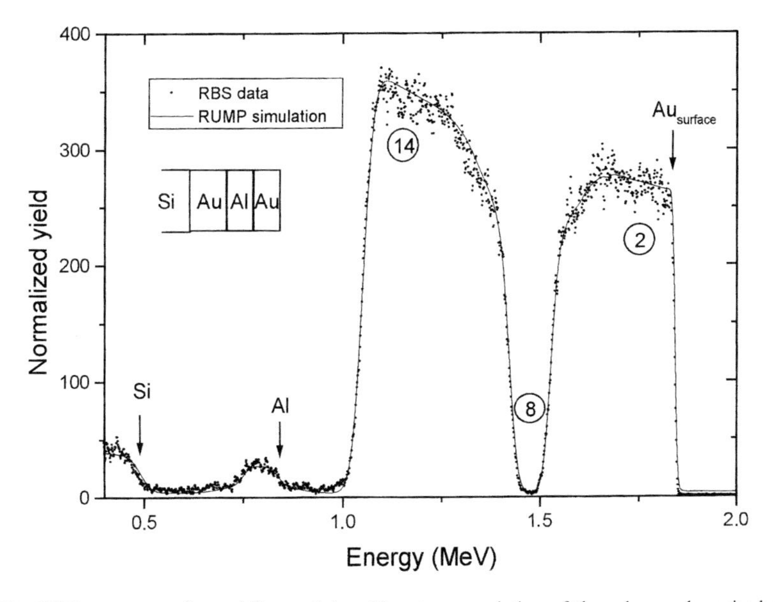

Fig. 10.21 RBS spectrum of a multilayered Au–Al system consisting of three layers deposited on single-crystalline silicon (layer system: Au/Al/Au/Si). The numbers indicate sublayers used for the RBS simulation [refer to Table 4 from Markwitz *et al.* (1997)].

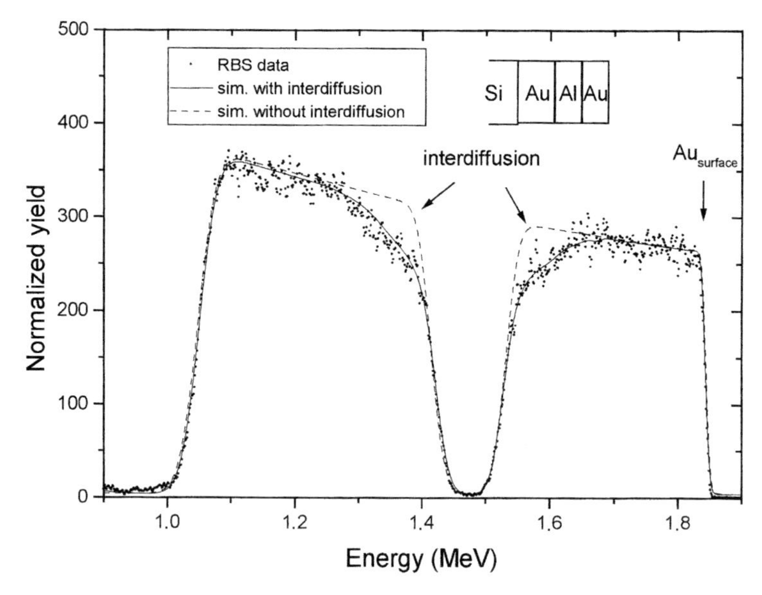

Fig. 10.22 Detection of the Au–Al interdiffusion by simulating the RBS spectrum with and without Au–Al interdiffusion zones (Markwitz *et al.* 1997).

Table 10.4 Data of the simulation used for evaluating the RBS spectrum shown in Fig. 10.22

	Layer system 1	
Sub-layers used in RBS simulation	Thickness (10^{15} atoms cm^{-2})	Component Au (%)
1	20	0
2	770	100
3	150	95
4	150	87
5	200	70
6	150	60
7	115	50
8	1115	0
9	115	50
10	150	60
11	200	70
12	150	87
13	150	95
14	930	100

Al, 150 nm Au). Glassy carbon was used as the substrate. It is noted in Figs 10.21 and 10.23 that the signals of Al are much less intense than the Au signals due to the Z^2 dependence of RBS cross-sections. The accuracy of the Al concentration is then very much achievable when measuring the absence of Au signals in the layered structure rather than the presence of Al signals.

The contribution of light elements like H and O in the layered structure may modify

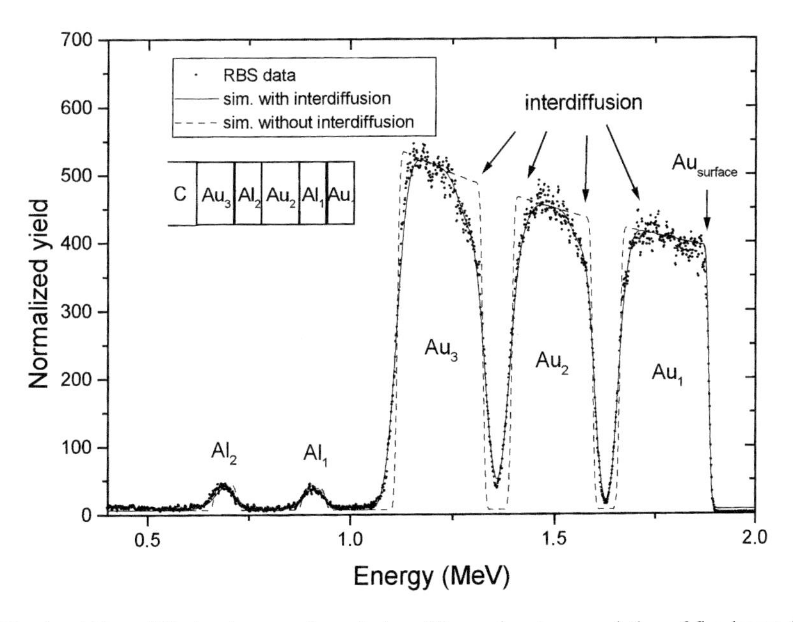

Fig. 10.23 Au–Al interdiffusion in an as-deposited multilayered system consisting of five layers (Au/Al/Au/Al/C). Glassy carbon was used as the substrate.

the Al–Au interdiffusion properties. The H concentration was determined by using a prompt resonant nuclear reaction induced by ^{15}N ions (Lanford *et al.* 1976). The O analysis was performed by using 7.6 MeV ^{4}He ions (backscattering angle of 171°), where the elastic cross-section is more than 100 times larger than the Rutherford cross-section (energy region 7.3–7.65 MeV) (Campisano *et al.* 1983; Cheng *et al.* 1993). By using this high ion energy, the oxides at the surface and interfaces can be measured in one single measurement, contrary to the 'classic' $^{16}O(\alpha,\alpha)^{16}O$ resonance reaction at 3.045 MeV, which requires an energy scan (Barbour & Doyle 1995; Markwitz *et al.* 1997). Furthermore, at 3.045 MeV, the σ/σ_R enhancement factor is only 24. The experimental spectra were evaluated by using the RUMP simulation program (Doolittle 1986), incorporating the non-Rutherford-enhanced cross-section data.

Open dots and the dotted line in Fig. 10.24 show a part of the spectrum of the HE–BS analysis (O region) of the as-deposited and irradiated Au–Al layer systems on C. In addition to the backscattering signals around 2.75 and 2.55 MeV (the kinetic factor for oxygen is indeed 0.36), two peaks appeared in the HE–BS spectrum at 2.3 and 2.87 MeV, originating from (α,p) reactions with Al. As can be seen in Fig. 10.24, significant differences in the O concentration at the specimen surfaces, the Au–Al interfaces, and the Al–substrate interfaces were measured. In the as-deposited state, only $5\pm2 \times 10^{15}$ O cm^{-2} were detected at the specimen surface. After irradiation of the specimen (300 nA, 2 MeV α particles, 10 min, beam area 0.5 mm^{2}) the O amount increased to $1.3\pm0.1 \times 10^{16}$ atoms cm^{-2} indicating the formation of Al_2O_3 at the surface (Fig. 10.24, full dots and continuous line). Quite different O concentrations were measured in the Au layer ($C_0 = 1\pm1$ at.%) and

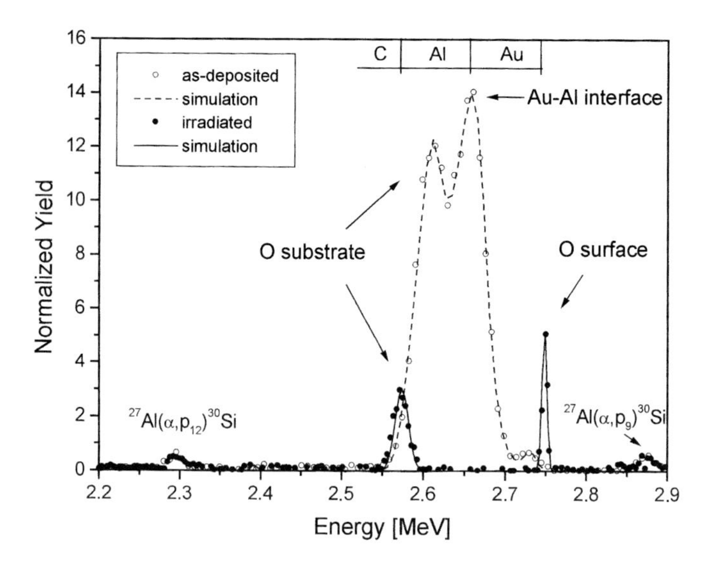

Fig. 10.24 A part of the high-energy backscattered spectrum (non-Rutherford) induced by 8 MeV α particles (Markwitz *et al.* 1998) and simulations in the O region of as-deposited and irradiated Au–Al layers deposited on C (schematic layer structure before irradiation is included).

the Al layer ($C_0 = 10 \pm 1$ at.%). The O atoms possibly originate from the residual gas in the evaporation chamber. However, the O content in the Au layer is 10 times lower, even though the vacuum conditions were identical during deposition. At the Au–Al interface, $2 \pm 0.3 \times 10^{16}$ O cm^{-2} were detected. In contrast to the increase of O at the specimen surface after ion-beam bombardment, O decreased at the Al–substrate interface from $3.5 \pm 0.3 \times 10^{16}$ O cm^{-2} to $5 \pm 2 \times 10^{15}$ atoms cm^{-2}. Remarkably, after ion-beam bombardment, oxygen was no longer found in the interdiffused Au–Al layer (detection limit 0.1 at.%). In addition, the carbon substrate is oxygen free after ion-beam bombardment. The oxygen loss in the interdiffused Au–Al layer can be attributed to an ion-beam-induced O diffusion, possibly supported by the presence of H (enhanced diffusion).

10.10.2 *RBS study of layers on spherical samples*

A series of ZSM 5 zeolites containing Ga, Al or both elements in variable amounts was synthesised involving methylamine as a mobilising–complexing agent that favours a quasi-homogeneous Ga distribution (Gabelica & Demortier 1998).

Simultaneous RBS and PIXE with 2 MeV protons were used to characterise the elemental concentration for species heavier than Si. The contribution to the RBS spectrum from traces of Ca and Fe (which occasionally appear in PIXE spectra) has been calculated by assuming that these elements are homogeneously distributed in depth. They could be present in the ingredients before they are mixed with the Ga source. RBS spectra, using 960 keV α particles, have been computed with the classical RUMP program (Doolittle 1986). The granulometry of the zeolite crystallites has been controlled and spherical particles have been selected. Grains were of nearly the same diameter for each sample (Fig. 10.25). The powder has been compressed to pellets using a gentle pressure in order to avoid any modification of the surface of individual grains (controlled by SEM). A large number of grains was irradiated by an incident broad beam of α particles (diameter 5 mm). The detection angle is 178°. For depth profiling adjustment, account was taken of the fact that incident α particles on the edge of a spherical grain would interact with layers apparently thicker than those in the centre of each grain (Fig. 10.26). The 956 keV incident α energy was chosen in order to avoid scattering on both sides of the grains (except at the edge). For depth profiling extending on thicknesses ranging from 5 to 25% of the grain radius, the ideal RBS had to be corrected using a geometrical factor, as given in Fig. 10.27. The classical rectangular shape of the RBS contribution of Ga in a surface layer of a spherical grain is indeed modified in the low-energy part owing to this geometrical effect.

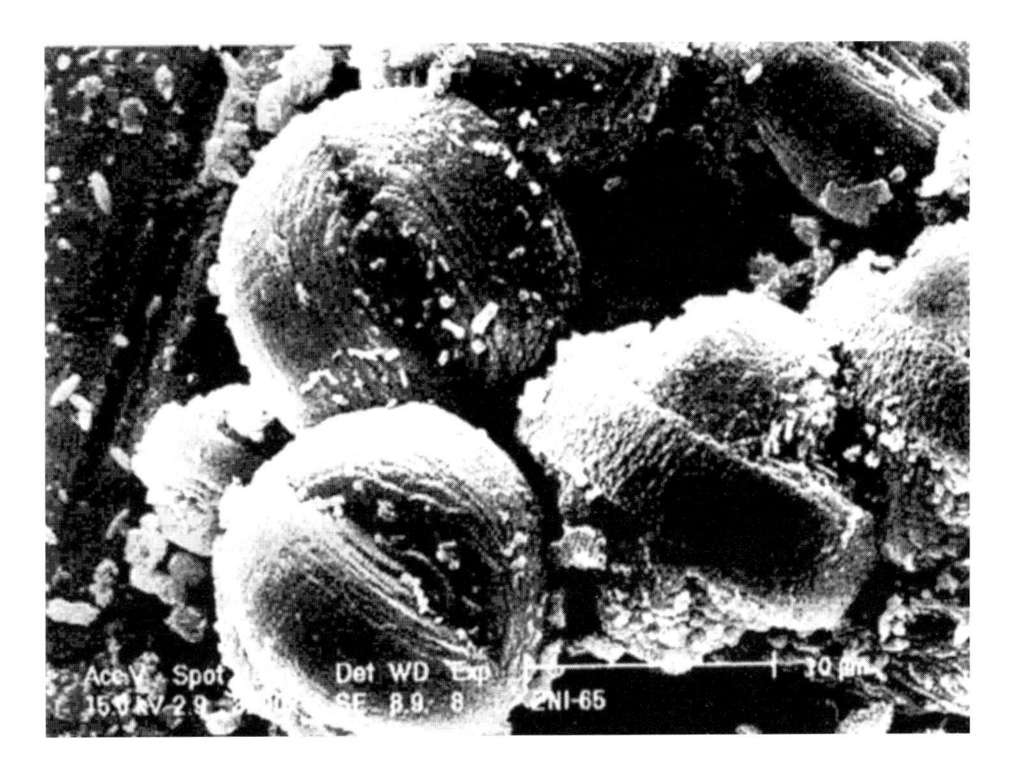

Fig. 10.25 SEM micrograph of zeolite crystallites ZN-2, showing the regular shape of quasi-spherical twinned (Ga_2Al_2)-ZSM-5 (from Gabelica and Demortier 1998).

10.10.3 *Bulk analysis of ancient gold artefacts*

Pre-Hispanic goldsmiths developed original techniques to manufacture and to gild ornaments and artefacts. Gold metallurgy in ancient America was mainly a surface technology. For changing the colour of the artefact, two kinds of gilding procedures were used. The first involved the addition of a gold alloy at the surface (fusion gilding,

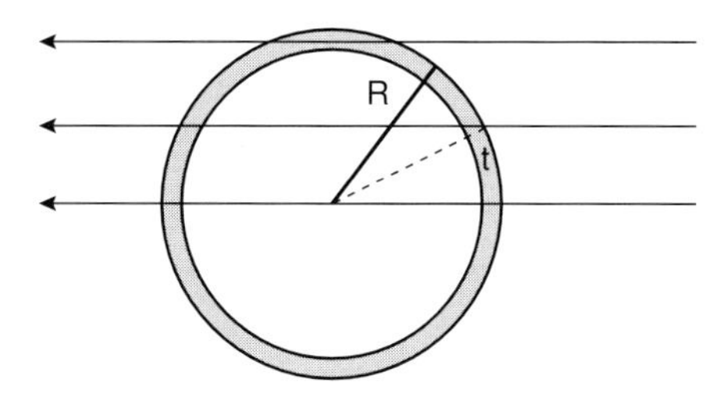

Fig. 10.26 Geometrical factors for the study of depth profiles in spherical granules.

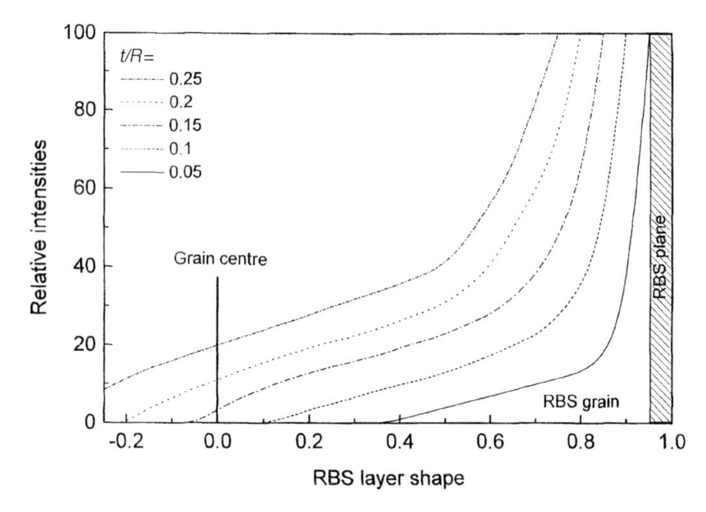

Fig. 10.27 Modification of the classical RBS layer shape for spherical grains for various t/R ratios (t = surface layer thickness; R = grain radius).

electrochemical plating), while the second removed less noble metals from the surface in order to obtain a gold-enriched surface (depletion gilding) (see Demortier & Ruvalcaba-Sil 1996).

Fusion gilding involves the application of a molten metal to the surface of a metallic artefact (usually made with a copper alloy containing no gold at all) by dipping the basic ornament into a bath containing a rich molten gold alloy. The resulting gold alloy layer strongly adheres to the substrate by melting the metal of the ornament at the interface. The coating is thick (10 μm or more) and not uniform. Precise control of the melting point of the gold alloys and of the core metal is required. In contrast, by using electrochemical plating, the ornament is coated with a very thin and uniform layer of a gold-rich alloy (0.5–2 μm). This process involves the dissolution of a gold alloy in a solution of corrosive minerals mixed with salts. Fusion gilding and electrochemical plating give rise to a sharp interface between the bulk and the gold alloy layer. Because of its depth profiling ability, RBS is suitable in rapidly distinguishing both procedures.

For depletion gilding, the ornament is to be made with a gold alloy containing a large amount of copper. The basic principle of this original process (exclusively of American origin) involves the removal of less noble metals from the surface but leaving the gold content unaffected. When the ornament is heated (at 500°C, for instance), various oxides are formed at its surface. If the ornament is dipped in a solution of salts and acids derived from plants, the copper and sometimes silver oxides are removed to leave a 'gold enriched' surface. This process is repeated several times, giving at each step an increase of gold at the surface. After each step, the colour of the alloy is modified and at the end of the procedure, the ornament has a lustrous 'golden' appearance.

Surface studies of artefacts from various regions of pre-Hispanic South America have been carried out using a 2 MeV proton microbeam (10 μm diameter). Several areas of such artefacts were scanned. RBS and PIXE were used simultaneously to determine the elemental composition and to characterise the homogeneity in depth and at the surface. RBS and PIXE spectra collected in four different microregions of a zoomorphic pendant (Fig. 10.28) of pre-Hispanic origin from Columbia (Ruvalcaba-Sil & Demortier 1997) are given in Fig. 10.29. Spectrum (a) shows a gold-depleted surface region with characteristic signals of C and O which are typical of the surface corrosion of copper. Spectra (b)–(d), in contrast, indicate gold enrichment at the surface. The most important gold enrichment is shown in spectrum (c), the less important in spectrum (d). The difference is also indicated in the corresponding Cu–Au ratios of the PIXE signals.

10.10.4 *Depth profiling light elements near an interface*

A knowledge of depth concentrations of atoms introduced by diffusion or implantation is essential, but it is difficult to measure. Resonant nuclear reaction analysis

Fig. 10.28 Zoomorphic pendant made in Tumbaga, Columbia, 500–1500 AD. Length: 71 mm.

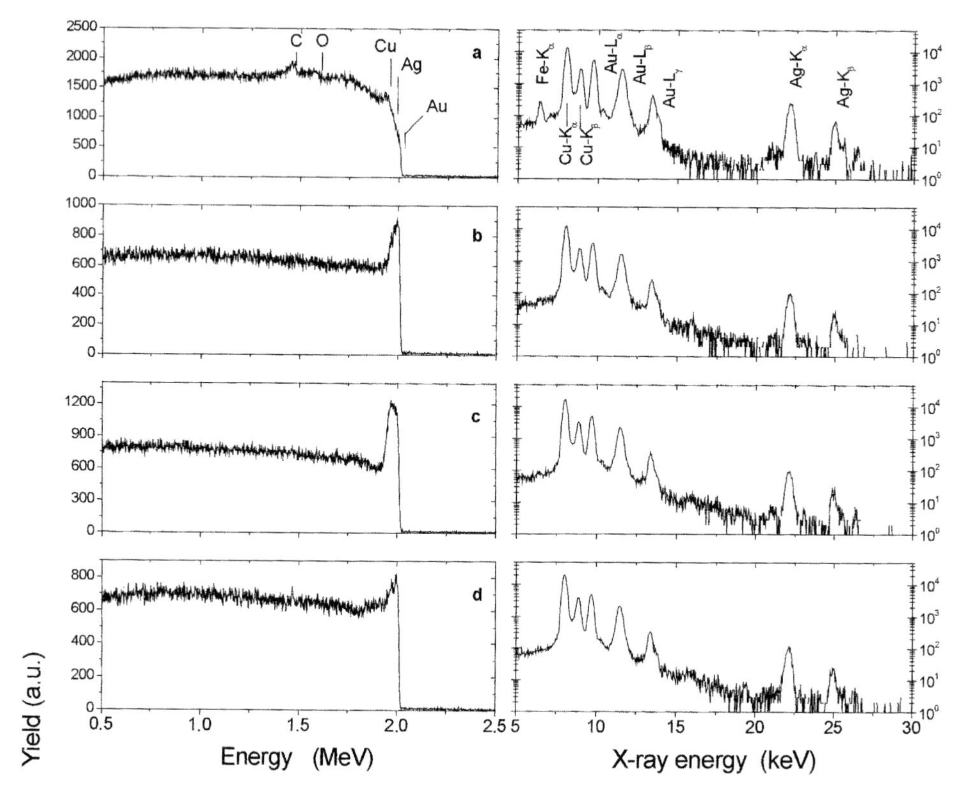

Fig. 10.29 Microbeam RBS and corresponding PIXE spectra recorded at four impacts on the zoomorphic gold artefact. Cu/Au PIXE ratio decreases when the gold concentration at the surface is higher (from Ruvalcaba-Sil & Demortier 1997).

(RNRA) is a powerful technique to determine the depth profile of light elements in a matrix. These reactions often occur in the low-energy range of protons or α particles. The depth profiles can be obtained by counting the number of γ-rays or emitted particles for various incident energy values increasing from an energy just below E_R (resonant energy) to an energy above E_R. At energies above E_R, the nuclear reaction takes place only in the depth. The depth resolution at the surface depends on the resonance width used and the beam energy resolution. The determination of the exact localisation of an interface remains a problem. The reason is that in the NRA technique the depth is calibrated using the energy loss of the particle. The energy scale should therefore be converted into atoms cm^{-2}. Moreover, in the case of depth profiling measurements, the density is not uniform. The different origins of inaccuracy in the determination of location are accuracy of the stopping power data, validity of Bragg's Rule [Equation (10.4)], exact composition of the surface layer and non-homogeneity in this layer. To considerably reduce these uncertainties, the energy loss of the incident particles is measured by using RBS simultaneously with NRA near the resonance energy. The depth profile at the interface is then calibrated from the direct observation of an RBS spectrum without using stopping power parameters. The method is also insensitive to concentration variations in the surface layer.

A typical application of the method is in the study of the fluorine and sodium migration in SnO_2 coatings on glass (Terwagne & Deconninck 1992). Fluorine is expected to be concentrated in the SnO_2 coatings, but sodium only is present in the glass. The depth profiles are measured by proton-induced resonant reactions and the depth calibration, which is essential to localise the interface, is done by proton backscattering. The depth profile is obtained by varying the energy of the incident particle by small energy steps over the whole energy range of the NRA. Figure 10.30 shows the experimental set-up of the detection system. The incident proton beam is impinging on the sample which is placed inside the NaI detector in order to obtain a maximum efficiency for γ-ray detection. The backscattered protons are detected at 177° relative to the incident beam direction in a passivated implanted planar silicon detector (PIPS). The beam was also rastered on the sample in order to avoid local overheating during irradiation (which could give rise to ion migration).

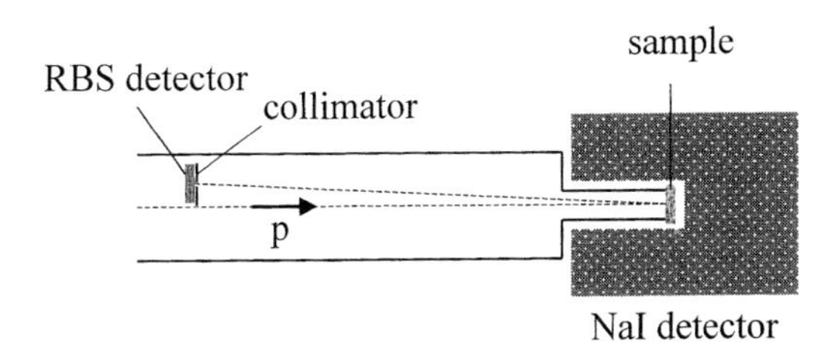

Fig. 10.30 Schematic view of the experimental set-up showing the NaI well detector, which is placed in the beam axis, and the PIPS detector positioned at 177° relative to the incident beam direction. The sample is located in the hole of the NaI detector (from Terwagne & Deconninck 1992).

10.10.4.1 *Sodium profiling* Profiling of sodium is achieved by using $^{23}Na(p,\alpha\gamma)^{20}Ne$ at a resonant energy $E_R = 1010.5$ keV ($\Gamma_R < 0.5$ keV) and detecting the 1634 keV γ-rays in the NaI detector. The experimental excitation curve is displayed in Fig. 10.31(b) showing a low surface concentration followed by a steep step and a constant concentration of 16 at.%, which is the glass bulk concentration. The energy loss, ΔE_p, is measured by proton backscattering. Figure 10.31(a) shows the Rutherford backscattering spectrum for the SnO_2 layer on glass in the same conditions as those used for depth profiling, i.e. $E_p = 1.0$ MeV. The width (29.2 keV) of the Sn peak is $2\Delta E_p$. As the kinematic factor for Sn is 0.97, the energy of particles scattered by Sn is almost the same as that of the incident particles. ΔE_p can be related directly to the sodium depth profile in order to localise exactly the interface as indicated in Fig. 10.31(b). The advantage of this procedure is that it is not necessary to deconvolute the experimental data. Of course, it is possible to obtain a depth scale (in nm) by using the stopping power of the SnO_2 layer and the stopping power of glass beyond the interface, but this is not essential for the analysis because the analyst would like to localise exactly the concentration step with respect to the interface.

10.10.4.2 *Fluorine profiling* Profiling of fluorine is achieved by using $^{19}F(p,\alpha\gamma)^{16}O$ at a resonant energy $E_R = 340$ keV ($\Gamma_R = 2.4$ keV). This isolated resonance allows the depth profiling of fluorine with a very low background on a large depth (1.5 μm). Another resonance also appears at 226.9 keV (Ajzenberg-Selove & Lauritsen 1959), but it does not interfere with the 340 keV resonance because the energy loss in the $SnO_2(F)$ layer is less than 100 keV and there is no fluorine in the glass. High-energy γ-

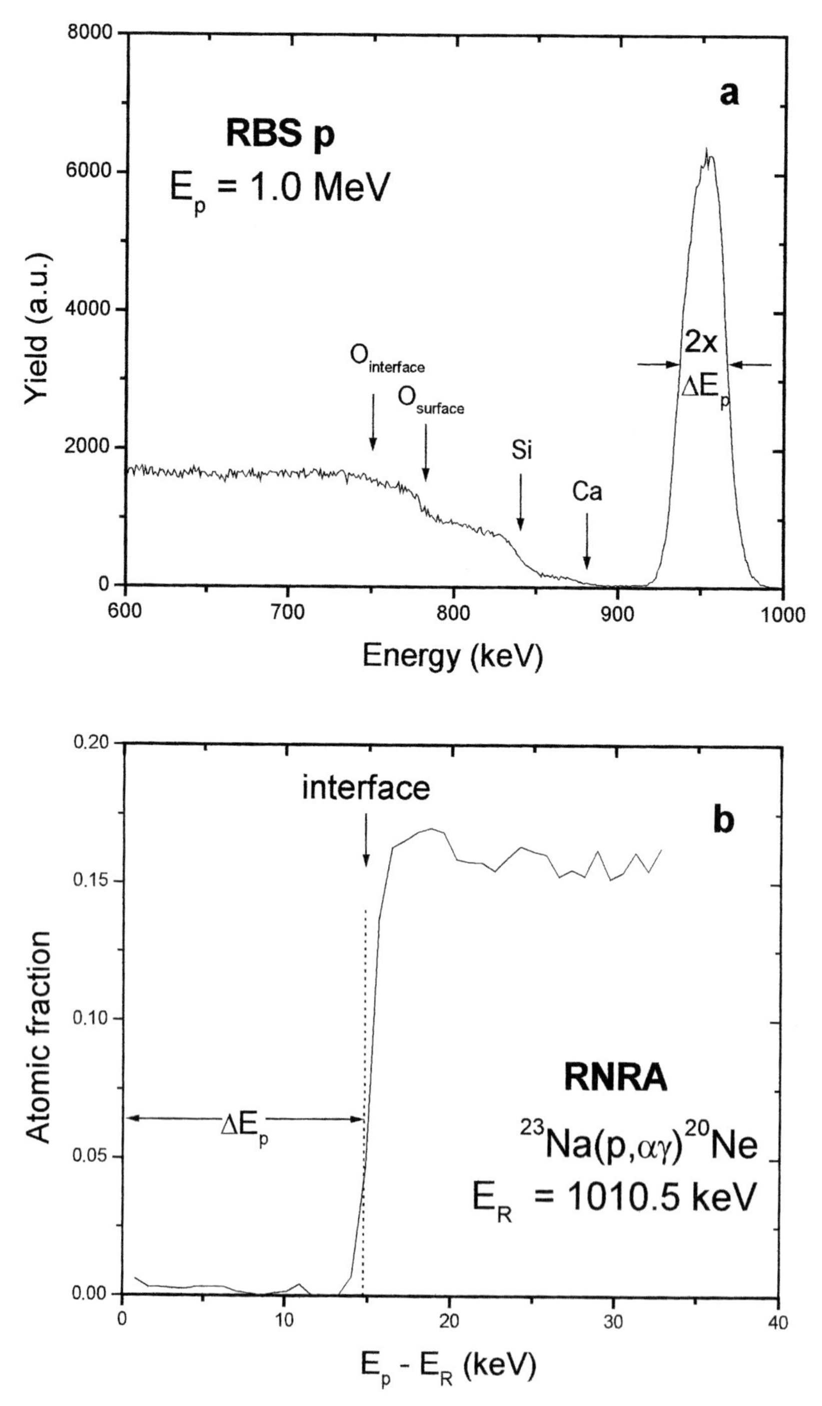

Fig. 10.31 Determination of the sodium profile by nuclear reactions: (a) interface depth measurement by 1.0 MeV proton backscattering on Sn; (b) sodium profile after deconvolution of the excitation curve using $^{23}Na(p,\alpha\gamma)^{20}Ne$ at $E_R = 1010.5$ keV. The interface position is obtained from ΔE_p as measured in (a) (from Terwagne & Deconninck 1992).

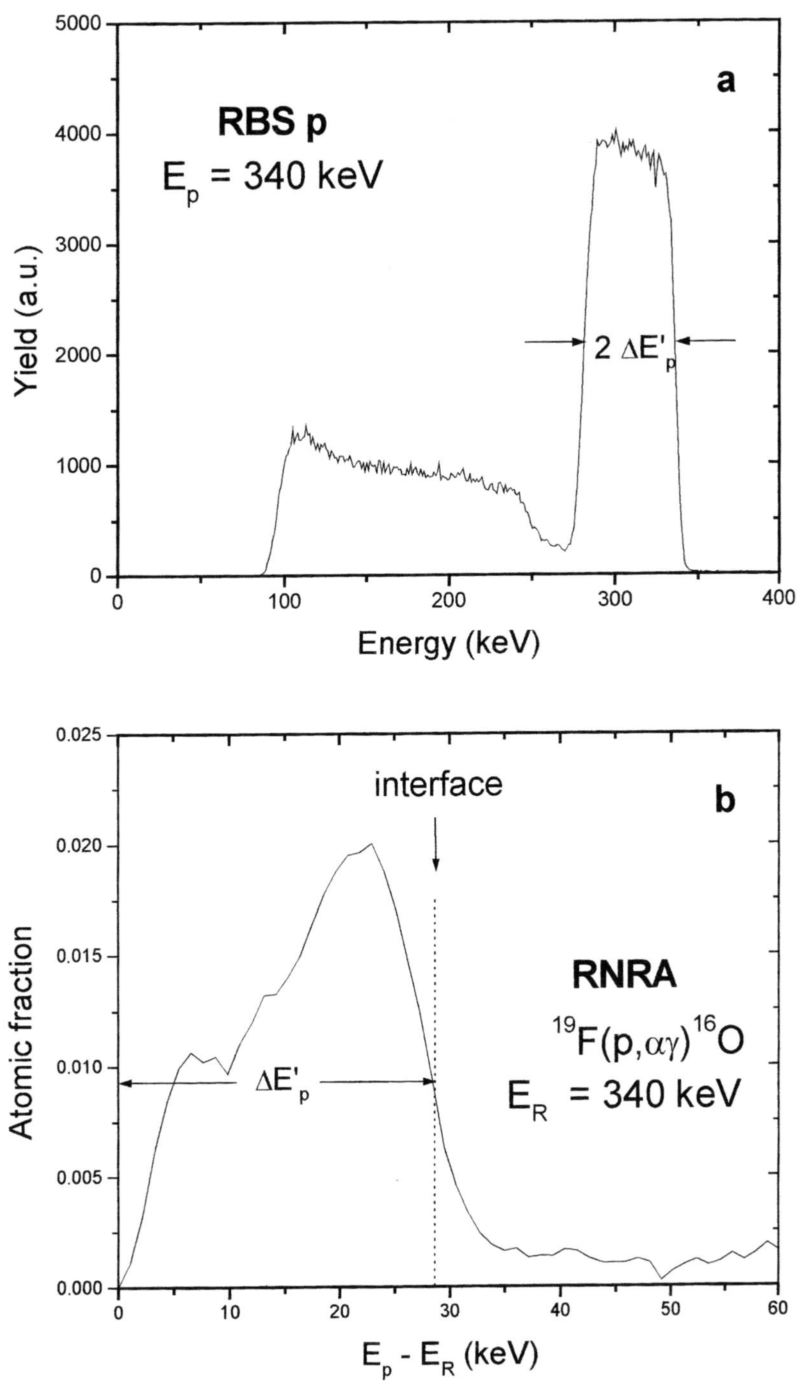

Fig. 10.32 Determination of the fluorine profile by nuclear reactions: (a) interface depth measurement by 340 keV proton backscattering on Sn; (b) fluorine profile after deconvolution of the excitation curve using $^{19}F(p,\alpha\gamma)^{16}O$ at E_R = 340 keV. The interface position is obtained from ΔE_p as measured in (a) (from Terwagne & Deconninck 1992).

rays (6–7 MeV) are detected in a 4 × 4 in^2 NaI detector with the same geometry as that shown in Fig. 10.30. The excitation curve is displayed in Fig. 10.32(b) showing a shallow concentration followed by a huge peak that is localised exactly with respect to the interface and the sodium step [Fig. 10.31(b)]. The energy loss, $\Delta E'_p$, corresponding to the interface is measured by proton backscattering, as shown in Fig. 10.32(a). The Sn signal in the RBS spectrum is nearly rectangular with a FWHM, $2\Delta E'_p$, which is found to equal 55.2 keV.

As a conclusion to this example, it is possible to localise exactly an interface and to depth profile light elements by combining RBS and RNRA techniques.

10.10.5 *Light elements analysis of nitrogen-implanted aluminium*

The growth of an oxide layer by an anodic process on synthesised aluminium nitride gives an interesting oxide on semiconductor material with surprising dynamic and decorative properties. During the anodic oxidation, ionic movements are involved in the near-surface region of the aluminium material; these ionic movements have been studied by RBS and RNRA on thin aluminium foils pre-implanted with nitrogen and post-oxidised in an ammonium pentaborate solution. In order to analyse quantitatively the light elements with the BS technique, we have prepared thin self-supported aluminium foils (700 nm) to accurately measure the BS cross-section. To avoid heat damaging the film during implantation, it is necessary to have a good heat conductor as the substrate. Copper has been chosen for this purpose, and Al was evaporated onto a 0.5 mm thick and 12 mm diameter Cu sample. The aluminium was implanted with 100 keV $^{15}N_2^+$ to different doses (10^{17}, 5×10^{17} and 10^{18} N$^+$ cm^{-2}) using the 600 kV implanter based in Århus. The copper was then dissolved and the aluminium foil was mounted on an aluminium frame. The oxidation was carried out at different voltages ranging from 50 to 300 V.

10.10.5.1 *Depth profiling measurements* The nitrogen distribution was measured using the $^{15}N(p,\alpha\gamma)^{12}C$ nuclear resonant reaction, which offers a very narrow resonance at $E_R = 429$ keV ($\Gamma_R < 120$ eV), and detecting the 4.43 MeV γ-rays in the NaI detector. The experimental set-up is the same as that shown in Fig. 10.30. The retained dose of nitrogen was obtained by integrating the nitrogen depth profile and normalising the data with a known standard (TiN). The evolution of the retained dose versus the applied voltage is shown in Fig. 10.33. For the lower implanted dose, the retained dose remains constant (9.8×10^{16} atoms cm^{-2}), and for the medium and the higher implanted doses, the same behaviour is observed: the retained dose of the nitrogen is first constant and then a sudden decrease of nitrogen concentration appears at a well-defined voltage.

10.10.5.2 *Rutherford backscattering measurements* RBS was performed with 2.4 MeV α particles impinging on the surface of the specimen at normal incidence. The backscattered particles were collected at 175° with respect to the incident beam direction. The backscattered spectra observed for the medium-dose-implanted foil (5×10^{17} atoms cm^{-2}) are shown in Fig. 10.34 for different preparation voltages. Light elements such as nitrogen and oxygen are still easily observable because self-

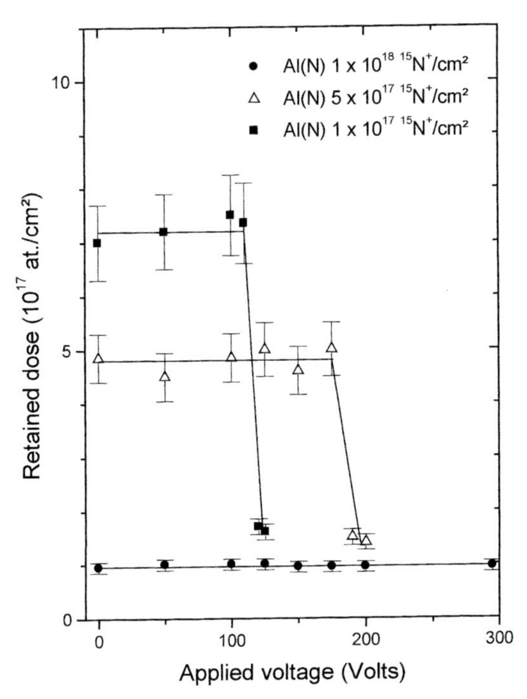

Fig. 10.33 Evolution as a function of the applied voltage during the anodic oxidation process of the nitrogen-retained dose for aluminium implanted to different doses. The full lines have been drawn as a guide for the eye (from Terwagne *et al.* 1990).

supported thin aluminium foils were used. Therefore, there is no interference between the Al and O signals and no contribution from the backing.

A complete analysis of the RBS spectra indicates that the relative concentrations of aluminium and oxygen do not correspond exactly to the stoichiometry of Al_2O_3. This can arise from the presence of boron and probably hydrogen in the oxide layer. In addition, since a BS measurement gives only the areal density [atoms cm^{-2}], the evolution in more conventional units [nm] cannot be considered.

Nevertheless, for the pure and the low-dose-implanted Al foils the thickness of the oxide layer is proportional to the applied voltage (Fig. 10.35). It is also observed that the growth of oxide is reduced for implanted specimens as long as the dose increases and aluminium nitride is present. This effect may result from a reduction in the amount of metallic aluminium atoms that are involved in the growth mechanism. It has been shown (Lucas *et al.* 1991) that when nitrogen is implanted into aluminium the width and the gradient of the nitrogen distribution both increase with the dose because nitrogen diffuses towards the surface. Thus, if the total amount of metallic aluminium between the oxide and the nitride layers is low (owing to the high-dose implantation), the growth of the oxide is reduced compared to the non-implanted sample.

For the high-dose-implanted samples, the sudden increase of the oxide layer is correlated to a sudden decrease of nitrogen (Fig. 10.33). This experiment confirms

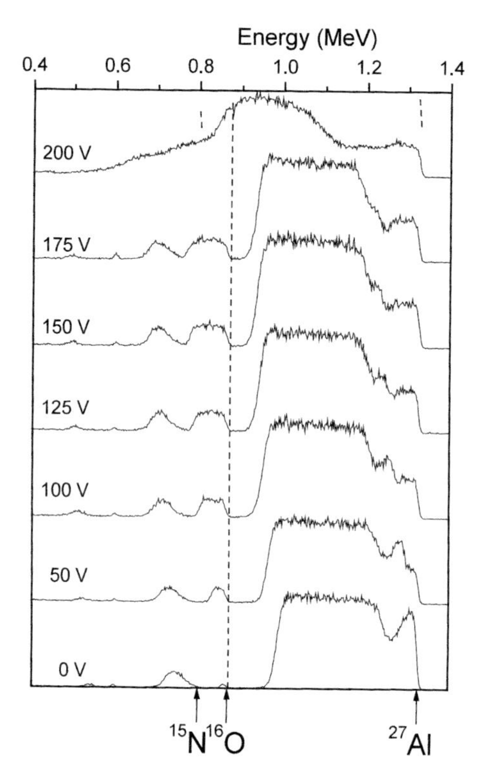

Fig. 10.34 Evolution of RBS spectra for thin aluminium films implanted with 5×10^{17} N^+ cm^{-2} and anodically oxidised at various voltages in an ammonium pentaborate solution (from Terwagne *et al.* 1990).

that nitrogen escapes from the aluminium nitride through the oxide layer i.e., N_2 bubbles were observed at the top of the sample. The film growth by anodic oxidation proceeds by Al^{3+} egress and OH^- and/or O^{2-} ingress trough a pre-existing air-formed film. These ionic movements have been explained elsewhere (Terwagne *et al.* 1990).

10.10.6 *High-resolution depth profiling of light and heavy elements in multilayers by using ^3He particles*

Profiling hydrogen using elastic recoil detection (ERD) induced by ^3He particles is not very usual. α Particles are often preferred to other particles owing to the fact that reactions induced by these particles, with an energy below 2 MeV, leads almost to elastic scattering. Nevertheless, ^3He particles can be used for hydrogen profiling because they induce nuclear reactions on light elements such as ^{12}C, ^{14}N or ^{16}O as well as elastic recoil on hydrogen and elastic scattering.

For example, carbon, hydrogen, copper and silicon from a $C_xH_y/Cu/C_xH_y/Si$ multilayer, obtained by D.C.-magnetron sputtering, can be simultaneously analysed if, respectively, the cross-section of the $^{12}C(^3He,p_0)^{14}N$ nuclear reaction (Kuan *et al.* 1964; Terwagne 1997), the $^1H(^3He,^1H)^3He$ elastic recoil reaction (Terwagne *et al.* 1996), and the elastic cross-section on copper and silicon are known. The first layer of

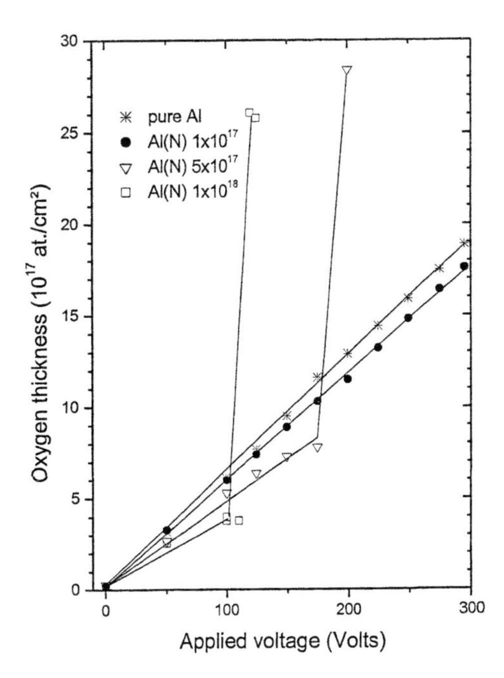

Fig. 10.35 Evolution as a function of the applied voltage during the anodic oxidation process of the total atomic concentration of oxygen for pure and implanted aluminium. The full lines have been drawn as a guide for the eye (from Terwagne *et al.* 1990).

the multilayer was C_xH_y deposited to a thickness of 50 nm on a silicon wafer; a second layer of copper (thickness 50 nm) and a third layer similar to the first were then deposited.

10.10.6.1 *Combining RBS and NRA techniques* Very high depth resolutions can be obtained for profiling carbon if the experimental set-up shown in Fig. 10.36 is used. All information concerning heavy and light elements in the multilayer (except for hydrogen) are detected in the RBS detector placed at a backward angle (160°). The

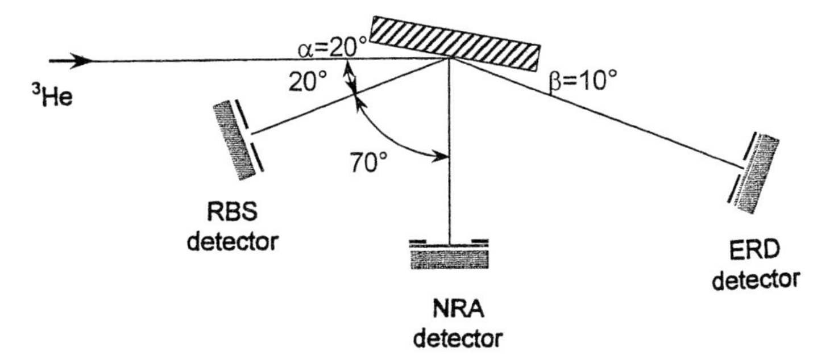

Fig. 10.36 Schematic view of the experimental set-up showing the three particle detectors. The incident angle, α, is 20° and the outgoing angle, β, for the ERD detector is set to 10°.

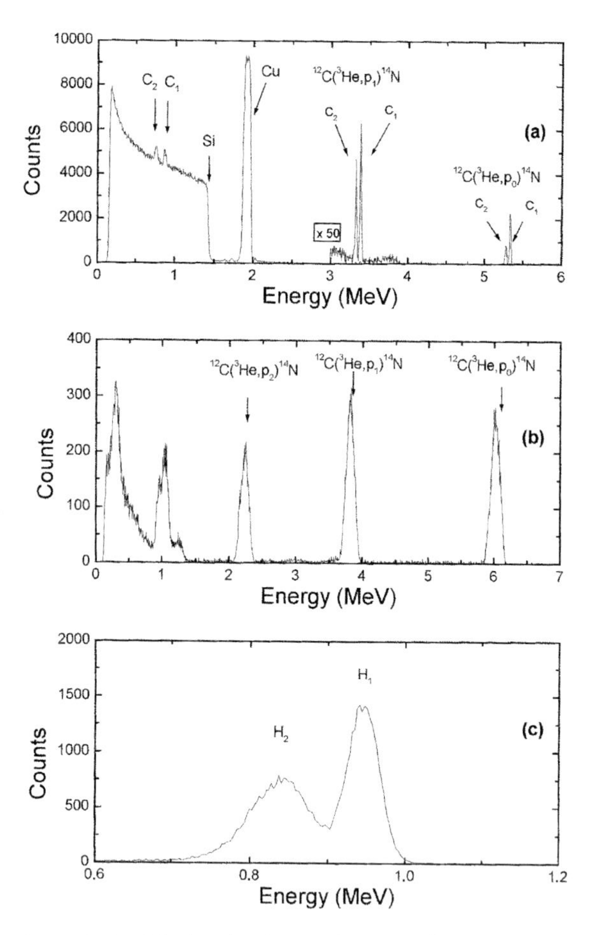

10.37 (a) RBS spectrum obtained by bombardment of 2.4 MeV ^{3}He particles on a C_xH_y/Cu/C_xH_y/Si multilayer. The peaks around the Cu signal and above 3 MeV are due to nuclear reactions induced by ^{3}He particles. The data above 3 MeV have been multiplied by a factor of 50. The peaks labelled C_1 and C_2 are due, respectively, to carbon in the two C_xH_y layers. (b) Spectrum recorded in the NRA detector under the same conditions as in (a). (c) ERD spectrum recorded under the same conditions as in (a) and (b). The peaks labelled H_1 and H_2 are due, respectively, to hydrogen in the two C_xH_y layers (from Terwagne *et al.* 1996).

spectrum represented in Fig. 10.37(a) was obtained with 2.4 MeV ^{3}He incident particles. The spectrum shows clearly the contribution of carbon present in the top layer and in the third layer labelled C_1 and C_2. The information relative to the $^{12}C(^3He,p_0)^{14}N$ nuclear reaction as well as the $^{12}C(^3He,p_1)^{14}N$ nuclear reaction can be observed. These peaks appear without background because the corresponding nuclear reactions are exo-energetic. Some contributions due to the $^{12}C(^3He,p_2)^{14}N$ and $^{12}C(^3He,\alpha_0)^{11}C$ nuclear reactions are also present just below the copper peak, which is due to elastic backscattering, but the area of these peaks is small compared to those of the copper and silicon contributions. In the low-energy part (below 1 MeV), it is observed that the C_1 and C_2 peaks are also due to elastic scattering on carbon. An RBS spectrum was simulated with the RUMP code (Doolittle 1986) using the data from Table 10.5. The carbon concentration in each C_xH_y layer was calculated by integrating the C_1 and C_2 peaks, which correspond to the $^{12}C(^3He,p_1)^{14}N$ nuclear reaction, and using the relationship

$$N_C = \frac{I_C \cos\theta}{\Omega_{RBS} \frac{d\sigma}{d\Omega} N_I} \tag{10.34}$$

where I_C is the area under carbon peak, Ω_{RBS} is the solid angle of the RBS detector, θ is the angle between the incident beam and the normal of the sample, $d\sigma/d\Omega$ is the differential cross-section corresponding to the $^{12}C(^3He,p_1)^{14}N$ nuclear reaction (Kuan *et al.* 1964) and N_I is the number of incident particles. Values of 2.4×10^{17} C cm^{-2} for the first layer and 2.2×10^{17} C cm^{-2} for the second layer containing carbon atoms are obtained. These results are in very good agreement with the results relative to the RBS simulation (Table 10.5).

Table 10.5 Analysis for the C_xH_y/Cu/C_xH_y/Si multilayer obtained using the RUMP simulation code (Doolittle 1986)

Layer number	Thickness (10^{15} atoms cm^{-2})	Composition and atomic fraction
1	414	$C_{0.54}H_{0.46}$
2	407	Cu
3	450	$C_{0.55}H_{0.45}$
4	∞	Si

The spectrum illustrated in Fig. 10.37(b) was recorded simultaneously in the NRA detector placed 90° relative to the incident beam (Fig. 10.36). The peaks relative to $^{12}C(^3He,p_i)^{14}N$ ($i = 0, 2$) are easily observable and are more important than those in the RBS spectrum [Fig. 10.37(a)] because the solid angle of the NRA detector was 14 times the solid angle of the RBS detector. Both contributions from carbon are not separated in spectrum (Fig. 10.37b) as a 12.2 μm mylar absorber was positioned in order to stop scattered particles.

10.10.6.2 *ERD technique* The spectrum represented in Fig. 10.37(c) was recorded simultaneously in the ERD detector placed 30° relative to the incident beam. Two peaks (H_1 and H_2) due to recoil of hydrogen from the C_xH_y layers were observed. The spectrum was stimulated with the ALEGRIA program (Schiettecatte & Ross 1997) and gives 1.8×10^{17} H cm^{-2} in the first layer and 2.0×10^{17} H cm^{-2} in the second, assuming a mean stopping power (Al) for the whole sample. These results reveal a very good agreement between the three complementary techniques (RBS, NRA and ERD).

10.10.7 *Steel surfaces modified by ion implantation*

The modification of steel surfaces by ion beams has been established as a favourable method of improving its mechanical properties, for example, by implantation of nitrogen ions. However, the wear-resistance behaviour of steel surfaces was found to be influenced not only by nitride formation but also favoured by the presence of C–N phases, where the carbon originates from the residual gas inside the implanter recipient. To correlate the mechanical steel properties with the chemical composition, an analysis of both nitrogen and carbon atoms is of interest. The elastic recoil detection (ERD) offers such a possibility and provides information on additional elements, e.g. oxygen and hydrogen, possibly incorporated within the nitride or carbonitride surface layers. Nitrogen ions (50 kV) were chosen to prepare a tool steel by implantation.

ERD with 30 MeV Cl^{6+} ions were used by a team in Rossendorf (Neelmeijer *et al.* 1992) to investigate H and N in layered structures on a Si substrate. Alternate layers of a Si–H and Si–N(H), each 20 nm thick, were produced by plasma CVD. The scattering angle was 30° and a mylar foil 9.2 μm thick was inserted between the target and the $\Delta E.E$ detector system. This telescope assembly associates a thin Si detector with a signal used in coincidence analysis with the pure energy signal, E, to identify the mass of the incident particles by the product ΔE–E. This product is indeed proportional to $M_1 Z^2{}_1$ as seen in Equation (10.1). The mylar foil completely stops the scattered Cl ions as well as the recoiling Si ions. The ERD spectrum only contains information from light elements H, C and N. Figure 10.38 illustrates the results.

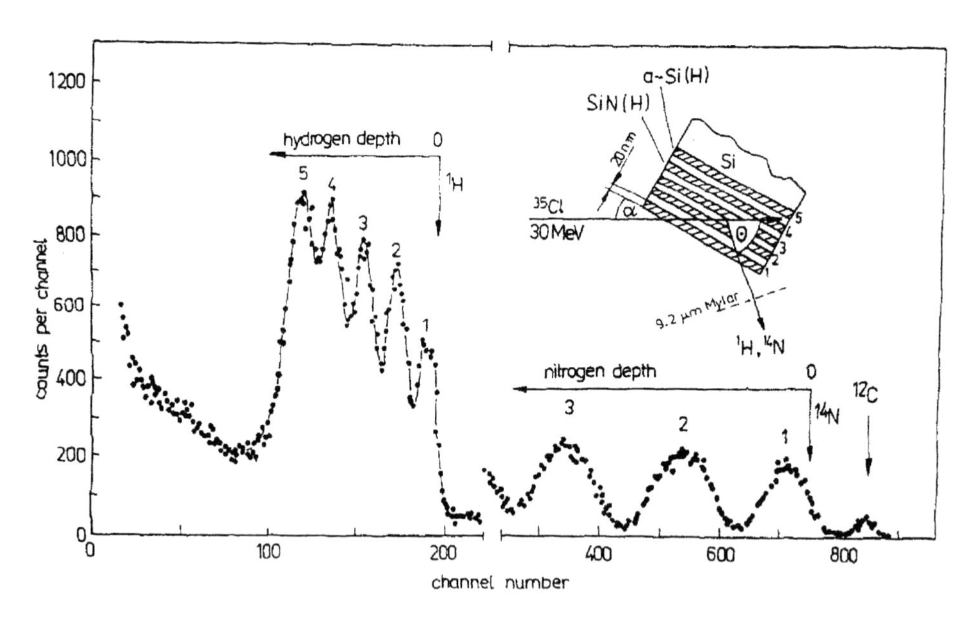

Fig. 10.38 ERD energy spectrum of a sandwich structure produced by deposition of SiN(H) and a-Si(H) layers of 20 nm. Note the geometry of the experiment in the inset (from Neelmeijer *et al.* 1992).

One clearly identifies the five layers containing hydrogen and three of the four layers also containing nitrogen. A surface peak due to carbon appears at the highest energy. The regions between the hydrogen peaks, corresponding to Si(H), contain nearly twice the hydrogen concentration compared with the Si–N(H) regions.

References

Ajzenberg-Selove, F. & Lauritsen, T., *Nucl. Phys.*, **11** (1959), 1.

Baglin, J.E.E., Kellock, A.J., Crokett, M.A. & Shih, A.H., *Nucl. Instru. Meth. Physics Res.*, **B64** (1992), 469–474.

Barbour, J.C. & Doyle, B.L., Elastic recoil detection, in *Handbook of Modern Ion Beam Analysis*, J. Tesmer and M. Nastasi (eds.). Material Research Society, Pittsburgh, PA (1995).

Bittner, J.W. & Moffat, R.D., *Phys. Rev.*, **96** (1954), 374.

Bozoian, M., Hubbard, K.M. & Nastasi, M., *Nucl. Instru. Meth. Physics Res.*, **B51** (1990), 311.

Bozoian, M., *Nucl. Instru. Meth. Physics Res.*, **B56/57** (1991a), 740.

Bozoian, M., *Nucl. Instru. Meth. Physics Res.*, **B58** (1991b), 127.

Cameron, J.R. *Phys. Rev.*, **90** (1953), 839.

Campisano, S.U., Chang, C.T., Logindice, A. & Rimini, E., *Nucl. Instru. Meth. Physics Res.*, **B209–210** (1983), 139.

Cheng, H., Shen, H., Tang, T. & Yang, F., *Nucl. Instr. Meth. Physics Res.*, **B83** (1993), 449.

Deconninck, G., *Introduction to Radioanalytical Physics*. Elsevier, Amsterdam (1978).

Demortier, G. & Ruvalcaba-Sil, J.L., *Nucl. Instr. Meth. Physics Res.*, **B118** (1996), 347.

Demortier, G. PIXE analysis of high *Z* complex matrices, *Nucl. Instr. Meth. Physics Res.*, **B150** (1999), 520–531.

Demortier, G., Nuclear reaction analysis, to be published in *Encyclopedia of Analytical Chemistry*, Wiley (2000).

Doolittle, L.R., A semi-automatic algorithm for Rutherford backscattering analysis, *Nucl. Instru. Meth. Physics Res.*, **B15** (1986), 227–231.

Gabelica, Z. & Demortier, G., *Nucl. Instr. Meth. Physics Res.*, **B136–138** (1998), 1312.

Hsin-Min Kuan, Bonner, T.W. & Risser, J.R., *Nucl. Phys.*, **51** (1964), 481.

Jarjis, R.A., Nuclear cross section data for surface analysis, thesis, University of Manchester (1979).

Knox, J.M., Non Rutherford scattering of proton by light elements, *Nucl. Instru. Meth. Physics Res.*, **B66** (1992), 31–37.

Lanford, W.A., Frautvetter, H.P., Ziegler, J.F. & Keller, J., *J. Applied Phys. Letters*, **28** (1976), 556.

Leavitt, J.A., McIntyre Jr., L.C., Champlin, R.S., Stoner Jr., J.O., Lin, Z., Ashbaigj, M.D., Cox, R.P. & Franck, J.D., *Nucl. Instru. and Meth. in Physics Research*, B85 (1994) 37.

L'Ecuyer, J., Davies, J.A. & Matsunami, N., *How accurate are absolute Rutherford backscattering yields*, *Nucl. Instru. Meth.*, **160** (1979), 337–346.

Lucas, S., Terwagne, G., Piette, M. & Bodart, F., *Nucl. Instru. Meth.*, **B59/60** (1991), 925.

Markwitz, A., Vandesteene, N., Waldschmidt, M. & Demortier, G., Characterization of the interdiffusion in Au–Al layers by RBS, *Fresenius' J. Anal. Chem.*, **358** (1997a), 59–63.

Markwitz, A., Ruvalcaba-Sil, J.L. & Demortier, G., *Nucl. Instr. Meth. Physics Res.*, **B122** (1997b), 685.

Markwitz, A. & Demortier, G., Ion current dependent interdiffusion in Au–Al layers during ion beam irradiation, *Defect Diffusion Forum*, **143–147** (1997), 1303–1308.

Markwitz, A. & Demortier, G., Light element impurities in ion beam bombarded Au–Al layers, *Nucl. Instru. Meth. Physics Res.*, **B136–138** (1998), 516–520.

Markwitz, A., Matz, W., Waldschmidt, M. & Demortier, G., Investigation of the atomic interdiffusion and phase formation in ion beam-irradiated thin Cu–Al and Ag–Al multilayers by in situ RBS and XRD, *Surface Interface Analysis*, **26** (1998), 160–174.

Mozer, F.N., *Phys. Rev.*, **104** (1956), 1386.

Neelmeijer, C., Grötzschel, R., Hentschel, E., Klabes, R., Kolitsch, A. & Richter, E., *Nucl. Instr. Meth. Physics Res.*, **B66** (1992), 242–249.

Ruvalcaba-Sil, J.L. & Demortier, G., *Nucl. Instr. Meth. Physics Res.*, **B130** (1997), 297.

Saarilahti, J. & Rauhala, E., Interactive personal computer data analysis of ion backscattering spectra, *Nucl. Instru. Meth. Physics Res.*, **B64** (1992), 734–738.

Schiettecatte, F. & Ross, G.G., *AIP Press Proceedings*, **392** (1997), 711.

Strathman, M.D. & Bauman, S., Angle resolved imaging of single crystal materials with MeV helium ions, *Nucl. Instr. Meth. Physics Res.*, **B64** (1992), 840–845.

Terwagne, G., Lucas, S., Bodart, F., Sorensen, G & Jensen, H., *Nucl. Instr. Meth.*, **B45** (1990), 95.

Terwagne, G. & Deconninck, G., *Nucl. Instr. Meth.*, **B64** (1992), 153.

Terwagne, G., Ross, G.G. & Leblanc, L., *J. Appl. Phys.*, **79** (1996), 8886.

Terwagne, G., *Nucl. Instr. Meth.*, **B122** (1997), 1.

Tirira, J., Serruys, Y. & Trocellier, P., *Forward Recoil Spectroscopy*. Plenum Press, New York (1996).

Vavilov, P.V., Ionization losses of high energy heavy particles, *Soviet Physics*, **JETP 5** (1957), 749–751.

Wenzel, W.A. & Whaling, W., *Phys. Rev.*, **87** (1952), 499.

11 Experimental data analysis

M. ABOUDY and Z.B. ALFASSI

11.1 Introduction

The aim of most experiments is to find the value of a physical quantity. This value may either result directly from the measurements or be derived from the values of measured quantities by a mathematical relationship. Indeed, we are *sampling* from a large *population* when we measure some quantity. For example, if someone is interested in measuring the vanadium content of steel it will be the vanadium content of steel samples, not the vanadium content in the whole steel, that is measured. The second stage after sampling and measuring is to make an inference about a population based on information contained in the samples. For making inferences about a population statistical methods are needed.

11.2 Properties of a population

11.2.1 *Random variable and distribution*

A real variable x that is defined for all possible outcomes of some experiment or some measurement (a real number is assigned to each possible outcome) is called a *random variable*. For example, the value of a thrown dice (the numbers 1, 2, 3, 4, 5, 6) or the vanadium content in a steel sample is a random variable. If we can assign a probability of occurrence to all possible outcomes of some experiment or measurement then it is possible to assign a *probability distribution* to any random variable. There are two kinds of random variable, discrete and continuous. A random variable x that takes discrete values only, $x_1, x_2, x_3, \ldots, x_n$ (like the values 1, 2, 3, 4, 5, 6 of the thrown dice) with appropriate probabilities $p_1, p_2, p_3, \ldots, p_n$ (1/6, 1/6, 1/6, 1/6, 1/6, 1/6 in the case of dice) is called a *discrete random variable*. A random variable that is defined for a continuous range of values between given limits (like the vanadium content in steel, which can take any values within limits) is called a *continuous random variable*.

11.2.2 *Probability function and probability density function*

We can define a function that assigns a probability of occurrence to each random variable with respect to the discussion of discrete and continuous random variables. In the case of a discrete variable X we define a *probability function* (PF) $f(x)$ that gives the probability $P(X = x)$ of obtaining x in the experiment or the measurement. For example in the example of the dice it is possible to define the probability function, as described below.

$$f(x) = P(X = x) = \begin{cases} \frac{1}{6} & \text{if } x = 1, 2, 3, 4, 5, 6 \\ 0 & \text{otherwise} \end{cases}$$

In the case of a continuous random variable X we can define the *probability density function* (PDF) $f(x)\mathrm{d}x$ that gives the probability that X lies in the interval $x < X \leq x + \mathrm{d}x$, namely $P(x < X \leq x + \mathrm{d}x) = f(x)\mathrm{d}x$. The probability must be a real positive number between 1 and 0, so the PF and the PDF must obey the following condition for every x: $0 \leq f(x) \leq 1$. Since the probabilities must sum to unity (because this is the sum of the probabilities of all possible outcomes and something must happen), we require:

$$\sum_x f(x) = 1 \quad \text{(PF)}, \qquad \int_x f(x)\mathrm{d}x = 1 \quad \text{(PDF)} \tag{11.1}$$

where the summation and the integration are over all allowed values of x. For example, in the case of dice:

$$\sum_x f(x) = \sum_{i=1}^{6} f(i) = f(1) + f(2) + \ldots + f(6) = \frac{1}{6} + \frac{1}{6} + \frac{1}{6} + \frac{1}{6} + \frac{1}{6} + \frac{1}{6} = 1$$

We can also define the *cumulative probability function* (CPF) $F(x)$ the value of which gives the probability of obtaining $X \leq x$:

$$F(x) = P(X \leq x) = \sum_{x_i \leq x} f(x_i) \quad \text{(in the case of PF)}, \; F(x) = P(X \leq x) = \int_a^x f(u)\mathrm{d}u \quad \text{(in the case of PDF)} \tag{11.2}$$

where a (in the case of PDF) is the minimum value of x and can be $-\infty$ (minus infinity). For example, in the case of dice, the probability of achieving a number less than or equal to 3 in one throw of the dice is given by:

$$F(3) = \sum_{x_i \leq 3} f(x_i) = \sum_{i=1}^{3} f(i) = f(1) + f(2) + f(3) = \frac{1}{6} + \frac{1}{6} + \frac{1}{6} = \frac{3}{6}$$

We may calculate the probability that X lies between two limits a_1 and a_2 by the mean of the CPF:

$$P(a_1 < x \leq a_2) = \sum_{a_1 \leq x_i \leq a_2} f(x_i) = \sum_{x_i \leq a_2} f(x_i) - \sum_{x_i \leq a_1} f(x_i) = F(a_2) - F(a_1) \quad \text{(in the case of PF)}$$

$$P(a_1 < x \leq a_2) = \int_{a_1}^{a_2} f(x)\mathrm{d}x = \int_{a_1}^{a} f(x)\mathrm{d}x + \int_{a}^{a_2} f(x)\mathrm{d}x = \int_{a}^{a_2} f(x)\mathrm{d}x - \int_{a}^{a_1} f(x)\mathrm{d}x = F(a_2) - F(a_1) \quad \text{(in the case of PDF)} \tag{11.3}$$

11.2.3 *Properties of distributions*

The kth *moment* of distribution about zero is defined by Equation (11.4). The *mean* (also known as *expectation value*) of the distribution is defined as the first moment of X about zero, see Equation (11.5):

$$E(X^k) = \sum_x x^k f(x) \quad \text{(in the case of PF)}, \quad E(X^k) = \int_x x^k f(x)\mathrm{d}x \quad \text{(in the case of PDF)} \tag{11.4}$$

$$\mu = E(X) = \sum_x xf(x) \quad \text{(in the case of PF)}, \mu = E(X) = \int_x xf(X)\mathrm{d}x \quad \text{(in the case of PDF)} \tag{11.5}$$

$$\mu = E(g(X)) = \sum_x g(x)f(x), \; u = E(g(X)) = \int_x g(x)f(x)\mathrm{d}x \quad \text{(in the case of PDF)} \tag{11.6}$$

We can also define the expectation value of any function $g(X)$ of the random variable X as shown above in Equation (11.6). From the definition of the expectation value it is easy to show that the expectation value has the following properties:

If a is a constant then

$$E(a) = a, \qquad E(aX) = aE(X) \tag{11.7}$$

If $g(X)$ and $h(x)$ are two functions then

$$E(g(X) + h(X)) = E(g(X)) + E(h(X)) \tag{11.8}$$

If X and Y are two random variables and a and b are constants then

$$E(aX + bY) = aE(X) + bE(Y) \tag{11.9}$$

These properties result directly from the fact that the expectation value $E(X)$ is a linear operator. For example, in the case of thrown dice the mean or the expectation value is given by:

$$\mu = E(X) = \sum_{i=1}^{6} if(i) = 1\cdot\frac{1}{6} + 2\cdot\frac{1}{6} + 3\cdot\frac{1}{6} + 4\cdot\frac{1}{6} + 5\cdot\frac{1}{6} + 6\cdot\frac{1}{6} = 3\frac{1}{2}$$

The *mode* of the distribution is defined as the value of random variable X for which the PF or the PDF has its maximum (or greatest) value. The mode is the most probable value from the viewpoint of occurrence. If there is more than one value for which the PF or the PDF reaches a maximum then each value may be called the mode of the distribution. For example, in the dice throw all numbers 1 through 6 may be called the mode.

The *median* of a distribution is defined as the value of the first random variable X for which the CDF is larger than or equal to 0.5 and in a condition that the previous value was less than 0.5.

The kth *central moment* of the distribution is defined as:

$$E\left((X-\mu)^k\right) = \sum_x (x-\mu)^k f(x) \quad \text{(in the case of PF)},$$
$$E\left((X-\mu)^k\right) = \int_x (x-\mu)^k f(x) \quad \text{(in the case of PDF)} \tag{11.10}$$

where $\mu = E(X)$ is the mean.

The *variance* of a distribution is defined as the second central moment of the distribution:

$$\sigma^2 = V(X) = E\left((X-\mu)^2\right) = \sum_x (x-\mu)^2 f(x) \text{ (in the case of PF)}$$
$$\sigma^2 = V(X) = E\left((X-\mu)^2\right) = \int_x^x (x-\mu)^2 f(x) \text{ (in the case of PDF)} \tag{11.11}$$

The *standard deviation* of the distribution is defined as the positive square-root of the variance: $\sigma = \sqrt{V(X)}$. Roughly speaking, the standard deviation (σ) is a measure of the spread of values in which X can assume a position about the mean. For example, in the case of the dice throw the standard deviation will be

$$\sigma = \sqrt{\sum_{i=1}^{6}(i-3.5)^2 f(i)} = \sqrt{\sum_{i=1}^{6}(i-3.5)^2 \frac{1}{6}} = 1.7078$$

From the definition of the variance we can deduce the following properties. If a and b are constants and if X and Y are two independent random variables then, using the properties of the expectation [Equations (11.5)–(11.9)] we can prove the following relations:

$$V(a) = 0, \quad V(aX+b) = a^2V(X), \quad V(aX+bY) = a^2V(X) + b^2V(Y) \tag{11.12}$$

$$\begin{aligned} V(X) &= E\left((X-\mu)^2\right) = E(X^2 - 2\mu X + \mu^2) \\ &= E(X^2) - 2\mu E(X) + \mu^2 = E(X^2) - 2\mu^2 + \mu^2 = E(X^2) - \mu^2 \end{aligned} \tag{11.13}$$

11.3 Important distributions

11.3.1 *Discrete distribution*

11.3.1.1 *Binomial distribution* Consider an experiment that has two possible outcomes, one of which will be defined as a 'success' and the other as a 'failure'. If we assume that the probability of success (p) is constant for each single trial, then the probability of observing a specified number of successes x in n trials is governed by the binomial distribution. The probability that in n trials we will obtain x successes followed by $n-x$ failures is given by:

$$\underbrace{ppp\cdots p}_{x\text{ times}} \cdot \underbrace{(1-p)(1-p)\cdots(1-p)}_{n-x\text{ times}} = p^x(1-p)^{n-x}$$

where p is the probability of observing a success in any one trial, $(1-p)$ the probability of observing a failure in any one trial, x is the number of successes in n trials and n is the total number of trials. This is just one permutation of n successes and $n-x$ failures. The total number of permutations of n objects, of which x are identical and $n-x$ are identical, is given by $n!/[x!\,(n-x)!]$. Thus, the total probability of obtaining x successes in n trials is given by:

$$f(x) = P(X = x) = \frac{n!}{x!(n-x)!}p^x(1-p)^{n-x} \tag{11.14}$$

where $x = 0, 1, 2, 3, \ldots, n$ which is the binomial probability function. For example, the probability of throwing a 3 on exactly four occasions with seven throws of a dice is given by:

$$(n = 7, x = 4), p = \frac{1}{6} \Rightarrow f(4) = \frac{7}{4!(7-4)!}\left(\frac{1}{6}\right)^4\left(1-\frac{1}{6}\right)^{7-4} = 0.0156$$

For $1 > p > 0$ it is clear that $f(x) \geq 0$ for every x in its domain, and by using the binomial expansion

$$(a+b)^n = \sum_{x=0}^{n}\frac{n!}{x!(n-x)!}b^x a^{n-x}$$

we can prove that

$$\sum_{x=0}^{n} f(x) = \sum_{x=0}^{n}\frac{n!}{x!(n-x)!}p^x(1-p)^{n-x} = (1-p+p)^n = 1$$

as expected from the probability function [see Equation (11.1)]. It can be shown (Hogg & Craig 1971, pp. 87–88) that the mean and the variance of the binomial distribution are given by:

$$\mu = E(X) = np, \qquad \sigma^2 = V(X) = np(1-p) \tag{11.15}$$

Let us define a random variable $\hat{p} = x/n$ as the proportion of successes in n trials. This random variable might also be presented as an estimate of the probability of success (p). Since x has the binomial distribution and n is a constant then by way of Equations (11.7), (11.12) and (11.15) we can prove that the mean and the variance of $\hat{p}$ are given by:

$$\mu = E(\hat{p}) = p, \quad \sigma^2 = V(\hat{p}) = \frac{p(1-p)}{n} \tag{11.16}$$

11.3.1.2 *Hypergeometric distribution* In an experiment in which one samples from a relatively small group of items where the items are divided into two classes, then the hypergeometric distribution can be defined. Suppose n items are sampled from a population of N items where N is not necessarily large. We also suppose that the population contain $N{\cdot}p$ items of class A and $N\,(1-p)$ items of class B where p is the fraction of items belonging to class A. Then the probability of taking a sample of n items containing x items of class A and $(n-x)$ items of class B will be given by the following steps:

(1) there are $\dfrac{N!}{n!(N-n)!}$ ways of selecting the samples;

(2) there are $\dfrac{(N \cdot p)!}{x!(N \cdot p - x)!}$ ways that x items can be chosen from $N{\cdot}p$ items;

(3) there are $\dfrac{(N(1-p))!}{(n-x)!(N(1-p)-(n-x))!}$ ways that $(n-x)$ items can be chosen from $N\,(1-p)$ items.

Combining steps 1 through 3 gives the desired probability:

$$f(x) = \frac{\left[\dfrac{(N \cdot p)!}{x!(N \cdot p - x)!}\right]\left[\dfrac{(N(1-p))!}{(n-x)!(N(1-p)-(n-x))!}\right]}{\left[\dfrac{N!}{n!(N-n)!}\right]}$$

$$[\text{where } x = 0, 1, 2, 3, \cdots, \min(n, Np)] \tag{11.17}$$

$f(x)$ is the probability function of hypergeometric distribution. For example, the probability of selecting five good items and two defective items in a sample of seven items from a box containing 150 items of which 120 are good and 30 are defective is given by:

$$N = 150, \quad n = 7, \quad p = \frac{30}{150}, \quad x = 2 \Rightarrow f(2) = 0.2819$$

When $n \to \infty$ (n approaches infinity) and b is constant we get:

$$\frac{n!}{(n-b)!} = \underbrace{n \cdot (n-1) \cdot (n-2) \cdots (n-b+1)}_{b\ \text{terms}} \underset{n \to \infty}{\approx} n^b \tag{11.18}$$

Using Equations (11.17) and (11.18) we can obtain the following result:

$$f(x) = \left[\frac{n!}{x!(n-x)!}\right] \frac{\left[\dfrac{(N \cdot p)!}{(N \cdot p - x)!}\right]\left[\dfrac{(N(1-p))!}{(N(1-p)-(n-x))!}\right]}{\left[\dfrac{N!}{(N-n)!}\right]}$$

$$\underset{N \to \infty}{\approx} \frac{n!}{x!(n-x)!} \frac{(N \cdot p)^x (N(1-p))^{n-x}}{N^n} = \frac{n!}{x!(n-x)!} p^x (1-p)^{n-x} \tag{11.19}$$

Equation (11.19) indicates that the hypergeometric distribution approaches the binomial distribution for an infinite population ($N \to \infty$) or, in other words, when the sample is much less than the population ($N \gg n$).

It can be shown (Kendaill & Stuart 1969, pp. 133–134) that the mean and the variance of the hypergeometric distribution are given by:

$$\mu = E(X) = np, \quad \sigma^2 = V(X) = \frac{N-n}{N-1} np(1-p) \tag{11.20}$$

11.3.1.3 *Poisson distribution* The binomial distribution describes the number of successful outcomes in a certain number of trials (n). The Poisson distribution describes the probability of obtaining a given number of successes for a situation in which the number of trials cannot be enumerated ($n \to \infty$). The Poisson distribution describes the situation in which discrete events occur in a continuum.

It can be shown (Brunk 1965, pp. 119–120) that if the total number of the trial (n) in the binomial distribution [see Equation (11.14)] approaches infinity ($n \to \infty$) in such a way that $\lim_{n\to\infty} (n \cdot p) = \lambda$ (constant) where p is the probability of success on any one trial, then the binomial distribution approaches the Poisson distribution:

$$\lim_{n\to\infty} \left(\frac{n!}{x!(n-x)!} p^x (1-p)^{n-x} \right) = \frac{\lambda^x \, e^{-\lambda}}{x!}$$

Thus, the probability function of the Poisson distribution will be:

$$f(x) = \frac{\lambda^x \, e^{-\lambda}}{x!} \quad \text{for } x = 1, 2, 3, \ldots \tag{11.21}$$

By using Maclaurin's expansion of e^{λ} $\left(e^{\lambda} = \sum_{x=0}^{\infty} \frac{\lambda^x}{x!} \right)$ it is easy to verify that the sum of Poisson probabilities is 1 $\left(\sum_{x=0}^{\infty} f(x) = \sum_{x=0}^{\infty} \frac{\lambda^x e^{-\lambda}}{x!} = 1 \right)$ as expected from the probability function [see Equation (11.1)]. It can be shown (Brunk 1965, pp. 118–119) that the mean and the variance of the Poisson distribution are given by:

$$\mu = E(X) = \lambda, \qquad \sigma^2 = \lambda \tag{11.22}$$

For example, if a radioactive isotope has an average of one alpha (α) disintegration in one hour, the probability of having exactly three disintegrations in two hours is given by:

$$\lambda = 1 \frac{\text{disintegrations}}{\text{hour}} \cdot 2(\text{hours}) = 2; \quad x = 3 \Rightarrow f(3) = \frac{2^3 \, e^{-2}}{3!} = 0.1804$$

11.3.2 *Continuous distribution*

11.3.2.1 *Normal distribution* The normal distribution is also called a *Gaussian* distribution and is the most important continuous distribution. There are many random variables of interest, in all areas of the physical sciences and beyond, which are described either exactly or approximately by a normal distribution. Moreover the normal distribution can be used to approximate other probability distributions.

Let us look at the function $f(y) = e^{-y^2/2}$. It can be shown (Hogg & Craig 1971, p.104) that the improper integral

$$I = \int_{-\infty}^{\infty} f(y) = \int_{-\infty}^{\infty} e^{-y^2/2} \, dy$$

is equal to $\sqrt{2\pi}$. Therefore, if we want to define a probability density function that satisfies Equation (11.1) we must to define it as:

$$\phi(z) = \frac{1}{\sqrt{2\pi}} e^{-z^2/2} \text{ for } -\infty < z < \infty \tag{11.23}$$

Here, $\phi(z)$ is defined as the *standard normal distribution*. The graph of $\phi(z)$ is shown in Fig. 11.1.

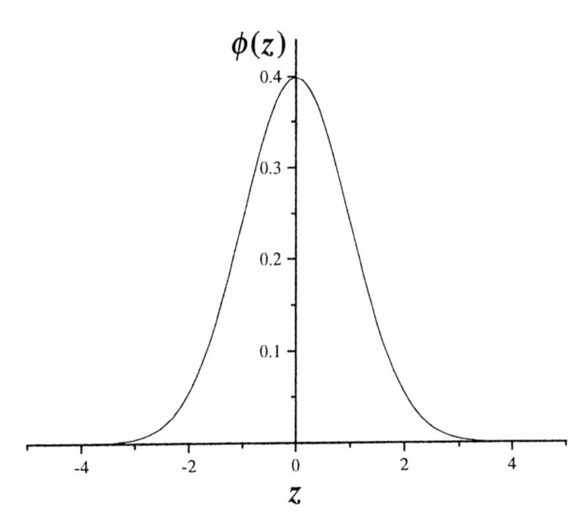

Fig. 11.1 Graph of the standard normal distribution.

The standard normal distribution is an even function ($\phi(z) = \phi(-z)$). This means that $\phi(z)$ is symmetrical about the y-axis ($z = 0$). Let us evaluate the mean of the standard normal distribution:

$$E(z) = \int_{-\infty}^{\infty} z\phi(z)\mathrm{d}z = \frac{1}{\sqrt{2\pi}} \int_{-\infty}^{\infty} z\,\mathrm{e}^{-z^2/2}\mathrm{d}z = -\frac{1}{\sqrt{2\pi}} \mathrm{e}^{-z^2/2}\Bigg|_{-\infty}^{\infty} = 0 \tag{11.24}$$

The variance of the standard normal distribution can be evaluated by partial integration:

$$\begin{aligned} V(z) &= \int_{-\infty}^{\infty} (z-\mu)^2\phi(z)\mathrm{d}z = \int_{-\infty}^{\infty} z^2\phi(z)\mathrm{d}z = \frac{1}{\sqrt{2\pi}} \int_{-\infty}^{\infty} z^2\mathrm{e}^{-z^2/2}\mathrm{d}z \\ &= -\frac{1}{\sqrt{2\pi}} z\,\mathrm{e}^{-z^2/2}\Bigg|_{-\infty}^{\infty} + \frac{1}{\sqrt{2\pi}} \int_{-\infty}^{\infty} \mathrm{e}^{-z^2/2}\mathrm{d}z = 0 + \frac{1}{\sqrt{2\pi}}\sqrt{2\pi} = 1 \end{aligned} \tag{11.25}$$

The cumulative probability function of the standard normal distribution is given by:

$$\Phi(z) = P(Z \leq z) = \int_{-\infty}^{z} \phi(u)\mathrm{d}u = \frac{1}{\sqrt{2\pi}} \int_{-\infty}^{z} \mathrm{e}^{-u^2/2}\mathrm{d}u \tag{11.26}$$

Unfortunately the indefinite integral in Equation (11.26), $\left(\int \mathrm{e}^{-u^2/2}\mathrm{d}u\right)$, cannot be evaluated analytically. Values of the function $\Phi(z)$ are tabulated in Table 11.1.

Table 11.1 The cumulative probability function $\Phi(z)$ of the standard normal distribution as given by Equation (11.26)

z	$\Phi(z)$	z	$\Phi(z)$	z	$\Phi(z)$	z	$\Phi(z)$	z	$\Phi(z)$
0.000	0.5000	0.700	0.7580	1.350	0.9115	1.960	0.9750	2.576	0.9950
0.005	0.5199	0.750	0.7734	1.400	0.9192	2.000	0.9772	2.600	0.9953
0.100	0.5398	0.800	0.7881	1.450	0.9265	2.050	0.9798	2.650	0.9960
0.150	0.5596	0.850	0.8023	1.500	0.9332	2.100	0.9821	2.700	0.9965
0.200	0.5793	0.900	0.8159	1.550	0.9394	2.150	0.9842	2.750	0.9970
0.250	0.5987	0.950	0.8289	1.600	0.9452	2.200	0.9861	2.800	0.9974
0.300	0.6179	1.000	0.8413	1.645	0.9500	2.250	0.9878	2.850	0.9978
0.350	0.6368	1.050	0.8531	1.650	0.9505	2.300	0.9893	2.900	0.9981
0.400	0.6554	1.100	0.8643	1.700	0.9554	2.326	0.9900	2.950	0.9984
0.450	0.6736	1.150	0.8749	1.750	0.9599	2.350	0.9906	3.000	0.9987
0.500	0.6915	1.200	0.8849	1.800	0.9641	2.400	0.9918	3.100	0.9990
0.550	0.7088	1.250	0.8944	1.850	0.9678	2.450	0.9929	3.150	0.9992
0.600	0.7257	1.282	0.9000	1.900	0.9713	2.500	0.9938	3.200	0.9993
0.650	0.7422	1.300	0.9032	1.950	0.9744	2.550	0.9946	3.250	0.9994

The function $\Phi(z)$ may be approximated by (Meier & Zünd 1993, pp. 240–241):

$$\Phi(z) = 1.0 - 0.5(y(z))^{-16} \quad \text{for } z \geq 0 \tag{11.27}$$

$$y(z) = 1.0 + 0.049867347z + 0.0211410061z^2 + 0.0032776263z^3 + 0.0000380036z^4 + 0.000488906z^5 + 0.0000053830z^6$$

$$z = \Phi^{-1}(z) = -0.9069066 - 3.64591x - 2.205586x^2 - 0.9623506x^3 - 0.2366192x^4 - 0.02921359x^5 - 0.001375013x^6 \tag{11.28}$$

$$x = \log_{10}(1 - \Phi(z)) \qquad \text{for } 0.5 \leq \Phi(z) < 1$$

The inverse function of the standard normal distribution may be approximated by (Meier & Zünd 1993, pp. 241–243)

$$P(Z \leq z) + P(Z > z) = 1 \Rightarrow P(Z > z) = 1 - P(Z \leq z)$$

Let us derive some properties of the standard normal distribution. Thus, by using Equation (11.26) we get:

$$P(Z > z) = 1 - P(Z \leq z) = 1 - \Phi(z) \tag{11.29}$$

$$\Phi(-z) = \frac{1}{\sqrt{2\pi}} \int_{-\infty}^{-z} \mathrm{e}^{-u^2/2}\mathrm{d}u \underset{r=-u}{\Rightarrow} \Phi(-z) = \frac{1}{\sqrt{2\pi}} \int_{z}^{\infty} \mathrm{e}^{-r^2/2}\mathrm{d}r = P(Z > z)$$

and by using Equation (11.29) we get:

$$\Phi(-z) = 1 - \Phi(z) \tag{11.30}$$

For example, the probability that Z will fall between -1.96 and 1.96 is given by [using Equations (11.3), (11.26), (11.30) and values from Table 11.1]

$$P(-1.96 < z \leq 1.96) = \Phi(1.96) - \Phi(-1.96) = \Phi(1.96) - (1 - \Phi(1.96))$$
$$= 2\Phi(1.96) - 1 = 2 \times 0.975 - 1 = 0.95$$

The interval (or the range) of Z in which 99% of the values will fall, is given by [using Equations (11.3), (11.26), (11.30) and values from Table 11.1]

$$P(-z < Z \leq z) = 0.99 \Rightarrow \Phi(z) - \Phi(-z) = \Phi(z) - (1 - \Phi(z)) = 2\Phi(z) - 1 = 0.99$$
$$\Rightarrow \Phi(z) = \frac{0.99 + 1}{2} = 0.995 \Rightarrow z = 2.576 \Rightarrow P(-2.576 < Z \leq 2.576) = 0.99$$

The mode of the standard normal distribution may be obtained by differentiation of Equation (11.23) and equating to zero:

$$\frac{\mathrm{d}\phi}{\mathrm{d}z} = -\frac{z}{\sqrt{2\pi}}\mathrm{e}^{-z^2/2} = 0 \Rightarrow z = 0 \Rightarrow \text{ mode} = 0$$

The median of the standard normal distribution may be obtained by using Table 11.1, namely $\Phi(z) = 0.5 \Rightarrow z = 0.0$. So, for the standard normal distribution, we get:

$$\text{mean} = \text{mode} = \text{median} = 0.0 \tag{11.31}$$

Let us define a random variable x by the following definition:

$$x = \sigma z + \mu \text{ or } z = \frac{x - \mu}{\sigma} \tag{11.32}$$

where z is the standard random variable of the standard normal distribution. By changing the variable in Equation (11.23) according to Equation (11.32), it follows that

$$f(x)\mathrm{d}x = \phi(z)\mathrm{d}z \Rightarrow f(x) = \phi(z)\frac{\mathrm{d}z}{\mathrm{d}x}$$

and we get the probability density function of the *normal distribution*:

$$f(x) = \frac{1}{\sigma\sqrt{2\pi}}\mathrm{e}^{-\frac{1}{2}[(x-\mu)/\sigma]^2} \tag{11.33}$$

By using Equations (11.7), (11.8), (11.12), (11.24) and (11.25) we can find the mean and the variance of the random variable x:

$$E(x) = E(\sigma z + \mu) = \sigma E(z) + E(\mu) = 0 + \mu = \mu \tag{11.34}$$

$$V(x) = V(\sigma z + \mu) = \sigma^2 V(z) = \sigma^2 \tag{11.35}$$

We can find the relation between the cumulative probability function of the standard normal distribution and the cumulative probability function of the normal distribution by changing the variable in the integral:

$$F(x) = \int_{-\infty}^{x} f(u)\,du = \frac{1}{\sigma\sqrt{2\pi}} \int_{-\infty}^{x} e^{-\frac{1}{2}[(u-\mu)/\sigma]^2} du \underset{v=(u-\mu)/\sigma}{=} \frac{1}{\sqrt{2\pi}} \int_{-\infty}^{(x-\mu)/\sigma} e^{-y^2/2} dv = \Phi\left(\frac{x-\mu}{\sigma}\right) \tag{11.36}$$

Thus, if a random variable X has a normal distribution with mean μ and standard deviation σ [for brevity we can write X is $n(\mu,\sigma)$] then the probability of obtaining $X \leq x$ is given by:

$$P(X \leq x) = F(x) = \Phi\left(\frac{x-\mu}{\sigma}\right) = P\left(Z \leq \frac{x-\mu}{\sigma}\right)$$

For example, if X has a normal distribution with mean ($\mu = 10$) and standard deviation of $\sigma = 2.5$ then the probability that X will fall in the interval $\mu \pm \sigma$ is given by [using Equations (11.3), (11.36) and Table 11.1]:

$$p(\mu - \sigma < X \leq \mu + \sigma) = F(\mu + \sigma) - F(\mu - \sigma) = \Phi\left(\frac{\mu + \sigma - \mu}{\sigma}\right) - \Phi\left(\frac{\mu - \sigma - \mu}{\sigma}\right) = \Phi(1) - \Phi(-1) = 2\Phi(1) - 1 = 2 \times 0.8413 - 1 = 0.6826$$

$$\Rightarrow P(10 - 2.5 < X \leq 10 + 2.5) = P(7.5 < X \leq 12.5) = 0.6826$$

It may be shown for the normal distribution in the same way as has been done with the standard normal distribution that

$$\text{mean} = \text{mode} = \text{median} = \mu \tag{11.37}$$

11.3.2.2 *Chi-square distribution* Let us define the *gamma function* as:

$$\Gamma(\alpha) = \frac{1}{\beta^{\alpha}} \int_{0}^{\infty} x^{\alpha-1} e^{-x/\beta} dx \quad \text{for } \alpha > 0,\ \beta > 0 \tag{11.38}$$

It can be shown (Hogg & Craig 1971, p. 99) that

$$\Gamma(1) = 1 \quad \text{and for } \alpha > 1 \quad \Gamma(\alpha) = (\alpha - 1)\Gamma(\alpha - 1) \tag{11.39}$$

and if α is a positive integer then:

$$\Gamma(\alpha) = (\alpha - 1)! \tag{11.40}$$

The gamma function may be approximated by (Press *et al.* 1992, pp. 206–207):

$$\Gamma(\alpha) \approx (\alpha + 5.5)^{\alpha+0.5} e^{-(\alpha+5.5)} \frac{\sqrt{2\pi}}{\alpha} \times \left[m_0 + \frac{m_1}{\alpha+1} + \frac{m_2}{\alpha+2} + \frac{m_3}{\alpha+3} + \frac{m_4}{\alpha+4} + \frac{m_5}{\alpha+5} + \frac{m_6}{\alpha+6}\right] \tag{11.41}$$

$m_0 = 1.000000000190015$, $m_1 = 76.18009172947146$, $m_2 = -86.50532032941677$,

$m_3 = 24.01409824083091$, $m_4 = -1.231739572450155$,

$m_5 = 0.1208650973866179 \times 10^{-2}$, $m_6 = -0.5395239384953 \times 10^{-5}$

If we want to calculate a probability density function from the gamma function that satisfies Equation (11.1) we must define it as:

$$f(x) = \frac{x^{\alpha-1}e^{-x/\beta}}{\Gamma(\alpha)\beta^{\alpha}} \text{ for } 0 < x < \infty \tag{11.42}$$

The PDF in Equation (11.42) is called the *gamma distribution*. It may be shown (see Hogg & Craig 1971, p. 101) that the mean and the variance of the gamma distribution are given by:

$$\mu = E(X) = \alpha \times \beta, \quad \sigma^2 = V(X) = \alpha \times \beta^2 \tag{11.43}$$

The *chi-square distribution* is a special case of the gamma PDF family in which $\alpha = n/2$ where n is a positive integer, and $\beta = 2$ [see Equation (11.42)]. Thus, the chi-square distribution is defined by:

$$f(x) = \frac{x^{\frac{n}{2}-1}e^{-\frac{x}{2}}}{\Gamma(n/2) \times 2^{\frac{n}{2}}} \; for \; 0 < x < \infty \tag{11.44}$$

The parameter n in Equation (11.44) is called the *number of degrees of freedom* of the chi-square distribution. For brevity we can write X is $\chi^2(n)$, meaning that X has a chi-square distribution with n degrees of freedom.

We can find the mean and the variance of the chi-square distribution by substituting $\alpha = n/2$ and $\beta = 2$ into Equations (11.43):

$$\mu = E(X) = \frac{n}{2} \times 2 = n, \quad \sigma^2 = V(X) = \frac{n}{2} \times 2^2 = 2n \tag{11.45}$$

The graph of chi-square distribution with 2 and 8 degrees of freedom is given in Fig. 11.2. We can find the mode of the chi-square distribution by differentiation of Equation (11.44) and equating it to zero:

$$\text{mode} = n - 2 \tag{11.46}$$

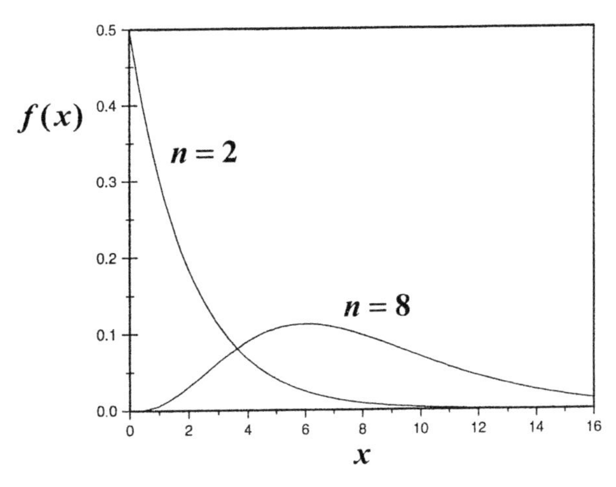

Fig. 11.2 The chi-square distribution with 2 and 8 degrees of freedom.

The cumulative probability function of the chi-square distribution is given by:

$$P(X \leq x) = F(x) = \int_0^x f(u)\mathrm{d}u = \int_0^x \frac{u^{n/2-1}\,\mathrm{e}^{-u/2}}{\Gamma(n/2) \times 2^{\frac{u}{2}}}\mathrm{d}u \qquad (11.47)$$

The cumulative probability function in Equation (11.47) may be approximated by the following series (Hogg & Craig 1971, pp. 135–136):

$$F(x) = \int_0^x \frac{u^{u/2-1}\,\mathrm{e}^{-u/2}}{\Gamma(n/2) \times 2^{\frac{u}{2}}}\mathrm{d}u = \left(\frac{x}{2}\right)^{n/2} \mathrm{e}^{-x/2} \sum_{i=0}^{\infty} \frac{x^i}{\Gamma(n/2+1+i)} \qquad (11.48)$$

Here, Γ is the gamma function as given by Equation (11.38).

The cumulative probability function in Equation (11.47) is tabulated in Table 11.2. For example, if X is $\chi^2(11)$ (X has a chi-square distribution with 11 degrees of free-

Table 11.2 The cumulative probability function of the chi-square distribution as given by Equation (11.47)

	$F(x) = P(X) \leq x)$								
n	0.01	0.025	0.050	0.500	0.75	0.90	0.95	0.975	0.99
1	0.000	0.001	0.004	0.455	1.320	2.710	3.840	5.020	6.630
2	0.020	0.051	0.103	1.390	2.770	4.610	5.990	7.380	9.210
3	0.115	0.216	0.352	2.370	4.110	6.250	7.810	9.350	11.300
4	0.297	0.484	0.711	3.360	5.390	7.780	9.490	11.100	13.300
5	0.554	0.831	1.150	4.350	6.630	9.240	11.100	12.800	15.100
6	0.872	1.240	1.640	5.350	7.840	10.600	12.600	14.400	16.800
7	1.240	1.690	2.170	6.350	9.040	12.000	14.100	16.000	18.500
8	1.650	2.180	2.730	7.340	10.200	13.400	15.500	17.500	20.100
9	2.090	2.700	3.330	8.340	11.400	14.700	16.900	19.000	21.700
10	2.560	3.250	3.940	9.340	12.500	16.000	18.300	20.500	23.200
11	3.050	3.820	4.570	10.300	13.700	17.300	19.700	21.900	24.700
12	3.570	4.400	5.230	11.300	14.800	18.500	21.000	23.300	26.200
13	4.110	5.010	5.890	12.300	16.000	19.800	22.400	24.700	27.700
14	4.660	5.630	6.570	13.300	17.100	21.100	23.700	26.100	29.100
15	5.230	6.260	7.260	14.300	18.200	22.300	25.000	27.500	30.600
16	5.810	6.910	7.960	15.300	19.400	23.500	26.300	28.800	32.000
17	6.410	7.560	8.670	16.300	20.500	24.800	27.600	30.200	33.400
18	7.010	8.230	9.390	17.300	21.600	26.000	28.900	31.500	34.800
19	7.630	8.910	10.100	18.300	22.700	27.200	30.100	32.900	36.200
20	8.260	9.590	10.900	19.300	23.800	28.400	31.400	34.200	37.600
21	8.900	10.300	11.600	20.300	24.900	29.600	32.700	35.500	38.900
22	9.540	11.000	12.300	21.300	26.000	30.800	33.900	36.800	40.300
23	10.200	11.700	13.100	22.300	27.100	32.000	35.200	38.100	41.600
24	10.900	12.400	13.800	23.300	28.200	33.200	36.400	39.400	43.000
25	11.500	13.100	14.600	24.300	29.300	34.400	37.700	40.600	44.300
26	12.200	13.800	15.400	25.300	30.400	35.600	38.900	41.900	45.600
27	12.900	14.600	16.200	26.300	31.500	36.700	40.100	43.200	47.000
28	13.600	15.300	16.900	27.300	32.600	37.900	41.300	44.500	48.300
29	14.300	16.000	17.700	28.300	33.700	39.100	42.600	45.700	49.600
30	15.000	16.800	18.500	29.300	34.800	40.300	43.800	47.000	50.900

dom) then the probability of finding X in the interval $3.82 \leq X \leq 21.9$ is given by (using Equation (11.3) and Table 11.2):

$$P(3.82 \leq X \leq 21.9) = F(21.9) - F(3.82) = 0.975 - 0.025 = 0.95$$

In a further example, if X is $\chi^2(5)$ then the a value that gives the probability $P(a < X) = 0.05$ is given by (using Table 11.2):

$$P(X \leq a) = 1 - P(X > a) = 1 - 0.05 = 0.95 \Rightarrow a = 11.1$$

It can be shown (van der Waerden 1969; Kendaill & Stuart 1969, pp. 246–247) that if $Z_1, Z_2, Z_3, \ldots, Z_n$ are independent normally distributed with mean zero and variance one [mean standard normal distribution; see Equations (11.23)–(11.25)] then the sum of the squares $Y = Z_1^2 + Z_2^2 + Z_3^2 + \ldots + Z_n^2$ has a chi-square distribution with n degrees of freedom. For brevity we can write:

$$\text{If } Z_1^2, Z_2^2, \ldots, Z_n^2 \text{ are } n(0,1) \text{ and independent then}$$
$$Y = Z_1^2 + Z_2^2 + Z_3^2 + \ldots + Z_n^2 \text{ is } \chi^2(n) \tag{11.49}$$

11.3.2.3 *t (Student's) distribution* The (Student's) t distribution is quite useful in certain problems of statistical inference. If Z is $n(0,1)$ and X is $\chi^2(n)$ (Z has standard normal distribution and X has chi-square distribution with n degrees of freedom), and X and Z are assumed stochastically independent, then the joint PDF of X and Z is the product of the PDF of X and that of Z [see Equations (11.23) and (11.44)]:

$$\theta(z, x) = \phi(z) \times f(x) = \frac{1}{\sqrt{2\pi}} e^{-z^2/2} \frac{x^{n/2-1} e^{-x/2}}{\Gamma\left(\frac{n}{2}\right) \times 2^{n/2}} \quad for\, 0 < x < \infty, -\infty < z < \infty \tag{11.50}$$

If we change the variable in Equation (11.50) according to the following equations

$$t = \frac{z}{\sqrt{\frac{x}{n}}} \text{ and } u = x \tag{11.51}$$

and integrate the resulting function over u we will get (Hogg & Craig 1971, pp.135–136):

$$g(t) = \frac{\Gamma\left(\frac{n+1}{2}\right)}{\sqrt{n\pi}\,\Gamma\left(\frac{n}{2}\right)} \left(1 + \frac{t^2}{n}\right)^{-(n+1)/2} \quad -\infty < t < \infty \tag{11.52}$$

where Γ is the gamma function as given by Equation (11.38). The PDF in Equation (11.52) is usually called a *t distribution* or *Student's distribution*. From the previous discussion we conclude that if Z is $n(0,1)$ and X is $\chi^2(n)$ and they are independent then $T = \frac{Z}{\sqrt{X/n}}$ has t distribution as defined by Equation (11.52). The graph of the t distribution for $n = 2$ is given in Fig. 11.3.

It may be shown (Kendaill & Stuart 1969, pp. 59–60, 375) that the mean and the variance of the t distribution are given by:

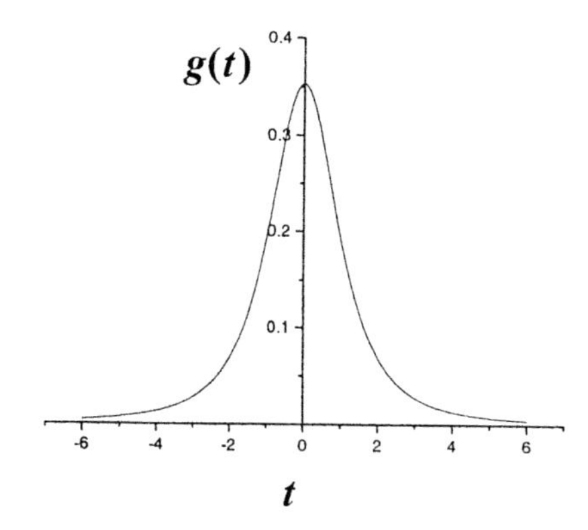

Fig. 11.3 t (Student's) distribution for $n = 2$.

$$\mu = E(T) = 0 \text{ for } n > 1, \quad \sigma^2 = V(T) = \frac{n \cdot \Gamma\left(\frac{n}{2} - 1\right)}{2\Gamma\left(\frac{n}{2}\right)} \text{ for } n > 2 \tag{11.53}$$

We can find the mode of the Student's t distribution by differentiating Equation (11.52) and equating it to zero:

$$\text{mode} = \mu = 0 \tag{11.54}$$

The cumulative probability function of the t distribution is then given by:

$$P(T \leq t) = G(t) = \int_{-\infty}^{t} g(u)\mathrm{d}u = \int_{-\infty}^{t} \frac{\Gamma\left(\frac{n+1}{2}\right)}{\sqrt{n\pi}\,\Gamma\left(\frac{n}{2}\right)} \left(1 + \frac{u^2}{n}\right)^{-(n+1)/2} \mathrm{d}u \tag{11.55}$$

Some values of $G(t)$ [Equation (11.55)] for selected values of t and n are given in Table 11.3.

The function $g(t)$ in Equation (11.52) is an even function ($g(t) = g(-t)$), so we can derive the following property of $G(t)$:

$$P(T \leq -t) = G(-t) = \int_{-\infty}^{-t} g(u)\,\mathrm{d}u \underset{v=-u}{=} \int_{t}^{\infty} g(v)\mathrm{d}v = P(T > t)$$
$$= 1 - P(T \leq t) = G(t) \tag{11.56}$$

For example, if T has a t distribution with 6 degrees of freedom then the t that gives a probability of 0.9 that we will find T in the interval $-t \leq T \leq t$ is given by [using Table 11.3 and Equations (11.3) and (11.56)]:

$$P(-t \leq T \leq t) = G(t) - G(-t) = G(t) - (1 - G(t)) = 2G(t) - 1 = 0.9 \Rightarrow$$
$$G(t) = 0.95 \Rightarrow t = 1.943 \Rightarrow P(-1.943 \leq T \leq 1.943) = 0.9$$

11.3.2.4 *F Distribution* Let us consider two independent chi-square random variables X and Y having n_1 and n_2 degrees of freedom, respectively. The joint PDF of X

Table 11.3 The cumulative probability function of the t distribution $G(t)$ [Equation (11.55)]

	$P(T \leq t) = G(t)$				
n	0.90	0.95	0.975	0.99	0.995
1	3.078	6.314	12.706	31.821	63.657
2	1.886	2.920	4.303	6.965	9.925
3	1.638	2.353	3.182	4.541	5.841
4	1.533	2.132	2.776	3.747	4.604
5	1.476	2.015	2.571	3.365	4.032
6	1.440	1.943	2.447	3.143	3.707
7	1.415	1.895	2.365	2.998	3.499
8	1.397	1.860	2.306	2.896	3.355
9	1.383	1.833	2.262	2.821	3.250
10	1.372	1.812	2.228	2.764	3.169
11	1.363	1.796	2.201	2.718	3.106
12	1.356	1.782	2.179	2.681	3.055
13	1.350	1.771	2.160	2.650	3.012
14	1.345	1.761	2.145	2.624	2.977
15	1.341	1.753	2.131	2.602	2.947
16	1.337	1.746	2.120	2.583	2.921
17	1.333	1.740	2.110	2.567	2.898
18	1.330	1.734	2.101	2.552	2.878
19	1.328	1.729	2.093	2.539	2.861
20	1.325	1.725	2.086	2.528	2.845
21	1.323	1.721	2.080	2.518	2.831
22	1.321	1.717	2.074	2.508	2.819
23	1.319	1.714	2.069	2.500	2.807
24	1.318	1.711	2.064	2.492	2.797
25	1.316	1.708	2.060	2.485	2.787
26	1.315	1.706	2.056	2.479	2.779
27	1.314	1.703	2.052	2.473	2.771
28	1.313	1.701	2.048	2.467	2.763
29	1.311	1.699	2.045	2.462	2.756
30	1.310	1.697	2.042	2.457	2.750

and Y will be given by the product of the PDF of X and that of Y [see Equation (11.44)]:

$$\theta(x, y) = \frac{x^{(n_1/2)-1} y^{(n_2/2)-1} e^{(x+y)/2}}{\Gamma\left(\frac{n_1}{2}\right)\Gamma\left(\frac{n_2}{2}\right) 2^{(n_1+n_2)/2}} \text{ for } 0 < x < \infty, 0 < y < \infty \tag{11.57}$$

It may be shown (Hogg & Craig 1971, pp. 137–138) that if we change the variable in Equation (11.57) according to

$$f = \frac{x/n_1}{y/n_2}, \quad z = y$$

and integrate the resulting function over z we get:

$$g(f) = \frac{\Gamma\left(\frac{n_1+n_2}{2}\right)\left(\frac{n_1}{n_2}\right)^{n_1/2}}{\Gamma\left(\frac{n_1}{2}\right)\Gamma\left(\frac{n_2}{2}\right)} \cdot \frac{f^{(n_1/2)-1}}{\left(1+\frac{n_1 f}{n_2}\right)^{(n_1+n_2)/2}} \text{ for } 0 < f < \infty \tag{11.58}$$

The distribution of the PDF in Equation (11.58) is called an *F distribution*. From the previous discussion we conclude that if X is χ^2 (n_1) and Y is χ^2 (n_2) and they are independent then $F = \dfrac{X/n_1}{Y/n_2}$ has F distribution as defined by Equation (11.58).

The mode and the mean and the variance of the F distribution are given by (Kendaill & Stuart 1969, p. 378):

$$\text{mode} = \frac{(n_1-2)n_2}{n_1(n_2+2)}, \mu = E(F) = \frac{n_2}{n_2-2} \text{ for } n_2 > 2, \tag{11.59}$$

$$\sigma^2 = V(F) = \frac{2n_2^2(n_1+n_2-2)}{n_1(n_2-2)^2(n_2-4)} \text{ for } n_2 > 4$$

The cumulative probability function of the F distribution is given by:

$$P(F \le f) = G(f) = \int_0^f g(u)\,\mathrm{d}u = \frac{\Gamma\left(\frac{n_1+n_2}{2}\right)\left(\frac{n_1}{n_2}\right)^{n_1/2}}{\Gamma\left(\frac{n_1}{2}\right)\Gamma\left(\frac{n_2}{2}\right)} \int_0^f \frac{u^{(n_1/2)-1}}{\left(1+\frac{n_1 u}{n_2}\right)^{(n_1+n_2)/2}}\mathrm{d}u \tag{11.60}$$

Some values of $G(f)$ in Equation (11.60) are given in Table 11.4.

For example, if F has an F distribution with $n_1 = 6$ and $n_2 = 12$ then a and b that give a probability of 0.9 that we will find F in the interval $a \le F \le b$ and give a probability of 0.95 that $F \le b$ are given by:

[using Equation (11.60) and Table 11.4]

$$P(F \le b) = G(b) = 0.95 \Rightarrow b = 3.00 \Rightarrow p(F \le 3) = 0.95$$

[using Equations (11.2) and (11.3)]

$$P(a \le F \le b) = G(b) - G(a) = 0.9 \Rightarrow G(a) = 0.05 \Rightarrow P(F \le a) = G(a) = 0.05$$

$$= P\left(\frac{1}{F} \ge \frac{1}{a}\right) = 1 - P\left(\frac{1}{F} \le \frac{1}{a}\right) \Rightarrow P\left(\frac{1}{F} \le \frac{1}{a}\right) = 1 - 0.05 = 0.95$$

[if F has an F distribution with $n_1 = 6$ and $n_2 = 12$ then $1/F$ also has F distribution $n_1 = 12$ and $n_2 = 6$ – see conclusion after Equation (11.58)]

$$P\left(\frac{1}{F} \le \frac{1}{a}\right) = 0.95 \Rightarrow \frac{1}{a} = 4.0 \Rightarrow a = \frac{1}{4} = 0.25 \Rightarrow P(0.25 \le F \le 3.00) = 0.9$$

11.4 Properties of a sample

As was mentioned in the Introduction, a finite sample is drawn from a large or infinite population. It is often useful to describe the properties of a sample by simple numbers

Table 11.4 The cumulative probability function of the F distribution $G(f)$ [Equation (11.60)]

		n_1											
$G(f)$	n_2	1	2	3	4	5	6	7	8	9	10	12	15
0.95	1	161	200	216	225	230	234	237	239	241	242	244	246
0.975		648	800	864	900	922	937	948	957	963	969	977	985
0.99		4052	4999	5403	5625	5764	5859	5928	5982	6023	6056	6106	6157
0.95	2	18.5	19.0	19.2	19.3	19.3	19.4	19.4	19.4	19.4	19.4	19.4	19.4
0.975		38.5	39.0	39.2	39.2	39.3	39.4	39.4	39.4	39.4	39.4	39.4	39.4
0.99		98.5	99.0	99.2	99.2	99.3	99.3	99.4	99.4	99.4	99.4	99.4	99.4
0.95	3	10.1	9.55	9.28	9.12	9.01	8.94	8.89	8.85	8.81	8.79	8.74	8.70
0.975		17.4	16.0	15.4	15.1	14.9	14.7	14.6	14.5	14.5	14.4	14.3	14.3
0.99		34.1	30.8	29.5	28.7	28.2	27.9	27.7	27.5	27.3	27.2	27.1	26.9
0.95	4	7.71	6.94	6.59	6.39	6.26	6.16	6.09	6.04	6.00	5.96	5.91	5.86
0.975		12.2	10.6	9.98	9.60	9.36	9.20	9.07	8.98	8.90	8.84	8.75	8.66
0.99		21.2	18.0	16.7	16.0	15.5	15.2	15.0	14.8	14.7	14.5	14.4	14.2
0.95	5	6.61	5.79	5.41	5.19	5.05	4.95	4.88	4.82	4.77	4.74	4.68	4.62
0.975		10.0	8.43	7.76	7.39	7.15	6.98	6.85	6.76	6.68	6.62	6.52	6.43
0.99		16.3	13.3	12.1	11.4	11.0	10.7	10.5	10.3	10.2	10.1	9.89	9.72
0.95	6	5.99	5.14	4.76	4.53	4.39	4.28	4.21	4.15	4.10	4.06	4.00	3.94
0.975		8.81	7.26	6.60	6.23	5.99	5.82	5.70	5.60	5.52	5.46	5.37	5.27
0.99		13.7	10.9	9.78	9.15	8.75	8.47	8.26	8.10	7.98	7.87	7.72	7.56
0.95	7	5.59	4.74	4.35	4.12	3.97	3.87	3.79	3.73	3.68	3.64	3.57	3.51
0.975		8.07	6.54	5.89	5.52	5.29	5.12	4.99	4.90	4.82	4.76	4.67	4.57
0.99		12.2	9.55	8.45	7.85	7.46	7.19	6.99	6.84	6.72	6.62	6.47	6.31
0.95	8	5.32	4.46	4.07	3.84	3.69	3.58	3.50	3.44	3.39	3.35	3.28	3.22
0.975		7.57	6.06	5.42	5.05	4.82	4.65	4.53	4.43	4.36	4.36	4.20	4.10
0.99		11.3	8.65	7.59	7.01	6.63	6.37	6.18	6.03	5.91	5.81	5.67	5.52
0.95	9	5.12	4.26	3.86	3.63	3.48	3.37	3.29	3.23	3.18	3.14	3.07	3.01
0.975		7.21	5.71	5.08	4.72	4.48	4.32	4.20	4.10	4.03	3.96	3.87	3.77
0.99		10.6	8.02	6.99	6.42	6.06	5.80	5.61	5.47	5.35	5.26	5.11	4.96
0.95	10	4.96	4.10	3.71	3.48	3.33	3.22	3.14	3.07	3.02	2.98	2.91	2.85
0.975		6.94	5.46	4.83	4.47	4.24	4.07	3.95	3.85	3.78	3.72	3.62	3.52
0.99		10.0	7.56	6.55	5.99	5.64	5.39	5.20	5.06	4.94	4.85	4.71	4.56
0.95	12	4.75	3.89	3.49	3.26	3.11	3.00	2.91	2.85	2.80	2.75	2.69	2.62
0.975		6.55	5.10	4.47	4.12	3.89	3.73	3.61	3.51	3.44	3.37	3.28	3.18
0.99		9.33	6.93	5.95	5.41	5.06	4.82	4.64	4.50	4.39	4.30	4.16	4.01
0.95	15	4.54	3.68	3.29	3.06	2.90	2.79	2.71	2.64	2.59	2.54	2.48	2.40
0.975		6.20	4.77	4.15	3.80	3.58	3.41	3.29	3.20	3.12	3.06	2.96	2.86
0.99		8.68	6.36	5.42	4.89	4.56	4.32	4.14	4.00	3.89	3.80	3.67	3.52

called *sample statistics* or *statistics*. The statistic is any quantity that depends on the sample alone. The aim in calculating these statistics is to ascertain information about the properties of the population from which the sample is drawn. The sample is random (one of all possible samples of the same size) and so also the statistic that depends on the sample is a random variable. Because of the random nature of the statistic we can talk about a statistic distribution or a statistic expectation value and statistic variance.

11.4.1 *Sample mean*

The *sample mean* is the statistic defined as

$$\bar{x} = \frac{1}{N}\sum_{i=1}^{N} x_i \tag{11.61}$$

where $x_1, x_2, \ldots, x_N$ are the sample values and N is the sample size. For example, if we have five measurements of stainless steel tensile strength, i.e. 85 000, 90 000, 93 000, 89 000, and 95 000 psi, then the sample mean will be:

$$\bar{x} = \frac{85\,000 + 90\,000 + 93\,000 + 89\,000 + 95\,000}{5} = 90\,400 \text{ psi}$$

If the sample is drawn from a population that has an expectation value (mean) μ and standard deviation σ (variance σ^2), then the expectation value and the variance of the sample mean are given according to Equations (11.8), (11.9) and (11.12):

$$\begin{aligned} E(\overline{X}) &= E\left(\frac{x_1}{N} + \frac{x_2}{N} + \cdots \frac{x_N}{N}\right) = \frac{1}{N}E(X) + \frac{1}{N}E(X) + \cdots + \frac{1}{N}E(X) \\ &= N\left(\frac{E(X)}{N}\right) = E(X)\,\mu \end{aligned} \tag{11.62}$$

$$\begin{aligned} V(\overline{X}) &= V\left(\frac{x_1}{N} + \frac{x_2}{N} + \cdots \frac{x_N}{N}\right) = \frac{1}{N^2}V(X) + \frac{1}{N^2}V(X) + \cdots + \frac{1}{N^2}V(X) \\ &= N\left(\frac{V(X)}{N^2}\right) = \frac{V(X)}{N} = \frac{\sigma^2}{N} \end{aligned} \tag{11.63}$$

Equations (11.62) and (11.63) can be interpreted as follows: the expectation value of the sample mean is the same as the population mean (expectation value) and the variance of the sample mean is the population variance divided by the sample size (N). If the sample size approaches infinity ($N \rightarrow \infty$), or the sample is large enough to contain the entire population, then the variance of the sample mean will approach zero

$$\lim_{N\rightarrow\infty} V(\overline{X}) = \lim_{N\rightarrow\infty} \frac{\sigma^2}{N} = 0$$

and there will be no difference between the sample mean and the population mean. The last result can be derived also from *Chebyshev's* (or Tchebycheff's) inequality:

$$P(|X - \mu| \geq b) \leq \frac{\sigma^2}{b^2} \tag{11.64}$$

Chebyshev's inequality states that the probability that a random variable X will differ absolutely from its expectation ($\mu = E(X)$) by b or more units is always less than or equal to the ratio of σ^2 (variance) to b^2. The proof of Chebyshev's inequality may be found in Brunk (1965, pp. 127–128) or Hogg and Craig (1971, pp. 54–55). If we now apply the inequality in Equation (11.64) to the random variable $\overline{X}$ (sample mean) and remember that

$$E(\overline{X}) = E(X) = \mu, \qquad V(\overline{X}) = \frac{V(X)}{N} = \frac{\sigma^2}{N}$$

then we obtain:

$$P(|\overline{X} - \mu| \geq b) \leq \frac{\sigma^2}{Nb^2} \tag{11.65}$$

This means that as the sample size approaches infinity ($N \rightarrow \infty$) the probability that $\overline{X}$ differs from $\mu = E(X)$ by any positive quantity tends to zero. In the case of the inequality given in Equation (11.65), we say that $\overline{X}$ converges in probability to $\mu = E(X)$. So, if the samples are large enough ($N \rightarrow \infty$) then we can use the sample mean $\overline{x}$ [Equation (11.61)] as an estimate of the population mean, $\mu = E(X)$. In general, if we have quantity I and if we can find random variable η, such that $E(\eta) = I$, then we say that η is an *unbiased estimator* of I. This is because $\overline{\eta}$ converges in probability to $E(\eta)$ [Equation (11.65)] and thus to I. For example, the sample mean is an unbiased estimator of the population mean.

11.4.2 *Sample variance*

The *sample variance* is a statistic defined as

$$s^2 = \frac{1}{N-1}\sum_{i=1}^{N}(x_i - \overline{x})^2 \tag{11.66}$$

where $x_1, x_2, \ldots, x_N$ are the sample values and N is the sample size and $\overline{x}$ is the sample mean as defined by Equation (11.61). In an analogy with the standard deviation of the population values, we can define the *standard deviation of the sample values* as s [the square-root of Equation (11.66)]. For example, six measurements of albumin concentration in blood serum gave the following results: 39.1, 40.2, 43.5, 42.6, 41.5, 39.8 $\text{g}\,\text{l}^{-1}$. The average of the measurements will be [using Equation (11.61)]:

$$\overline{x} = \frac{39.1 + 40.2 + 43.5 + 42.6 + 41.5 + 39.8}{6} = 41.1\,\text{g}\,\text{l}^{-1}$$

The variance of the measurements will be [using Equation (11.66)]:

$$s^2 = \frac{(39.1-41.1)^3 + (40.2-41.1)^2 + (43.5-41.1)^2 + (42.6-41.1)^2 + (41.5-41.1)^2 + (39.8-41.1)^2}{6-1}$$
$$= 2.934\,(\text{g}\,\text{l}^{-1})^2$$

The standard deviation of the measurements will be: $s = \sqrt{2.934} = 1.713\,\text{g}\,\text{l}^{-1}$.

Let us now find the expectation value of the sample variance. Using Equations (11.7), (11.8), (11.13) and (11.61) we obtain:

$$E(S^2) = E\left(\frac{1}{N-1}\sum_{i=1}^{N}(x_i - \overline{x})^2\right) = \frac{1}{N-1}E\left(\sum_{i=1}^{N}(x_i^2 - 2x_i\overline{x} + \overline{x}^2)\right)$$

$$= \frac{1}{N-1}E\left(\sum_{i=1}^{N}x_i^2 - 2\overline{x}\sum_{i=1}^{N}x_i + N\overline{x}^2\right) = \frac{1}{N-1}E\left(\sum_{i=1}^{N}x_i^2 - 2\overline{x}N\overline{x} + N\overline{x}^2\right)$$

$$= \frac{1}{N-1}E\left(\sum_{i=1}^{N}x_i^2 - N\overline{x}^2\right) = \frac{1}{N-1}\left[\sum_{i=1}^{N}E(x_i^2) - NE(\overline{x}^2)\right]$$

[and from Equation (11.13)]

$$V(X) = E(X^2) - [E(X)]^2 \Rightarrow \sigma^2 = E(X^2) - \mu^2 \Rightarrow E(X^2) = \sigma^2 + \mu^2$$

The last result is valid for any random variable, so we get for x_i^2 and $\overline{x}^2$:

$$E(x_i^2) = V(x_i) + [E(x_i)]^2 = V(X) + [E(X)]^2 = \sigma^2 + \mu^2$$

$$E(\overline{x}^2) = V(\overline{X}) + [E(\overline{X})]^2 = \frac{\sigma^2}{N} + \mu^2$$

In the last equation we used Equations (11.62) and (11.63). Substituting the previous results into the previous equation for $E(S^2)$ gives:

$$E(S^2) = \frac{1}{N-1}\left[\sum_{i=1}^{N}E(x_i^2) - NE(\overline{x}^2)\right] = \frac{1}{N-1}\left[\sum_{i=1}^{N}(\sigma^2 + \mu^2) - N\left(\frac{\sigma^2}{N} + \mu^2\right)\right]$$

$$= \frac{1}{N-1}[N\sigma^2 + N\mu^2 - \sigma^2 - N\mu^2] = \sigma^2 = V(X) \quad (11.67)$$

Equation (11.67) indicates that the sample variance [s^2 in Equation (11.66)] is an unbiased estimator of the variance of the population [σ^2 in Equation (11.11)] from which the sample is drawn. So we can use s^2 to estimate σ^2. In scientific hand calculations and some textbooks we can find Equation (11.66) for calculating s^2 with N instead of $N-1$ in the denominator. If we use N instead of $N-1$ in calculating s^2, then s^2 will be a *biased estimator* (instead of an unbiased estimator) of σ^2 and will give a poorer estimation (underestimation) of σ^2. Obviously, for large values of N (large sample size) the difference between N and $N-1$ will be negligible.

11.5 Confidence interval

In the previous section we discussed how the population mean and the population variance may be estimated by the sample mean and by the sample variance. The sample mean and the sample variance are random variables so the sample mean and the sample variance ordinarily will not be equal to the population mean and to the population variance correspondingly. It is necessary to qualify how much our estimates differ from the real values (population mean and population variance). Usually this is done by showing a *confidence interval*, which is an estimated range of values with a given probability of covering the true population value. The extreme values of the confidence interval are called the *confidence limits*. The size of the confidence interval will depend on how certain we want to be that it includes the true value: the greater the certainty, the greater the interval required.

11.5.1 *Confidence interval for the mean with known population variance (σ^2)*

The mean of a sample of size N drawn from a normally distributed population of expectation value (mean) μ and variance σ^2 is normally distributed with expectation value (mean) μ and variance σ^2/N. In other words, if the independent random variables $x_1, x_2, x_3, \ldots, x_N$ are $n(\mu,\sigma)$ (normally distributed with mean μ and standard deviation σ) and they are independent, then:

$$\overline{x} = \frac{1}{N}\sum_{i=1}^{N} x_i$$

is $n\left(\mu, \dfrac{\sigma}{\sqrt{N}}\right)$ (normally distributed with mean μ and standard deviation $\dfrac{\sigma}{\sqrt{N}}$).

The proof of the last theorem may be found in Brunk (1965, pp. 228–229) or Hogg and Craig (1971, pp. 158–159, 163).

Another theorem, called the *central limit theorem*, states that the mean of a sample of size N drawn from any distribution having positive variance ($\sigma^2 > 0$) (and hence finite mean μ) has a limiting normal distribution (as N approaches infinity) with mean μ and variance σ^2/N. In other words, the central limit theorem states that the distribution of the sample mean tends to the normal distribution with population mean μ and standard deviation $\sigma/\sqrt{N}$ (σ is the population standard deviation, N is the sample size) as N increases even if the original population is not normal. The proof of the central limit theorem may be found in Kendaill and Stuart (1969, pp. 193–194) or Hogg and Craig (1971, pp. 182–184).

If the sample mean ($\overline{x}$) has normal distribution (exactly if the population has normal distribution or approximately for large sample size if the population is not normal) with population mean μ and standard deviation $\sigma/\sqrt{N}$ (σ is the population standard deviation, N is the sample size) then the probability that

$$z = \frac{\overline{x} - \mu}{\sigma/\sqrt{N}}$$

will lie between two limits $-a_\alpha$ and a_α will be given by [using Equations (11.3), (11.26), (11.30), (11.32) and (11.36)]

$$P\left(-a_\alpha \leq \frac{\overline{x} - \mu}{\sigma/\sqrt{N}} \leq a_\alpha\right) = \Phi(a_\alpha) - \Phi(-a_\alpha) = \Phi(a_\alpha) - (1 - \Phi(a_\alpha))$$

$$= 2\Phi(a_\alpha) - 1 = 1 - \alpha$$

where Φ is the cumulative probability function of the standard normal distribution as given by Equation (11.26) and Table 11.1. The last equation can be rearranged (within the parentheses of the probability) to give:

$$P\left(-a_\alpha \leq \frac{\bar{x}-\mu}{\frac{\sigma}{\sqrt{N}}} \leq a_\alpha\right) = P\left(-a_\alpha \frac{\sigma}{\sqrt{N}} \leq \bar{x}-\mu \leq a_\alpha \frac{\sigma}{\sqrt{N}}\right) = 2\Phi(a_\alpha) - 1 = 1-\alpha$$

$$P\left(-a_\alpha \frac{\sigma}{\sqrt{N}} \leq \bar{x}-\mu \leq a_\alpha \frac{\sigma}{\sqrt{N}}\right) = P\left(a_\alpha \frac{\sigma}{\sqrt{N}} \geq \mu-\bar{x} \geq -a_\alpha \frac{\sigma}{\sqrt{N}}\right)$$
$$= 2\Phi(a_\alpha) - 1 = 1-\alpha$$

$$P\left(\bar{x}+a_\alpha \frac{\sigma}{\sqrt{N}} \geq \mu \geq \bar{x}-a_\alpha \frac{\sigma}{\sqrt{N}}\right) = 2\Phi(a_\alpha) - 1 = 1-\alpha \quad (11.68)$$

From Equation (11.68) we can say that the interval $\bar{x} \pm a_\alpha(\sigma/\sqrt{N})$ is a $100(1-\alpha)\%$ confidence interval for μ where $\bar{x}+a_\alpha(\sigma/\sqrt{N}), \bar{x}-a_\alpha\ (\sigma/\sqrt{N})$ are the confidence limits. For example, a pin diameter is distributed normally with a standard deviation of 0.001 cm in a process of manufacturing steel pins. A random sample of 20 pins was drawn from the process. The mean of the sample was [according to Equation (11.61)] $\bar{x} = 1.92$ cm. Then a confidence interval of 95% for the mean of the pin diameter in the process (an interval that includes the mean pin diameter of the process with a probability of 95%) will be given by [using Equation (11.68) and Table 11.1]:

$$\text{confidence interval of } 95\% \Rightarrow 0.95 = 1-\alpha \Rightarrow \alpha = 0.05$$

$$2\Phi(a_\alpha) - 1 = 1-\alpha \Rightarrow \Phi(a_\alpha) = 1-\frac{\alpha}{2} = 1-\frac{0.05}{2} = 0.975 \Rightarrow a_\alpha =$$
$$\Phi^{-1}(0.975) = 1.96$$

$$N = 20, \quad \bar{x} = 1.92, \quad \sigma = 0.001 \Rightarrow$$

$$P\left(\bar{x}+a_\alpha \frac{\sigma}{\sqrt{N}} \geq \mu \geq \bar{x}-a_\alpha \frac{\sigma}{\sqrt{N}}\right) = P\left(1.92+1.96\frac{0.001}{20} \geq \mu \geq 1.92-1.96\frac{0.001}{20}\right)$$
$$= 0.95$$

$$P(1.92+0.000098 \geq \mu \geq 1.92-0.000098) = 0.95$$

So, a confidence interval of 95% for the mean pin diameter of the process will be:

$$\mu = 1.92 \pm 0.000098 \text{ cm}$$

11.5.2 *Confidence interval for the mean with unknown population variance*

In the previous section we derived a confidence interval for the mean when the variance of the population was known. However, we do not usually know the variance of the population and we can only estimate (or approximate) the population variance by sample variance [s^2; see Equation (11.66)]. In this section we will derive a confidence interval for the population mean without knowing the population variance.

Let us look at a sample of size N drawn from a normally distributed population of expectation value (mean) μ and variance σ^2. In the following paragraph we will prove that the statistic $(N-1)s^2/\sigma^2$ [N is the sample size, σ^2 is the population variance, s^2 is the sample variance as given by Equation (11.66)] has chi-square distribution with

$N-1$ degrees of freedom, if the sample is drawn from a normally distributed population. Using Equation (11.66) we get:

$$\frac{(N-1)s^2}{\sigma^2} = \frac{1}{\sigma^2}\sum_{i=1}^{N}(x_i - \overline{x})^2$$

where $x_1, x_2, \ldots, x_N$ are the sample values and $\overline{x}$ is the sample mean as defined by Equation (11.61). The last expression may be written as:

$$\frac{(N-1)s^2}{\sigma^2} = \frac{1}{\sigma^2}\sum_{i=1}^{N}(x_i - \overline{x})^2 = \frac{1}{\sigma^2}\left[\sum_{i=1}^{N}((x_i - \mu) - (\overline{x} - \mu))^2\right]$$

where μ is the population mean. We can expand the last expression further:

$$\begin{aligned}\frac{(N-1)s^2}{\sigma^2} &= \frac{1}{\sigma^2}\left[\sum_{i=1}^{N}((x_i - \mu) - (\overline{x} - \mu))^2\right] \\ &= \frac{1}{\sigma^2}\left[\sum_{i=1}^{N}\left((x_i - \mu)^2 - 2(x_i - \mu)(\overline{x} - \mu) + (\overline{x} - \mu)^2\right)\right]\end{aligned}$$

Since $\overline{x}-\mu$ does not involve the summation index i, we have:

$$\frac{(N-1)s^2}{\sigma^2} = \frac{1}{\sigma^2}\left[\sum_{i=1}^{N}(x_i - \mu)^2 - 2(\overline{x} - \mu)\sum_{i=1}^{N}(x_i - \mu) + N(\overline{x} - \mu)^2\right]$$

But [using Equation (11.61)]

$$\sum_{i=1}^{N}(x_i - \mu) = \sum_{i=1}^{N}x_i - \sum_{i=1}^{N}\mu = N\overline{x} - N\mu = N(\overline{x} - \mu)$$

so that

$$\begin{aligned}\frac{(N-1)s^2}{\sigma^2} &= \frac{1}{\sigma^2}\left[\sum_{i=1}^{N}(x_i - \mu)^2 - 2N(\overline{x} - \mu)^2 + N(\overline{x} - \mu)^2\right] \\ &= \sum_{i=1}^{N}\left(\frac{x_i - \mu}{\sigma}\right)^2 - \left(\frac{\overline{x} - \mu}{\sigma/\sqrt{N}}\right)^2\end{aligned}$$

Let us now look at the last result. The first term at the right is the sum of the squares of N independent random variables, where each has standard normal distribution [$n(0,1)$; see Equations (11.32), (11.33) and (11.36)]. We saw in Section 11.3.2.2, Equation (11.49), that the sum of the squares of N independent standard normal random variables has a chi-square distribution with N degrees of freedom. Thus, $\sum_{i=1}^{N}[(x_i - \mu)/\sigma]^2$ has a chi-square distribution with N degrees of freedom ($\chi^2(N)$). We also saw in Section 11.5.1 that

$$\overline{x} = \frac{1}{N}\sum_{i=1}^{N}x_i \text{ is } n\left(\mu, \frac{\sigma}{\sqrt{N}}\right)$$

(normally distributed with mean μ and standard deviation $\sigma/\sqrt{N}$), so

$$\frac{\overline{x} - \mu}{\sigma/\sqrt{N}}$$

has standard normal distribution ($n(0,1)$). We can conclude that

$$\left(\frac{\overline{x} - \mu}{\sigma/\sqrt{N}}\right)^2$$

has chi-square distribution with one degree of freedom ($\chi^2(1)$) from Equation (11.49) in Section 11.3.2.2 (because we have only one member in the sum of the squares, $N = 1$). Putting these facts together, we have:

$$\frac{(N-1)s^2}{\sigma^2} + \underbrace{\left(\frac{\overline{x} - \mu}{\sigma/\sqrt{N}}\right)^2}_{is\,\chi^2(1)} = \underbrace{\sum_{i=1}^{N}\left(\frac{x_i - \mu}{\sigma}\right)^2}_{is\,\chi^2(N)}$$

It may be shown (Brunk 1965, pp. 82, 232) that the sample mean $\overline{x}$ and the sample variance s^2 are independent random variables in sampling from the normal distribution. Because $\overline{x}$ and s^2 are independent random variables,

$$\left(\frac{\overline{x} - \mu}{\sigma/\sqrt{N}}\right)^2 \text{ and } \frac{(N-1)s^2}{\sigma^2}$$

are also. It also may be shown (Hogg & Craig 1971, p. 159) that if a random variable has chi-square distribution with n_1 degrees of freedom ($\chi^2(n_1)$), and an independent random variable has chi-square distribution with n_2 degrees of freedom ($\chi^2(n_2)$), then the new random variable formed from the sum of these variables has chi-square distribution with $n_1 + n_2$ degrees of freedom ($\chi^2(n_1 + n_2)$).

Using these two theorems and the previous results of our expansion we can conclude that if we are sampling from a normal distribution then the random variable $[(N-1)s^2]/\sigma^2$ has chi-square distribution with $N-1$ degrees of freedom ($\chi^2(N-1)$).

Let us define a new statistic t to be:

$$t = \frac{\dfrac{\overline{x} - \mu}{\sigma/\sqrt{N}}}{\sqrt{\dfrac{(N-1)s^2}{\sigma^2(N-1)}}} = \sqrt{N}\frac{\overline{x} - \mu}{s} \tag{11.69}$$

We stated previously that when we are sampling from the normal distribution, $(\overline{x} - \mu)/(\sigma/\sqrt{N})$ has standard normal distribution ($n(0,1)$) (see Sections 11.5.1 and 11.5.2) and $[(N-1)s^2]/\sigma^2$ has chi-square distribution with $N-1$ degrees of freedom ($\chi^2(N-1)$ (see Section 11.5.2) and they are independent. From these facts we conclude that when we are sampling from a normal distribution, the statistic t has t distribution with $N-1$ degrees of freedom as has been shown in Section 11.3.2.3. With this knowledge in hand we can determine a confidence interval for the population mean. The probability that $t = \sqrt{N}[(\overline{x} - \mu)/s]$ lies between two limits $-t_{N-1,\alpha}$ and $t_{N-1,\alpha}$ is given by [using Equations (11.3), (11.55), (11.56)]

$$P\left(-t_{N-1,\alpha} \leq \sqrt{N}\frac{\overline{x} - \mu}{s} \leq t_{N-1,\alpha}\right) = G(t_{N-1,\alpha}) - G(-t_{N-1,\alpha}) = G(t_{N-1\alpha})$$
$$-(1 - G(t_{N-1,\alpha})) = 2G(t_{N-1,\alpha}) - 1 = 1 - \alpha$$

where $G(t)$ is the cumulative probability function of the t distribution as given by Equation (11.55) and Table 11.3. The last equation can be rearranged (within the parentheses of the probability) to give:

$$P\left(-t_{N-1,\alpha} \leq \sqrt{N}\frac{\overline{x} - \mu}{s} \leq t_{N-1,\alpha}\right) = 2G(t_{N-1,\alpha}) - 1 = 1 - \alpha$$

$$P\left(\frac{-s \cdot t_{N-1,\alpha}}{\sqrt{N}} \leq \overline{x} - \mu \leq \frac{s \cdot t_{N-1,\alpha}}{\sqrt{N}}\right) = 2G(t_{N-1,\alpha}) - 1 = 1 - \alpha$$

$$P\left(\overline{x} + \frac{s \cdot t_{N-1,\alpha}}{\sqrt{N}} \geq \mu \geq \overline{x} - \frac{s \cdot t_{N-1,\alpha}}{\sqrt{N}}\right) = 2G(t_{N-1,\alpha}) - 1 = 1 - \alpha \tag{11.70}$$

From Equation (11.70) we can say that the interval $\overline{x} \pm [(s \cdot t_{N-1,\alpha})/\sqrt{N}]$ is the $100(1-\alpha)\%$ confidence interval for the population mean μ. For example, a pin diameter is distributed normally in a process for manufacturing steel pins. A random sample of five pins was drawn from the process with the following diameters: 1.13, 1.11, 1.14, 1.20, 1.16 cm. A confidence interval of 99% for the mean of the pin diameter in the process will be given by the following calculation:

$$N = 5$$

$$\text{sample mean [Equation (11.61)]} = \overline{x} = \frac{1}{N}\sum_{i=1}^{N} x_i = 1.148\,\text{cm}$$

$$\text{sample variance [Equation (11.66)]} = s^2 = \frac{1}{N-1}\sum_{i=1}^{N}(x_i - \overline{x})^2 = 0.00117\,\text{cm}^2$$

$$\text{sample standard deviation} = s = \sqrt{0.00117} = 0.0342\,\text{cm}$$

$$\text{confidence interval of 99\%} \Rightarrow 1 - \alpha = 0.99 \Rightarrow \alpha = 0.01$$

$$\text{Equation (11.70)} \Rightarrow 2G(t_{N-1,\alpha}) - 1 = 1 - \alpha \Rightarrow G(t_{N-1,\alpha}) = 1 - \frac{a}{2} = 0.995$$

$$\Rightarrow \text{Table 11.3 with } N - 1 = 4 \text{ degrees of freedom } t_{N-1,\alpha} = G^{-1}(0.995) = 4.604$$

$$\text{confidence interval [Equation (11.70)]} \Rightarrow P\left(\overline{x} + \frac{s \cdot t_{N-1,\alpha}}{\sqrt{N}} \geq \mu \geq \overline{x} - \frac{s \cdot t_{N-1,\alpha}}{\sqrt{N}}\right)$$
$$= 1 - \alpha = 0.99$$

$$P\left(1.148 + \frac{0.0342 \times 4.604}{\sqrt{5}} \geq \mu \geq 1.148 - \frac{0.0342 \times 4.604}{\sqrt{5}}\right)$$
$$= P(1.148 + 0.0704 \leq \mu \leq 1.148 - 0.0704)$$

So, a confidence interval of 99% for the mean pin diameter of the process will be $\mu = 1.148 \pm 0.070$ cm.

11.5.3 *Confidence interval for the difference of means with known population variances*

Occasionally we need to compare two sample means $\overline{x}_1$ and $\overline{x}_2$, for example, when comparing two methods of measurement or two populations. In assessing that the two samples (or two methods of measurement, or two samples from two populations) give the same results within statistical deviation, we need to test whether $\overline{x}_1 - \overline{x}_2$ differs significantly from zero. In other words we must give a confidence interval for the difference of population means from which the samples were drawn.

It can be shown (Hogg & Craig 1971, pp. 158–159) that any linear combination of some n independent random variables $x_1, x_2, \ldots, x_n$, each normally distributed, is normally distributed random variable. From this theorem or from the theorem in Section 11.5.1 we conclude that the mean of the sample $\overline{x}$, drawn from the normal population is normally distributed. We also know from the central limit theorem (see Section 11.5.1) that for a large sample the sample mean tends to normal distribution even if the original population is not normal.

Let us assume that we have two independent samples. The first sample of size N, is drawn from a normally distributed population number 1 with expectation value (mean) μ_1 and variance σ_1^2. The second sample of size N_2 is drawn from a normally distributed population number 2 with expectation value (mean) μ_2 and variance σ_2^2. Let us denote $\overline{x}_1$ and $\overline{x}_2$ as the sample means [according to Equation (11.61)] of the first and second sample, respectively. When we are sampling from a normal distribution or for a large sample size $\overline{x}_1$ and $\overline{x}_2$ have normal distribution ($\overline{x}_1$ is $n(\mu_1, \sigma_1/\sqrt{N_1})$, $\overline{x}_2$ is $n(\mu_2, \sigma_2/\sqrt{N_2})$) (see Sections 11.5.1 and 11.5.2).

As stated previously, any linear combination of independent random variables each normally distributed, is also normally distributed, so $\overline{x}_1 - \overline{x}_2$ is also normally distributed. We can find the expectation value (mean) and the variance of $\overline{x}_1 - \overline{x}_2$ by using Equations (11.9) and (11.12):

$$E(\overline{X}_1 - \overline{X}_2) = E(\overline{X}_1) - E(\overline{X}_2) = \mu_1 - \mu_2 \tag{11.71}$$

$$V(\overline{X}_1 - \overline{X}_2) = V(\overline{X}_1) + V(\overline{X}_2) = \frac{\sigma_1^2}{N_1} + \frac{\sigma_2^2}{N_2} \tag{11.72}$$

Using Equations (11.32), (11.36), (11.71) and (11.72) we can state that

$$\frac{\overline{x}_1 - \overline{x}_2 - (\mu_1 - \mu_2)}{\sqrt{\dfrac{\sigma_1^2}{N_1} + \dfrac{\sigma_2^2}{N_2}}}$$

has standard normal distribution ($n(0,1)$). Using this fact and Equations (11.3), (11.26), (11.30), (11.32), and (11.36) we can obtain a confidence interval for the difference of means of two populations:

$$P\left(-a_\alpha \leq \frac{\bar{x}_1 - \bar{x}_2 - (\mu_1 - \mu_2)}{\sqrt{\dfrac{\sigma_1^2}{N_1} + \dfrac{\sigma_2^2}{N_2}}} \leq a_\alpha\right) = \Phi(a_\alpha) - \Phi(-a_\alpha) = \Phi(a_\alpha) - (1 - \Phi(a_\alpha))$$

$$= 2\Phi(a_\alpha) - 1 = 1 - \alpha$$

where Φ is the cumulative probability function of the standard normal distribution as given by Equation (11.26) and Table 11.1. The last equation can be rearranged (within the parentheses of the probability) to give:

$$P\left(\bar{x}_1 - \bar{x}_2 - a_\alpha\sqrt{\frac{\sigma_1^2}{N_1} + \frac{\sigma_2^2}{N_2}} \leq \mu_1 - \mu_2 \leq \bar{x}_1 - \bar{x}_2 + a_\alpha\sqrt{\frac{\sigma_1^2}{N_1} + \frac{\sigma_2^2}{N_2}}\right)$$
$$= 2\Phi(a_\alpha) - 1 = 1 - \alpha \tag{11.73}$$

From Equation (11.73) we can say that the interval

$$\mu_1 - \mu_2 = \bar{x}_1 - \bar{x}_2 \pm a_\alpha\sqrt{\frac{\sigma_1^2}{N_1} + \frac{\sigma_2^2}{N_2}}$$

is the $100(1-\alpha)\%$ confidence interval for the difference in two population means. If we can assume that the two populations have the same variance or that the two samples were drawn from the same population then ($\sigma_1^2 = \sigma_2^2 = \sigma^2$) and the confidence interval for the difference in two population means are given by:

$$\mu_1 - \mu_2 = \bar{x}_1 - \bar{x}_2 \pm a_\alpha\sigma\sqrt{\frac{1}{N_1} + \frac{1}{N_2}} \tag{11.74}$$

11.5.4 *Confidence interval for the difference of means with unknown population variances*

In the previous section we derived a confidence interval for the difference of means when the variances of the populations were known. However, we do not usually know the variances of the populations and we can only estimate (or approximate) the population variances by the variances of samples [s_1^2 and s_2^2: see Equation (11.66)]. In this section we will derive a confidence interval for the difference of means without knowing the variances of populations. In Section 11.5.3 we showed that the random variable

$$\frac{\bar{x}_1 - \bar{x}_2 - (\mu_1 - \mu_2)}{\sqrt{\dfrac{\sigma_1^2}{N_1} + \dfrac{\sigma_2^2}{N_2}}}$$

has standard normal distribution ($n(0,1)$). In Section 11.5.2 we showed that the random variable $[(N-1)s^2]/\sigma^2$ has chi-square distribution with $N-1$ degrees of freedom ($\chi^2\,(N-1)$) when the sample is drawn from a normal population. Using this fact we can state that the random variables $[(N_1-1)s_1^2]/\sigma_1^2$ and $[(N_2-1)s_2^2]/\sigma_2^2$ have chi-square distribution with N_1-1 and N_2-1 degrees of freedom, respectively. Their sum

$$\frac{(N_1-1)s_1^2}{\sigma_1^2}+\frac{(N_2-1)s_2^2}{\sigma_2^2}$$

has chi-square distribution with N_1+N_2-2 degrees of freedom, according to the theorem in Section 11.5.2 (see also Hogg & Craig 1971, p. 159). If we assume that $\sigma_1^2 = \sigma_2^2 = \sigma^2$ then the random variable

$$T=\frac{\dfrac{\bar{x}_1-\bar{x}_2-(\mu_1-\mu_2)}{\sqrt{\dfrac{\sigma^2}{N_1}+\dfrac{\sigma^2}{N_2}}}}{\sqrt{\dfrac{\dfrac{(N_1-1)s_1^2}{\sigma^2}+\dfrac{(N_2-1)s_2^2}{\sigma^2}}{N_1+N_2-2}}}$$

$$=\frac{\bar{x}_1-\bar{x}_2-(\mu_1-\mu_2)}{\sqrt{\dfrac{(N_1-1)s_1^2+(N_2-1)s_2^2}{N_1+N_2-2}}\sqrt{\dfrac{1}{N_1}+\dfrac{1}{N_2}}}$$

has t distribution with N_1+N_2-2 degrees of freedom according to Section 11.3.2.3. Using the last fact and Equations (11.3), (11.55) and (11.56) we can ascertain a confidence interval for the difference of means of two populations when the variance of populations is not known, namely

$$P\left(-t_{N_1+N_2-2,\alpha}\le\frac{\bar{x}_1-\bar{x}_2-(\mu_1-\mu_2)}{\sqrt{\dfrac{(N_1-1)s_1^2+(N_2-1)s_2^2}{N_1+N_2-2}}\sqrt{\dfrac{1}{N_1}+\dfrac{1}{N_2}}}\le t_{N_1+N_2-2,\alpha}\right)$$
$$=G(t_{N_1+N_2-2,\alpha})-G(-t_{N_1+N_2-2,\alpha})=G(t_{N_1+N_2-2,\alpha})-(1-G(t_{N_1+N_2-2,\alpha}))$$
$$=2G(t_{N_1+N_2-2,\alpha})-1=1-\alpha$$

where $G(t)$ is the cumulative probability function of the t distribution as given by Equation (11.55) and Table 11.3. The previous equation can be rearranged (within the parentheses of the probability) to give:

$$P\left(\begin{array}{l}\bar{x}_1-\bar{x}_2-t_{N_1+N_2-2,\alpha}\sqrt{\dfrac{(N_1-1)s_1^2+(N_2-1)s_2^2}{N_1+N_2-2}}\sqrt{\dfrac{1}{N_1}+\dfrac{1}{N_2}}\le\mu_1-\mu_2\\ \le\bar{x}_1-\bar{x}_2+t_{N_1+N_2-2,\alpha}\sqrt{\dfrac{(N_1-1)s_1^2+(N_2-1)s_2^2}{N_1+N_2-2}}\sqrt{\dfrac{1}{N_1}+\dfrac{1}{N_2}}\end{array}\right)$$
$$=2G(t_{N_1+N_2-2,\alpha})-1=1-\alpha \tag{11.75}$$

From Equation (11.75) we can say that the interval

$$\mu_1 - \mu_2 = \overline{x}_1 - \overline{x}_2 \pm t_{N_1+N_2-2,\alpha}\sqrt{\frac{(N_1-1)s_1^2 + (N_2-1)s_2^2}{N_1+N_2-2}}\sqrt{\frac{1}{N_1}+\frac{1}{N_2}}$$

is a $100(1-\alpha)\%$ confidence interval for the difference in two population means. In the derivation of Equation (11.75) we assume that $\sigma_1^2 = \sigma_2^2 = \sigma^2$ (population variances). If we cannot assume equal population variances then an approximate method (Hays & Winkler 1971; Miller & Miller 1994) is to modify Equation (11.75):

$$P\left(\overline{x}_1 - \overline{x}_2 - t_{df,\alpha}\sqrt{\frac{s_1^2}{N_1}+\frac{s_2^2}{N_2}} \le \mu_1 - \mu_2 \le \overline{x}_1 - \overline{x}_2 + t_{df,\alpha}\sqrt{\frac{s_1^2}{N_1}+\frac{s_2^2}{N_2}}\right)$$

$$= 2G(t_{df,\alpha}) - 1 = 1 - \alpha \qquad (11.76)$$

$$df = \left\{\frac{\left(\frac{s_1^2}{N_1}+\frac{s_2^2}{N_2}\right)^2}{\frac{\left(\frac{s_1^2}{N_1}\right)^2}{N_1+1}+\frac{\left(\frac{s_2^2}{N_2}\right)^2}{N_2+1}}\right\}$$

The degrees of freedom (df) in Equation (11.76) must be rounded to the nearest whole number. For example, two samples of plastic sheet were selected from different batches of the same production line. The thicknesses of five plastic sheets in the first sample were 1.47, 1.49, 1.42, 1.45, 1.46 cm. The thicknesses of four plastic sheets in the second sample were 1.47, 1.48, 1.43, 1.46 cm. If we assume that the thickness of the plastic sheet is distributed normally and if we assume that the population variances are equal because the samples were taken from the same production line, then by using Equation (11.75) we obtain a confidence interval for the difference of the means of the two batches:

$$N_1 = 5,\ N_2 = 4$$

$$\overline{x}_1 = 1.458, \overline{x}_2 = 1.46 \text{ [Equation (11.61)]}$$

$$s_1^2 = 0.00067, s_2^2 = 0.000467 \text{ [Equation (11.66)]}$$

$$95\% \text{ confidence interval} \Rightarrow \mathbf{1} - \alpha = 0.95 = 2G(t) - 1 \Rightarrow G(t) = 0.975$$

$$G(t) = 0.975 \text{ with } N_1 + N_2 - 2 = 7 \text{ degrees of freedom (using Table 11.3)}$$
$$\Rightarrow t = 2.365$$

$$\mu_1 - \mu_2 = \overline{x}_1 - \overline{x}_2 \pm t_{N_1+N_2-2,\alpha}\sqrt{\frac{(N_1-1)s_1^2 + (N_2-1)s_2^2}{N_1+N_2-2}}\sqrt{\frac{1}{N_1}+\frac{1}{N_2}}$$

$$\mu_1 - \mu_2 = 1.458 - 1.46 \pm 2.365\sqrt{\frac{(5-1)0.00067 + (4-1)0.000467}{5+4-2}}\sqrt{\frac{1}{5}+\frac{1}{4}}$$

$$\mu_2 - \mu_1 = 0.002 \pm 0.0383$$

11.5.5 *Confidence interval for population variance*

For some industrial applications, product uniformity is of primary importance. The population variance may be used to characterize uniformity. Usually the population variance is estimated by the sample variance [Equation (11.66)]. Let us derive the confidence interval for population variance using the sample variance. In Section 11.5.2 we verified that the random variable $[(N-1)s^2]/\sigma^2$ (N is sample size, s^2 is sample variance, σ^2 is population variance) has chi-square distribution with $N-1$ degrees of freedom ($\chi^2(N-1)$) when sampling from normal distribution. Using this fact and Equations (11.3) and (11.47) we can derive a confidence interval for population variance

$$P\left(a_\alpha \le \frac{(N-1)s^2}{\sigma^2} \le b_\alpha\right) = 1 - \alpha = F(b_\alpha) - F(a_\alpha)$$

where $F(x)$ is the cumulative distribution function of the chi-square distribution (with $N-1$ degrees of freedom) as given by Equation (11.47) and Table 11.2. The last equation can be rearranged (within the parentheses of the probability) to give:

$$P\left(\frac{(N-1)s^2}{b_\alpha} \le \sigma^2 \le \frac{(N-1)s^2}{a_\alpha}\right) = 1 - \alpha = F(b_\alpha) - F(a_\alpha) \tag{11.77}$$

From Equation (11.77) we can say that the interval

$$\left(\frac{(N-1)s^2}{b_\alpha}, \frac{(N-1)s^2}{a_\alpha}\right)$$

is a $(1-\alpha)$ 100% confidence interval for the population variance σ^2. For example, the thicknesses of five plastic sheets drawn from a normal population were 1.47, 1.49, 1.42, 1.45, 1.46 cm. According to Equation (11.66) the sample variance will be $s^2 = 0.00067\ \text{cm}^2$. Using Table 11.2 we can find the confidence limit for a 95% confidence interval with $N-1 = 5-1 = 4$ degrees of freedom:

$$F(b_\alpha) - F(a_\alpha) = 0.95 = 0.975 - 0.025$$

$$F(b_\alpha) = 0.975 \Rightarrow b_\alpha = 11.1, \quad F(a_\alpha) = 0.025 \Rightarrow a_\alpha = 0.484$$

Thus, a 95% confidence interval for population variance will be [using Equation (11.77)]:

$$P\left(\frac{(N-1)s^2}{b_\alpha} \le \sigma^2 \le \frac{(N-1)s^2}{a_\alpha}\right) =$$

$$P\left(\frac{(5-1)\times 0.00067}{11.1} \le \sigma^2 \le \frac{(5-1)\times 0.00067}{0.484}\right) = F(11.1) - F(0.484) = 0.95$$

$$P(0.000241 \le \sigma^2 \le 0.005537) = 0.95$$

11.5.6 *Confidence interval for ratio of variances*

In many situations it is important to compare the variances of two sets of data. For example, we need to know sometimes if two populations have the same variance in the case of finding the confidence interval for the difference of means (see Sections 11.5.3 and 11.5.4) or in the case of comparing two industrial processes that produce the same product.

Let us assume that we have two independent samples. The first sample of size N_1 is drawn from a normally distributed population number 1 with variance σ_1^2. The second sample of size N_2 is drawn from a normally distributed population number 2 with variance σ_2^2. From Section 11.5.2 we know that the random variables $[(N_1-1)s_1^2]/\sigma_1^2$ and $[(N_2-1)s_2^2]/\sigma_2^2$ have chi-square distribution with N_1-1 and N_2-1 degrees of freedom, respectively, where s_1^2 and s_2^2 are the sample variances [see Equation (11.66)] of the first and the second sample, respectively. From Section 11.3.2.4 we can conclude that the ratio

$$\frac{s_1^2/\sigma_1^2}{s_2^2/\sigma_2^2}$$

has F distribution with N_1-1 and N_2-1 degrees of freedom:

$$\left.\begin{array}{l} \dfrac{(N_1-1)s_1^2}{\sigma_1^2} \text{ is } \chi^2(N_1-1) \\ \dfrac{(N_2-1)s_2^2}{\sigma_2^2} \text{ is } \chi^2(N_2-1) \end{array}\right\} \Rightarrow \frac{\dfrac{(N_1-1)s_1^2}{\sigma_1^2(N_1-1)}}{\dfrac{(N_2-1)s_2^2}{\sigma_2^2(N_2-1)}} = \frac{s_1^2/\sigma_1^2}{s_2^2/\sigma_2^2}$$

is F with N_1-1 and N_2-1 degrees of freedom.

Using this fact and Equations (11.3) and (11.60) we can calculate a confidence interval for the ratio of two population variances, i.e.

$$P\left(a_\alpha \le \frac{s_1^2/\sigma_1^2}{s_2^2/\sigma_2^2} \le b_\alpha\right) = 1-\alpha = G(b_\alpha) - G(a_\alpha)$$

where $G(f)$ is the cumulative distribution function of the F distribution (with N_1-1 and N_2-1 degrees of freedom) as given by Equation (11.60) and Table 11.4. The previous equation can be rearranged (within the parentheses of the probability) to give:

$$P\left(a_\alpha \frac{s_2^2}{s_1^2} \le \frac{\sigma_2^2}{\sigma_1^2} \le b_\alpha \frac{s_2^2}{s_1^2}\right) = 1-\alpha = G(b_\alpha) - G(a_\alpha) \tag{11.78}$$

From Equation (11.78) we can say that the interval

$$\left(a_\alpha \frac{s_2^2}{s_1^2}, b_\alpha \frac{s_2^2}{s_1^2}\right)$$

is a $(1-\alpha)100\%$ confidence interval for the ratio of population variances σ_2^2/σ_1^2. For example, two samples drawn from different processes for manufacturing steel rods gave the following results for rod diameter: $N_1 = 11$, $s_1^2 = 0.501$ cm^2; $N_2 = 6$, $s_2^2 = 0.547$ cm^2, where N is the sample size and s^2 is the sample variance. Using Table 11.4, Equation (11.78) and assuming that the rod diameter is normally distributed, we can find the confidence limit for the 95% confidence interval for the ratio of population variances with $N_2-1=5$ and $N_1-1=10$ degrees of freedom:

$$95\% \Rightarrow 1-\alpha = 0.95 = 0.975 - 0.025 = G(b_\alpha) - G(a_\alpha) \Rightarrow$$

$$G(b_\alpha) = 0.975 \Rightarrow b_\alpha = 4.24;\ G(a_\alpha) = 0.025 = P(f \le a_\alpha) = P\left(\frac{1}{f} \ge \frac{1}{a_\alpha}\right) =$$

$$1 - P\left(\frac{1}{f} \le \frac{1}{a_\alpha}\right) \Rightarrow P\left(\frac{1}{f} \le \frac{1}{a_\alpha}\right) = 1 - 0.025 = 0.975 \Rightarrow \text{from Table 11.4 with}$$

$$n_1 = N_1 - 1 = 10 \text{ and } n_2 = N_2 - 1 = 5 \Rightarrow \frac{1}{a_\alpha} = 6.62 \Rightarrow a_\alpha = 0.15106 \Rightarrow$$

[Equation(11.78)]

$$P\left(a_\alpha \frac{s_2^2}{s_1^2} \le \frac{\sigma_2^2}{\sigma_1^2} \le b_\alpha \frac{s_2^2}{s_1^2}\right) = 1 - \alpha \Rightarrow P\left(0.15106 \frac{0.547}{0.501} \le \frac{\sigma_2^2}{\sigma_1^2} \le 4.24 \frac{0.547}{0.501}\right) = 0.95$$

$$\Rightarrow P\left(0.165 \le \frac{\sigma_2^2}{\sigma_1^2} \le 4.629\right) = 0.95$$

11.6 Hypothesis and significance tests

In Section 11.4 we estimated the population properties (mean and standard deviation) by calculating statistics (sample mean and sample variance) from samples drawn from the population. In Section 11.5 we derived the confidence interval for the population property θ by finding a statistic u the distribution of which is known for each value of θ. A $100(1-\alpha)\%$ confidence interval for θ will be given by two numbers a and b, which will generally depend on θ, $a = a(\theta)$, $b = b(\theta)$, such that $P(a(\theta) \le u \le b(\theta)) = 1-\alpha$. In this section we will try to make decisions (or inferences) about a population from only the information we get from the sample (statistics and confidence intervals). The first stage of making a decision is to state a *hypothesis* about some property of the population. The initial hypothesis (or assumption) would be defined as a *null hypothesis* and symbolized as H_0. There is an *alternative hypothesis* for each null hypothesis and is symbolized as H_1. Usually the hypothesis will take one of the following forms (θ is some population property or parameter):

Form:	1	2	3
H_0:	$\theta = \theta_0$	$\theta \le \theta_0$	$\theta \ge \theta_0$
H_1:	$\theta \ne \theta_0$	$\theta > \theta_0$	$\theta < \theta_0$
Name of the test :	two-tailed test	upper-tailed test	lower-tailed test

From the previous forms of the hypothesis, it is obvious that if we decide that the null hypothesis is true then the alternative hypothesis will be false and, vice versa, if we decide that the null hypothesis is false then the alternative hypothesis will be true.

The second stage after stating the hypothesis is to test the truth of the hypothesis. In stating a test of statistical hypothesis we mean establishing a rule which, when the experimental sample values (statistics) have been obtained, leads to a decision to accept or to reject the hypothesis under consideration. We saw that assuming a population parameter θ_0 with statistic u with known distribution for θ_0, we can give an interval with probability of $1-\alpha$ that u will lay within it (confidence interval). We can also define a complementary interval to the confidence interval with probability α that u will lie within it. This interval will be called the *critical region* or *rejection region*. If the calculated value of u from a given sample falls into the region of rejection then the hypothesis (or the assumption) that the population parameter is θ_0 will be rejected. In these circumstances, we say that we reject the hypothesis at a *significance level* of $100\alpha\%$. In other words, if the statistic is in a region of rejection (or the critical region) then it *departs significantly from the expectation* under the hypothesis. In carrying out this decision process, we incur two possible types of errors. First, we can say the hypothesis is false (reject the hypothesis) when it is really correct (*Type-I error*) and second we can say the hypothesis is correct (accept the hypothesis) when it is really false (*Type-II error*). The probability of making a Type-I error is given by α (the significance level). In the following sections, we will give several examples of hypothesis testing.

11.6.1 *Test of hypothesis for population mean (μ)*

Steel wires have a yield strength of 34.9 lb in^{-2}. A new process was suggested for the production of these steel wires. A sample of 21 wires from the new process was selected randomly and the yield strength of these wires was tested. The results of the test were sample mean $\overline{x}$ = 35.4 lb in^{-2} [Equation (11.61)] and standard deviation of the sample s = 2.03 lb in^{-2} [Equation (11.66)]. Production management wants to determine whether the yield strength of the steel wires has changed under the new conditions. The null hypothesis in this case will be H_0: μ = 34.9 lb in^{-2}. The alternative hypothesis in this case will be H_1: $\mu \neq 34.9$ lb in^{-2}. To test the null hypothesis we assume that the yield strength of the wires is normally distributed. With this assumption the statistic $t = \sqrt{N}[(\overline{x}-\mu)/s]$ has t distribution with $N-1$ degrees of freedom [N is the sample size; see Section 11.5.2 and Equation (11.70)]. If we choose to test the hypothesis with a significance level of 5% (α = 0.05) then the critical values (for the critical region) in the two-tailed test and $N-1=21-1=20$ degrees of freedom [using Table 11.3 and Equation (11.70)] are given by:

$$P\left(-t_{N-\alpha} \leq \sqrt{N}\frac{\overline{x}-\mu}{s} \leq t_{N-1,\alpha}\right) = 2G(t_{N-1,\alpha}) - 1 = 1-\alpha$$

$$P\left(\frac{-s \cdot t_{N-1,\alpha}}{\sqrt{N}} + \mu \leq \overline{x} \leq \frac{s \cdot t_{N-1,\alpha}}{\sqrt{N}} + \mu\right) = 2G(t_{N-1,\alpha}) - 1 = 1-\alpha = 1-0.05 =$$

$$0.95 \Rightarrow G(t_{N-1,\alpha}) = 0.975 \Rightarrow t_{21-1,0.05} = 2.086 \Rightarrow$$

$$P\left(\frac{-s \cdot t_{N-1,\alpha}}{\sqrt{N}} + \mu \leq \overline{x} \leq \frac{s \cdot t_{N-1,\alpha}}{\sqrt{N}} + \mu\right) =$$

$$P\left(\frac{-2.03 \cdot 2.086}{\sqrt{21}} + 34.9 \leq \overline{x} \leq \frac{2.03 \cdot 2.086}{\sqrt{21}} + 34.9\right) =$$

$$P(33.976 \leq x \leq 35.824) = 0.95$$

We will accept H_0: $\mu = 34.9$ lb in^{-2} if the sample average $\overline{x}$ falls in the interval $33.976 \leq \overline{x} \leq 35.824$ and we will reject H_0 (accept H_1: $\mu \neq 34.9$ lb in^{-2}) if the sample average $\overline{x}$ falls in the complementary intervals $\overline{x} \leq 33.976$ or $\overline{x} \geq 35.824$. In this case, $\overline{x} = 35.4$ so we accept H_0. We can say that the yield strength of the steel wires has not changed under the new conditions (with a significant level of 5%).

The production management could also ask if the yield strength of the steel wires has increased under the new conditions (beyond a reasonable doubt). The null hypothesis in this case will be H_0: $\mu \leq 34.9$ lb in^{-2}. The alternative hypothesis in this case will be H_1: $\mu > 34.9$ lb in^{-2}. If we again choose to test the hypothesis with a significance level of 5% ($\alpha = 0.05$) then the critical values for the upper-tailed test and $N-1=21-1=20$ degrees of freedom [using Table 11.3 and Equation (11.55)] are:

$$P\left(\sqrt{N}\frac{\overline{x}-\mu}{s} \leq t_{N-1,\alpha}\right) = G(t_{N-1,\alpha}) = 1-\alpha = 0.95 \Rightarrow G(t_{N-1,\alpha}) =$$

$$0.95 \Rightarrow t_{21-1,0.05} = 1.725 \Rightarrow P\left(\overline{x} \leq \frac{s \cdot 1.725}{\sqrt{N}} + \mu\right) =$$

$$P\left(\overline{x} \leq \frac{2.03 \cdot 1.725}{\sqrt{21}} + 34.9\right) = P(\overline{x} \leq 35.664) = 0.95$$

In our case $\overline{x} = 35.4$, so we accept the null hypothesis H_0: $\mu \leq 34.9$ lb in^{-2} and we conclude that improvement in wire strength has not been demonstrated beyond a reasonable doubt.

11.6.2 *Test of hypothesis for difference of means*

Two steel rod samples were drawn randomly from the same production line. The yield strength of the first sample (of size 13) was measured by method 1 and gave the following results: sample average $\overline{x}_1 = 36.4$ lb in^{-2} [Equation (11.61)]; sample standard deviation $s_1 = 1.82$ lb in^{-2} [Equation (11.66)]. The yield strength of the second sample (of size 9) was measured by method 2 and gave the following results: sample average $\overline{x}_2 = 34.2$ 1b in^{-2} [Equation (11.61)]; sample standard deviation $s_2 = 1.93$ lb in^{-2}. The production manager wants to know if the two methods of measurement give the same results within statistical deviation. To check if the two methods give the same results we will test the null hypothesis H_0: $\mu_1 = \mu_2$ against the alternative hypothesis H_1: $\mu_1 \neq \mu_2$ where μ is the population mean. In other words, we test if the two methods of measurement give the same population mean. If we assume that the two samples were drawn from the same normally distributed population (assuming $\sigma_1^2 = \sigma_2^2$ where σ^2 is the population variance; see Section 11.6.5) then the statistic has t distribution (see Section 11.5.4):

$$T = \frac{\overline{x}_1 - \overline{x}_2 - (\mu_1 - \mu_2)}{\sqrt{\frac{(N_1-1)s_1^2 + (N_2-1)s_2^2}{N_1+N_2-2}}\sqrt{\frac{1}{N_1}+\frac{1}{N_2}}}$$

So the critical region for the two-tailed test and significance level of 5% ($\alpha = 0.05$) and with 2 degrees of freedom ($N_1 + N_2 - 2 = 13 + 9 - 2 = 20$) will be given by Equation (11.75):

$$P\left(-t_{N_1+N_2-2,\alpha} \le \frac{\overline{x}_1 - \overline{x}_2 - (\mu_1 - \mu_2)}{\sqrt{\frac{(N_1-1)s_1^2 + (N_2-1)s_2^2}{N_1+N_2-2}}\sqrt{\frac{1}{N_1}+\frac{1}{N_2}}} \le t_{N_1+N_2-2,\alpha}\right) =$$

$$2G(t_{N_1+N_2-2,\alpha}) - 1 = 1 - \alpha = 1 - 0.05$$

$$G(t_{N_1+N_2-2,\alpha}) = 0.975 \Rightarrow t_{20,\,0.05} = 2.086 \Rightarrow$$

$$P\left(-2.086 \le \frac{\overline{x}_1 - \overline{x}_2 - 0}{\sqrt{\frac{(13-1)\times 1.82^2 + (9-1)\times 1.93^2}{13+9-2}}\sqrt{\frac{1}{13}+\frac{1}{9}}} \le 2.086\right) =$$

$$P(1.687 \le \overline{x}_1 - \overline{x}_2 \le -1.687) = 0.95$$

In our case $\overline{x}_1 - \overline{x}_2 = 36.4 - 34.2 = 2.2$ lay out of the interval ± 1.687, so we reject the null hypothesis H_0: $\mu_1 = \mu_2$ with significance level of 5%. From the last test we conclude that the two methods of measurement yield different results.

11.6.3 *Test of hypothesis for population variance*

A manufacturer has claimed that he can produce steel rods with diameter of 10.1 mm with standard deviation of 0.04 mm. A sample of 20 steel rods drawn randomly from the production line yields the following results for the rod diameter: sample average $\overline{x}$ = 10.12 mm [Equation (11.61)], sample standard deviation s = 0.06 mm [Equation (11.66)]. If we want to check if the manufacturer's claim is correct about the precision of the rod diameter then we must test the following null hypothesis H_0: σ = 0.04 mm against the alternative hypothesis H_1: $\sigma > 0.04$ mm (we do not care if $\sigma < 0.04$ mm). If we assume that the rod diameter is normally distributed then the statistic $(N-1)s^2/\sigma^2$ (N is the sample size, s^2 is the sample variance, σ^2 is the population variance) has chi-square distribution with $N-1$ degrees of freedom (see Sections 11.5.2 and 11.5.5). So the critical value for this upper-tailed test with $N-1 = 20-1 = 19$ degrees of freedom and 5% significance level [α = 0.05, using Equation (11.77) and Table 11.2]:

$$P\left(\frac{(N-1)s^2}{\sigma^2} \le b_\alpha\right) = F(b_\alpha) = 1 - \alpha = 1 - 0.05 = 0.95 \Rightarrow$$

$$F(b_\alpha) = 0.95 \Rightarrow b_{0.05} = 30.1 \Rightarrow$$

$$P\left(\frac{(20-1)/s^2}{0.04^2} \le 30.1\right) = P(s^2 \le 0.00253) = 0.95$$

In our case $s^2 = 0.06^2 = 0.0036$ which is greater than the critical value 0.00253, so we reject the null hypothesis H_0: σ = 0.04 with significance level of 5%. We conclude that the manufacturer's claim about the precision of the rod diameter is not correct as a result of the last test.

11.6.4 *Test of hypothesis for comparing two population variances*

In Section 11.6.2 we assume that two samples of steel rods are drawn from the same population and the corresponding population variances are equal, $\sigma_1^2 = \sigma_2^2$. In order to check this assumption we will test the null hypothesis H_0: $\sigma_1^2 = \sigma_2^2$ against the alternative hypothesis H_1: $\sigma_1^2 \neq \sigma_2^2$. The statistic ratio $(s_1^2/\sigma_1^2)/(s_2^2/\sigma_2^2)$ has F distribution with N_1-1 and N_2-1 degrees of freedom [where s^2 is the sample variance (Equation 11.66), σ^2 is the population variance and N is the sample size] if the samples are drawn from normally distributed populations (see Section 11.5.6). Using Equation (11.78) and Table 11.4 we obtain the critical values for the two-tailed test with a significance level of 5% ($\alpha = 0.05$):

$$P\left(a_\alpha \leq \frac{s_1^2/\sigma_1^2}{s_2^2/\sigma_2^2} \leq b_\alpha\right) = 1 - \alpha = 1 - 0.05 = 0.95 = G(b_\alpha) - G(a_\alpha) \Rightarrow$$

$$N_1 - 1 = 13 - 1 = 12, \quad N_2 - 1 = 9 - 1 = 8 \Rightarrow$$

$$G(b_\alpha) = 0.975 \Rightarrow b_{0.05,12,8} = 5.67$$

$$G(a_\alpha) = 0.025 = P(F_{12,8} \leq a_{0.05}) = P\left(\frac{1}{F_{12,8}} \geq \frac{1}{a_{0.05}}\right) = 1 - P\left(\frac{1}{F_{12,8}} \leq \frac{1}{a_{0.05}}\right) =$$

$$1 - P\left(F_{8,12} \leq \frac{1}{a_{0.05}}\right) \Rightarrow P\left(F_{8,12} \leq \frac{1}{a_{0.05}}\right) = 1 - 0.025 = 0.975 \Rightarrow \frac{1}{a_{0.05}} =$$

$$3.51 \Rightarrow a_{0.05} = 0.285 \Rightarrow \frac{\sigma_1^2}{\sigma_2^2} = 1 \Rightarrow P\left(0.285 \leq \frac{s_1^2}{s_2^2} \leq 5.67\right)$$

In our case $s_1^2/s_2^1 = 1.82^2/1.93^2 = 0.889$ so it lies within the interval (0.285, 5.67). As a result of the last fact we accept the null hypothesis H_0: $\sigma_1^2 = \sigma_2^2$ and we conclude that the two samples came from the same population.

11.7 Outliers

11.7.1 *Introduction*

An important question every experimentalist meets quite often concerns one or two data points that deviate from the others and hence influences quite significantly the standard deviation, though not so much the mean. The tendency is usually to say that some exceptional error was involved in this measurement and to discard the data points concerned. However, is this justified? It may be that the deviation is just due to random error. We refer to measurements like these as outliers and this section discusses the question of whether or not they can be disregarded.

Outliers can result, for example, from blunders or malfunctions of the methodology or from unusual losses or contaminations. Assume that we measure aliquots from the same sample and obtain the values for manganese content as 5.65, 5.85, 5.94, 6.52, 5.73, 5.80. At first sight we will tend to reject the fifth result and use only five values to calculate the mean and the standard deviation. Is this rejection justified? We will test this rejection assuming that the whole population is normally distributed.

11.7.2 *Dixon's Q test*

See Dixon (1953). Q is calculated from the equation

$$Q = | \text{ suspected value} - \text{nearest value } | /(\text{largest value} - \text{smallest value}) \qquad (11.79)$$

i.e. in our case $Q = (6.52 - 5.94)/(6.52 - 5.65) = 0.666$. Table 11.5 gives the values of the critical Q. For a risk of 5% false rejection and for a sample size of six the critical value of Q is 0.560, and since our Q exceeds the critical value, the value of 6.52 should be rejected. If we did not perform the last two procedures the same Q still applies, but since the sample size is four the critical value becomes 0.765 and thus $Q < Q_{\text{critical}}$ and hence the 6.52 measurement should be retained.

Table 11.5 Values for use in Dixon's Q test for outliers

Number of observations, n	Risk of false rejection			
	0.5%	1%	5%	10%
3	0.994	0.988	0.941	0.886
4	0.926	0.889	0.765	0.679
5	0.821	0.780	0.642	0.557
6	0.740	0.698	0.560	0.482
7	0.680	0.637	0.507	0.434
8	0.725	0.683	0.554	0.479
9	0.677	0.635	0.512	0.441
10	0.639	0.597	0.477	0.409
11	0.713	0.679	0.576	0.517
12	0.675	0.642	0.546	0.490
13	0.649	0.615	0.521	0.467
14	0.674	0.641	0.546	0.492
15	0.647	0.616	0.525	0.472
16	0.624	0.595	0.507	0.454
17	0.605	0.577	0.490	0.438
18	0.589	0.561	0.475	0.424
19	0.575	0.547	0.462	0.412
20	0.562	0.535	0.450	0.401

Actually the Q value calculated by Equation (11.79) is correct only for sample size n in the range $3 \leq n \leq 7$. For larger sample sizes other ratios are calculated. If we arrange the x_i in either increasing or decreasing order such that the outlier is the last one, x_n, the following ratios are calculated:

$3 \leq n \leq 7 \qquad Q = | (x_n - x_{n-1})/(x_n - x_1 | \qquad 8 \leq n \leq 10 \qquad Q = | (x_n - x_{n-1})/(x_n - x_2) |$

$11 \leq n \leq 13 \qquad Q = | (x_n - x_{n-2})/(x_n - x_2) | \qquad 14 \leq n \leq 25 \qquad Q = | (x_n - x_{n-2})/(x_n - x_3) |$

However, n is rarely larger than seven and Equation (11.79) is commonly used.

11.7.3 *Grubb's test for outliers*

See Grubb (1972). The following procedure should be followed:

(1) Rank the data either in ascending or describing order such that x_n is the

outlier, i.e. if the outlier is the largest then $x_n > x_{n-1} > \ldots x_1$ but if the outlier is the smallest then $x_n < x_{n-1} \ldots < x_2 < x_1$.

(2) Calculate the mean and the standard deviation of all the data points.

(3) Calculate the statistics T from the equation

$$T = \frac{|x_n - \overline{x}|}{s} \tag{11.80}$$

(4) Compare the calculated T with the critical values given in Table 11.6.

Table 11.6 Values for Grubb's T test for outliers

Number of observations, n	Risk of false rejection				
	0.1%	0.5%	1%	5%	10%
3	1.155	1.155	1.155	1.153	1.148
4	1.496	1.496	1.492	1.463	1.425
5	1.780	1.764	1.749	1.672	1.602
6	2.011	1.973	1.944	1.822	1.729
7	2.201	2.139	2.097	1.938	1.828
8	2.358	2.274	2.221	2.032	1.909
9	2.492	2.387	2.323	2.110	1.977
10	2.606	2.482	2.410	2.176	2.036
15	2.997	2.806	2.705	2.409	2.247
20	3.230	3.001	2.884	2.557	2.385
25	3.389	3.135	3.009	2.663	2.486
50	3.789	3.483	3.336	2.956	2.768
100	4.084	3.754	3.600	3.207	3.017

For the example considered previously, we have $\overline{x} = 5.915$, $s = 0.313$ and hence

$$T = \frac{6.52 - 5.915}{0.313} = 1.933$$

For $p = 0.05$ this value is larger than the critical value in Table 11.6. (1.822) and should be rejected.

11.7.4 *Youden's test for outlying laboratories*

Youden (1982) developed a statistical test applicable to a group of laboratories analysing the same samples, as in, for instance, the purposes of certifying the contents of standard reference materials, in order to reject the data of laboratories that are outliers. Each laboratory receives several different samples to analyse. For each sample the laboratories are ranked such that the one obtaining the highest value ranked 1, the second highest 2, etc. At the end the different rankings of each laboratory are summed up. The expected ranges for the cumulative scores, for a random distribution, at the 0.05 level of significance (95% confidence) are given in Table 11.7. Any laboratory which is outside this range is treated as an outlier.

Let us look at the case of seven laboratories each analysing five different samples of ores for their copper content (in %). The data are displayed in Table 11.8.

Table 11.7 shows that with a 95% confidence limit the range is expected to be

Table 11.7 Values for Youden's range to identify outlying laboratories (5% two-tailed limits)

Number of participants	Number of materials												
	3	4	5	6	7	8	9	10	11	12	13	14	15
3		4	5	7	8	10	12	13	15	17	19	20	22
		12	15	17	20	22	24	27	29	31	33	36	38
4		4	6	8	10	12	14	16	18	20	22	24	26
		16	19	22	25	28	31	34	37	40	43	46	49
5		5	7	9	11	13	16	18	21	23	26	28	31
		19	23	27	31	35	38	42	45	49	52	56	59
6	3	5	7	10	12	15	18	21	23	26	29	32	35
	18	23	28	32	37	41	45	49	54	58	62	66	70
7	3	5	8	11	14	17	20	23	26	29	32	36	39
	21	27	32	37	42	47	52	57	62	67	72	76	81
8	3	6	9	12	15	18	22	25	29	32	36	39	43
	24	30	36	42	48	54	59	65	70	76	81	87	92
9	3	6	9	13	16	20	24	27	31	35	39	43	47
	27	34	41	47	54	60	66	73	79	85	91	97	103
10	4	7	10	14	17	21	26	30	34	38	43	47	51
	29	37	45	52	60	67	73	80	87	94	100	107	114
11	4	7	11	15	19	23	27	32	36	41	46	51	55
	32	41	49	57	65	73	81	88	96	103	110	117	125
12	4	7	11	15	20	24	29	34	39	44	49	54	59
	35	45	54	63	71	80	88	96	104	112	120	128	136
13	4	8	12	16	21	26	31	36	42	47	52	58	63
	38	48	58	68	77	86	95	104	112	121	130	138	147
14	4	8	12	17	22	27	33	38	44	50	56	61	67
	41	52	63	73	83	93	102	112	121	130	139	149	158
15	4	8	13	18	23	29	35	41	47	53	59	65	71
	44	56	67	78	89	99	109	119	129	139	149	159	169

Table 11.8 Application of Youden's test to seven laboratories analysing the copper content of five different ore samples

Laboratory	Measured results					Score					Sum of scores
	1	2	3	4	5	1	2	3	4	5	
A	9.8	15.8	21.3	18.6	24.9	4	6	3	4	5	22
B	10.0	16.1	20.5	18.7	25.3	2	4	7	3	1	17
C	10.2	17.2	29.7	18.8	25.2	1	1	1	2	2	7
D	9.7	15.9	21.5	18.9	25.0	5	5	2	1	4	17
E	9.6	16.8	20.7	18.3	25.1	6	2	5	5	3	21
F	9.9	16.3	20.8	18.1	24.8	3	3	4	7	6	23
G	9.3	15.4	20.6	18.2	24.0	7	7	6	6	7	33

within 8–32. Accordingly, laboratory C (Table 11.8) is considered to provide higher results and laboratory G lower results than the other laboratories. In calculating the certified value of Reference Standard Materials the results of these laboratories will be rejected, at least for Cu.

11.8 Calibration curves

Very few of the instrumental methods are absolute and usually the instrument signals which are used for calculation of the amount of analyte in the analysed sample depends on several of the instrument's parameters. The quantitative determination of the analysed sample is achieved by comparison with standards. In most cases more than one concentration of standards is used and a calibration curve is plotted, a curve which gives the concentration versus the electronic signal measured by the instrument. These calibration curves also provide the certainty (confidence limits) of our measurement and the range of concentrations for which the method can be applied accurately.

11.8.1 *Standards for calibration curves*

Standards can be purchased commercially or can be prepared by the analyst. There are commercially available standards for single elements or multielement standards in various concentrations. However, an analyst will often prepare standards or some part of them.

Great care should be taken in the preparation of standards used for calibration purposes. Only chemicals with exactly known stoichiometry should be used, and it is crucial that the amount of water of hydration is known accurately. It is preferable to use chemicals without any water of hydration. The standard should be dried, at sufficient temperature, until a constant weight is reached. A standard solution is prepared by weighing a known mass of this dried chemical into a known volume. Care should be taken that other standard solutions of lower concentration are not prepared by dilution of a principal standard solution, as an error in the preparation of the first solution will propagate to all other standard solutions. It is recommended that 3–4 different solutions are prepared by weighing and that these solutions are diluted. Since weighing is usually more accurate than pipetting, it is recommended that weighing is used whenever possible. If a standard solution is prepared by dilution of another standard solution it is recommended that a burette and at least 25 ml of the original solution are used. If two chemicals with exactly known stoichiometry are available for the same element it is recommended that a standard of each one of them is used, thus verifying that both stoichiometries are actually known. *It should always be remembered that the accuracy of results cannot be better than the accuracy of the standards.*

11.8.2 *Derivation of an equation for the calibration curves*

Calibration curves and/or functions are obtained by measuring several standard solutions or samples with varying elemental concentrations with an instrument and thus obtaining a set of data points each describing the concentration of the element (x_i) and the corresponding instrument signal (y_i). We can plot the data on millimetric paper and connect the points either by a broken line, a line which seems to us to be the best, or a curve using a curved ruler and then interpolate the concentration of the unknown sample (the x axis) from the instrument's signal (the y axis). However this method requires longer time in routine work and may also lead to large and subjective errors depending on the analyst.

The best and most objective way to obtain a calibration curve is to use a set of coupled values to calculate a function $y(x)$ and obtain from it the concentrations of the analysed samples according to their signal, y. Many analysts today are not bothered how this function is calculated since, in contrast to even 10–20 years ago, when the calibration function was calculated by the analyst, instruments today, all of them PC based, are supplied with programs to calculate the calibration function. If not calculated by the instrument PC, the functions are derived usually by a spreadsheet program (e.g. Quatropro or Excel).

To understand how this is done on the computer, we will give here the principles. Several scientific calculators also have the ability to calculate this function assuming it is linear.

Finding the best functional equation which fits the experimental data is referred to as *regression*. Most calibration curves are linear and hence the search for the best linear function is called *linear regression*. How do we choose the best line? The basic assumption is that the values of x are exact (although actually this is not true since there will be at least a random error in the volumes of the standard solutions taken for calibration) while the values of y are subject to a normal distribution. In linear regression we are interested in finding a function $y = ax + b$ which best fits our data. What is the criterion for best fit? We cannot request the sum of deviations between our observed values (y_{obs}) and our calculated values from the fitted equation ($y_{cal} = ax + b$) to be a minimum [minimal value of $\sum_{i=1}^{n} y_i - (ax_i + b)$, where n is the number of observations (x_i, y_i) used to calculate the calibration equation] since in this case negative deviations will cancel positive deviations and minimal values of the sum do not warrant small deviations. In order to cancel the effect of sign, instead of summing the deviations, the summing of either the absolute values of the deviations or their squares can be done. These deviations are referred to as residuals so we have two alternatives: to look for the minimum sum of the absolute residuals (SAR); or to sum the squares of residuals (SSR). Which of the two alternatives is the better one? Without referring to an optimum quality the prevailing use is of the minimum SSR. The disadvantage of SSR, however, is that it has a larger influence from larger deviations, and many larger deviations occur at the limits of measurement of the instrument and hence will have larger associated errors. The main advantage of the SSR is that the parameters leading to the minimum values can be calculated by a simple calculus methods. Simple equations for calculating a and b can be obtained in this way. This method of minimizing the SSR is called the *least squares method*:

$$\mathrm{SSR} = \sum_{i=1}^{n} (ax_i + b - y_i)^2 \qquad (11.81)$$

Differentiating SSR with respect to a and b and equating the derivatives to zero leads to the following equations where the summation is done for the n calibration points.

$$\sum x_i[y_i - (ax_i + b)] = 0 \Rightarrow a\sum x_i^2 + b\sum x_i = \sum x_i y_i$$
$$\sum y_i - (ax_i + b) = 0 \quad \Rightarrow a\sum x_i + b \cdot n = \sum y_i \tag{11.82}$$

n being the number of experimental pairs of values measured for the calibration. Solving these two equations for a and b yields:

$$a = \frac{n\sum x_i y_i - \sum x_i \sum y_i}{n\sum x_i^2 - (\sum x_i)^2}; \quad b = \frac{\sum x_i^2 \sum y_i - \sum x_i \sum x_i y_i}{n\sum x_i^2 - (\sum x_i)^2} \tag{11.83}$$

Another form in which to write the equations for a and b is by using the means of x_i and y_i:

$$a = \frac{\sum(x_i - \overline{x})(y_i - \overline{y})}{\sum(x_i - \overline{x})^2}; \quad b = \overline{y} - a\overline{x} \tag{11.84}$$

Using a calculator it is usually simpler to use Equation (11.83) since all scientific calculators give $\sum x_i$ and $\sum x_i^2$.

The variance of the calculated y is given by the SSR divided by the number of degrees of freedom. Since n data points were used to calculate the two constants a and b, the number of degrees of freedom (independent variables) is $n-2$, thus

$$s_0^2 = \frac{\sum_{i=1}^{b}(y_{\text{obs},i} - y_{\text{cal},i})^2}{n-2} \quad \Rightarrow \quad s_0^2 = \frac{\sum_{i=1}^{n}(y_i - ax_i - b)^2}{n-2} \tag{11.85}$$

where a and b are those calculated by Equation (11.83) or (11.84). From the total variance s_0^2 given in Equation (11.85) we can calculate the variance of the coefficients a and b:

$$s_a^2 = \frac{n \cdot s_0^2}{n\sum x_i^2 - (\sum x_i)^2} = \frac{s_0^2}{\sum(x_i - \overline{x})^2} = \frac{s_0^2}{\sum x_i^2 - n(\overline{x})^2} \tag{11.86}$$

$$s_b^2 = \sum x_i^2 \cdot s_a^2 \tag{11.87}$$

The *confidence limits of the regression parameters* are calculated as for the measurement of the mean by the Student's t statistics (Table 11.3) with the appropriate level of significance (probability level) using $n - 2$ degrees of freedom. Thus the confidence limits of the parameters are:

$$a \pm ts_a \text{ and } b \pm ts_b \tag{11.88}$$

11.8.3 *Test for linearity*

In the previous section we assumed that the calibration curve is linear and calculates its parameters, but is this assumption correct and how do we check it? The most common method for testing the linear correlation is by the statistic R, called the product-moment linear correlation coefficient. Often it is named just the correlation coefficient since this is most used, although there are other types of correlation coefficients. The correlation coefficient is described by:

$$R^2 = \frac{[\sum (x_i - \overline{x})(y_i - \overline{y})]^2}{\sum (x_i - \overline{x})^2 \sum (y_i - \overline{y})^2} = \frac{[n \sum x_i y_i - \sum x_i \sum y_i]^2}{[n \sum x_i^2 - (\sum x_i)^2] \cdot [n \sum y_i^2 - (\sum y_i)^2]} \quad (11.89)$$

For absolute linearity, R^2 is equal to 1. The smaller R^2 is the larger the deviation from linearity. However, quite large R^2 is also obtained for non-linear curves as, e.g. with one side of a parabola, and it is advisable to check the obtained linearity by observation of the calculated line versus the observed data points. Testing the significance of R can be done using the t test, where t is calculated from Equation (11.90):

$$t = \left[\frac{R^2 (n-2)}{\sqrt{1 - R^2}} \right]^{\frac{1}{2}} \quad (11.90)$$

The calculated value of t is compared with the tabulated value (Table 11.3; a two-tailed test) with $n-2$ degrees of freedom. The null hypothesis in this case is that there is no correlation between x and y. If the calculated value of t is larger than the critical tabulated value then the null hypothesis is rejected and we conclude that within this level of confidence there is a correlation.

Another test for linearity is the F test. In this test we compare the variance of the linear regression s_0^2 with the standard deviation for a single measurement, obtained by repeating measurements of the same standard solution and calculating the single measurement variance s_s^2. F is the calculated from the ratio $F = s_0^2 / s_s^2$. If the calculated F value is larger than the tabulated critical value (Table 11.4) then the null hypothesis of a linear fit is rejected. If F is less than the tabulated value we retain the hypothesis of a linear correlation. But which of the single points of the calibration line do we repeat? It is best to repeat several of them (or even all of them) and pooled together the standard deviations of all of them. The pooling together of the standard deviations is done by their weighted arithmetic average when the weights are their degrees of freedom, i.e.

$$s_s^2 = \frac{\sum_{i=1}^{n} \mathrm{d}f_i \cdot s_i^2}{\sum_{i=1}^{n} \mathrm{d}f_i}$$

where $\mathrm{d}f_i$ and s_i are for each calibration point that was repeated more than once; n is the number of calibration points (different concentrations).

A further test of linearity is the comparison of the least squares linear regression parameter of dependence on x with its standard deviation. The smaller the ratio S_a/a the better is the linearity fit.

All curve-fitting and spreadsheet programs calculate both the parameters, their standard deviations and R^2, and actually nowadays there is no need to calculate these values ourselves. The equations are given only so that we will understand the meanings of the program's output.

In the calibration curve, signal $= a \times$ concentration $+ b$, the parameter b is called the 'blank' of the method as this is the reading obtained for zero concentration. The parameter a is called the sensitivity of the method. As we will see in the next section a influences the error of the determination, mainly for large concentrations. On the other hand, b influences mainly the limit of determination and the errors at lower concentrations. Ideally b should be zero, but in many applications there is a non-zero signal even for a zero concentration, which is due to impurities in the standards, the supposedly pure matrix or owing to instrumental reasons.

11.8.4 *Calculation of a concentration*

Once we have the calibration equation $y = ax + b$, which actually stands for the equation

$$\text{concentration} = (1/a) \times (\text{instrument's signal} - b) \tag{11.91}$$

then it is very simple to calculate the concentration from the instrument's signal. The calculation becomes more complicated when we want to state the confidence limit (interval) of the concentration. Remember that there are confidence limits for a and for b and those uncertainties in their values is reflected by uncertainty in the concentration. In many cases an approximate equation is used for the calculation of the standard deviation of the calculated concentration from Equation (11.91). If the experimental signal for the unknown solution is y_{u} and the linearization parameter a was calculated from n points (x_i, y_i) then the variance of the concentration, s_{c}^2 (the square of the standard deviation), is given by:

$$s_{\mathrm{c}}^2 = \frac{s_0^2}{a^2}\left[1 + \frac{1}{n} + \frac{(y_{\mathrm{u}} - \overline{y})^2}{a^2 \sum_i (x_i - \overline{x})^2}\right] \tag{11.92}$$

Instead of using $\sum(x_i - \overline{x})^2$ we can substitute it with $\sum_i x_i^2 - n(\overline{x})^2$.

If m readings were completed on the unknown solution and y_{u} is their mean, then the first term in the brackets is changed from 1 to $1/m$:

$$s_{\mathrm{c}}^2 = \frac{s_0^2}{a^2}\left[\frac{1}{m} + \frac{1}{n} + \frac{(y_{\mathrm{u}} - \overline{y})^2}{a^2 \sum_i (x_i - \overline{x})^2}\right] \tag{11.93}$$

If the concentration obtained by substituting y_{u} in Equation (11.91) is c then the confidence limits of the concentration are $c \pm ts_{\mathrm{c}}$.

Another way of calculating the confidence limits for the concentration is by using the confidence limits for the parameters using Equation (11.88). Treating a and b independently we have four possibilities of extreme values of concentrations, i.e. using firstly $a + ts_a$ and $b + ts_b$ in Equation (11.91) and then changing to $a + ts_a$ *and* $b - ts_b$, and the same with $a - ts_a$. Calculation of these four extreme concentrations give us the confidence limit around c, calculated from a and b. The disadvantage of the previous method is that it does not show how the confidence limit is changed if the signal for the unknown solution are read more than once. In this case for each reading the confidence limit is calculated and then the extreme values are averaged and the averaged confidence limit is divided by $\sqrt{m}$. However, this method requires many calculations although they can be easily added to the least squares program or written separately with a, b, s_a and s_b as the input values of the program.

The confidence limits can be narrowed (i.e. improved) by increasing either n, the number of points on the calibration line, or m, the number of repeating experiments on the unknown sample, or both.

11.8.5 *Weighted least squares linear regression*

We mentioned before that occasionally the points at the end of the calibration range are more erroneous than the others since we are tending to the limits of the instrument's capabilities. These points influence the fit more than the others since their deviations are larger. In order to overcome the disadvantage that the usual least squares linear regression gives similar importance to all points, more accurate and less accurate, a weighted regression line can be calculated. In this regression method each point receives a weighting that depend on its accuracy. How do we assess the accuracy, or relative accuracy of each point? The common method to give a weight to a point is to measure each concentration point several times, and then the weight is taken as inversely proportional to the variance of the various measurements of y for this x:

$$w_i = \frac{1}{s_i^2} \tag{11.94}$$

The weighted least squares linear regression minimizes the sum $\sum_i^n w_i(y_{\text{obs}} - y_{\text{cal}})^2$. Then each sum in Equation (11.83) also has w_i, and n is replaced by Σw_i:

$$\begin{aligned} a &= \frac{\sum w_i \sum w_i x_i y_i - \sum w_i x_i \sum w_i y_i}{\sum w_i \sum w_i x_i^2 - (\sum w_i x_i)^2}; \\ b &= \frac{\sum w_i x_i^2 \sum w_i y_i - \sum w_i x_i \sum w_i x_i \sum w_i x_i y_i}{\sum w_i \sum w_i x_i^2 - (\sum w_i x_i)^2} \end{aligned} \tag{11.95}$$

Some analysts prefer to normalize the weights so that their sum is equal to n, such that in Equation (11.95) $\sum w_i$ is replaced by n, which is similar to Equation (11.83). In this instance w_i is defined by

$$w_i = \frac{\dfrac{n}{s_i^2}}{\sum_i \dfrac{1}{s_i^2}} \tag{11.96}$$

11.8.6 *Non-linear calibration equations*

Several analytical methods give non-linear calibration plots, or they are linear only at limited range and at higher concentrations they become curved. Usually the curves are curved downward, reducing the sensitivity of the method in this range. Thus, if possible, it is recommended to dilute the analysed solution to the linear range. However, there are cases where accuracy is sacrificed for shorter times of analysis, prohibiting any additional step of dilution. So, we would like to fit our data also to non-linear functions. There are computer programs that can fit the data to any function we choose. Most curved calibration lines are fitted to a quadratic equation $y = ax^2 + bx + c$ or to a third-power equation $y = ax^3 + bx^2 + cx + d$. The principle of least squares fitting to these polynomials is the same as for the linear function. Thus, for a quadratic regression, we want to minimize the SSR:

$$\sum_i [y_i - (ax_i^2 + b_i x_i + c_i)]^2 \tag{11.97}$$

By differentiating the SSR with respect to a, b and c and equating the derivatives to zero we will obtain three equations with the unknowns a, b and c. From these three equations we can derive formulas for a, b and c, similar to Equation (11.83).

It is also possible to perform these polynomial fittings with spreadsheets such as Quatropro, which does not have a polynomial regression facility. Although they normally have only linear regression, such programs can perform multilinear regression where the regression assumes that y depends linearly not only on x but also on z, etc. For example in bilinear regression the program minimizes the SSR of the function $y = az + bx + c$. However, if in the column of z_i values we insert a column of x_i^2 values we can perform a least squares analysis of the quadratic equation $y = ax^2 + bx + c$.

11.8.7 *Calibration curves in activity measurements*

The efficiency, ε, of an HPGe detector (i.e. the fraction of the γ-ray photons that produces a photopeak signal in the detector) as a function of the energy of the photons E can be approximated by the equation (Debertin & Helmer 1988; Knoll 1989):

$$\varepsilon = a \times E^b \tag{11.98}$$

The experimental data ε_i versus E_i can be fitted to the calibration curve either by non-linear regression or by writing the equation in a linear form by using the logarithm of the variable.

$$\log \varepsilon = \gamma + b \log \mathrm{E} \quad (\gamma = \log a) \tag{11.99}$$

Here we have a linear equation in the parameters γ and b, and the equation of linear regression can be used.

In non-linear regression we have to find a and b values which will minimize the SSR:

$$\mathrm{SSR} = \sum_i (\varepsilon_i - aE_i^b)^2 \tag{11.100}$$

The values of a and b, which minimize the SSR, cannot be found by substitution in equations as was done for linear regression and the appropriate values must be found by iteration. The most common method for non-linear regression is the one suggested by Marquardt (1963). This method is used in many of the new computer programs that deal with curve plotting and fitting as for example, Sigmaplot, Origin, Table Curve and several others. However, most people are used to linear regression, especially that it is packaged with most of the data-sheet programs such as Excel, Quatropro and Lotus.

A slightly better approximating equation is:

$$\log \varepsilon = a + b \log E + c (\log E)^2 \tag{11.101}$$

The data can be fitted to this equation either by polynomial regression or by multi-parameter linear regression as discussed previously. In this instance, the minimization of the SSR is also done by substituting the data in the equations for the parameters and there is no need for iteration.

11.8.8 *Calibration curves for neutron capture radiography*

Neurtron capture radiography is used to measure instrumentally the concentration of elements which have high cross-section for the (n, α) reaction, e.g. $^{10}B(n,\alpha)^7Li$ and $^6Li(n,\alpha)T$. This method uses selectively the emitted α tracks induced by irradiation with thermal neutrons and their registration by an α-sensitive film. The observed calibration curve of the number of the tracks as a function of boron concentration is sub-linear. The explanation for the sub-linearity is that existing tracks limit the area of the film that is available for new tracks.

For infinitesimally small tracks the rate of formation of tracks dn/dt is proportional to the concentration c of the activatable nuclide (^{10}B or ^6Li).

$$dn/dt = \alpha c$$

α is a proportionality constant dependent on the flux of the neutrons and the cross-section for the formation of the α particle. For final-size tracks the area of unavailable film is proportional to the number of already existing tracks and thus the fraction of available area is $(1-\beta n)$ where β is a proportionality parameter. Taking these two factors together we obtain

$$\frac{dn}{dt} = \alpha c(1 - \beta n) \Rightarrow \ln(1 - \beta n) = -\beta\alpha c t \Rightarrow c = -\frac{1}{\gamma}\ln(1 - \beta n) \tag{11.102}$$

where $\gamma = \alpha\beta t$ and t is the irradiation time.

The final calibration equation for n has two unknown parameters that cannot be fitted by a linear regression, but can be fitted by non-linear regression analysis. One way to overcome the need for non-linear regression is by iteration of the linear regression. The linear equation chosen is $c = a_0 + a_1 \ln(1-\beta n)$. The value of β is changed iteratively in order to obtain $a_0 = 0$. The initial value of $1/\beta$ was selected as slightly above the largest value of n_i. The value of a_0 was found to be positive and $1/\beta$ was increased gradually until a_0 become zero. For larger $1/\beta$ values a_0 will be negative (Alfassi & Probst 1999).

11.8.9 *Limit of detection*

Almost the first question asked by the analyst about an analytical method or an instrument is about its limit of detection, which means the minimum amount that can be detected by the instrument, within a confidence level that this detection is not a false one. The limit of detection does not reveal what are the confidence limits of the concentration calculated, but just the confidence level indicating that the element is really present in the sample. The limit of detection is not changed by repeated measurements, but the confidence limits of the measured concentration will be narrowed by repetition of the measurement. The limit of detection of the analyte is determined by the limit of detection of the signal. If the blank signal was zero with standard deviation of zero any signal measured by the instrument would be significant, but actual measurements usually have blank signals and are not accurate numbers, though they have distribution. So how do we recognize if a signal is really due to the presence of the element and not just a blank signal? Different people use different criteria for this limit of detection. The suggestion made by the National Institute of Standards and Technology of the USA is to use the 95% level of confidence. So if the blank is distributed normally with standard deviation σ_0 then the limit of detection of the signal is:

$$s_D = 3.29\sigma_0 \quad (11.103)$$

This means that only if our signal is $b + 3.29\sigma_0$ can we say that we have a true signal of the analyte with false positive and false negative risks each of 5%. The 3.29 value is taken from $2z_{0.95}$ where z is the value of the probability of the normal distribution. The factor 2 comes from the fact that not only the blank has a normal distribution but also our measured signal has a normal distribution. If we neglect the factor 2 we will obtain what is called a limit of decision. It will guarantee (with 95% confidence) that we do not claim the presence of the analyte when it is actually absent, but it will not assure us that it is absent while actually present. Many people use the factor 3 instead of 3.29. This is rounding-off the factor. Using the normal distribution table shows that this factor ($z = 1.5$) means a risk of 7% of either false positive or false negative answers. In most cases we do not know the real value of σ_0 and we estimate it from the calculation of s, obtained from repeated measurements of the blank. In this case the Student's t distribution should be used rather than the normal one and

$$s_D = 2 \times t \times s \quad (11.104)$$

where t is taken for df $= n-1$ where n is the number of repeated measurements.

If the blank value is obtained from a least squares linear regression then σ_0 is equal to $\sqrt{2} \times s_b$. Some people suggest using s_0 as σ_0 instead.

The analyte detection limit, c_D is related to the signal detection limit by the sensitivity of the measurement, and the parameter which transforms signals to concentration, i.e. the parameter a in the calibration linear equation:

$$c_D = s_D/a \quad (11.105)$$

Limit of determination and *Limit of quantitation* (synonyms) are sometimes used for the lower limit of a precise quantitative measurement in contrast to limit of detection which deals with qualitative detection. A value of $10s_b$ was suggested as the limit of determination of the signal and $10s_b/a$ as the limit of determination of the concentration. It should be remembered that the accuracy of the measurement can be improved by repeated measurements of the sample. In any case very few reports include this term and usually only the limit of detection is given. It should always be borne in mind that it is actually almost impossible to measure quantitatively samples with a concentration close to c_D. We can only detect the presence of the element, but its real value will be considerably in error (large confidence intervals).

References

Alfassi, Z.B. & Probst, T.U., *Nucl. Instrum. Methods. Phys. Res.* **428A**, 502 (1999).

Brunk, H.D., *An Introduction to Mathematical Statistics*, 2nd edn. Blaisdell Publishing Company, Massachusetts, 1965.

Debertin, K. & Helmer, R.G., *Gamma and X-ray Spectrometry with Semiconductor Detectors*. North Holland, Amsterdam, 1988, p. 220.

Dixon, W.J., *Biometrics, BIOMA*, **9**, 1953, p. 74.

Grabb, F.E. & Beck, G., *Technometrix TCMTA*, **14**, 1972, p. 847.

Hays, W.L. & Winkler, R.L., *Statistics, Probability, Inference, and Decision*. Holt, Rinehart and Winston, New York, 1971, p. 347.

Hogg, R.V. & Craig, A.T., *Introduction to Mathematical Statistics*, 3rd edn. Collier-Macmillan, London, 1971.

Kendaill, M.G. & Stuart, A., *The Advanced Theory of Statistics*, 3rd edn. Hafner Publishing Company, New York, 1969, Vol. 1.

Knoll, G.F., *Radiation Detection and Measurement*, 2nd edn. Wiley, New York, 1989.

Marquardt, D.W., *J. Soc. Indus. Appl. Math.* **11**, 431 (1963).

Meier, P.C. & Zünd, R.E., Statistical methods in analytical chemistry, in: *Chemical Analysis*, eds. J.D. Winefordner & I.M. Kolthoff, Vol. 123. Wiley, New York, 1993, pp. 240–241.

Miller, J.C. & Miller, J.N., *Statistics for Analytical Chemistry*, 3rd edn. Ellis Horwood and Prentice Hall, New York, 1994, p. 57.

Press, W.H., Teukolsky, S.A., Vetterling, W.T. & Flannery, B.P., *Numerical Recipes in FORTRAN, The Art of Scientific Computing*, 2nd edn. Cambridge University Press, New York, 1992, pp. 206–207, 209–210, 215.

van der Waerden, B.L., *Mathematical Statistics*, George Allen & Unwin, London, 1969, pp. 96–97.

Youden, W.J., *J. Qual. Technol.*, **4**, 1982, p. 1.

Index